Advances in Conservation Agriculture

Volume 1: Systems and Science

It is widely recognised that agriculture is a significant contributor to global warming and climate change. Agriculture needs to reduce its environmental impact and adapt to current climate change whilst still feeding a growing population, i.e. become more 'climate-smart'. Burleigh Dodds Science Publishing is playing its part in achieving this by bringing together key research on making the production of the world's most important crops and livestock products more sustainable.

Based on extensive research, our publications specifically target the challenge of climate-smart agriculture. In this way we are using 'smart publishing' to help achieve climate-smart agriculture.

Burleigh Dodds Science Publishing is an independent and innovative publisher delivering high quality customer-focused agricultural science content in both print and online formats for the academic and research communities. Our aim is to build a foundation of knowledge on which researchers can build to meet the challenge of climate-smart agriculture.

For more information about Burleigh Dodds Science Publishing simply call us on +44 (0) 1223 839365, email info@bdspublishing.com or alternatively please visit our website at www.bdspublishing.com.

Related titles:

Advances in Conservation Agriculture Volume 2: Practice and Benefits
Print (ISBN 978-1-78676-268-9); Online (ISBN 978-1-78676-270-2, 978-1-78676-271-9)

Assessing the environmental impact of agriculture
Print (ISBN 978-1-78676-228-3); Online (ISBN 978-1-78676-230-6, 978-1-78676-231-3)

Improving organic crop cultivation
Print (ISBN 978-1-78676-184-2); Online (ISBN 978-1-78676-186-6, 978-1-78676-187-3)

Managing soil health for sustainable agriculture Volume 1: Fundamentals
Print (ISBN 978-1-78676-188-0); Online (ISBN 978-1-78676-191-0, 978-1-78676-190-3)

Managing soil health for sustainable agriculture Volume 2: Monitoring and management
Print (ISBN 978-1-78676-192-7); Online (ISBN 978-1-78676-195-8, 978-1-78676-194-1)

Chapters are available individually from our online bookshop:
https://shop.bdspublishing.com

climate•SMART•publishing
IN AGRICULTURAL SCIENCE

BURLEIGH DODDS SERIES IN AGRICULTURAL SCIENCE
NUMBER 61

Advances in Conservation Agriculture

Volume 1: Systems and Science

Edited by Professor Amir Kassam, University of Reading, UK and Moderator, Global Conservation Agriculture Community of Practice (CA-CoP), FAO, Rome, Italy

burleigh dodds
SCIENCE PUBLISHING

Published by Burleigh Dodds Science Publishing Limited
82 High Street, Sawston, Cambridge CB22 3HJ, UK
www.bdspublishing.com

Burleigh Dodds Science Publishing, 1518 Walnut Street, Suite 900, Philadelphia, PA 19102-3406, USA

First published 2020 by Burleigh Dodds Science Publishing Limited

Library of Congress Control Number: 2019951731

British Library Cataloguing in Publication Data
A catalogue record for this book is available from the British Library

ISBN 978-1-78676-264-1 (Print)
ISBN 978-1-78676-267-2 (PDF)
ISBN 978-1-78676-266-5 (ePub)
ISSN 2059-6936 (print)
ISSN 2059-6944 (online)

DOI 10.19103/AS.2019.0048

Typeset by Deanta Global Publishing Services, Dublin, Ireland
Printed and bound by CPI Group (UK) Ltd, Croydon, CR0 4YY

Contents

Series list

Title	Series number
Achieving sustainable cultivation of maize - Vol 1 From improved varieties to local applications Edited by: Dr Dave Watson, CGIAR Maize Research Program Manager, CIMMYT, Mexico	001
Achieving sustainable cultivation of maize - Vol 2 Cultivation techniques, pest and disease control Edited by: Dr Dave Watson, CGIAR Maize Research Program Manager, CIMMYT, Mexico	002
Achieving sustainable cultivation of rice - Vol 1 Breeding for higher yield and quality Edited by: Prof. Takuji Sasaki, Tokyo University of Agriculture, Japan	003
Achieving sustainable cultivation of rice - Vol 2 Cultivation, pest and disease management Edited by: Prof. Takuji Sasaki, Tokyo University of Agriculture, Japan	004
Achieving sustainable cultivation of wheat - Vol 1 Breeding, quality traits, pests and diseases Edited by: Prof. Peter Langridge, The University of Adelaide, Australia	005
Achieving sustainable cultivation of wheat - Vol 2 Cultivation techniques Edited by: Prof. Peter Langridge, The University of Adelaide, Australia	006
Achieving sustainable cultivation of tomatoes Edited by: Dr Autar Mattoo, USDA-ARS, USA & Prof. Avtar Handa, Purdue University, USA	007
Achieving sustainable production of milk - Vol 1 Milk composition, genetics and breeding Edited by: Dr Nico van Belzen, International Dairy Federation (IDF), Belgium	008
Achieving sustainable production of milk - Vol 2 Safety, quality and sustainability Edited by: Dr Nico van Belzen, International Dairy Federation (IDF), Belgium	009
Achieving sustainable production of milk - Vol 3 Dairy herd management and welfare Edited by: Prof. John Webster, University of Bristol, UK	010
Ensuring safety and quality in the production of beef - Vol 1 Safety Edited by: Prof. Gary Acuff, Texas A&M University, USA & Prof. James Dickson, Iowa State University, USA	011
Ensuring safety and quality in the production of beef - Vol 2 Quality Edited by: Prof. Michael Dikeman, Kansas State University, USA	012
Achieving sustainable production of poultry meat - Vol 1 Safety, quality and sustainability Edited by: Prof. Steven C. Ricke, University of Arkansas, USA	013
Achieving sustainable production of poultry meat - Vol 2 Breeding and nutrition Edited by: Prof. Todd Applegate, University of Georgia, USA	014

Dedication

This book is dedicated to the global Conservation Agriculture movement and all the stakeholders engaged in this movement who have successfully generated and sustained an international momentum to transform the unsustainable and out-of-date conventional tillage agriculture into the alternate paradigm of Conservation Agriculture, representing an ecosystem approach to regenerative sustainable agriculture. The book is particularly dedicated to all the Conservation Agriculture pioneer and champion farmers and their families, researchers and extension agents as well as all the champions in the national and international public, private and civil sectors who have dedicated themselves to mobilizing policy and institutional support for making Conservation Agriculture a global mainstream reality for sustainable development in the 21st century.

Acknowledgements

This book builds on the vast amount of innovative and pioneering development-oriented work done over the past several decades in the public, private and civil sectors by those individuals and organizations who make up the global Conservation Agriculture (CA) Community of Practice. These CA stakeholders include farmers, their families and organizations, service providers along the value chain, scientific and empirical researchers, educators and trainers at all levels in schools, colleges, universities and extension services, technical assistance and development organizations and their experts at the national, regional and international levels, manufacturers of inputs, policy analysts and policy-makers, politicians and all the innovators and champions. I express my very special and sincere gratitude to all the stakeholders of this CA Community who are making sustainable agriculture real in all continents through the successful adoption and spread of the new paradigm of CA and the institutional capacity to sustain the transformation.

I would like to recognize the enormous assistance and cooperation extended to me in the preparation of this book from a large numbers of colleagues and organizations. First of all I would like to extend my sincere thanks to the Publisher Burleigh Dodds Scientific Publishing for inviting me to edit this book, and specifically to Francis Dodds who was actively engaged throughout in overseeing the project, with the able assistance from Amanda Renwick and Rennie Alphonsa at Deanta who oversaw production of the book.

Most importantly, I would like to express my deep gratitude to every lead author and all the co-authors for their valuable contributions and for spending their precious time and resources to bring together the relevant and latest global knowledge and information on CA science and systems, practice and benefits, and adoption and spread.

I acknowledge with many thanks the external reviewers of the different chapters of the book for their help to ensure that the chapters comprise the best quality material. They include: Theodor Friedrich, Rolf Derpsch, Sjoerd Duiker, Jeff Mitchell, Ivo Mello, Rafael Fuentes, Paulo Sérgio Pavinato, Denizart Bolonhezi, Telmo Amado, Enamul Haque, Frank Place, D. Pedro Freitas, Robelio Marchão, Julio César Salton, Luis Carlos Balbino, Brian Sims, Jacky Desbiolles, Willem Hoogmoed, Andrés Sylvestre Begnis, Sebastián Senesi, Gottlieb Basch, Hendrick Smith, Don Reicosky, Timothy Reeves, Yashpal Singh Saharawat, Vicenzo Tabaglio, Gualbert Gbehounou, Kevin Gallagher, Johnnie Van den Berg, Mário Carvalho, Jean-Pierre Sarthou, Emilio Gonzalez, Daryl Domitruk, David Coates, Tom Goddard, Julian Dumanski, Michele Pisante, Francis Shaxson, Ademir de Oliveira Ferreira and Zete Stumpf.

Finally, I express my very special thanks to Dr. Qu Dongyu, the Director General of the Food and Agriculture Organization of the United Nations, for his inspiring and illuminating Foreword. Dr. Dongyu's contribution is all the more important because the promotion CA in China is a national priority that has led to a very rapid growth in the area under CA and in the institutional capacity to sustain it.

Foreword

These volumes are a timely celebration of the most progressive change in farming practices that has been set in motion during the past 60 years and which is gathering momentum around the world at an extraordinary pace.

For thousands of years, soil inversion, whether by hoe or plough, has been almost universally applied by farmers and seen as essential for the successful growing of annual crops. Tilling the soil buries weeds, loosens the soil to let roots penetrate easily and allows rainfall to sink in and become available to the crop. By burying crop residues, it may also interrupt the life cycles of crop diseases and pests.

However, as the frequency and depth of tillage has increased, the negative effects have become more obvious. Every time soil is dug or ploughed, its structure is broken up and it becomes increasingly at risk to water and wind erosion. The speed with which soil organic matter content falls is accelerated, causing the surface to become susceptible to crusting, thereby reducing rainfall infiltration and increasing run-off while also restricting moisture retention capacity in the plant rooting zone below the surface. These processes tend to make crops and soils more vulnerable to drought and can ultimately lead to farmland becoming so degraded that it is abandoned.

As we understand more about the processes of climate change, we are becoming increasingly aware of the extent to which frequent tillage also contributes to global warming. As the organic matter content of soil falls this reduces the capacity of farmed land to serve as a carbon sink. Moreover, soil inversion, whether manual or mechanical, is very heavy in its energy requirements, with tractor use for ploughing accounting for a large share of the fossil fuel consumption in food production.

We are also belatedly learning that frequent soil inversion, especially when associated with heavy applications of pesticides, reduces soil biotic activity and undermines soil health with a corresponding fall in productivity.

The 'Dust Bowl' in the United States in the 1930's awoke farmers and scientists to the damaging effects of excessive tillage but it was not until the 1960s that American farmers began to adopt various no-till systems to reduce wind and water erosion. These set a precedent for the progressive emergence in the following decades of Conservation Agriculture (CA) which consists of a combination of continuous no tillage with year-round biomass soil cover and crop associations and/or rotations involving cover crops – often legumes.

The USA continues to be the country with the largest area under CA (43.5 million ha in 2015/16 or 35.1% of its arable farming area). Its farmers' associations are very active in pioneering new technologies, especially those related to the better use of cover crops in CA systems to increase Nitrogen availability and to cut herbicide dependence

One could claim that CA was ahead of its time in that it contains the main elements of what is now termed 'sustainable agriculture'. Indeed, there is much to be learnt from the CA story of the past 60 years which is relevant to inducing the global shift that must urgently be made from the currently unsustainable food production and consumption systems to ones that are truly sustainable.

Five of such lessons from the CA experience are:

Firstly, CA shows that fundamental changes in farming technologies - in this case, getting rid of the plough - can spread very rapidly throughout the world, in both developed and developing countries in which it is being taken up by large and small-scale farmers alike. The area of arable land under CA has grown from about 2 million ha in the USA in the early 1970s to 180 million ha (12.5% of global cropland) in 78 countries in 2015-16.

Secondly, the growth in CA uptake and the adaptation of CA methods to different ecological and societal conditions has been driven largely by practitioners, especially innovative farmers and machinery and hand-tool manufacturers. Approaches to CA are constantly evolving and new developments are being openly shared between all those involved. The more that farmers find that CA can boost their incomes, the faster will be the pace of change.

Thirdly, formal research has been important in developing new approaches to CA but most studies have been focussed on identifying the impact of the shift to CA on crop performance; the physical and biological conditions of soils; global warming, and farm incomes. Such validation studies have help boost the case for policy support for CA.

Fourthly, in most countries, however, CA has so far spread between farmers without explicitly supportive government policies. It seems certain that the rate of diffusion can be accelerated by targeted incentives (for instance, subsidies on appropriate machinery and equipment, payments to farmers for soil carbon accumulation and enhancement of water resources) and improvements in agricultural extension.

Lastly, although international agencies such as the FAO have only invested quite small resources in CA, they have played a valuable catalytic role, mainly by promoting exchanges of experiences between practitioners, countries and regions and nurturing the emergence of regional promotional institutions. FAO has placed CA at the core of its vision for sustainable food and agriculture which calls for 'a world in which food is nutritious, safe and accessible for everyone, where natural resources are managed sustainably, and where rural dwellers have decent livelihoods and contribute actively to economic development'. The FAO has also sponsored the foundation of a CA Community of Practice (COP) which shares new developments between its many members.

Although my own country, China, carried out trials on zero tillage from the 1980s, it was only after the turn of the century that the promotion of CA was adopted as a national priority. As a result, there has been a very rapid growth in the area under CA from just a few ha in 2002 to 9 million ha in 2015. The results have been good in terms of yield increases, especially for maize, and of reductions in erosion and river sedimentation. The government, therefore, intends to take additional measures to boost farmer uptake of CA including the recent creation of the China Institute for Conservation Tillage, technical support for the manufacture of improved machinery, subsidies for the provision of eco-systems services by farmers, and improvements in agricultural extension. It also intends to enhance the benefits of CA by linking its promotion to other aspects of its sustainable development agenda including Integrated Pest Management (IPM).

This edifying book is a compendium of much of what we have learnt across the world about CA. Its authors have all been deeply and enthusiastically involved in one way or another in the remarkable evolution and spread of Conservation Agriculture in recent decades. It will serve as an immensely valuable source of reference – and inspiration - for all those who are committed to putting the world's food systems on a truly sustainable footing.

It is with great pleasure, therefore, that I commend this book to you,

Qu Dongyu,
Director General, UN Food and Agriculture Organization (FAO),
Rome

Preface

Conservation Agriculture (CA) and its community of practice have made tremendous strides in the last two decades, particularly since the modern version of CA as we know it today was defined by the FAO at its first Regional Workshop on CA in Harare, Zimbabwe in 1998. The term CA was globalized through the 1st World Congress on Conservation Agriculture organized by the ECAF in partnership with the FAO in 2001 in Madrid, Spain. In 1999, the global spread of CA stood at about 45 M ha of cropland with some 10 countries in which CA was being practiced and promoted. In 2008/09, the global spread of CA was about 106 M ha of cropland across some 36 countries, corresponding to an annual rate of expansion of about 6.1 M ha. In 2015/16, the global spread of CA was some 180 M ha of cropland, split about equally between the Global South and the Global North, involving 78 countries. This corresponds to an annual rate of increase of some 10.5 M ha, a spectacular rate of transformation, led mainly by farmers and their associations with support from national and international champions and enthusiasts, many on a voluntary basis, and some national and international institutions. Although the spread of CA in the 1990s and 2000s was led by countries in North and South America and Australia, the spread of CA since 2008/09 across Europe (including Russia and Ukraine), Asia and Africa has been accelerating, making the transformation from conventional tillage agriculture to CA a truly global phenomenon.

The Green Revolution of the 1960s and 1970s was based on production intensification of wheat and rice, and later maize, relying on the unsustainable intensive tillage-based agriculture and expensive production inputs which brought short-term benefits to some types of farmers only in a handful of countries. The Green Revolution agriculture lost its effectiveness and appeal in the 1990s due to the unacceptably high negative economic, environmental and social impact as well as due to the loss of control by farmers of their own affairs related to production management and capital investment. The alternate CA revolution on the other hand has involved all types of smallholder and large-scale farmers, men and women, and rich and poor farming households, in all major land-based agro-ecologies in all continents. It has brought to the rural communities and society at large a wide range of productivity, economic, environmental and social benefits which cannot be harnessed with the conventional Green Revolution agriculture.

The expansion in area of CA across the world and the increase in number of smallholders adopting CA has benefitted from the growing support of the international research and development community including the FAO, the IFAD, the World Bank, the EU, CIRAD, and CGIAR as well as from many national and local level research and extension systems including NGOs, farmer associations and private sector service providers. Overall, the spread of CA since the very beginning has largely been led by farmers. Initially this was in response to the need to minimize soil erosion and land degradation, but soon this became a strategy to build and maintain soil health and productivity, reduce the high cost of production and diminishing returns, harness ecosystem services for society and nature, address climate change, and support pro-poor sustainable agricultural development strategies.

The CA Community of Practice is made up of many stakeholders and champions including farmers and their families, extension workers, development experts, researchers and academics, heads of institutions, policy analysts and decision-makers, as well as national and regional CA associations. Collectively, they have provided the pioneers and champions and support that keeps generating and sustaining the momentum to what has now become a global phenomenon, transforming conventional tillage-based agriculture into CA as the basis for sustainable agriculture and land use intensification. At the same time, they have generated enormous amounts of new knowledge, formal and experiential, as well as scientific and empirical evidence, regarding CA science and systems, CA practice and benefits and CA adoption and spread. All this knowledge and evidence constitutes the new understanding about regenerative and sustainable agriculture and represents the desire by the CA Community of Practice to move away from the degrading paradigm of tillage-based Green Revolution agriculture to the alternative paradigm of CA.

In light of the above, I was easily persuaded by the publisher Burleigh Dodds Science Publishing to take up the challenge of editing this book 'Advances in Conservation Agriculture' to bring together the latest state-of-the-art global knowledge and development-oriented information about CA science and systems, practice and benefits and adoption and spread. This book is purposely not designed to be a theoretical debate about what scientists and academics with no practical experience of real farming or of CA think about CA and how CA is performing internationally. Global scientific and empirical evidence speaks volumes about the productivity, economic, environmental and social attractiveness of CA to farmers in all continents. Globally, the rate of adoption of CA is accelerating but much remains to be explained in terms of the superior performance of CA, and also much remains to be innovated in the coming decades to maximize the wide range of benefits offered by CA to farmers, society and the natural world.

The past and current research on CA reviewed in this book is aimed at being of value to all CA stakeholders, including students, especially in the practical context of addressing global challenges related to sustainable development with effective solutions. This is particularly true in the adoption of strategies dealing with: sustainable production intensification, climate smart agriculture, regenerative agriculture, agroecology, and restoration of degraded lands including biodiversity and land-mediated ecosystem services. Thus, this book is a contribution to making sustainable agriculture development real globally. This would not have been possible without the extraordinary help of colleagues and field experts from the global CA Community of Practice who have been involved over many years in championing and bringing about the ongoing and accelerating global CA revolution. More than 120 authors based in more than 30 countries have made this book possible, building their respective chapters on the vast amount of global evidence and knowledge that is now available regarding the superior global performance of the alternate ecologically sustainable paradigm of CA.

In the long run, the expanding knowledge and education system globally must become fully engaged in generating the future human and institutional capacity and appropriate new mindset to underpin and sustain the process of mainstreaming the CA paradigm as a core component of the much-needed sustainable global food and agriculture system. Mainstreaming of any alternate paradigm in any field requires that all the relevant public, private and civil sector institutions and policies must align themselves behind it and ensure the necessary strategic and practical support to sustain its evolution. It is hoped that the information contained in this book will contribute to more rapid mainstreaming of CA globally by inspiring and generating increasing number of stakeholders and champions, particularly youngsters with practical training of CA, to become engaged in the promotion of CA as a foundation for sustainable agricultural development.

Chapter 1

The need for Conservation Agriculture

Amir Kassam, University of Reading, UK; and Laila Kassam, Animal Think Tank, UK

1 Introduction

The ability of agriculture to meet future demand is generally analysed by mainstream scientists and policy analysts in terms of available resources and production inputs to supply the required level of agricultural products. Similarly, production systems are commonly assessed on the efficiency and effectiveness of different combinations of inputs, technologies and/or practices to produce certain agricultural outputs. Despite the continuous attention to ensure that advances in agriculture are based on science, technology and innovation (STI), it is only relatively recently that analyses have begun to address externalities of production systems, such as environmental damage; the associated input factor inefficiencies and sub-optimal yield ceilings; losses in agroecological production potentials due to land degradation; and poor resilience against major external biotic and abiotic challenges (Brisson et al., 2010; FAO, 2011a, 2016; Li et al., 2016; Nkonya et al., 2016; Vlek et al., 2017; González-Sánchez et al., 2017, 2018). However, relatively rarely do mainstream researchers question the actual agricultural paradigm itself (characterised here as conventional tillage agriculture) in terms of its continuing ecological appropriateness for the sustainable development agenda and for the environmental and land degradation challenges faced by agriculture and societies globally. Equally, the delivery of supportive, regulatory, provisioning and cultural ecosystem services to society by conventional tillage agriculture has not been an area of serious

http://dx.doi.org/10.19103/AS.2019.0048.01

mainstream research concern (MEA, 2005; Beddington, 2011; Lal and Stewart, 2013; Kassam et al., 2013; Palm et al., 2014). Of even greater longer-term concern is the fact that much of the mainstream agricultural education globally continues to promote STI and knowledge to support conventional tillage agriculture – the current dominant agricultural paradigm – which represents 'business as usual'.

Thus, in general, mainstream approaches to agricultural assessments are simplistic, conservative and limited in scope. As a result, they are unable to identify and address the root causes of the damage caused to land resources, the environment and human health by the current dominant agricultural paradigm. Such assessments are also decoupled from the human and ethical consequences of the demands and pressures placed upon agricultural production by the food and agriculture system, including consumer demand, diets, industry, government and the economy.

In general, over the past several millennia, agricultural land use globally has led to soil physical, chemical, biological and hydrological degradation. This state of affairs continues unabated in most farmlands (MEA, 2005; Montgomery, 2007; FAO, 2011a; Kassam et al., 2013). This is true on small and large farms, on farms using mechanised or manual farm power, in developing countries and in industrialised countries, in the tropics and outside the tropics. The dominant farming systems paradigm globally is based on mechanical tillage of various types to control weeds (often with herbicides), soften the seedbed for crop establishment and loosen compacted sub-soil. At the centre of this paradigm there are farming practices for crop, soil, nutrient, water and pest management that are considered by most agricultural stakeholders to be 'modern, good and normal'. However, the same farming practices have also forced farmers to accept that, supposedly, any accompanying soil degradation and loss of ecosystem services are inevitable and 'natural' consequences of farming – consequences which can be kept under control but not avoided altogether. This view is increasingly being challenged and considered to be outdated, and the inherited farming practices are considered unable to deliver the multifunctional objectives of productivity with ecosystem services now being demanded from agricultural land and producers who use it for farming. Further, an unabated and continued degradation of the production resources, namely soil, water and biological, cannot secure a sustained survival of humanity on earth considering that the degradation processes are advancing globally and that 90% of the food for humankind originates from soil-based production systems.

In the past three decades, ideas and concepts, as well as an ecosystem approach to sustainable production intensification have led to the emergence of an alternative approach to farming across all continents. Not only how and what crops are grown matters, but also the interactions of the two in space and time lead to effects and consequences which influence the system performance and the delivery of ecosystem services. Some ecosystem services involve processes such as the hydrological, carbon and nutrient cycling that operate at the level of the fields on the farm, the landscape, the watershed and beyond. In addition, agricultural soil management is undertaken within different farming systems for producing biological products for the markets, and a range of production inputs, equipment and machinery, and management skills are needed to operate successfully (Kassam et al., 2009, 2013).

This chapter illustrates and discusses the inherently destructive and inefficient nature of the conventional tillage agriculture paradigm and therefore its inability to contribute sustainably and meaningfully to present and future societal needs. It highlights the role of conventional tillage agriculture in causing soil, landscape and agroecological degradation, and its consequent inability to function optimally at maximum output and profit with maximum efficiency and resilience at any level of agricultural and economic development, and to adequately deliver ecosystem societal services. The chapter elaborates the alternate production paradigm of Conservation Agriculture (CA) which can support sustainable production intensification to meet future food and agricultural needs (Kassam et al., 2009, 2013).

It is beyond the scope of this chapter to address the ethical issues related to sustainability noted earlier. However, CA's ability to reduce production inputs and increase outputs getting 'more for less' (i.e. raise factor and total productivity) and thus provide a truly pro-poor solution for poor farmers, unlike the current paradigm, is most certainly an issue of ethics, as is the ability of CA to protect and sustain the natural resource base in agricultural ecosystems (i.e. sustainable land management) and stop and reverse land degradation, regenerate agroecosystem functions and sustainably deliver societal services for all communities.

In Volume 2, Chapter 1 describes how greater crop and land productivity potentials are being mobilised under CA which has been spreading rapidly in all continents, particularly since the 1990s (Goddard et al., 2006; Friedrich et al., 2013; Jat et al., 2014; Farooq and Siddique, 2014; Kassam et al., 2009, 2010, 2013, 2015, 2017a,b).

2 The 'hidden' reality and societal cost of conventional tillage agriculture

2.1 Conventional tillage agriculture

Conventional tillage-based production systems are often referred to as the Green Revolution agriculture paradigm. Particularly since WWII, these systems have led to a paradigm for production intensification that is based on the intensification of tillage and the notion that more output can only come from applying more purchased inputs, especially of modern seeds, agrochemicals (for crop nutrition and protection), energy for tillage, and water. Conventional tillage-based systems have generally become unsustainable due to the degradation they cause (Montgomery, 2007; Kassam et al., 2013). This degradation includes loss of agricultural land, productivity and ecosystem and societal services (Montgomery 2007; Goddard et al., 2006; Kassam et al., 2009, 2013; Lindwall and Sonntag, 2010; Basch et al., 2012; Jat et al., 2014; Farooq and Siddique, 2014). Yet, agricultural science that continues to be supported by most governments and institutions deals mainly with tillage agriculture, whether inorganic or organic. However, the situation has been changing and increased attention is being directed towards generating and applying scientific and empirical knowledge and experience for CA, particularly since the end of the 1990s and the beginning of this century (Kassam et al., 2009, 2010, 2013, 2015, 2019).

The Green Revolution approach does not seem to be going anywhere now, even in nations such as India and Pakistan where it is claimed to have made a special impact in the 1960s and the 1970s, nor in the West where it all began after the end of WWII. For example, it is often stated that countries in Asia were the first to benefit from the Green Revolution paradigm in the developing world, but the question that arises is, why did the Green Revolution not continue to spread across India and Pakistan, or across Asia to benefit more and more smallholder farmers? (FAO, 2011b, 2016). In fact, the conventional 'modern' approach to crop production intensification, based on expensive inputs of intensive tillage, modern seeds, high agrochemicals and energy is often not affordable by resource-poor smallholder farmers. Where inputs are affordable or subsidised, the situation invariably leads to increased soil erosion and land degradation and sub-optimal yield ceilings.

The input-intensive Green Revolution mindset also includes the indoctrination and creation of a certain behavioural culture in agri-*culture*. This culture promotes the notion that farmers and their service providers and governments do not need to worry about the negative externalities that arise from the production practices being applied (Pretty, 2002; Beddington, 2011). The approach does not even call for an understanding by producers, dealers and their extension advisors, of the key ecological elements, functions and processes in the agroecosystem that should be managed and sustained to serve as the ecological foundation of sustainable production intensification (Kassam et al., 2009, 2010, 2013). In addition, the science and technology related to intensification under the Green Revolution paradigm has led to the application of economic models such as commodity specialisation that have led to extended landscapes with monocropping (Pretty, 2002).

Thus, the question of how the ecological foundation of agriculture should be managed to enhance and deliver both the desired output and ecosystem services[1] to society, while performing at the highest possible levels of efficiency and resilience, including coping with the climate change, does not receive the attention it deserves. Nor is there any serious concern being expressed in the Green Revolution approach about agricultural land area continuing to be severely degraded and abandoned in the North and the South due to the negative impact of the conventional tillage-based production paradigm (Juniper, 2013; Montgomery, 2007; Nkonya et al., 2016; Vlek et al., 2017). Indeed, many areas which in human history were the cradle of culture and intensive tillage-based agriculture are deserts today (Montgomery, 2007).

2.2 Agricultural soil degradation

Soil is a non-renewable resource that mediates and ensures crucial environmental, social and economic functions, and it has a central role in any approach aimed at defining the principles and practices of sustainable agriculture (Warkentin, 1995). Agricultural soil degradation is generally understood as the loss in the quality or productivity of soil because of human activities, leading therefore to less intensive usages or even land abandonment for agricultural use. In the 'Guidelines for General Assessment of the Status of Human-induced Soil Degradation' (Oldeman, 1988) the different forms of human-induced soil degradation are distinguished comprehensively between two main categories, which are: (a) the displacement of soil material through water and wind erosion; and (b) the chemical and physical deterioration, such as the depletion of soil nutrients and organic matter, salinisation, acidification and pollution, but also compaction, sealing and crusting, truncation of the soil profile or waterlogging. Despite this distinction between the two categories, there is a strong relationship between them, once the occurrence and degree of soil displacement is appreciated as being a consequence of chemical and physical deterioration of the soil. In addition, both categories of agricultural soil degradation may lead to severe off-site effects such as the sedimentation of reservoirs, harbours or lakes; flooding, river-bed filling and riverbank erosion, eutrophication of water bodies, etc.

In such earlier definitions and descriptions of agricultural soil degradation, the soil is treated mainly as a physical entity. However, a productive agricultural soil is a living system in which biological processes

1 Ecosystem services are provided to society by nature. Such services include edible and nonedible biological products, clean drinking water, processes that decompose and transform organic matter, and cleansing processes that maintain air quality. Several categories of ecosystem services are recognised: provisioning, regulating, cultural, and supporting (Millennium Ecosystem Assessment – MEA, 2005). In agricultural landscapes, provisioning ecosystem services can be delivered effectively and efficiently when the linked regulatory and supporting services are also allowed to operate normally. Ecosystem functions that protect and enhance regulatory and supporting ecosystem services in the soil and landscape in which crops are grown appear, in general, to offer an effective way of harnessing the best productivity, as well as ecological and economic performances (Kassam et al., 2013). Chapter 13 in Volume 2 is about harnessing ecosystem services with CA.

carried out by soil microorganisms and meso-fauna are key elements in the creation, maintenance and enhancement of soil health and its productive capacity. Soil health represents the soil's physical, chemical, hydrological and biological status and its ability to respond to agricultural production inputs and to climatic variability including extreme weather events. For example, soil physical, hydrological and chemical characteristics such as soil structure and porosity, soil aeration, water infiltration and drainage, soil water and nutrient retention capacity, total exchange capacity, pH are greatly influenced by soil biological properties such as soil organic matter turnover and the dynamics of soil biodiversity which has an intimate relationship with plant roots, affecting its phenotypic expression and stress behaviour below and above the ground. The deterioration of soil biological health, and the consequent loss in soil productive capacity, is often not given much prominence in agricultural soil management and degradation research or in farming system management. Thus, the role of soil microorganisms and meso-fauna and the soil organic matter they require to function effectively and self-sustainably in the maintenance of soil health and the important role they play in crop phenotypic expression, resilience and crop performance, are overlooked. This includes the diverse kinds of symbiotic relationships that exist between soil biodiversity and plants about which we know very little (Uphoff et al., 2006), presumably because of the difficulty in establishing through scientific experimentation the causal relationships with productivity and ecosystem services.

According to The Global Assessment of Human-Induced Soil Degradation (GLASOD) up to half the world's agricultural land is degraded to some degree (Oldeman et al., 1991). Degradation of cropland is most extensive in Africa, affecting 65% of cropland areas, compared with 51% in Latin America and 38% in Asia (CA, 2007). Loss of organic matter and physical degradation of soil, that not only reduces nutrient availability and has significant negative impacts on infiltration and porosity that consequently impact local and regional water productivity, the resilience of agroecosystems, and global carbon cycles. Accelerated on-farm soil erosion leads to substantial yield losses and contributes to downstream sedimentation and the degradation of water bodies and infrastructure (Vlek et al., 2010). Nutrient depletion and chemical degradation of soil are a primary cause of decreasing yields and result in low on-site water productivity and off-site water pollution. Globally, agriculture is the main contributor to non-point-source water pollution. Water quality problems can often be as severe as those of water availability. Secondary salinisation and waterlogging in irrigated areas threaten productivity gains. According to the Millennium Ecosystem Assessment (MEA, 2005) some two-thirds of our ecosystems are degraded. According to FAO (2011a), only some 10% of the global agricultural land is under improving condition, the rest has suffered some degree of degradation, with 70% being characterised as being moderately to highly degraded.

Unfortunately, the problem of agricultural soil degradation is often considered to be unique to tropical and sub-tropical regions (Greenland and Lal, 1977) or confined in the developing regions, which is now recognised to be not so. Soil mismanagement and the traditional physical view of soils have led to serious soil degradation in temperate agroecologies in the industrialised countries (Pretty, 2002; Montgomery, 2007). For example, in 2002 the European Union initiated the so-called 'Thematic Strategy for Soil Protection', as it recognised that 'Soil is a vital and non-renewable resource and had not been the subject of comprehensive EU action'. At that time, the Commissioner of Environment even said that 'for too long, we have taken soil for granted. However, soil erosion, the decline in soil quality and the sealing of soil are major problems across the EU.' The ensuing discussion in the frame of this strategy identified eight major soil threats, which are soil erosion, decline in soil organic matter, soil contamination, soil sealing, soil compaction, decline in soil biodiversity, salinisation, floods and landslides. Notwithstanding this, the approach to understanding the root causes of soil degradation in any agricultural environment has remained relatively narrow, lacking the fuller appreciation of the role of soil biology in the maintenance of soil health, the role of symbiotic relationships between soil microorganisms and crop performance, and the disruptive effect of mechanical soil disturbance on soil health and productive capacity, and on production system resilience (Kassam et al., 2009, 2010, 2013).

2.3 Causes of degradation of agricultural soils

The root cause of soil degradation in agricultural land use and of decreasing productivity – as seen in terms of loss of soil health – is the low soil-carbon and soil-life disrupting paradigm of mechanical soil tillage which, to create conditions for improved crop performance, debilitates many important soil-mediated ecosystem functions (FAO, 2008). For the most part agricultural soils are becoming de-structured; our landscape is exposed and unprotected; and soil life is starved of organic matter, reduced in biological activity and deprived of habitat. The loss of soil biodiversity, damaged structure and its self-recuperating capacity or resilience, increased topsoil and sub-soil compaction, runoff and erosion, and greater infestation by pests, pathogens and weeds indicate the current poor state of the health of many of our soils. In the developing regions, this is a major cause for inadequate food and nutrition security.

In industrialised countries, the poor condition of soils due to excessive disturbance through mechanical tillage is being exacerbated by (a) the over-reliance on the application of mineral fertilisers, as the main source of plant nutrients, on to farmland that has been losing its ability to respond to nutrient inputs due to degradation in biological soil health – related to declining stocks of soil carbon – including loss/destruction of adequate soil porosity and reduced soil moisture storage and increased runoff, leading to poor root system, nutrient loss and decreased in nutrient uptake, and (b) reducing or doing away with crop diversity and rotations including legumes and pastures (which were largely in place around the time of WWII) facilitated by high levels of agrochemical inputs, standardised fixed agronomy, and commodity-based market forces that are insensitive to on-farm and landscape ecosystem functions.

The situation in the industrialised nations is now leading to further problems of increased threats from insect pests, diseases and weeds against which farmers are forced to apply ever more pesticides and herbicides, and which further damages biodiversity and pollutes the environment.

It seems that with mechanical tillage (intensive or otherwise) and with low soil input of atmospheric carbon and nitrogen and exposed soil surfaces as a basis of the current agriculture production and intensification paradigm, we have now arrived at a 'dangerous' point in soil and agroecosystem degradation globally, including in the industrialised North. However, we also know that the solution for sustainable soil management for farming has been known for a long time, at least since the mid-thirties when the mid-west of USA suffered massive dust storms and soil degradation due to intensive ploughing of the prairies. Dust bowls and large-scale soil degradation continue to occur in vast regions and in developed and developing countries (Baveye et al., 2011), despite the recognition of soil health being critical to life on earth.

For instance, in 1943, Edward Faulkner wrote a book *Ploughman's Folly* in which he provocatively stated that there is no scientific evidence for the need to plough. More recently, David Montgomery in his well-researched book *Dirt: The Erosion of Civilizations* shows that generally with any form of tillage, including non-inversion tillage, the rate of soil degradation (loss of soil health) and soil erosion is generally by orders of magnitude greater than the rate of soil formation, rendering agroecosystems unsustainable. According to Montgomery's research, tillage has caused the destruction of the agricultural resource base and of its productive capacity nearly everywhere, and continues to do so (Montgomery, 2007).

For these natural science writers as far back as 1945, tillage, regardless of type and intensity, is not compatible with sustainable agriculture. We only have to look at the various international assessments of the large-scale degradation of our land resource base and the loss of productivity globally to reach a consensus as to whether the further promotion of any form of tillage-based agriculture is a wise development strategy. We contend that to continue with intensive tillage agriculture now verges on irresponsibility towards society and nature. Thus, we maintain that with tillage-based agriculture in all agroecologies, no matter how different and unsuitable they may seem for no-till farming, crop productivity (efficiency) and output *cannot be optimised* to the full potential. Further, agricultural land under tillage is not fully able to deliver the needed range and quality of environmental services that are mediated by ecosystem functions in the soil, terrain, climate and biodiversity system. Obviously, something must change.

3 Agricultural intensification based on the Green Revolution agricultural paradigm

3.1 Intensification since WWII

The post-WWII agricultural intensification placed increasing reliance upon breeding 'new' high yielding seeds, more intensive tillage of various types pulled by heavy and more powerful machines, combined with even more chemical fertilisers, pesticides and herbicides, supposedly making crop rotations superfluous and promoting apparent efficiency through specialisation with monocropping. This was complemented with the notion that with more mineral nitrogen input came the need for new, more responsive cultivars because the traditional cultivars were not capable of responding to higher doses of mineral nitrogen. A slogan of that era, coined by DuPont, was 'Better Living through Chemistry'. Agroindustry and the Land Grant Colleges joined forces in promoting an industrial model for agriculture that was based on the use of chemical inputs and large volumes of output. Even FAO launched in 1961 the Freedom From Hunger Campaign (FFHC), which was partly financed by the world fertiliser industry. The FFHC's main target was to encourage the use of fertilisers by small-scale farmers through education, effective means of distribution and credit. The overall idea was that agricultural production could not be significantly increased in the developing countries of the world without improving the nutrient status of most soils. In the late 1970s, the FFHC was replaced by FAO´s Fertiliser Programme. Concurrently, rapid urbanisation and land consolidation in the industrialised countries forced agriculture 'labour' to be substituted by 'capital', particularly in the form of agricultural equipment and machinery. Large tractors with large ploughs became common in the 1980s and symbolised modern farming. This technological 'interventionist' approach became the accepted paradigm for production intensification, and was promoted globally including in the developing regions – referred to as the Green Revolution paradigm of the 1950s, '60s and '70s – and which, despite boosting crop yields, increased the likelihood of soil erosion and degradation, loss in ecosystem functions and in productivity and output, leading to abandoned and desertified farmland and landscapes.

Smallholder farmers in developing regions using manual labour to till the land and burning or removing all crop residues from the field also experience the above consequences and remain trapped in a degrading vicious cycle that cannot be broken just by applying mineral fertiliser and replacing traditional varieties with the latest breeding results. This also applies to farms in industrialised regions where the voices demanding more sustainable farming practices, both environmentally and economically, are getting louder. As soil degradation advances, the need for purchased inputs increases until the point where the compensatory effect is no longer possible, forcing farmers to use even higher inputs with equally higher environmental impact.

Examples of large-scale agricultural soil degradation from different parts of the world appear to share several common experiences as can be seen from the cases of South America, China and Australia reported in Kassam et al. (2013). These cases reflect contrasting agricultural environments ranging from the tropical and sub-tropical environment with summer rainfall in Brazil to sub-tropical environment with winter precipitation in Western Australia to temperate environment with winter precipitation in northern China from east to west.

Various reports state that between 7 and 12 million ha of agricultural land are lost or abandoned every year due to land degradation. We believe this area includes 0.4 to 0.5 billion ha of agricultural cropland that was once suitable but has been degraded and abandoned over the years (Dregne and Chou, 1992; Pimentel et al., 1995; Montgomery, 2007; Gibbs and Salmon, 2015), particularly since WWII. This abandonment is due to the severe degradation and erosion arising from tillage-based agriculture systems, and the science, education and development community that supports them through the public, private and civil sector institutions, in both industrialised and less-industrialised countries (Montgomery, 2007). A recent study puts the annual global cost of agricultural land degradation due to land use and cover change at 300 billion USD (Nkonya et al., 2016). Other reports indicate much higher costs, and in cases where priceless ecosystem services are lost, it is not possible to put a cost value (Juniper, 2013).

Additionally, it is assumed that yield gaps can continue to be filled based on the current practice of intensive tillage-based soil management and increased application of costly and excessive production inputs, assuming the same or even higher production increase rates than in the past. In other words, the current agricultural paradigm assumed to meet future food demand in the scenarios of the FAO and of their collaborators such as the IFPRI, World Bank and others, is the land degrading and land destroying 'business as usual' tillage agriculture (FAO, 2012a). This 'more of the same' approach to intensification and maintenance of sub-optimal yields, factor productivities and overall performance can no longer be considered to be economically, environmentally, socially or developmentally sustainable anywhere; not in industrialised nations nor in emerging economies (Montgomery, 2007; Beddington, 2011; Gibbs and Salmon, 2015; Nkonya et al., 2016). In low-income countries, tillage agriculture based on the use of hoes and animal traction to pull simple ploughs and tillage equipment also leads to land degradation and loss of top soil and soil functions, to the point where land is eventually abandoned. Often, the lack of mineral fertilisers and fertility-enhancing and soil-health-promoting cropping systems accelerates the loss in crop and land productivity (Montgomery, 2007; Gibbs and Salmon, 2015; Nkonya et al., 2016).

Further, in many important high yielding production areas, crop yields have reached sub-optimal ceilings. For example, national-level yields of wheat crops appear to have stagnated at about 7 t ha^{-1} since around 1996 across several countries in Europe (Fig. 1) with inputs and input costs going up, and diminishing returns setting in (Brisson et al., 2010). In some countries in Europe such as Switzerland and Spain, wheat yields seem to have stagnated over the last 25 years (Table 1).

3.2 Agricultural land use unsustainability at a structural level

As indicated earlier, 0.4–0.5 billion ha of agricultural lands are reported to have been abandoned since WWII due to severe soil and land degradation, and loss of biodiversity and resilience. Yields of staple cereals in industrialised and

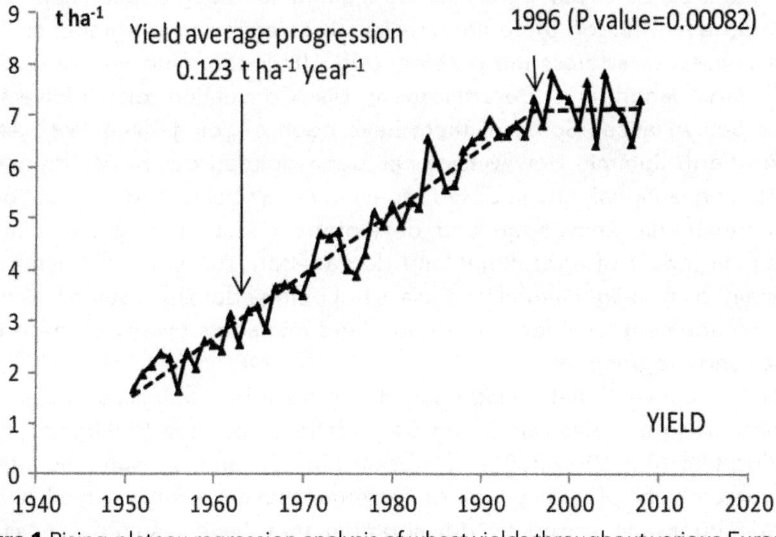

Figure 1 Rising-plateau regression analysis of wheat yields throughout various European countries. Source: adapted from Brisson et al. (2010).

Table 1 Year of stagnation in wheat yields in countries in Europe

Country	Year of stagnation
Denmark	1995*
France	1996*
Germany	1999
Italy	1994
The Netherlands	1993*
Spain	1989
Switzerland	1990*
United Kingdom	1996*

Year of stagnation: * very significant $P < 0.01$; no star $P > 0.05$.
Source: adapted from Brisson et al. (2010).

less-industrialised regions appear to have stagnated under tillage agriculture (Kassam et al., 2009, 2013, 2017b; Montgomery, 2007; Brisson et al., 2010, Jat et al., 2014; Farooq and Siddique, 2014; Gibbs and Salmon, 2015). These are signs of unsustainability and the institutionalisation of the supporting systems of agricultural research to generate new technology and knowledge. Thus, it is at the structural level, for both the supply and demand sides of food and agricultural security, that we need transformed mindsets about production, consumption and distribution, and policies and institutional capacities for science and development to support an alternative agricultural paradigm described in the next section.

What is surprising though is that agricultural land degradation continues unabated even though there are several UN treaties and programmes that are supposed to address the problem (e.g. UN Programme on Combating Land Degradation and Desertification; UN Convention on Biodiversity, Global Soil Alliance). Some of them have been ongoing since 1992, after the Rio Earth Summit. However, as has been pointed out by Montgomery (2007), conventional tillage-based agriculture globally, and the national and international knowledge and development systems that maintain it, is the main driver of agricultural land degradation. This structural nature of unsustainability in agricultural land use is not being addressed at the practical land management level for farmers and land managers by any of these UN treaties and programmes.

Thus, conventional tillage-based agricultural practices have all contributed to more soil and land degradation, decreased infiltration and waterlogging (Fig. 2), runoff and erosion (Fig. 3a and b), water pollution, and vulnerability of agriculture to extreme climatic events. At all levels of agricultural and economic development they have resulted in loss of agricultural land, decrease in attainable yields and in input factor productivity

Figure 2 Soil compaction and loss in water infiltration ability caused by regular soil tillage lead to impeded drainage and flooding after a thunderstorm in the ploughed field (right) and no flooding in the no-till field (left). Photograph taken on June 2004 in a plot from a long-term field trial 'Oberacker' at Zollikofen close to Berne, Switzerland, started in 1994 by SWISS NO-TILL. The three water-filled 'cavities' in the no-till field derive from soil samples taken for 'spade tests' prior to the thunderstorm. Source: Wolfgang Sturny.

(Montgomery, 2007). They have also contributed to excessive use of seeds, agrochemicals, water, energy and labour, all leading to increased cost of production. They have led to poor production system resilience, dysfunctional agroecosystems, degraded ecosystem services to society, including lower water quality and quantity, poor nutrient and carbon cycling, sub-optimal water, nutrient and carbon provisioning and regulatory water services, and loss of soil and landscape biodiversity. They all constitute the high, real cost of food production, of agricultural products for industry, and of environmental management being passed on to the public, and to future generations (Kassam et al., 2009, 2013).

On the other hand, if we are to (i) mobilise greater crop and land productivity potentials sustainably to meet future food, agriculture and environmental demands; (ii) maintain the highest levels of productivity, efficiency and resilience (getting 'more from less'); and (iii) rehabilitate degraded and abandoned agricultural land and ecosystem services we would need to replace the faltering production 'engine' of the conventional tillage-based production paradigm. We need to transform all the components of the food and agriculture systems, including the supporting science, education and policy that are built

Figure 3 (a) Erosion and runoff on conventionally tilled bare soil near Cordoba, Spain (Source: Emilio Gonzalez-Sanchez). (b) Erosion and runoff on conventionally tilled bare soil in southern Brazil (Source: Rolf Derpsch).

upon it. This transformation is now ongoing as far as agriculture production is concerned and needs to be accelerated (Goddard et al., 2006; Kassam et al., 2009, 2013, 2015, 2017b, 2019; Lindwall and Sonntag, 2010; FAO, 2011b, 2016; Jat et al., 2014; Farooq and Siddique, 2014).

The following sections elaborate on an alternative agricultural paradigm that is being applied by farmers to: minimise and reverse the agricultural soil and landscape degradation trends; and mobilise greater agricultural land potentials.

4 Replacing conventional tillage agricultural with Conservation Agriculture

4.1 Nurturing soils as living biological systems

A productive soil is a living biological system and its health and productivity depend on managing it as a complex biological system, not as a geological entity. This is because soil's productive capacity is derived from its many components (including physical, biological, chemical, hydrological, climate, cropping system, management, development level) all of which interact dynamically in space and time within cropping systems and within agroecological and socioeconomic environments (Kassam et al., 2013).

Alongside the concern for soil erosion and the destruction of soil structure and soil life caused by frequent and intensive tillage has been the growing understanding of the important role soil life and soil biology play in the maintenance of soil health. According to Doran and Zeiss (2000):

> Soil health is the capacity of soil to function as a living system with ecosystem and land use boundaries, to sustain plant and animal productivity, maintain or enhance water and air quality, and promote plant and animal health.

According to Peter Trutmann, quoted in FAO (2008), the above emphasises, a unique property of biological systems, since inert components cannot be sick or healthy. Management of soil health thus becomes synonymous with management of the living portion of the soil to maintain essential functions of soil to sustain plant and animal productivity, maintain or enhance water and air quality, and promote plant and animal health (Kassam et al., 2013). According to David Wolfe, quoted in FAO (2008), healthy soils maintain a diverse community of soil organisms that help to control plant disease, insect and weed pests, form beneficial symbiotic associations with plant roots (e.g. nitrogen-fixing bacteria and mycorrhizal fungi); recycle essential plant nutrients; improve soil structure (e.g. aggregate stability) with positive repercussions or soil water and nutrient holding capacity, and ultimately improve crop production. More recently, it has become clear that mycorrhiza also play an important role in carbon sequestration by producing glomalin, and inert sugar-protein that also acts as a biological cement in the creation of stable soil micro- and macro-aggregates.

In this context, ongoing supply of carbon-rich organic matter for soil organisms is essential, from which they source both energy and nutrients. Examples of management practices for maximising soil health include maintaining vegetative cover of the land year-round to increase organic matter input and minimise soil erosion, more reliance on biological as opposed to chemical approaches to maintain crop productivity (e.g. rotations with legumes and disease-and weed-suppressive cover crops), and avoiding physical (mechanical) interventions which might compact, alter or destroy the biologically created porous structural arrangements of soil components (FAO, 2008).

A key factor for sustainability in any production system, in contrast to sustainable intensification, was described in FAO (2008) as follows:

> Sustainability of land's capacities to continue yielding both plant products and water year after year depends primarily on maintaining the soil in fit condition for active life processes of the whole soil/plant system. This relates to the ongoing generation and re-generation of the porous soil architecture – the soil's 'self-recuperation capacity' – with respect to the repair of damaged soil and to its physical resilience in the face of adverse shocks of weather and/or of poor management.

It is now recognised more widely that a productive agricultural soil, together with its inhabiting plants and other biota, is a living biological system (Tikhonovich and Provorov, 2011; Doran and Zeiss, 2000; Doran, 2002) which is made up of a complex web of interactions between a large diversity of microorganism and meso-fauna and between microorganisms and plants roots as well as above-ground parts. Relatively little is known about this complex agrobiodiversity or soil biota and its ecosystem functions, as its role in crop productivity has been generally ignored, even in recent decades.

In addition to the need to feed soil food web with organic matter, it is equally necessary to recognise the role organic matter play in establishing and sustaining food webs at the ground level and above the ground. These food webs form part of the pathogen and insect pest-predator dynamics that maintain a balance between pests and their natural enemies so that crops are not readily attacked by insect pests and diseases.

The above-described soil biological processes and ecosystem functions as well as above-ground pest-predator dynamics cannot be performed adequately in soils and on agricultural lands that are mechanically disturbed by tillage regularly and whose structure and porosity are repeatedly impaired as a result. Soil biological health is further hindered by the inadequate amount of organic substrate being supplied, to feed and maintain soil microorganisms and natural enemies of pests and their functions, at rates equal to, or faster than, its rate of oxidation following tillage.

As FAO's 'Save and Grow' approach and other reviews have shown (FAO, 2011b, 2016; Kassam et al., 2013; Jat et al., 2014; Kassam et al., 2011, 2017a,b), to harness the conditions that are sufficient for achieving sustainable production intensification, agriculture must literally return to its roots and rediscover the importance of healthy soils, landscapes and ecosystems. At the same time, it must conserve resources, enhance natural capital and the flow of ecosystem and societal services at all levels – field, farm, community, landscape, territory and national (and beyond). The no-till agricultural production paradigm, known as CA, is totally compatible with the above multidimensional and multifunctional goal.

4.2 Principles and description of Conservation Agriculture

According to Derpsch (2004), research on 'conservation' or reduced tillage with early versions of a chisel plough was initiated in the Great Plains in the 1930s to alleviate wind erosion. Stubble mulch farming was also developed and can be seen as a forerunner of no-tillage farming. This collection of practices led to what became known as conservation tillage which includes a range of tillage practices from high soil disturbance tillage to low soil disturbance that maintains at least 30% soil cover.

The *Ploughman's Folly* (Faulkner, 1943) was an important milestone in the development of sustainable soil management for agriculture. As noted earlier, Faulkner questioned the wisdom of ploughing and explained the destructive nature of soil tillage. Further research in the UK, the USA and elsewhere during the late-1940s and 1950s and the development of herbicide technology made no-tillage farming possible, and the practice began to spread in the USA in the 1960s, and in Brazil, Argentina, Paraguay and Australia in the 1970s. In 1973, Shirley Phillips and Harry Young published the book *No-tillage Farming* (Phillips and Young, 1973), the first of its kind in the world, and this was followed in 1984 by the book *No-Tillage Agriculture: Principles and Practices* by E. R. Phillips and S. H. Phillips (Phillips and Phillips, 1984).

The modern successor of no-till farming – now generally known as CA – goes much further as elaborated in the following section. It involves the simultaneous application of the following three practical principles based on locally formulated practices, along with complementary good agricultural practices of crop and production management, all contributing to enhancing and sustaining optimum system performance and resilience, namely (Hobbs, 2007; FAO, 2008, 2011b; Friedrich et al., 2009; Kassam et al., 2009, 2013, 2019; www.fao.org/conservation-agriculture).

a) **Continuous no or minimum mechanical soil disturbance:** implemented by the practice of no-till seeding or broadcasting of crop seeds, and direct placing of planting material into untilled soil; no-till weeding; causing minimum soil disturbance from any cultural operation, harvest operation or farm traffic. Sowing seed or planting crops directly into untilled soil and no-till weeding: reduces runoff and soil erosion; minimises the loss of soil organic matter through oxidation; reduces disruptive mechanical cutting and smearing of pressure faces; promotes soil microbiological processes; protects and builds soil structure and connected pores; avoids impairing movement of gases and water through the soil; and promotes overall soil health.

b) **Maintaining a permanent mulch cover on the soil surface:** implemented by retaining crop biomass, rootstocks and stubbles and biomass from cover crops and other sources of biomass from ex-situ sources. Use of crop residues (including stubbles) and cover crops: reduces runoff and soil erosion; protects the soil surface; conserves water and nutrients; supplies organic matter and carbon to the soil system; promotes soil microbiological activity to enhance and maintain soil health including structure and aggregate stability (resulting from glomalin production by mycorrhiza); and contributes to integrated weed, insect pest and pathogen management and to integrated nutrient and water management.

c) **Diversification of species in the cropping system:** implemented by adopting a cropping system with crops in rotations, and/or sequences and/or associations involving annuals and perennial crops, including a balanced mix of legume and non-legume crops and cover crops. Use of diversified cropping systems: contributes to diversity in rooting morphology and root compositions; enhances microbiological activity; enhances crop nutrition and crop protection through the suppression of pathogens, diseases, insect pests and weeds; and builds up soil organic matter. Crops can include annuals, short-term perennials, trees, shrubs, nitrogen-fixing legumes and pastures, as appropriate.

The mindset that is driving the CA community of practice defines CA as an ecosystem approach to regenerative sustainable agriculture and land management based on the practical application of context-specific and locally adapted three interlinked principles described earlier. CA systems are present in all continents, involving rainfed and irrigated systems including annual cropland systems, perennial systems, orchards and plantation systems, agroforestry systems, crop-livestock systems pasture and rangeland system, organic production systems and rice-based systems. Conservation Tillage, Reduced Tillage and Minimum Tillage are not CA, nor is No-Till on its own (Derpsch et al., 2014; Kassam et al., 2013, 2019; Reicosky, 2015). A practice such as No-Till can only be referred to as being a CA practice if it is part of an actual CA system as per the above definition. Similarly, for soil mulch practice and crop diversification practice, both of which can only be considered to be

CA practices if they are part of a CA system based on the application of the three interlinked principles.

Thus, CA along with complementary practices pay particular attention to establishing dynamic ecological conditions in the whole soil-plant-landscape system that offers resilient and climate-smart performance with maximum productivity (output and efficiency) and flow of ecosystem and societal services. This is possible because CA systems enhance soil health by nurturing soil and landscape as living biological systems, promote good functional relationships between root system and soil microorganisms, and generate system resilience by integrating and promoting biodiversity in the cropping system, in the soil and in the landscape (Kassam et al., 2013).

CA is considered to be climate-smart because of its ability to mitigate climate change, including through carbon sequestration (FAO, 2012b), and to be more adaptable to increased climatic variations in rainfall and temperature (González-Sánchez et al., 2017, 2018, 2019). Thus, CA is increasingly being considered as a suitable core component of climate-smart agriculture (Kassam et al., 2017b).

4.3 Integrating Conservation Agriculture principles into farming systems

Principles of CA can be integrated into most if not all types of land-based production or farming systems. This is because they provide ecological underpinnings to production and farming systems to generate greater productivity and environmental benefits. For example:

CA cropping systems with annual crops now exist in all continents and major agroecologies exhibiting productivity, economic, environmental and social benefits resulting from improvement of soil health, reduced input requirement and more efficient use of all production inputs including seeds, nutrients, water, pesticides, energy, labour and capital. These CA systems replace the tillage-based arable systems and involve cropping system diversification based on intercropping and mixed cropping as well as rotations and sequences (Jat et al., 2014; Farooq and Siddique, 2014; Kassam et al., 2017b).

Organic agriculture based on CA can lead to greater soil health and productivity, increased efficiency of use of organic matter, and reduction in use of energy. Organic CA farming is already being practised on a smaller scale in the USA, Brazil and Germany, as well as by subsistence CA farmers in Africa and elsewhere. Tillage-based organic farming is often characterised more by what practices it excludes from its production systems than by what it actually does to harness sustainable production intensification and ecosystem services. Introducing CA principles into organic farming would reduce soil disturbance, improve weed control with mulch cover and crop diversification, and generate greater amounts of organic matter from *in situ* sources within a more diversified cropping system involving legumes (Altieri et al., 2011).

Perennial systems such as orchards and plantations are taking advantage of CA principles in all continents. In Europe, olive groves and vineyards, as well as fruit tree orchards, are being transformed into CA systems. Spain and Italy have the largest area of CA olive groves involving ground covers of different kinds. Similarly, plantation systems are also being maintained as CA systems or converted to CA systems. These include shaded coffee plantations in El Salvador and oil palm and rubber plantations in Malaysia (Kassam et al., 2013).

Agroforestry systems involve the cultivation of woody perennials and annual crops together in a sustainable manner and are increasingly practised in degraded areas with perennial legumes (Saha et al., 2010). CA works well with trees and shrubs and within agroforestry and related systems. In fact, several tree crop systems in the developing and developed regions already practice some form of CA, but these systems can be further enhanced with improved crop associations including legumes, and integration with livestock. Alley cropping has been one innovation in this area that is beginning to offer productivity, economic and environmental benefits to producers (Sims et al., 2009).

CA with trees has now become an important option for many farming situations, particularly in the tropics. These CA systems incorporate varying densities of fertiliser trees in order to enhance biological nitrogen fixation, increase biomass production for surface residue, and conserve moisture. They have become the basis for major scaling-up programmes with hundreds of thousands of farmers in Zambia, Malawi, Niger, and Burkina Faso (Garrity et al., 2010; Garrity, 2011). The incorporation of the indigenous acacia species *Faidherbia albida* into maize-based CA in Zambia on a large scale is a noteworthy example. These programmes have demonstrated the practical opportunities for combining fertiliser trees with CA in both small-scale and commercial-scale farming systems.

Shifting agriculture, also referred to as 'swidden' or 'slash and burn', entails the clearing of land to prepare a cultivation plot and subsequently returning this to re-growth and eventual natural reforestation, during which damaged soil structure and depleted 'indigenous' plant nutrients are restored. Shifting cultivation has acquired a negative connotation, particularly because of the burning of vegetation. However, for sustainable intensification, such systems can be adapted to follow CA principles, changing from slash and burn systems into *slash and mulch* systems with diversified cropping (including legumes and perennial crops) that reduce the need for extra land clearing.

Rice-based systems based on CA principles have taken root in a number of countries including Brazil, Argentina, India, Pakistan, Bangladesh, Laos, Cambodia, Vietnam, China and North Korea. Trained farmers have shown CA to offer higher income and productivities (use efficiencies) of inputs of labour, nutrients and water, and as requiring less seeds, water, energy, fertiliser and labour compared with conventional irrigated or rainfed flooded rice production systems. CA advantages have been shown to apply to traditional as well as modern cultivars which also appear to be less susceptible to pest and disease damage. The no-till CA wheat-rice cropping system in the Indo-Gangetic Plains in India involves some 3.5 million ha with biomass from the previous rice crop providing soil cover and changes in irrigation practice of flooding to those maintaining aerobic soil conditions such as sub-surface drip irrigation.

Integrated crop-livestock systems including trees and pasture have long been part of agriculture. In recent decades, there have been practical innovations that harness synergies between the production sectors of crops, livestock and forestry that ensure economic and ecological viability while providing a flow of valued ecosystem services to society. System integration increases environmental and livelihood resilience through increased biological diversity, more efficient nutrient cycling and recycling, improved soil health, enhanced forest preservation and contributes to adaptation and mitigation of climate change. The integration of production sectors can enhance livelihood diversification and efficiency of use of production inputs including labour, offer resilience to economic stresses and reduce risks (Landers, 2007; FAO, 2010).

However, in most situations livestock integration into CA systems is not necessary in an absolute sense to capture the ecological and production benefits. In general, livestock production is not an efficient or optimum way of producing food, nor is it the best way to ensure food security particularly in areas where plant-based food can be grown reliably in diversified cropping systems. Further, the environmental footprint of livestock production tends to be much greater than its pant-based equivalent in terms of calories and proteins. In the case of ruminants, there is an added problem of methane emission, contributing to global warming. Thus, it is increasingly being realised internationally that integration of livestock into CA systems should not be thought of as being an essential part of CA or even a necessary part of a CA production system.

In many parts of the world animal flesh and secretions have become a central part of the diet due to various factors including colonisation, the globalisation of Western diets, manufactured demand of the corporate food regime and the 'industrial grain-oilseed-livestock complex' which is integral to it (Nibert, 2013; Weis 2013). The fact that animal products have become such a central part of the diet, especially since WWII, means it is difficult for many, including some scientists and policy-makers, to imagine that livestock production is not an essential component of sustainable land use and environmental management. It is even more difficult for many to accept that livestock products in the diet are not as healthy as we have been led to believe (Willett et al., 2019). In any case, if and when livestock is produced, it is beneficial to integrate it into CA systems so that the fodder base is produced close to the livestock production and the nutrient recycling avoids pollution problems resulting from intensive livestock production.

5 Advantages of Conservation Agriculture at the field and landscape level

Evidence from different parts of the world suggests that it may not be possible to separate sustainable soil management from sustainable production system management. Both are inextricably linked in ways that sustainable crop production systems must first be ecologically sustainable. This means that any production system that permits the mechanical disruption of soil life and biology and soil structure and quality, and therefore ecosystem services, cannot be considered to be sustainable ecologically. The aim of 'sustainable soil management' should be to reverse the trends indicated by the items listed above, via the inducing of improvements in the quality of the soil as a rooting-environment for plants. Also, an agricultural soil system is of no value if the crops grown are attacked by weeds, insects and pathogens. In other words, sustainable soil management is not enough for sustainable production as an outcome, and certainly not where sustainable production intensification is the objective in which care must be taken of crop, soil, nutrient, water, pest and farm power management in space and time to remain ecologically and economically viable.

The agroecological principles that underpin sustainable production systems for small and large farmers from an eco-commercial viewpoint relate to resource conservation and efficiency of resource use, both natural and purchased, while profitably managing sustainable production intensification and ecosystem services. At its core and based upon a large amount of empirical evidence from farmers themselves in all continents, we can say that sustainable production derives from a number of practical principles that can be applied simultaneously through combined crop-soil-water-nutrient-pest-ecosystem management practices. These practices are locally devised and adapted to capture a range of productivity, socioeconomic and environmental co-benefits of agriculture and ecosystem services at the farm, landscape (watershed) and provincial or national scale (Pretty, 2008; Baig and Gamache, 2009; Kassam et al., 2009, 2013; Godfray et al., 2010; FAO, 2011b; Pretty et al., 2011).

However, different from the tillage-based *intrusive approach* to farming described earlier, there are now many production systems with a predominantly *ecosystem or agroecological approach* generally characterised by minimal disturbance of the ecosystem, with both natural and managed biodiversity in order to provide food, raw materials and other ecosystem services. Biologically healthy soils underpin these systems. Thus, in order to achieve sustainable intensification, a production system must be able to support and maintain the ecosystem functioning, and services derived from it, by limiting interventions (which may appear necessary for intensifying the production) to levels which do not disrupt these functions.

Sustainable production systems based on ecosystem approaches offer a range of productivity, socioeconomic and environmental benefits to producers and to society at large. To achieve the increased productivity required to meet 2050 food demands and the range of ecosystem services expected by society, sustainable production systems should be based on five technical principles:

- Simultaneous achievement of increased agricultural productivity and enhanced ecosystem services;
- Enhanced input-use efficiency, where key inputs include water, nutrients, pesticides, energy, land and labour;
- Reduced dependency from external inputs derived from fossil fuels (such as mineral fertiliser and pesticides) and preference for alternatives (such as biological nitrogen fixation and integrated pest management);
- Protection of soil, water and biodiversity through use of minimum disturbance of natural systems – interventions must not have accumulative effects but must have an impact and frequency lower than the natural recovery capacity of the ecosystem;
- Use of managed and natural biodiversity to build and/or re-build system resilience to abiotic, biotic and economic stresses.

Systems following these objectives based on the application of interlinked CA principles will show over time increasing production levels and decreasing levels of input use as soil health and functions improve. In many degraded situations, better retention of incoming water – its capture, infiltration and in-soil storage at plant-available tensions – is an important achievement which makes possible the optimum functioning of the entire soil-plant system.

Soils formed on different landform facets will have different risks and fragility characteristics related to crop conditions. Soil biologic processes will occur differentially as well by landscape position because of the variable microclimate conditions and soil development (or degradation). Land managers need to recognise the range of soil health and functional characteristics associated with landscapes in order to develop CA systems as well as monitoring and evaluating performance and risks.

Soil quality strongly affects agricultural land use and thus the shaping of the landscape. Any change in soil quality whether through degradation processes or soil health improvement will have consequences not only on field or farm level but also on a greater scale, the landscape. In addition, landscape normally consists of a combination of different ecosystems that are interlinked more or less closely with each other. The healthier the soil under agricultural use, the less off-site effects can be expected upon adjacent ecosystems of the same landscape. When contiguous fields are managed under CA systems, then there is a positive effect on landscape health and function such that large-scale ecosystem services that are soil mediated can be harnessed at the watershed scale and beyond. Examples of such societal services based on landscape health include watershed services in the Parana Basin III in Brazil delivering good quality water to the ITAIPU dam reservoir, and the carbon offset trading scheme in Alberta, Canada (Kassam et al., 2013).

6 Conclusion

Essentially, we have two farming paradigms operating globally, and both aspire to manage the soil and landscape sustainably. The two paradigms are:

(1) The tillage-based farming systems, including intensive tillage with inversion ploughing during the last century, aim at modifying soil structure to create a clean seedbed for planting seeds and to bury weeds or incorporate residues. This is the *intrusive paradigm* as represented by Green Revolution agriculture in which most aspects of crop production are controlled by technological interventions such as soil tilling; genetically engineered cultivars; protective or curative insect pest-, pathogen- and weed control with agrochemicals; and the application of mineral fertilisers for plant nutrition. This is still the predominant agricultural paradigm around the world but it is increasingly giving way to the alternate no-till systems paradigm which is considered to be ecological, regenerative and climate-smart as described next.

(2) With the development of no-tillage technologies from the 1940s onwards, and the discovery of specific farming systems since the 1970s, many of those have taken a predominantly ecosystem approach, and are productive and ecologically sustainable at field and landscape level. This is the *agroecological paradigm* as represented by CA. Crop production based on an ecosystem or agroecological approach can sustain the health of farmland already in use and can regenerate land left in poor condition by past misuse. The core agroecological elements of CA, and of production intensification based on it, are the practices that implement the three interlinked principles - continuous no or minimum mechanical soil disturbance, permanent soil mulch cover, and species diversification - plus other production and management best practices dealing with crop management including integrated pest control, plant nutrition, water management, labour, energy, etc., as well as the integration of pastures and trees into the production system and

supported by adequate and appropriate farm power and equipment. This concept and associated practical implications of CA must be placed at the centre of any effort to intensify production at any farm scale and to transform the unsustainable tillage agriculture into a sustainable alternate agriculture.

It is possible to develop ecologically and economically sustainable production system based on how and what crops are grown but always following CA principles. This would allow the maintenance of the underpinnings of ecological sustainability of production systems in good order so that sustainable production of biological products, including food, as well as other ecosystem services becomes the norm. This transformational change from tillage-based systems to CA systems in annual croplands is now occurring worldwide on all continents and ecologies at an annual rate of some 10.5 million ha, and in 2015/16 CA cropland systems covered more than 180 million ha (12.5% of the global annual cropland).

7 Where to look for further information

The need for CA is a topic that has been elaborated in several books and websites particularly since 2000. Books include: Goddard et al. (2006), Baker et al. (2007), FAO (2011b, 2016), Jat et al. (2014), Farooq and Siddique (2014), Chan and Fantle-Lepczyk (2015) and Kassam et al. (2017b).

Nationally oriented information on the need for CA is published for several countries including Australia (Crabtree, 2010), Canada (Lindwall and Sonntag, 2010), North America (Lessiter, 2011, 2018), Brazil (Junior et al., 2012; de Freitas and Landers, 2014) and Argentina (Peiretti and Dumanski, 2014).

FAO website provides general information on the need for CA at: www.fao.org/conservation-agriculture. More information regarding the need for CA can be found on websites of national and regional CA organizations.

Special issues of journals also provide information on the need for CA. For example, the International Soil and Water Conservation Research published a special issue on Pioneers in soil conservation and Conservation Agriculture which provides good information on the need for CA in several countries (Dumanski et al., 2014).

8 References

Altieri, M. A., Lana, M. A., Bittencourt, H. V., Kieling, A. S., Comin, J. J. and Lovato, P. E. 2011. Enhancing crop productivity via weed suppression in organic no-till cropping systems in Santa Catarina, Brazil. *Journal of Sustainable Agriculture* 35, 1–15. doi:10.1080/10440046.2011.588998.

Baig, M. N. and Gamache, P. M. 2009. *The Economic, Agronomic and Environmental Impact of No-Till on the Canadian Prairies*. Alberta Reduced Tillage Linkages, Canada.

Baker, C. J., Saxton, K. E., Ritchie, W. R., Chamen, W. C. T., Reicosky, D. C., Ribeiro, M. F. S., Justice, S. E. and Hobbs, P. R. (Eds). 2007. *No-Tillage Seeding in Conservation Agriculture* (2nd edn.). CABI Publishing, Wallingford, UK.

Basch, G., Kassam, A., Friedrich, T. F., Santos, F. L., Gubiani, P. I., Calegari, A., Reichert, J. M. and dos Santos, D. R. 2012. Sustainable soil water management systems. In: Lal, R. and Stewart, B. A. (Eds), *Soil Water and Agronomic Productivity, Advances in Soil Science*. CRC Press, Taylor and Francis Group, Boca Raton, FL, pp. 229–89.

Baveye, P. C., Rangel, D., Jacobsen, A. R., Laba, M., Darnault, C., Otten, W., Radulovich, R. and Camargo, F. A. O. 2011. From Dust Bowl to Dust Bowl: soils are still very much a frontier of science. *Soil Science Society of America Journal* 75, 2037–48.

Beddington, J. 2011. *The Future of Food and Farming Futures: Challenges and Choices for Global Sustainability*. Government Office for Science, UK.

Brisson, N., Gate, P., Gouache, D., Charmet, G., Oury, F. X. and Huard, F. 2010. Why are wheat yields stagnating in Europe? A comprehensive data analysis for France. *Field Crops Research* 119, 201–12.

CA. 2007. *Water for Food, Water for Life: A Comprehensive Assessment of Water Management in Agriculture*. Earthscan, London and International Water Management Institute, Colombo, 645pp.

Chan, C. and Fantle-Lepczyk, J. 2015. *Conservation Agriculture in Subsistence Farming: Case Studies from South-Asia and Beyond*. CABI Publishing, Wallingford, 278pp.

Crabtree, W. L. 2010. *Search for Sustainability with No-Till Bill in Dryland Agriculture*. Crabtree Agricultural Consulting, Western Australia, 204pp. Available at: www.no-till.com.au.

de Freitas, P. L. and Lander, J. N. 2014. The transformation of agriculture in Brazil through development and adoption of Zero Tillage Conservation Agriculture. *International Soil and Water Conservation Research* 1(2), 35–46.

Derpsch, R. 2004. History of crop production, with and without tillage. *Leading Edge* 3, 150–4.

Derpsch, R., Franzluebbers, A. J., Duiker, S. W., Reicosky, D. C., Koeller, K., Friedrich, T., Sturny, W. G., Sá, J. C. M. and Weiss, K. 2014. Why do we need to standardize no-tillage research? *Soil and Tillage Research* 137, 16–22. doi:10.1016/j.still.2013.10.002.

Doran, J. W. 2002. Soil health and global sustainability: translating science into practice. *Agriculture, Ecosystems and Environment* 88(2), 119–27. doi:10.1016/S0167-8809(01)00246-8.

Doran, J. W. and Zeiss, M. R. 2000. Soil health and sustainability: managing the biotic component of soil quality. *Applied Soil Ecology* 15(1), 3–11. doi:10.1016/S0929-1393(00)00067-6.

Dregne, H. E. and Chou, N. T. 1992. Global desertification dimensions and costs. In: Dregne, H. E. (Ed.), *Degradation and Restoration of Arid Lands*. Texas Technical University, Lubbock, TX, pp. 73–92.

Dumanski, J., Reicosky, D. C. and Peiretti, R. A. 2014. Pioneers in soil conservation and Conservation Agriculture. *International Soil and Water Conservation Research* 2(1).

FAO. 2008. Investing in sustainable crop intensification: the case for soil health. Report of the *International Technical Workshop*, FAO, Rome, July. Integrated Crop Management (vol. 6). FAO, Rome. Available at: http://www.fao.org/ag/ca/.

FAO. 2010. An international consultation on integrated crop- livestock systems for development – the way forward for sustainable production intensification. Integrated Crop Management (vol. 13). FAO, Rome.

FAO. 2011a. The state of the world's land and water resources for food and agriculture (SOLAW): managing systems at risks. Summary report. FAO, Rome, 47pp.

FAO. 2011b. Save and grow. A policymakers' guide to the sustainable intensification of smallholder crop production. FAO, Rome, 98 pp. Available at: www.fao.org/ag/save-and-grow/.

FAO. 2012a. World agriculture towards 2030/2050: the 2012 revision. Alexandratos, N. and Bruinsma, J. (Eds). ESA working paper no. 12-30. FAO, Rome.

FAO. 2012b. Soil organic carbon accumulation and greenhouse gas emission reductions from Conservation Agriculture: a literature review. Integrated Crop Management (vol. 16). FAO, Rome.

FAO. 2016. *Save and Grow in Practice: Maize, Rice and Wheat*. FAO, Rome.

Farooq, M. and Siddique, K. H. M. (Eds). 2014. *Conservation Agriculture*. Springer International, Switzerland.

Faulkner, E. H. 1943. *Ploughman's Folly*. Michael Joseph, London, 142pp.

Friedrich, T., Kassam, A. H. and Shaxson, T. F. 2009. Conservation Agriculture. *Agriculture for Developing Countries. Science and Technology Options Assessment (STOA) Project*. European Technology Assessment Group, Karlsruhe, Germany.

Friedrich, T., Derpsch, R. and Kassam, A. 2013. *Overview of the global spread of Conservation Agriculture. Facts Reports* Special Issue 6, 1–7.

Garrity, D. P. 2011. Making Conservation Agriculture ever green. *5th World Congress of Conservation Agriculture Incorporating 3rd Farming Systems Design Conference*, September 2011, Brisbane, Australia.

Garrity, D. P., Akinnifesi, F. K., Ajayi, O. C., Weldesemayat, S. G., Mowo, J. G., Kalinganire, A., Larwanou, M. and Bayala, J. 2010. Evergreen agriculture: a robust approach to sustainable food security in Africa. *Food Security* 2(3), 197–214. doi:10.1007/s12571-010-0070-7.

Gibbs, H. K. and Salmon, J. M. 2015. Mapping the world's degraded lands. *Applied Geography* 57, 12–21.

Goddard, T., Zoebisch, M., Gan, Y., Ellis, W., Watson, A. and Sombatpanit, S. (Eds). 2006. No-Till Farming Systems. Special Pub. No. 3. Bangkok: World Association of Soil and Water Conservation (WASWAC), Bangkok.

Godfray, H. C., Beddington, J. R., Crute, I. R., Haddad, L., Lawrence, D., Muir, J. F., Pretty, J., Robinson, S., Thomas, S. M. and Toulmin, C. 2010. Food security: the challenge of feeding 9 billion people. *Science* 327(5967), 812–8. doi:10.1126/science.1185383.

González-Sánchez, E. J., Moreno-Garcia, M., Kassam, A., Holgado-Cabrera, A., Trivino-Tradas, P., Carbonell-Bojollo, R., Pisante, M., Veroz-Gozalez, O. and Basch, G. 2017. *Conservation Agriculture: Making Climate Change Mitigation and Adaptability Real in Europe*. European Conservation Agriculture Federation (ECAF), 154pp.

González-Sánchez, E. J., Mkomwa, S., Conway, G., Kassam, A., Ordonez-Fernandez, R., Moreno-Garcia, M., Repullo-Ruiberriz de Torres, M., Gil-Ribes, J., Basch, G., Veroz-Gozalez, O., Trivino-Tradas, P., Holgado-Cabrera, A., Miranda-Fuentes, A. and Carbonell-Bojollo, R. 2018. *Making Climate Change Mitigation and Adaptability Real in Africa with Conservation Agriculture.* European Conservation Agriculture Federation (ECAF) and African Conservation Tillage Network (ACT), 143pp.

González-Sáncheza, E. J., Veroz-Gonzalez, O., Conway, G., Moreno-Garcia, M., Kassam, A., Mkomwa, S., Ordoñez-Fernandez, R., Triviño-Tarradas, P. and Carbonell-Bojolloe, R. 2019. Meta-analysis on carbon sequestration through Conservation Agriculture in Africa. *Soil and Tillage Research* 190, 22-30. doi:10.1016/j.still.2019.02.020.

Greenland, D. and Lal, R. (Eds). 1977. *Soil Conservation and Management in the Humid Tropics*. John Wiley and Sons, Chichester, 283pp.

Hobbs, P. R. 2007. Conservation agriculture: what is it and why is it important for future sustainable food production? *Journal of Agricultural Science* 145, 127-37.

Jat, R. A., Sahrawat, K. L. and Kassam, A. H. 2014. *Conservation Agriculture: Global Prospects and Challenges*. CABI Publishing, Wallingford, UK, 393pp.

Junior, R. C., de Araujo, A. G. and Llanillo, R. F. 2012. No-till Agriculture in Southern Brazil: factors that facilitated the evolution of the system and the development of the mechanization of conservation farming. FAO and IAPAR.

Juniper, T. 2013. *What Has Nature Ever Done for Us? How Money Really Does Grow on Trees*. Profile Books Book Company, London, UK, 324pp.

Kassam, A. H., Friedrich, T., Shaxson, F. and Pretty, J. N. 2009. The spread of Conservation Agriculture: justification, sustainability and uptake. *International Journal of Agricultural Sustainability* 7(4), 292-320. doi:10.3763/ijas.2009.0477.

Kassam, A. H., Friedrich, T. and Derpsch, R. 2010. Conservation Agriculture in the 21st Century: a paradigm of sustainable agriculture. *European Congress on Conservation Agriculture*, 4-6 October 2010, Madrid, Spain.

Kassam, A., Friedrich, T., Shaxson, T. F., Reeves, T., Pretty, J. and de Moraes Sà, J. C. 2011. Production systems for sustainable intensification: integrated productivity with ecosystem services. *Technikfolgenabschatzung - Theorie und Praxis* 2, 39-45.

Kassam, A., Basch, G., Friedrich, T., Shaxson, F., Goddard, T., Amado, T., Crabtree, B., Hongwen, L., Mello, I., Pisante, M. and Mkomwa, S. 2013. Sustainable soil management is more than what and how crops are grown. In: Lal, R. and Stewart, R. A. (Eds), *Principles of Soil Management in Agro-ecosystems. Advances in Soil Science*. CRC Press, Boca Raton, FL, pp. 337-400.

Kassam, A., Friedrich, T., Derpsch, R. and Kienzle, J. 2015. Overview of the worldwide spread of conservation agriculture. Field actions science reports 8. Available at: http://factsreports.revues.org/3966.

Kassam, A., Basch, G., Friedrich, T., Gonzalez, E., Trivino, P. and Mkomwa, S. 2017a. Mobilizing greater crop and land potentials sustainably. *Hungarian Geographical Bulletin* 66(1), 3-11. doi:10.15201/hungeobull.66.1.1.

Kassam, A., Mkomwa, S. and Friedrich, T. 2017b. *Conservation Agriculture for Africa: Building Resilient Farming Systems in a Changing Climate*. CABI Publishing, Wallingford, UK.

Kassam, A., Friedrich, T. and Derpsch, R. 2019. Global spread of Conservation Agriculture. *International Journal of Environmental Studies* 76(1), 29-51. doi:10.1080/00207233.2018.1494927.

Lal, R. and Stewart, R. A. (Eds). 2013. *Principles of Soil Management in Agro-ecosystems*. Advances in Soil Science. CRC Press, Boca Raton, FL.

Landers, J. 2007. Tropical crop-livestock systems in Conservation Agriculture: the Brazilian experience. Integrated Crop Management (vol. 5). FAO, Rome.

Lessiter, F. 2011. *40 Legends of the Past: 40th Anniversary Issue of No-till Farmer*, November 2011.

Lessiter, F. 2018. *From Maverick to Mainstream: A History of No-Till Farming*. No-Till Farmer, 416pp.
Li, H., He, J., Bharucha, Z. P., Lal, R. and Pretty, J. 2016. Improving China's food and environmental security with conservation agriculture. *International Journal of Agricultural Sustainability* 14(4), 377–91. doi:10.1080/14735903.2016.1170330.
Lindwall, C. W. and Sonntag, B. (Eds). 2010. *Landscape Transformed: The History of Conservation Tillage and Direct Seeding, Knowledge Impact in Society*. University of Saskatchewan, Saskatoon, Canada.
MEA. 2005. *Ecosystems and Human Well-Being: Synthesis*. Millennium Ecosystem Assessment. Island Press, Washington DC.
Montgomery, D. R. 2007. *Dirt: the Erosion of Civilizations*. University of California Press, Berkeley and Los Angeles, 285pp.
Nibert, D. 2013. *Animal Oppression and Human Violence: Domesecration, Capitalism, and Global Conflict*. Columbia University Press, NY, 352pp.
Nkonya, E., Mirzabaev, A. and von Braun, J. (Eds). 2016. *Economics of Land Degradation and Improvement – A Global Assessment for Sustainable Development*. IFPRI and ZEF. Springer Open.
Oldeman, L. R. 1988. *Guidelines for General Assessment of the Status of Human-Induced Soil Degradation*. International Soil Reference and Information Centre (ISRIC), Wageningen, the Netherlands.
Oldeman, L. R., Hakkeling, R. T. A. and Sombroek, W. G. 1991. *World Map of the Human-Induced Soil Degradation: an Explanatory Note*. International Soil Reference and Information Centre (ISRIC), Wageningen, the Netherlands.
Palm, C., Blanco-Canqui, H., Declerck, F., Gatere, L. and Grace, P. 2014. Conservation agriculture and ecosystems services. An overview. *Agriculture, Ecosystems and Environment* 187, 87–105.
Peiretti, R. and Dumanski, J. 2014. The transformation of agriculture in Argentina through soil conservation. *International Soil and Water Conservation Research* 1(2), 14–20.
Phillips, E. R. and Phillips, S. H. 1984. *No-Tillage Agriculture: Principles and Practices*. University of Michigan, Ann Arbor, MI, 306pp.
Phillips, S. H. and Young, H. M. 1973. *No-Tillage Farming*. Reiman Associates, Milwaukee, WI, 224pp.
Pimentel, D., Harvey, C., Resosudarmo, P., Sinclair, K., Kurz, D. and McNair, M. 1995. Environmental and economic costs of soil erosion and conservation benefits. *Science* 267(5201), 1117–23.
Pretty, J. 2002. *Agri-Culture: Reconnecting People, Land and Nature*. Earthscan Publications, London, 261pp.
Pretty, J. 2008. Agricultural sustainability: concepts, principles and evidence. *Philosophical Transactions of the Royal Society of London. Series B, Biological Sciences* 363(1491), 447–65. doi:10.1098/rstb.2007.2163.
Pretty, J. N., Toulmin, C. and Williams, S. 2011. Sustainable intensification in African agriculture. *International Journal of Agricultural Sustainability* 9(1), 5–24. doi:10.3763/ijas.2010.0583.
Reicosky, D. C. 2015. Conservation tillage is not Conservation Agriculture. *Journal of Soil and Water Conservation* 70(5), 103A–8A.
Saha, R., Ghosh, P. K., Mishra, V. K., Majumdar, B. and Tomar, J. M. S. 2010. Can agroforestry be a resource conservation tool to maintain soil health in the fragile

ecosystem of north-east India? *Outlook on Agriculture* 39(3), 191–6. doi:10.5367/oa.2010.0004.

Sims, B., Friedrich, T., Kassam, A. H. and Kienzle, J. 2009. Agroforestry and Conservation Agriculture: complementary practices for sustainable agriculture. *2nd World Congress on Agroforestry*, August 2009, Nairobi, Kenya.

Tikhonovich, I. A. and Provorov, N. A. 2011. Microbiology is the basis of sustainable agriculture: an opinion. *Annals of Applied Biology* 159(2), 155–68. doi:10.1111/j.1744-7348.2011.00489.x.

Uphoff, N., Ball, A. S., Fernandes, E., Herren, H., Husson, O., Laing, M., Palm, C., Pretty, J., Sanchez, P., Sanginga, N. and Thies, J. 2006. Issues for more sustainable soil system management. In: Uphoff, N., Ball, A. S., Fernandes, E., Herren, H., Husson, O., Laing, M., Palm, C., Pretty, J., Sanchez, P., Sanginga, N. and Thies, J. (Eds), *Biological Approaches to Sustainable Soil Systems*. CRC Press, Boca Raton, FL, pp. 715–27.

Vlek, P. L. G., Le, Q. B. and Tamene, L. 2010. Assessment of land degradation, its possible causes and threat to food security in sub-Saharan Africa. In: Lal, R. and Stewart, B. A. (Eds), *Advances in Soil Science - Food Security and Soil Quality*. CRC Press, Boca Raton, FL, pp. 57–86.

Vlek, P. L. G., Khamzina, A. and Tamene, L. (Eds). 2017. Land degradation and the sustainable development goals: threats and potential remedies. CIAT publication no. 440. International Center for Tropical Agriculture (CIAT), Nairobi, Kenya, 67p. Available at: http://hdl.handle.net/10568/81313.

Warkentin, B. P. 1995. The changing concept of soil quality. *Journal of Soil and Water Conservation* 50, 226–8.

Weis, T. 2013. *The Ecological Hoofprint: the Global Burden of Industrial Livestock*. Zed Books Ltd, London, UK.

Willett, W., Rockström, J., Loken, B., Springmann, M., Lang, T., Vermeulen, S., Garnett, T., Tilman, D., DeClerck, F., Wood, A., Jonell, M., Clark, M., Gordon, L. J., Fanzo, J., Hawkes, C., Zurayk, R., Rivera, J. A., De Vries, W., Majele Sibanda, L., Afshin, A., Chaudhary, A., Herrero, M., Agustina, R., Branca, F., Lartey, A., Fan, S., Crona, B., Fox, E., Bignet, V., Troell, M., Lindahl, T., Singh, S., Cornell, S. E., Srinath Reddy, K., Narain, S., Nishtar, S. and Murray, C. J. L. 2019. Food in the Anthropocene: the EAT-Lancet Commission on healthy diets from sustainable food systems. *The Lancet* 393(10170), 447–92. doi:10.1016/S0140-6736(18)31788-4.

Chapter 2

Development of Conservation Agriculture systems globally

Amir Kassam, University of Reading, UK; Rolf Derpsch, Consultant, Paraguay; and Theodor Friedrich, Food and Agriculture Organization of the United Nations (FAO), Italy

1 Introduction

Reducing soil disturbance by tillage in agricultural land began in the Great Plains in the United States in the 1930s in response to the devastation caused by intensive tillage with mouldboard plough accompanied by prolonged drought (Derpsch, 2004). This period became known as the 'Dust Bowl'. In response, a major soil and water conservation programme was established by USDA. Initial research on 'conservation' or reduced tillage involved early versions of a chisel plough, by which plant residues could be retained on the soil surface to alleviate wind and water erosion (Duley and Fenster, 1954; Mannering, 1979). Stubble mulch farming was also developed (Fenster, 1960), and this became a forerunner of no-tillage farming. This collection of practices led to what became known as conservation tillage, which by definition is any tillage leaving at least 30% of the soil surface covered with crop residues for erosion control, while no-till systems by definition avoid soil disturbance by no-till seeding and weeding and maintaining a biomass mulch cover on the soil surface (King and Holcomb, 1985; Kassam et al., 2009). Since 1990, there has been a steady development of no-till systems globally, and it was in 1997,

http://dx.doi.org/10.19103/AS.2019.0048.02

that the term 'Conservation Agriculture' (CA) was adopted and defined in terms of the three interlinked principles to represent the modern version of no-tillage farming. Since then, the development of CA systems globally, particularly in annual croplands, has continued, led by North and South America and Australia. Since 2008, CA systems with annual crops have been spreading in Europe, Asia and Africa. More recently, CA systems also include perennial systems including orchards, plantations and agroforestry. Overall, CA principles are being applied to rainfed and irrigated systems, including rice-based systems, and there is increasing interest being shown in the CA approach by conventional tillage-based organic farming systems.

This chapter provides an overview of the evolution of CA systems globally since its early development in the United States, from where it spread to South America in the early 1970s, Canada and Australia in the later 1980s and more recently to Europe (including Russia and Ukraine), Asia and Africa since the late 1990s.

2 Development of no-till farming and some of the pioneers and champions

The book, *Ploughman's Folly* by Edward Faulkner (Faulkner, 1943), an extension agronomist in Ohio, was an important milestone in the development of conservation agricultural practices. Faulkner questioned the wisdom of inversion ploughing and explained the destructive nature of soil tillage. He stated: 'No one has ever advanced a scientific reason for plowing' and 'There was nothing wrong with our soils except our interference'. Around the same time, in Japan, Masanobu Fukuoka started to question the appropriateness and sustainability of conventional tillage-based farming (Fukuoka, 1975). Further research in the United Kingdom, United States and elsewhere during the late-1940s and 1950s made no-tillage farming possible. The practice began to spread in the United States in the 1960s, in Brazil in the 1970s and in Argentina, Paraguay and Uruguay in the 1980s and 1990s, with farmers such as Herbert Bartz, Manoel Henrique (Nono) Pereira, Frank Dijkstra and John Landers in Brazil, Heri Rosso, Rogelio Fogante, Victor Trucco and Mario Gilardoni in Argentina, Carlos Crovetto in Chile and Akinobu Fukamiin in Paraguay championing the transformation of tillage farming into no-till farming systems. In Brazil, no-till research was pioneered in Londrina in 1971 with initiatives from Rolf Derpsch, collaborating with Brazilian colleagues such as Ademir Calegari on plots often visited by Herbert Bartz; in 1972 Rolf Derpsch sent his no-till wheat seeder to Herbert Bartz's farm 'Rhenania' at Rolandia, Paraná (30 km from Londrina), to seed a demonstration plot of half hectare of wheat after soybean.

In the United States in 1973, Shirley Phillips and Harry Young published the book *No-tillage Farming* (Phillips and Young, 1973), the first of its kind in the world. This was followed in 1984 by the book *No-Tillage Agriculture: Principles and Practices* by E. R. Phillips and S. H. Phillips (Phillips and Phillips, 1984). In Southern Rhodesia (now Zimbabwe), Tom Borland, a weed control specialist, published an article in 1974 in the Rhodesian Agriculture Today: *Which way weed control and tillage?* (Borland, 1974), after a study tour to the United States where he met with most of the no-till pioneers including Harry Young. He also published a series of articles on no-till in *The Rhodesian Farmer* magazine over the period from 1976 to 1979 and were later reprinted in *South African Farmer's Weekly* (Borland, 1980). This was followed in 1984 by Brian Oldreive in Zimbabwe designing an approach called Farming God's Way (subsequently called Foundation for Farming) comprising no-till, mulch cover and rotation (Oldreive, 2006). In West Africa, research on no-till farming was started in 1970 at the International Institute for Tropical Agriculture (IITA), Nigeria, and a series of articles and bulletins on mulch farming techniques and no-till farming were published in the 1970s and 1980s (e.g. Lal, 1973, 1974a,b, 1975, 1976a,b,c, 1983).

In addition to these pioneers, there have been many other innovators in no-till farming since the early seventies who have made tremendous contributions to its growth and popularity. To mark the 40th Anniversary of the publishing of *No-Till Farmer*, its editor Frank Lessiter published a list of '40 Legends of The Past from North America' (Lessiter, 2011). They include the internationally popular names such as Edward Faulkner, Dwayne Beck, Jill Clapperton, Steve Groff, Guy Lafond, Bill Richards, Shirley Phillips and Harry Young. However, the North American list of champions is longer than 40 and includes global promoters such as Keith Saxton, Don Reicosky, Tom Goddard, Scott Day, Sjoerd Duiker, Julian Dumanski and David Montgomery. Elsewhere, other no-till farming champions[1] have included John Baker and W. R. Ritchie in New Zealand, Rolf Derpsch, Terry Wiles, David Sharp, Ivo Mello, Joao Carlos de Moraes Sa, Ademir Calegari, Augusto Guilherme de Araujo, Rafael Fuentes Lanillo in Brazil, Gino Minucci, Mario Nardone, Jose Araya, Roberto Peiretti, Hugo Ghio, Jorge Romagnoli, Horacio Aguero, Cesar Belloso and Luis Giraudo in Argentina, Ken Moriya in Paraguay, Bill Crabtree, Jean-Francois Rochecouste, Steven Powles, Allen Postlethwaite, Jeff Tullberg, Jeff Esdaile, Tim Reeves and Richard Bell in Australia, Gao Huanwen and Li Hongwen in China, Wolfgang Sturny in Switzerland, Emilio Gonzales-Sanchez in Spain, Gottlieb Basch and Gabriela Cruz in Portugal, Benoit Lavier, Gerrad Rass, Frederique Thomas and Sara Singla in France, Tony Reynolds, Tony Gent and John Cherry in the United Kingdom, Michele Pisante in Italy and Soren Ilsoe in Denmark. Champions at FAO have included José Benites, Theodor Friedrich, John Ashburner, Amir

1 There have been many champions of no-till farming in different parts of the world. The aim here is not to offer a comprehensive list but to mention some names of 'early' pioneers.

Kassam, Francis Shaxson, Brian Sims and Josef Kienzle. From different parts of Africa, there have been several champions such as Ken Fowler, Richard Findlay, Hendrik Smith, Rachid Mrabet, Martin Bwalya and Saidi Mkomwa; in the Consultative Group for International Agricultural Research (CGIAR), Jeff Sayer, Patrick Wall, Muratbek Karabayev, Mekhlis, Suleimanov, Aziz Nurbekov, Peter Hobbs, Harminder Singh Sidhu, Mangi Lal Jat, Yashpal Singh Saharawat, Raj Gupta, Christian Thierfelder, Colin Piggin and Stephen Loss; and in the Center for International Research for Agricultural Development (CIRAD), Lucien Séguy, Olivier Husson, Stephane Boulakia and Jean-Claud Legopil. In addition to those working for CGIAR, other champions in South Asia include Raj Paroda, Inder Abrol, Mushtaq Gill and Enamul Haque. In the global CA community, there are many more champions from different parts of the world who deserve to be mentioned here. Indeed, the list is long and honourable, and growing (Dumanski et al., 2014).

3 The modern concept of CA

The modern concept of no-till farming or no-till system is now generally known as Conservation Agriculture (CA). The term was first used in 1997 at a regional workshop organized by FAO on the topic of Soil Management and Conservation – Efficient Tillage Methods for Soil Conservation in Ibadan, Nigeria, although the participants still talked more about conservation tillage than CA (Barbosa dos Anjos, 2000). Documented evidence shows that the term 'Conservation Agriculture' was first coined in Spanish at the IV RELACO meeting in Morelia, Michoacán, Mexico, in the year 1997 (RELACO, 1997). In English, RELACO stands for 'Latin-American Network for Conservation Tillage'. The functioning and purpose of RELACO was described by Benites (1997) at the IV RELACO meeting. At that forum in Morelia, Rolf Derpsch presented a paper on 'New paradigms of agricultural production' (Derpsch, 1997a) and 'Development and diffusion of sustainable agricultural production systems in Paraguay' (Derpsch, 1997b) which incited a lot of discussion. Only two other papers (Veiga, 1997; Amado and Reinert, 1997) from a total of 28 papers presented also focussed on no tillage and did not use the words 'conservation tillage'. Only one paper used the wording 'Conservation Agriculture' in the title (Friedrich, 1997). Conservation tillage reflected the old paradigm and mainstream thinking until the 1990's. During the RELACO meeting, Rolf Derpsch and Theodor Friedirich initiated a discussion to change the extended name of the organization from 'Red Latinoamericana de Labranza Conservacionista' (Latin-American Network for Conservation Tillage) to 'Red Latinoamericana de Agricultura Conservacionista' (Latin-American Network for Conservation Agriculture) arguing that one cannot conserve the soil while at the same time tilling it.

Conservation tillage is a contradiction, an oxymoron in the words of Don Reicosky (2015) because worldwide experience has clearly shown that tillage has nothing to do with conservation. In fact, tillage is a root cause of agricultural land degradation (Montgomery, 2007). The term 'conservation tillage' was developed in the United States as a consequence of the soil and water conservation practices that followed the 'Dust Bowl' in the 1930's. It was probably the right terminology for that time, but with the successful adoption of no-tillage since 1962 in the United States and its expansion to more than 180 M ha worldwide, the term has become obsolete.

After lengthy discussions at the RELACO meeting, the organizers (among them Theodor Friedrich and José Benites from FAO, Ramón Claveran, CENAPRO, Mexico) adopted the term 'Conservation Agriculture'. One reason for the relatively quick acceptance was that there was no need to change the acronym RELACO, which could stay the same for 'Red Latinoamericana de Agricultura Conservacionista'. An additional benefit for this denomination was the ease to translate it into English and even French. This name was also more appealing than no-tillage because it had a broader meaning and could be easily understood by any layman.

From there on, the two FAO representatives attending that meeting, Theodor Friedrich and José Benites, decided that CA was the term the FAO should use for sustainable agriculture systems in the future. An interdisciplinary informal workgroup on CA was created in the FAO with members from all divisions of the agricultural department. In 1998, the FAO organized its first regional workshop on CA in Harare, where a code of practice for CA was drafted describing the three interlinked CA principles as we know them today and their practical application. Also, at the workshop, the African Conservation Tillage (ACT) Network was founded with support from the GTZ. Yet, the ACT still used the term 'Conservation Tillage', although the FAO at that time had already started talking about CA, and only many years later did the ACT adopt CA as its focus.

A year later, in 1999, both Theodor Friedrich and José Benites were approached to join the founding of the European Conservation Agriculture Federation (ECAF). Between the ECAF and the FAO they then came up with the idea of having a World Congress on CA in 2001 – and that is how the world congresses on CA started, and how the three interlinked principles of CA were globalized.

RELACO over the following years struggled to survive with ever less support and a growing incidence of regional organizations promoting CA, such as the American Confederation of Associations for Sustainable Agriculture (CAAPAS). The first World Congress in Madrid concluded with an official Congress Declaration and brought the term 'Conservation Agriculture' along with the concept to a global audience and after that date the terminology 'Conservation

Agriculture' prevailed, gaining wider and wider acceptance with each new Congress being held and creating a global CA movement. Acceptance has been so strong and widespread, that some scientists are suggesting doing away with the term no-tillage (or its synonyms) and replacing it with 'Conservation Agriculture' when thinking of systems. The fact is that minimum soil disturbance with no-tillage as a practice is and will continue to be one of the main pillars or core principles to be applied in CA.

In 2001, the First World Congress on Conservation Agriculture was organized by the ECAF and the FAO in Madrid (Spain). Since then, there have been six more World Congresses, in Foz do Iguassu (Brazil, 2003), Nairobi (Kenya, 2005), New Delhi (India, 2009), Brisbane (Australia, 2011), Winnipeg (Canada, 2015) and Rosario (Argentina, 2017). The eighth World Congress will be held in Europe under the aegis of ECAF in 2020. The World Congress process is now well established and operates on an informal basis with regional CA organizations taking an active part in ensuring that the process continues to function on a voluntary and collegial basis with decisions being made on a consensus basis. At every World Congress, bids are invited for the next Congress to be held in 3 years' intervals. Bids are considered by a committee made up of the representative of all the regional CA organizations. The general rule is that the world congresses should alternate between the Global North and the Global South in succession.

Alongside the process of holding world congresses, meetings on CA have been held regularly by sub-national, national and regional CA organizations in all continents. The FAO regional and sub-regional offices also promote and support regional training and/or review workshops on CA. All these meetings are held to review progress, share experiences and learnings, address constraints, help with strategy and programme formulation and promote policy and institutional support decisions and actions. Also, CA organizations along with the FAO and several other bilateral and multilateral development agencies support the adoption of CA by farmers, generally working in groups and through farmer field schools (FFS) and farmer cooperatives, associations and networks. Indeed, much of the CA success in all regions is a result of hard work of many farmer-led activities.

In 2007, Amir Kassam and Francis Shaxson organized an international workshop on The Importance of Improving Soil Conditions for Water, Plant Nutrients and Biological Productivity to Sustain Agricultural Growth under Rising Population Pressure in a Changing Climate at Newcastle University under the aegis of the Tropical Agricultural Association (TAA) and in collaboration with the University of Newcastle, the University of Reading, the University of Nottingham, the University of Durham, the Association of Applied Biologists and ICRAF (World AgroForestry Centre). Several donor agencies were present at the meeting, including the Gates, Rockefeller, Ford and Syngenta

Foundations and DFID (Department for International Development, UK). Several international organizations participated including the FAO, ICRISAT, ICARDA, IWMI and CIAT (International Center for Tropical Agriculture). The FAO was represented by Theodor Friedrich, and several CA champions including from South America, Asia and Africa also attended. Organizations including Rothamsted, the NRI (Natural Resources Institute, Greenwich), ICAR (Indian Council for Agricultural Research), EMBRAPA (Brazilian Enterprise for Agricultural and Livestock Research), Kilimo Trust, MSSRF (M. S. Swaminathan Research Foundation), FARA (Forum for Agricultural Research in Africa), ACSAD (Arab Centre for the Studies of Arid Zones and Dry Lands), GFAR (Global Forum for Agricultural Research), CIRAD (Centre for Cooperation on International Research on Agricultural Development), SEI (Stockholm Environment Institute) and universities of Exeter, Durham, Free University of Amsterdam, Queensland, Kingston, Cornell, Mekerere, Kwa-Zulu Natal and Zimbabwe also participated. The main outcome of the workshop was the recommendation that CA was the best option for making sustainable agriculture a reality and for the promotion of sustainable agricultural intensification. Further, it was recognized that there was an emergent Community of Practice (CoP) that needed to be linked up to promote CA globally. Also, there was a need to hold a follow-up of the full international workshop at ICRAF to discuss how this could be achieved (TAA, 2007). Indeed, there were indications that a quiet revolution with no-till was underway (Huggins and Reganold, 2008).

Having failed to secure a venue at ICRAF due to political riots in Kenya, the FAO in Rome agreed to host the workshop. The *'International Technical Workshop on Investing in Sustainable Crop Intensification: The Case for Improving Soil Health'* was held at the FAO in the Plant Production and Protection Division in July 2008 (FAO, 2008) led by Amir Kassam, Theodor Friedrich, Francis Shaxson, Eric Kueneman and Andrew McMillan. Most of the CA champions around the world, some 100 individuals, including those in the FAO, attended to discuss and review the basis for successful adoption and spread of CA in different countries and regions, and how CA could be globalized.

The Workshop prepared a global action plan to mainstream CA and recommended that a multi-stakeholder Conservation Agriculture Community of Practice (CA-CoP) communication platform, hosted by the FAO, should be launched to facilitate the adoption and mainstreaming of CA globally (FAO, 2008). Justification for CA as the best choice for sustainable production intensification was published in 2009 based on the outcome of the international Workshop (Kassam et al., 2009). The CA-CoP platform was launched in January 2009 with Amir Kassam serving as the Moderator. This platform has been running continuously since then and now reaches several thousands of subscribers around the world. Through the CA-CoP platform, all kinds of information about CA, including journal articles, official documents and news, is distributed practically every day. When CA-CoP was established, the global CA cropland area was 106 M ha. In 2015/16 it was 180 M ha, an increase of some 69% (Kassam et al., 2018).

In 2011, the FAO published *Save and Grow: a Policymaker's Guide to Sustainable Intensification of Smallholder Crop Production*, which was based on CA as the new paradigm of agriculture (FAO, 2011). With that publication, the sustainable intensification of crop production, later integrating other agricultural production sectors, became a strategic objective of the FAO and a guideline for the FAO's work on sustainable agriculture systems.

CA is significantly different from the conventional tillage agriculture (Hobbs, 2007; Shaxson et al., 2008; Friedrich et al., 2009; Kassam et al., 2009). It represents a fundamental change in the agricultural production system paradigm. CA involves the simultaneous application of three inter-linked principles based on locally formulated adapted practices, along with several other complementary good agricultural practices of integrated crop, nutrient, pests (weeds, insects, pathogens), water, labour, farm power and energy management practices (Friedrich et al., 2009; Kassam et al., 2009, 2011, 2018) as described in Chapter 1 of this volume. CA along with the complementary practices are at the heart of the FAO's new sustainable agricultural intensification strategy for smallholder farmers (FAO, 2011, 2016; Kassam et al., 2011; Freidrich, 2013) which takes an ecosystems approach to enhancing and sustaining productivity and resilience as well as the flow of ecosystem services while reducing greenhouse gas emissions that come from the agriculture sector (Kassam et al., 2009, 2011, 2013; Gonzalez-Sanchez et al., 2012, 2017, 2018). These characteristics are also an integral part of climate-smart agriculture that seeks to increase productivity in an environmentally and socially sustainable way, strengthen farmers' resilience to climate change and reduce GHG emissions and sequester carbon (World Bank, 2012; IPCC, 2014). At the heart of sustainable agricultural intensification, or sustainable land management, is a strong focus on soil health and biology and the integration of soil and water conservation practices in agricultural production, with concurrent objectives of enhanced economic returns and environmental management (Kassam and Brammer, 2012; Kassam et al., 2013).

The description of CA used by the Moderator of the CA-CoP Communication Platform hosted by FAO is as follows:

> *'Conservation Agriculture is an ecosystem approach to regenerative sustainable agriculture and land management based on the practical application of context-specific and locally adapted three interlinked principles of: (i) Continuous no or minimum mechanical soil disturbance (no-till seeding/planting and weeding, and minimum soil disturbance with all other farm operations including harvesting); (ii) permanent maintenance of soil mulch cover (crop biomass, stubble and cover crops); and (iii) diversification of cropping system (economically, environmentally*

and socially adapted rotations and/or sequences and/or associations involving annuals and/or perennials, including legumes and cover crops), along with other complementary good agricultural production and land management practices. Conservation Agriculture systems are present in all continents, involving rainfed and irrigated systems including annual cropland systems, perennial systems, orchards and plantation systems, agroforestry systems, crop-livestock systems, pasture and rangeland systems, organic production systems and rice-based systems. Conservation Tillage, Reduced Tillage and Minimum Tillage are not Conservation Agriculture, and nor is No-Till on its own'.

4 History and adaptability of CA

Tillage, as a soil management concept, was questioned for the first time in the 1930s, when the 'Dust Bowl' devastated wide areas of the mid-west United States (Derpsch, 1998). Ideas for reducing tillage and keeping soil covered with crop biomass followed and the term 'conservation tillage' was introduced for practices aimed at erosion control. Seeding machinery developments allowed then, in the 1940s, to seed directly without any soil tillage. At the same time, theoretical concepts resembling today's CA principles were elaborated by Edward Faulkner in his book *Ploughman's Folly* (Faulkner, 1943) and Masanobu Fukuoka with the 'One Straw Revolution' (Fukuoka, 1975). But only in the 1960s did no-tillage enter into farming practice in the United States (Derpsch, 2004; Kassam et al., 2010, 2014a).

In the early 1970s and as the result of uncontrollable erosion problems in the southern states, no-tillage reached Brazil, where farmers together with scientists transformed the technology into the system which today is called CA. Yet it took another 20 years before CA reached significant adoption levels. During this time, farm equipment and agronomic practices in no-tillage systems were improved and developed to optimize the performance of crops, machinery and field operations. This process continues; the creativity of farmers and researchers is still producing improvements to the benefits of the production system, the soil and the farmer. While tillage-based agriculture has been researched for several centuries, CA is only half a century old and the functioning of CA systems can only be understood as the agro-ecosystems evolve under the new production management. From the early 1990s, the uptake of CA started growing exponentially, leading to a revolution in the agriculture of southern Brazil, Argentina and Paraguay and Uruguay.

During the 1990s, this development increasingly attracted attention from farmers and researchers in Europe, Asia, Africa and Australia, and from development and international research organizations such as FAO, World Bank, IFAD, GIZ, NORAD, CIRAD, ACIAR and the CGIAR system. Study tours to Brazil for farmers and policy makers, regional workshops, development and research projects were organized in different parts of the world. These produced increased levels of awareness and adoption in African countries such as South Africa, Zambia, Zimbabwe, Mozambique, Tanzania and Kenya as well as in Asia, particularly in Kazakhstan, India, Pakistan and China. The improvement of conservation and no-tillage practices within an integrated farming concept such as CA also led to a cropping system diversification and increased adoption of CA in industrialized countries after the end of the millennium, particularly Canada, United States, Australia, Spain, Italy, Finland, Ukraine and Russia (Derpsch et al., 2010). The spread of CA has continued to more countries in Europe, Asia and Africa (Kassam et al., 2014b, 2015, 2018).

CA crop production systems are popular worldwide. There are few countries where CA is not practised by at least some farmers and where there are no local research results about CA available (Jat et al., 2014). The total cropland area under CA in 2008/09 was estimated to be 106 M ha (Kassam et al., 2009; Derpsch and Friedrich, 2009a,b). By 2010/11, the global spread had increased to 145 M ha (Friedrich et al., 2012), and by 2013/14, 157 M ha (Kassam et al., 2014a,b, 2015, 2018). As reported below, the latest global estimate for CA cropland for 2015/16 is about 180 M ha.

CA systems are widely adaptable. Their presence extends from the equatorial tropics (e.g. Kenya, Tanzania, Uganda) to the arctic circle (e.g. Finland) North and to about 50° latitude South (e.g. Falkland Islands); from sea level in several countries of the world to 3000 m altitude (e.g. Bolivia, Colombia); from heavy rainfall areas with 2000 mm a year (e.g. Brazil) or 3000 mm a year (e.g. Chile) to extremely dry conditions in the Mediterranean environments with 250 mm or less a year (e.g. Morocco, Syria, Western Australia) (Derpsch et al., 2010; Jat et al., 2014). In some countries, CA has even allowed expansion of agriculture to marginal soils in terms of rainfall or fertility (e.g. Australia, Argentina). In southern Brazil, CA has facilitated the restoration of the degraded savanna and forest soils – the *cerrados* – to productive agricultural lands (Derpsch et al., 2010; Jat et al., 2014).

CA is practised on soils that vary from 90% sand (e.g. North Africa, southern Mediterranean zone, coastal zones in tropical Africa, Australia) to 80% clay (e.g. Brazil's Oxisols and Alfisols). Soils with high clay content in Brazil or in Europe are extremely sticky but this has not been a hindrance to no-till adoption when appropriate equipment is available. Soils that are readily prone to crusting and surface sealing under tillage farming do not exhibit this problem under established CA systems. This is because minimum soil disturbance, mulch cover, diversified cropping and increased soil organic matter all contribute to enhancement of soil quality that avoids the formation of surface crusts and soil compaction.

No-tillage in CA is practised on all farm sizes from less than half a hectare or a few hectares (e.g. India, China, Zambia, Zimbabwe, Brazil and Paraguay) to thousands of hectares (e.g. Argentina, Brazil, Canada, South Africa, Australia and Kazakhstan). All crops can be grown adequately in CA systems and to the authors' knowledge there has not yet been a crop that would not grow and produce under this system, including root and tuber crops (Derpsch and Friedrich, 2009a,b).

Despite the existence of several constraints to adoption, farmers in different parts of the world are continuing to find local solutions to support the spread of CA as well as to innovate with new practices and management methods to maximize the benefits. Major constraints to the adoption of CA practices continue to be knowledge about the existence of CA and on how to do it (know how), mindset (tradition, prejudice), inadequate policies, for example, commodity-based subsidies (EU, USA) and direct farm payments (EU), unavailability of appropriate equipment and machines (many countries of the world, including EU), and of suitable management strategies to facilitate weed and vegetation management, including mechanical, biological and chemical options as herbicides (especially for larger farms in low-income countries) (Friedrich and Kassam, 2009; Jat et al., 2014; Farooq and Siddique, 2014). Other area-specific constraints in semi-arid areas during the transformation to CA system relate to initial low supply of crop and vegetation biomass for soil mulch cover development; to initial short-term competition for crop residue as livestock feed; and to initial adoption of new manual weed management practices when the soil mulch cover and integrated weed management practice is being established.

Yet farmers who do become seriously interested in adopting CA develop local solutions to all these barriers. Many such cases have been reported for smallholder and large-scale farms in all continents (see list of publications at: http://www.fao.org/conservation-agriculture). Further, more international and national organizations have increased their support for CA as they have increased their awareness of its effectiveness in sustainable production intensification. These organizations include FAO, International Fund for Agricultural Development (IFAD), World Bank, European Union (EU), African Union - New Partnership for Africa's Development (AU-NEPAD), CIRAD, ACT, some CGIAR Centres (International Maize and Wheat Improvement Centre - CIMMYT, International Centre for Agricultural Research in the Dry Areas - ICARDA, International Crops Research Institute for Semi-Arid Tropics - ICRISAT, ICRAF), NGOs, some governments in the Global North and the Global South, national and multinational corporations, the growth of no-till/CA organizations worldwide, farmer to farmer support, even across continents and bilateral and multilateral donors. Thus, the continuing spread of CA globally is creating a need for effective national and regional policy and institutional support (Kassam et al., 2014c; FAO, 2017) and the need to ensure that the core elements of CA systems are respected in conducting research and in education and training

(Derpsch et al., 2012, 2013) to safeguard the quality and relevance of CA knowledge and information.

5 Global adoption and regional spread

5.1 Drivers of adoption

Looking back to the days of the 'Dust Bowl' and what we now know of how CA has been spreading globally, we consider the following to be major drivers that have and are contributing to farmers adopting CA in different regions of the world.

The initial driver as has been indicated was the wind erosion in the 1930s in the mid-west and elsewhere that was caused by intensive mouldboard ploughing for several years on large expanse of the prairies in the mid-west of the United States. When this is combined with several years of severe drought, the intensely disturbed and pulverized top soil took off with the wind, creating a dust storm that carried the soil all the way to the east coast. In addition, the opening up of the prairies also led to run-off and water erosion of the top soil on a vast scale. Similar situations elsewhere have also driven farmers to adopt CA in other parts of the world. Most notable example is that of Brazil, which led to the initial adoption of no-till farming in the early 1970s, in Canada, Ukraine, Russia, China and Australia where dust storms forced farmers to change from tillage to no-till farming. Because of the large areas being brought under CA in Canada and Australia, dust storms have disappeared, since about 1989. The situation is still not under control completely in Russia, Ukraine and China where dust storms are still reported. A recent case of a dust storm was in Germany in April 2011 which led to an 81 car pile-up on the highway and the death of eight people. The cause of the dust storm turned out to be intensive tillage combined with drought and high winds.

Severe droughts have been a major driver of adoption of CA. Well-known examples are the semi-arid area in western and southwestern Australia, northern Kazakhstan, Mongolia, northern China and southern Africa. The realization that tillage eventually leads to decrease in water infiltration and water retention by soil, due to loss of soil structure, only few days of dry weather lead to crops suffering from water stress. In dry semi-arid areas with moderate-to-high rainfall variability, drought can be a common feature of the agricultural environment.

In the early 1970s, there was a steep rise globally in the cost of fossil fuel and energy in general. This led to a direct increase in cost of agricultural field operations and in the cost of production inputs. Mechanized farmers in both industrialized and lower-income countries were forced to move away from intensive ploughing. More farmers, particularly in North and South America and in Australia, moved into no-till technology for crop establishment, whereas other farmers particularly in Europe moved into min-till technology which pulverized 100% of the top soil up to 100–150 cm deep. In both cases, there was a significant reduction in cost of fuel use, but in the former it was possible to evolve into no-till cropping system with mulch cover (i.e. CA system) and improve soil health, productivity and profit but in the latter, it was not possible to evolve into CA system. Instead, areas under min-till in Europe and elsewhere are causing erosion and operating suboptimally in terms of productivity and profit, often reverting back to full inversion tillage when economy allows.

In the 1980s and 1990s, it became more generally appreciated that agricultural land globally was degrading and causing loss in productivity and ecosystem services. This was documented by GLASOD and by the FAO and was confirmed by the Millennium Ecosystem Assessment in 2005 (MEA, 2005). Tillage was recognized by some farmers and countries as being the root cause of soil and land degradation globally. In some cases, such as Alberta, Canada, a carbon offset trading scheme was established to encourage farmers to move away from tillage agriculture and adopt CA to allow carbon sequestration for which they could be paid by industries that were emitting greenhouse gases beyond a certain limit. In Brazil, a programme was initiated in the Parana Basin III to enable farmers to adopt CA to generate clean water for the ITAIPU hydroelectric dam which earlier was being fed by water that was high in sediments because of tillage agriculture. Land degradation and dysfunctional ecosystem services became the driving force behind the change to CA in countries including Spain, Italy, Kazakhstan, China, India, Pakistan, Russia and Ukraine, while the need for increased output and farm economy continues to drive the expansion of CA in other countries such as Brazil, Argentina, Paraguay, Uruguay, United States, Canada and Australia.

During the early-2000s, in addition to land degradation and loss in ecosystem services in tillage-based agricultural land, climate change became an additional global challenge to be addressed along with the need for sustainable agricultural intensification. Both these challenges provided further reasons to promote CA. This is because CA systems are to be climate-smart because they have a high level of adaptability to climate change and can mitigate climate change, whereas conventional tillage agriculture is known to be poor in climate change adaptability and mitigation. Additionally, whereas tillage agriculture was becoming known for being unable to intensify sustainably without causing more degradation, CA systems were already known to be achieving intensification sustainably. More countries began to get interested in CA systems in the 2000s.

Within the context of the sustainable agricultural intensification goal, it was also becoming clear, based on the experience gained from different parts of the world, that CA systems fitted much better into a pro-poor development strategy. This is because CA works well for smallholder farmers in any land-based ecology. Also, CA systems can be used to improve local agriculture with minimum inputs, saving time and labour while improving productivity, even without any purchased inputs if necessary.

Today all the above drivers in various combinations are enabling CA systems to be promoted and adopted by smallholders, large-scale farmers, poor or rich farmers, men or women farmers in the Global South and in the Global North. There is clearly a need to mainstream CA globally. This requires that governments and public, private and civil institutions should all be aligning themselves towards promoting CA and supporting its adoption and spread. However, for now, major constraints continue to be the fact that the education, research and development efforts continue to direct their attention to the development of conventional tillage agriculture. Indeed, hardly any of the education systems globally, particularly at the tertiary level, are directing resources to integrate CA into their curricula and teach it at both the theoretical and practical level. Much talk continues to be promoted in many academic and development circles about sustainable agricultural intensification, climate-smart agriculture, regenerative agriculture and more recently about the Sustainable Development Goals (SDGs), but without the understanding that CA is the functional principle of all these concepts. The CA community has much to contribute to such debate and practical action to meet several of the SDGs as can be seen from the various chapters of this volume and volume 2.

5.2 Adoption and spread

Farmer-led transformation of agricultural production systems based on CA is progressing globally. It is in all world regions the initial and principal driver and reason for success in the adoption and expansion of CA. Since 2008/09, the adoption has increased exponentially with the impulse of the need for a new paradigm for 'sustainable intensification of crop production' including the delivery of ecosystem services, and as a base for 'climate-smart agriculture'.

The information on the adoption of CA in 2015/16 presented in this chapter applies only to annual cropland and is based on several sources: government statistics (e.g. Canada and United States); survey estimates by no-till farmer organizations (e.g. Australia, Brazil, Argentina, Paraguay and Uruguay), by Ministry of Agriculture (e.g. China, Malawi, Zimbabwe), NGOs (e.g. Europe, Russia, Madagascar, Zambia), well-informed individuals from research and development organizations (e.g. Pakistan, India, Kazakhstan, Ukraine). The database is up to date for 2015/16 for most countries with exceptions including Ukraine, India, DPR Korea, Lebanon, Azerbaijan, Chile, Colombia, Mexico, Venezuela, Kenya, Ghana and Sudan. For these countries, the information is from 2013/14. Besides, since 2013/14, CA annual cropland systems have been recorded in more countries such as Uganda, Swaziland and Algeria in Africa; in Asia, Pakistan, Tajikistan, Iran, Bangladesh, Laos, Vietnam and Cambodia; and in Europe, Austria, Bulgaria, Croatia, Cyprus, Czech Republic, Denmark, Estonia, Latvia, Lithuania, Luxemburg, Poland, Romania, Slovenia and Sweden.

There are reports of adoption of CA in individual countries in 2008/2009 (Kassam et al., 2009; Derpsch and Friedrich, 2009a), in 2010/11 (Kassam et al., 2010; Friedrich et al., 2012) and in 2013/14 (Kassam et al., 2015). There was an interim record of the global spread of CA in 2015/16 (Kassam et al., 2017a) which was refined and expanded to include data for more countries and corrected any previous errors, particularly related to information from countries in Africa (Kassam et al., 2018). There are also global state-of-the-arts reviews of CA systems and practices in Jat et al. (2014) and Farooq and Siddique (2014), and for Africa in Kassam et al. (2017b).

Global and regional data of CA adoption and spread in cropland in the following sections are based on the above-mentioned sources, including the latest global and regional CA data for 2015/16 as published in Kassam et al. (2018).

5.3 Global

It was estimated that the global extent of CA cropland in 2008/09 covered about 106 M ha (7.5% of global cropland) (Kassam et al., 2009). In 2013/14, it was about 157 M ha (11% of global cropland), representing a difference of some 51 M ha (some 47%) over the 5-year period (Table 1) (Kassam et al., 2015). In 2015/16, CA cropland was about 180 M ha (12.5% of global cropland), representing a difference of some 74 M ha (69%) over the 7-year period since 2008/09 or about 23 M ha (some 15%) over the 2 years since 2013/14.

CA in recent years has become a fast-growing production system for many reasons including greater factor productivity and farm output, reducing cost of production and improving profitability, greater resilience to biotic and abiotic stresses, minimizing soil erosion and degradation, building soil health, improving biodiversity, adapting to and mitigating climate change (Kassam et al., 2013, 2017a; Jat et al., 2014; Farooq and Siddique, 2014). Whereas in 1973/74 CA was applied on only 2.8 M ha worldwide (Fig. 1), the area had grown to 6.2 M ha in 1983/84 and to 38 M ha in 1996/97 (Derpsch, 1998). In 1999, worldwide adoption was 45 M ha (Derpsch, 2004), and by 2003 the area had grown to 72 M ha (Benites et al., 2002). During the period from 1999 to 2013, CA cropland area had expanded at an average rate of about 8.3 M ha per year, from 72 to 157 M ha (Kassam et al., 2015). The growth of the area under CA has been especially significant in South America where the MERCOSUR countries (Argentina, Brazil, Paraguay and Uruguay) are using the system on more than 70% of their total cropland area (Kassam et al., 2018).

Table 1 Extent ('000 ha) of adoption of CA worldwide by country in 2008/09, 2013/14 and 2015/16

No	Country	CA area (2008/09)	CA area (2013/14)	CA area (2015/16)
1	USA	26500.00	35613.00	43204.00
2	Brazil	25502.00	31811.00	32000.00
3	Argentina	19719.00	29181.00	31028.00
4	Canada	13481.00	18313.00	19936.00
5	Australia	12000.00	17695.00	22299.00
6	Paraguay	2400.00	3000.00	3000.00
7	Kazakhstan	1300.00	2000.00	2500.00
8	China	1330.00	6670.00	9000.00
9	Bolivia	706.00	706.00[a]	2000.00
10	Uruguay	655.10	1072.00	1260.00
11	Spain	650.00	792.00	900.00
12	South Africa	368.00	368.00[a]	439.00
13	Germany	354.00	200.00	146.00
14	Venezuela	300.00	300.00[a]	300.00[b]
15	France	200.00	200.00[a]	300.00
16	Finland	200.00	200.00	200.00
17	Chile	180.00	180.00[a]	180.00[b]
18	New Zealand	162.00	162.00[a]	366.00
19	Colombia	102.00	127.00	127.00[b]
20	Ukraine	100.00	700.00	700.00[b]
21	Italy	80.00	380.00	283.92
22	Zambia	40.00	200.00	316.00
23	Kenya	33.10	33.10[a]	33.10[b]
24	United Kingdom	24.00	150.00	362.00
25	Portugal	25.00	32.00	32.00[b]
26	Mexico	22.80	41.00	41.00[b]
27	Zimbabwe	15.00	90.00	100.00
28	Slovakia	10.00	35.00	35.00[b]
29	Sudan	10.00	10.00[a]	10.00[b]
30	Mozambique	9.00	152.00	289.00
31	Switzerland	9.00	17.00	17.00[b]
32	Hungary	8.00	5.00	5.00[b]
33	Tunisia	6.00	8.00	12.00
34	Morocco	4.00	4.00	10.50
35	Lesotho	0.13	2.00	2.00
36	Ireland	0.10	0.20	0.20
37	Russia	–	4500.00	5000.00

No	Country	CA area (2008/09)	CA area (2013/14)	CA area (2015/16)
38	India	–	1 500.00	1 500.00[b]
39	Malawi	–	65.00	211.00
40	Turkey	–	45.00	45.00
41	Moldova	–	40.00	60.00
42	Ghana	–	30.00	30.00[b]
43	Syria	–	30.00	30.00[b]
44	Tanzania	–	25.00	32.60
45	Greece	–	24.00	24.00[b]
46	Korea, DPR	–	23.00	23.00[b]
47	Iraq	–	15.00	15.00[b]
48	Madagascar	–	6.00	9.00
49	Uzbekistan	–	2.45	10.00
50	Azerbaijan	–	1.30	1.30[b]
51	Lebanon	–	1.20	1.20[b]
52	Kyrgyzstan	–	0.70	50.00
53	Netherlands	–	0.50	7.35
54	Namibia	–	0.34	0.34[b]
55	Belgium	–	0.27	0.27
56	Pakistan	–	–	600.00
57	Romania	–	–	583.82
58	Poland	–	–	403.18
59	Iran	–	–	150.00
60	Estonia	–	–	42.14
61	Czech Republic	–	–	40.82
62	Austria	–	–	28.33
63	Lithuania	–	–	19.28
64	Croatia	–	–	18.54
65	Bulgaria	–	–	16.50
66	Sweden	–	–	15.82
67	Latvia	–	–	11.34
68	Uganda	–	–	7.80
69	Algeria	–	–	5.60
70	Denmark	–	–	2.50
71	Slovenia	–	–	2.48
72	Bangladesh	–	–	1.50
73	Swaziland	–	–	1.30
74	Tajikistan	–	–	1.20
75	Vietnam	–	–	1.00
76	Cambodia	–	–	0.50
77	Laos	–	–	0.50

(Continued)

Table 1 *(Continued)*

No	Country	CA area (2008/09)	CA area (2013/14)	CA area (2015/16)
78	Luxembourg	-	-	0.44
79	Cyprus	-	-	0.27
	Total	*106 505.23*	*156 738.96*	*180 438.64*
	% difference		*47.17 since 2008/09*	*69.42 since 2008/09* *15.12 since 2013/14*

[a] 2013/14 values taken from 2008/09.
[b] 2025/16 values taken from 2013/14.
Source: 2008/09 and 2013/14 data from FAO AquaStat as reported in Kassam et al. (2015). 2015/16 data collected by authors as reported in Kassam et al. (2018).

Since 2008/09, the annual rate of change has increased to about 10.5 M ha, from 106 to 180 M ha in 2015/16, showing the increased interest of farmers in the CA farming systems approach to sustainable production and agricultural land management. Earlier, this expansion was mainly in North and South America and in Australia and New Zealand. More recently, it is also occurring in Asia, particularly in Kazakhstan and China with large farms, and in India and Pakistan with small farms. Wheat-based CA cropping systems have been spreading in these countries in recent years. In Kazakhstan and China, rainfed wheat systems are being transformed into CA systems. Crop diversity is increasing as research has shown the feasibility of integrating legumes in the rotations. In India and Pakistan, wheat-rice cropping systems in the Indo-Gangetic Plains are being transformed into CA systems, referred to as 'double no-till' rice-wheat systems, and in some areas it has been possible to add a short-season legume crop such as mung bean in the cropping system.

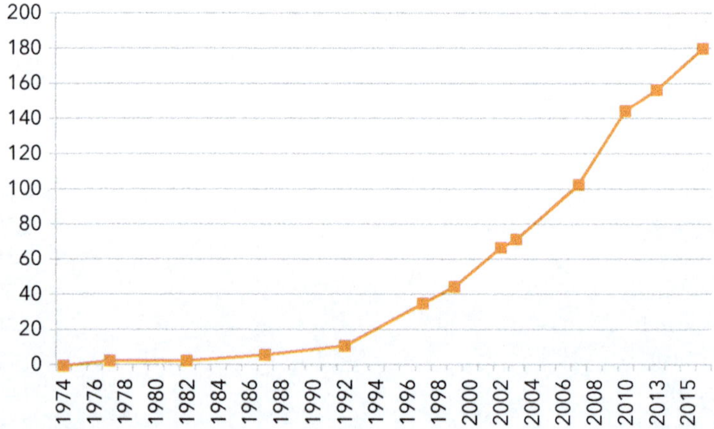

Figure 1 Global uptake of CA in M ha of cropland. Source: adapted from Kassam et al. (2017a).

Major increases in the adoption of CA cropping systems are expected across Asia in the coming decades. The situation is also changing in Africa where more smallholder farmers are taking up CA, particularly in eastern and southern Africa, and medium-scale farmers in North Africa. CA in Europe, and in Russia and Ukraine, has been expanding steadily during the past decade. These trends are expected to continue as farmers continue to share their experiences among themselves and other stakeholders, particularly no-till seeder machine companies and researchers.

Since 2008/09, the number of countries where CA adoption and uptake is occurring has increased from 36 to at least 55 in 2013/14 and to 78 in 2015/16, as shown in Table 1. The table does not include several countries where CA is known to be practised, but either at very small levels or without being reported in any systematic form. They is a significant list: Ethiopia, Rwanda, Burkina Faso, Senegal and Cameroon in Africa, Jordan, Nepal, Timor Este, the Philippines and Mongolia in Asia and Cuba, Costa Rica, Honduras, El Salvador, Guatemala and Nicaragua in Central America.

Further, the area of CA systems based on perennial crops such as in orchards and plantations, or mixture of annual and perennial crops such as trees in association with annual crops, or agroforestry systems, or crop-livestock-tree systems, or pasture systems, *are not included* in the total CA area reported in this chapter. Such CA systems with perennial crops are on the increase in all inhabited continents in most agroecologies.

In semi-arid areas with winter rainfall in the temperate and subtropical environments, CA orchards and vines include crops such as olive, grape, fruit and nut trees. Such CA systems can also be tailored with the direct sowing of field crops, cover crops and sown or natural pastures and vegetation beneath or between rows, giving permanent cover and improved soil aeration and biodiversity. The common complaint made by farmers to practising this latter type of inter-cropping or association is competition for soil water between trees and other crops. However, careful selection of deep rooting tree species and shallow rooting annuals resolves this. Functional CA systems do not replace but should be integrated with current good land husbandry practices (chapter 9, this volume).

In sub-humid and humid tropical environments, CA plantation systems include crops such as oil palm, cocoa, rubber, tea, coffee, coconut but also sugar cane with cover crops or natural or sown pastures or natural vegetation underneath (chapter 9, this volume). Thus, the CA cropland areas reported in this chapter are *conservative estimates* of global CA land use.

As Table 2 shows some 69.9 M ha (38.7%) of the total global area under CA is in South America, corresponding to some 63.2% of the cropland in the region, and some 63.2 M ha (35.0%) is in North America, corresponding to 28.1% of the cropland of the region. Some 22.7 M ha (12.6%) is in Australia

Table 2 Cropland area under CA (M ha) by region in 2008/09 and 2015/16, and the difference in %; CA area as % of global total cropland in 2015/16, and CA area as % of cropland of each region in 2015/16 (derived from Table 1)

Region	CA cropland area (2008/09)	CA cropland area (2015/16)	% Difference since (2008/09)	% of global CA cropland area (2015/16)	% of CA cropland area in the region (2015/16)
South America	49.56	69.90	41.0	38.7	63.2
North America	40.00	63.18	57.9	35.0	28.1
Australia and New Zealand	12.16	22.67	85.8	12.6	45.5
Asia	2.63	13.93	429.7	7.7	4.1
Russia and Ukraine	0.10	5.70	5600.0	3.2	3.6
Europe	1.56	3.56	127.4	2.0	5.0
Africa	0.49	1.51	211.0	0.8	1.1
Global total	*106.5*	*180.44*	*69.4*	*100*	*12.5*

and New Zealand, corresponding to 45.4% of the cropland, and some 13.9 M ha (7.7%) is in Asia, corresponding to 4.1% of the cropland in the region. Some 10.8 M ha (6.0%) of the total global CA area is in the rest of the world, comprising 5.7 M ha in Russia and Ukraine, 3.6 M ha in Europe and 1.5 M ha in Africa, corresponding to about 3.6%, 5.0% and 1.1% of their total cropland area, respectively. Thus, some 47.3% of the total CA area is located in the Global South and 53% in the Global North, involving millions of farmers particularly in the Global South.

In terms of CA adoption and uptake, Europe and Africa are the developing continents. Nevertheless, CA area in Europe of 3.6 M ha estimated in 2015/16 is greater by some 127.4% than the 1.56 M ha that was estimated in 2008/09. For Africa, the CA area of 1.5 M ha in 2015/16 corresponds to some 211% increase from 0.48 M ha in 2008/09. There has been this significant increase in CA area in Europe and Africa in recent years because many years of research in these continents have shown positive results for CA systems. There has also been the incentive of increased interest in CA systems shown by the New Partnership for Africa's Development (NEPAD), governments, European Commission (EC), NGOs such as ACT and ECAF and the private sector, international organizations and donors.

Besides other drivers mentioned earlier, the effects of climate change, which are more and more visible not only in the Global South, but also in the Global North, for example in Europe, are leading to change. For resource-poor smallholder farmers in low-income countries, CA fits better in pro-poor development strategies across most land-based farming systems (FAO, 2011, 2016), and there are several examples of successful CA scaling involving smallholder farmers in Africa with little-or-no purchased inputs (e.g. Owenya et al., 2011; Lalani et al., 2016, 2017).

The continuous growth of CA systems as shown in Fig. 1 is largely a result of the initiative of farmers and their organizations. This is augmented by technical and financial support from governments, donor agencies and international organizations for CA research and development in Africa and Asia in recent years (FAO, 2012, 2013, 2014; ACT, 2014, 2018; Kassam et al., 2017b). The uptake of CA in Africa and Asia is expected to accelerate in the coming years. When government policies support base-level initiatives, as in Kazakhstan and China, rapid growth rates occur (Nurbekov et al., 2014; Li et al., 2016).

In many countries such as the United States, Canada, Australia, New Zealand, Brazil, Argentina, Paraguay, Uruguay, South Africa, Zambia, Zimbabwe, Malawi, Mozambique, Namibia, Kazakhstan, India and China, attempts have been ongoing for CA to be 'mainstreamed' through agricultural development programmes or in some cases backed by suitable policies and institutional support in certain areas within countries depending on local policy makers and institutional leaders. However, countries such as the United States, Canada, Australia, New Zealand, Brazil and Argentina, despite a good progress in CA adoption, still lack national policies to support the spread and further development of CA. They either have state policies in support of CA such as Alberta in Canada or Paraná and Santa Catarina in Brazil, but unsuitable national policies, for example promoting certain commodities in monocropping systems, which could even endanger ongoing CA development, such as in the United States, Argentina and more recently in Brazil. Yet, the total area under CA worldwide has increased by 69.4% since 2008/09, from 106 M ha (7% of global cropland) to 180 M ha (12.5% of global cropland) in 2015/16. The adoption of CA globally since 1990 has been growing mainly in North and South America and in Australia, and more recently in Asia in particularly Kazakhstan, China, India and Pakistan, and in Europe especially in Spain, Italy, UK and France, and in Russia; and in Africa including in Zambia, Zimbabwe, Malawi, Mozambique, South Africa, Tunisia and Morocco. Thus, the area under CA is expanding in all regions of the world, and large areas of global agricultural land, including those under orchards and plantation systems, agroforestry systems and crop-pasture-tree systems are expected to transform to CA systems in the coming years and decades.

So far most of the CA development has been in rainfed annual cropping systems and some in irrigated crops in combination with rainfed crops such as the rice-wheat cropping system in the Indo-Gangetic Plains. The same CA principles apply for strengthening the ecological and economic sustainability of *irrigated systems*, including those in arid and semi-arid areas, with the additional benefit of improving water-use efficiency and avoiding or minimizing salinization problems (Basch et al., 2012; Kassam and Brammer, 2012). This is happening in the tropics and subtropics with irrigated rice-based systems in Brazil, Argentina, Pakistan, India, Bangladesh, and with other cropping systems such as irrigated cotton-based systems in Uzbekistan, and in irrigated systems in Spain and Italy. However, for rice-based CA systems, it is important also to change the rice irrigation and water management to maintain mainly moist soil with aerobic conditions and not interfere with the building up of healthy soils permitting the development of large and strong root system along with many more head bearing tillers. This water management practice so far is not yet widespread and is mostly introduced in India, Pakistan and Bangladesh, and in some South-east Asian countries, while Latin American rice growers still maintain flood irrigation in no-till rice systems.

If CA is to spread in Europe, Asia and Africa, it must be understood that in the context of sustainable agricultural mechanization it is more than just a technique, such as no-tillage and direct seeding. CA represents a fundamental change in the soil-crop-landscape system management and in the cropping system design and management, which in turn lead to consequential changes in the required operations and mechanization solutions. This involves a major shift in the current mix of mechanical technologies, some of which will remain but with only marginal use in future, and there will be the development of completely new set of mechanical technologies, changes in farm power requirements and in land-use suitability for sustainable intensification (Baker et al., 2007). This change process has been ongoing now in Europe, Asia and Africa.

In Europe, a range of machine companies over the past 5 years or so have become involved in manufacturing no-till seeders, including the angled disc seeders which are gaining in popularity, particularly in the United Kingdom. Equally, attempts are being made to innovate no-till weed management without the use of herbicides but based on equipment that can kill weeds using heat stress.

In Asia, there are now no-till seeders being manufactured in a number of countries including Turkey, Iran, India, Pakistan, Bangladesh and China. Two-wheel tractor-mounted seeders have a potential in Asia, particularly for smallholder farmers and an exciting seeder that is gaining popularity is the versatile multi-crop planter (VMP) from Bangladesh (Haque et al., 2016).

Africa too has launched a sustainable agricultural mechanization initiative supported by the African Union and the FAO which is laying a collaborative foundation among stakeholders for mechanization including for CA production systems and value chains (ACT, 2018).

As indicated earlier, CA principles and practices are also applicable for orchards, plantations and vine crops with the direct sowing of associated field crops, cover crops and pastures beneath or between rows, giving permanent ground cover and biomass production, controlling soil erosion, improving soil health and biodiversity, water infiltration and retention and soil aeration. In the dry areas of Africa, there is an increase of agroforestry systems integrating nitrogen fixing trees such as *Faidherbia albida* with CA systems (Garrity et al., 2010; see chapter 9, this volume). Orchard crops and vines are being converted into CA systems in Europe (Franco and Calatrava, 2006; Leyva et al., 2007; Martinez, 2009; Gomez et al., 2009; see chapter 9, this volume). Plantation tree crops such as oil palm, rubber, cocoa, citrus and coconut are also being successfully managed under CA systems in several countries such as Malaysia (Othman et al., 2012; chapter 9, this volume). In India, the area under CA rice-wheat and rice-maize cropping systems has significantly increased during the last 10 years or so (Saharawat et al., 2010; Farooq and Siddique et al., 2014; Jat et al., 2009a,b, 2010).

5.4 South America

The adoption levels of CA farming in Argentina, Paraguay, Uruguay and Southern Brazil are approaching 100%. Since 2008/09, the area under CA in the South America region has changed by some 41% from 49.6 M ha to 69.9 M ha in 2015/16. But there are serious concerns about the quality of some of the CA adoption. Following market pressures, which are partly increased by government policies, a considerable number of farmers are opting for soya monocropping, even without any cover crops between two soya crops. This approach, despite applying the no-till practice, has the bad results of erosion and soil degradation. Accordingly, the area under good-quality CA could be argued to be, particularly in Argentina, Uruguay and Brazil, significantly lower than the total area under no-till cropping. The problem was being solved in Brazil with strengthened extension, which more recently has again reverted, and in Uruguay with legal regulations for cover crops in the specific case of soya and subsidy programmes for good-quality CA. The problem has also been reported and is being solved in the recent no-till adoption report for Argentina, where recent policy changes have again opened up opportunities for farmers to grow crops other than soya (Nocelli, 2018).

Brazil has the longest experience in CA in South America, and in 2015/16 it had 32 M ha under various forms of CA. Since its first appearance in 1972, many useful lessons have originated from Brazil and from neighbouring Argentina and Paraguay, which now have respectively 31 M ha and 3 M ha of CA. They have also set important precedents for the engagement of farmers as principal actors in the development and adaptation of new technologies and practices in including the integration of pasture, trees and livestock (Mello and van Raij, 2006; Landers, 2007).

The first set of no-tillage experiments in Brazil were started in April 1971 at the Brazilian Institute for Agricultural and Livestock Research (IPEAME, later EMBRAPA), in Londrina, Paraná, by Rolf Derpsch, one of the co-authors of this chapter. The following year, Herbert Bartz, the first farmer to try the technology in Latin America, was already introducing the system on his farm. From there it took Brazil almost 20 years to reach the first million ha of no-tillage being applied by farmers, but after this milestone the practice has experienced an exponential growth.

Brazil took the initiative when herbicides (paraquat and diquat) and direct-drilling equipment became available in the United States, and the realization that conventional ploughing was leading to a severe environmental and economic crisis for farmers in southern Brazil. Progressive and wealthy farmers led the way, some travelling to the United States to learn about soil conservation and management systems there and to purchase direct-drilling equipment. Common interest groups were then formed among large-scale farmers and subsequently among small-scale farmers, joining to the Brazilian Federation for Direct Seeding into Straw (FEBRAPDP). The spread of CA in Brazil is mainly the result of farmer innovation, giving the main impulse for subsequent problem-solving support from input supply companies, state and federal research and extension organizations, universities, as well as long-term funding commitments from international donors such as the World Bank and GTZ. However, the momentum for innovation and adoption is still with farmers and their organizations, and the farmer-led organizations enhanced and reinforced the educational components required for understanding CA principles and concepts (de Freitas and Landers, 2014).

Apart from enabling their land to be cropped more intensively without risk of degradation, CA attracted Brazilian farmers because it increased crop yields (at least 10–25%), greatly reduced surface run-off and soil erosion and cut tractor use, resulting in big savings in fuel and production costs (Sorrenson and Montoya, 1984, 1991; Sorrenson, 1997). Such benefits explain why today, Latin American farmers practice CA systems on a continuous basis on some than 69.9 M ha. These systems are among the most competitive production systems anywhere in the world and they are sustained without any government subsidies.

In the early 1970's Argentina also began its first research and farm trials with no-till. Several farmers started with the system and then gave up because of the lack of adequate herbicides and machinery that, together with lack of know-how, constituted the main constraint for early adopters. A milestone in the development and spread of no-till in Argentina was the foundation in 1989 of the Argentinean Association of No-till Farmers (AAPRESID), based in Rosario. Since 1992, AAPRESID has been organizing no-till conferences in August of every year which have been attended by more than 1000 farmers at the beginning and nowadays exceed 2000 farmers. Since the founding of AAPRESID, Argentina also experienced an exponential growth of the no-till farming.

Argentina experienced a paradigm shift with the advent of the no-tillage practice and finally discarding the idea that tillage was necessary to grow crops. In Argentina, the concept of 'arable' soils has been abandoned after recognizing that soils that cannot be ploughed can be directly seeded. According to AAPRESID (2010) in 2007/08, there were 25.8 million ha of no-tillage being practiced in Argentina (http://www.aapresid.org.ar), making it one of the most successful countries in terms of no-till adoption. The first group of farmers started using no-till in 1977/78 after exchanging ideas with Carlos Crovetto, one of the most renowned no-till experts from Chile, as well as with Shirely Phillips and Grant Thomas from the United States. At the beginning, growth was slow because of lack of experience, knowledge on how to do it, machines and limitations on the availability of herbicides. It took 15 years until 1992/93 when about 1 million ha under no-tillage were reached. Since then adoption increased year by year as a result of the intensive activities of AAPRESID so that in 2008/09 about 79% of all cropland in Argentina was under no-tillage system. In 2015/16, it was 31 M ha, about 95% of all cropland.

One of the main factors that made the rapid growth of no-tillage possible in Argentina was the fact that machine manufacturers quickly responded to the increasing demand in no-till seeders. Among the many big and small no-till seeders manufacturers in Argentina, there are at least 15 that are in good conditions to export their equipment. No-tillage in Argentina is almost exclusively performed with disc seeders.

Similar to other countries in South America, farmers in Argentina prefer to do permanent no-tillage once they have started with the system. At the beginning cover crops were not an issue for no-till farmers in this country because it was believed that these crops would take too much moisture out of the soil. This has changed in recent years when research could show that water-use efficiency can be enhanced when using appropriate cover crops. A milestone in no-tillage in Argentina was reached on 7 May 2010 when G. Cabrini with the help of AAPRESID became the first farmer to certify his no-till production system. The certification protocol is based on principles and criteria developed from international initiatives that focus on sustainability (see chapter 12, this volume).

Paraguay has experienced a continuous and steady growth of CA adoption, almost all of it over the past 10 years. Tillage practices have disappeared almost completely. In tractor mechanized farming systems, about 90% (of the total 2.4 M ha in 2008) of all crop area was under CA (Derpsch and Friedrich, 2009b). In 2015/16, some 3 M ha (nearly all of the cropland) was in CA. Similarly, in small farmer production systems with animal traction or manual systems, no-till practices have increased to about 30 000 ha covering 22 000 small farmers. The increased interest in small-scale farmer CA systems has been a result of government support that provides extension services and grants for buying no-till equipment.

In Bolivia, CA practices increased in the last 10 years especially in the lowlands in the east of the country where the main crop is soybeans whose area has increased from around 240 000 ha in the year 2000 to 706 000 ha in the year 2007 (Derpsch and Friedrich, 2009b), and to 2 M ha in 2015/16. The occurrence of wind erosion in conventional tillage systems has been one of the major driving forces for adoption. Also, farmers value the increased water-use efficiency with no-till system in this low and erratic rainfall region. However, CA is only promoted in the eastern plains around Santa Cruz by the Association of Oilcrop and Wheat Producers (ANAPO). Early experiments with no-till systems were introduced by CIMMYT in the 1990s on the ANAPO experimental station and ANAPO is the only organization providing technical assistance in CA cropping systems. Outside the ANAPO area, no-till is practiced to a minor extent but permanent no-till systems such as CA are unknown. This is the case for the tropical eastern lowlands as well as for the Andean regions. Only recently, due to severe soil degradation, alarmingly low production levels and increasing pressures from climate change, the government is becoming seriously interested in CA.

In Uruguay, about 82% of cropland, that is 655 000 ha, was under no-till systems in the 2006/07 growing season according to the Uruguayan No-till Farmers Association (AUSID). This was a great progress compared to the 2000/01 season when only 119 000 ha of no-tillage were reported, corresponding to 32% adoption. In 2015/16, CA area covered was 1.26 M ha (52% of total arable land) and the total cropland was much bigger, some 2.41 M ha, due to expansion of no-till cropland into pasture areas. Some 65% of arable crops are seeded on rented land for which contracts are renewed every year, and this hinders the planning of medium-term crop rotation and investment strategies. In Uruguay, the integration of crops with livestock is very popular and CA systems fit well into the requirements for crop-livestock production systems. Pastures are grown for several years until they show signs of 'degradation'. Crops are then grown for several years according to the needs of the farmers and the market situation.

Venezuela, Chile, Colombia and Mexico each have modest amounts of their land under no-till systems, ranging from some 41 000 ha in Mexico and 127 000 ha in Colombia to 180 000 ha in Chile and 300 000 ha in Venezuela (Derpsch and Friedrich, 2009b; Kassam et al., 2018).

The main crops grown under CA in Latin America include soybean, maize, wheat, sunflower, canola as well as cassava (Howeler et al., 2013), and a number of horticultural and cover crops, while CA potato is marginal. CA practices are also being applied to perennial crops and to tree crops. Soil cover is achieved by growing cash crops and cover crops either in association or sequentially. Main cover crops include oats, oilseed-radish, rye, lupine, vetch, *Mucuna* (velvet bean), *Dolichos* and pigeon pea. In recent years, mixtures of up to ten different cover crops are sown to address specific needs of the soil or subsequent crop in terms of cover, nitrogen release, weed suppression or other pest control effects (Tilman et al., 1997). In some cases, especially among small-scale farmers, herbicide use is reduced by direct-drilling of the seed into a cover crop that has been flattened with a knife roller. Specialized no-till equipment has been developed in Brazil and the Americas, including tractor-mounted, animal-drawn and hand tools (including jab planters). These are being exported to Africa and Asia and being adapted there for local use and manufacture.

5.5 North America

CA adoption is the highest in the North-Western Parts of North America and in the southern parts of South America with adoption levels above 50%. Since 2008/09, the area under CA in the North America region has changed by 57.9% from 40.0 M ha to 63.2 M ha in 2015/16.

In Canada, CA is estimated to be practised on 19.9 M ha (38.2% of the crop area) in 2015/16, although no-till technology is used over a much larger area, 67% of crop area (Derpsch and Friedrich, 2009b). In 2008/09, CA was estimated to be 13.5 M ha (25.9% of the crop area) and the regions with highest percentage of adoption of no-tillage were Saskatchewan (60.1%), Alberta (47.8%), Ontario (31.2%), Manitoba (21.3%) and British Columbia (19.0%). Canada has had a similar development as the United States, with heavy erosion problems in the 1930's and the subsequent focus on conservation tillage. However, after the year 2000, more importance was given to a systems approach, not only focussing on reduced or zero tillage and chemical fallows but including factors such as soil organic cover and crop rotations. As a consequence, between 1999 and 2004 the amount of wheat grown in Canada went down by 6.4%, while the oil crops increased by 48.7% and pulses by 452.7%. At the same time, the use of fallow went down by 58.7% (Yuxia and Chi, 2007). These developments are parallel to the recent increase in the application of CA in Canada since the year 2000 (Goddard et al., 2006, 2009; Lindwall and Sonntag, 2010). This long-term and wide adoption of CA, mostly in the western provinces, has resulted in visible environmental benefits, including the absence of dust storms and a greater biodiversity. The main co-benefits from CA in Canada have been documented by Baig and Gamache (2009, 2011). There was also excellent technical research done on crop diversification and on integrated weed management (Blackshaw et al., 2007). Environmental services provided through CA are increasingly recognized, for example through a voluntary carbon offset trading scheme as in Alberta (Haugen-Kozyra and Goddard, 2009) that encourages industry under a 'cap and trade' regulation to purchase carbon offsets from farmer associations whose members are practicing a production system based on the government-approved no-till protocol (similar to the CA concept) to sequesters additional soil carbon and reduce greenhouse gas emissions carbon payment schemes as in Alberta (see chapter 13, volume 2).

More recently, research in Alberta, Canada, has shown that production cost savings and energy-use efficiency are best achieved if precision farming technology and variable rate technology (VRT) for fertilizer application are built upon no-tillage systems. No-tillage already forms a basis for operating an agricultural carbon offset trading scheme in Alberta, as well as for reducing greenhouse gas emissions from agriculture. For example, according to the research supported by the Agricultural Research and Extension Council of Alberta (ARECA), producers throughout Alberta can reduce their fuel use and become more energy-efficient by:

- converting from conventional tillage-based practices to zero-till practices;
- operating energy-efficient equipment technologies;
- adopting precision farming techniques and VRT for application of fertilizers.

Agriculture in Alberta, as elsewhere in the industrialized countries, is a major user of energy and conventionally tilled farms spend about 24% of their energy inputs on fuel and about 60% on fertilizer. Converting to zero-till practices alone increases energy efficiency, energy conservation and profitability. Fuel savings from converting from conventional tillage to zero-tillage averages around 38% (across all crop rotations). During the period of 2001–2006, zero-till practices increased by 1.6 M ha and during the same time period diesel fuel consumption fell by 70.2 M L. This led to decreased CO_2 emissions and improved soil conservation (see www.areca.ab.ca). Since 2007, Alberta has been operating an agricultural carbon offset scheme in which the protocol that defines the production system compliance characteristics is based on no-till (Haugen-Kazyra and Goddard, 2009; Goddard et al., 2009).

As noted earlier, no-till agriculture systems in the modern sense originated in the United States in the 1960s, and from then up until 2007 the United States had the largest area under no-till worldwide. In 2008/09, CA was 26.5 M ha (21.5% of cropland) and in 2015/16, it was 43.2 M ha (35.1% of cropland), despite long time experience with no-till farming. Conventional agriculture with tillage remains in the majority even if CA is a valid option for farmers, as compared with southern Latin America where no-till has become the majority agricultural system with 63.3% of the crop area. The awareness about crop rotations and cover crops as well as the additional benefits of permanent no-till systems is growing because of organized farmers' associations at the state, and at the regional level. Research has shown that it can take more than 20 years of continuous no-till to reap the full benefits of CA. Farmers that practice rotational tillage (plough their soils occasionally) will not experience the full benefits of the system (Derpsch, 2004).

A particularly exciting development in no-till system in the United States is the practice of 'planting green'. This operation allows the establishment of a crop, following a cover crop, without using any herbicides but instead using a roller crimper to subdue the cover crop (Duiker, 2017; Gullickson, 2018). In the United States, much of CA cropping has used maize, soybean and canola crops, but more recently cotton systems also.

5.6 Australia and New Zealand

The area of CA in Australia and New Zealand combined increased from 12.16 M ha in 2008/09 to 22.67 M ha in 2015/16, an increase of 85.8%.

In Australia, CA has been widely and quickly embraced by farmers, making it the country with the fourth largest area of 22.3 M ha in 2017, and increase from 12.0 M ha in 2008/09. It has improved weed control, time of sowing, given drought tolerance and has enabled dry regions to use water most efficiently (Crabtree, 2004, 2010; Flower et al., 2008; Llewellyn et al., 2009; Jat et al., 2014). In 2017, the adoption of no-till by farmers in Australia, according to the Australian Bureau of Statistics, varied from 84.5% in northern New South Wales to 93.5% in South Australia and 96.3% in Western Australia. Overall adoption of no-till in Queensland is approximately 72%. CA methods have led to large increases in profitability, sustainability and positive environmental impact in the Australian cropping belt. Also, the use of cover crops is getting popular among no-till farmers. Because of the water, time and fuel savings with no-till systems as well as the other advantages, cropland under no-tillage is expected to continue to grow. Another complementary technology used in Australia on no-tillage farms is controlled traffic farming to avoid soil compaction. Particularly in Eastern Australia, the adoption of CA is needed to reduce the contamination of water bodies with sediments, nutrients and pesticides, which is threatening the existence of the Great Barrier Reef on the eastern coast of Australia.

New Zealand has about 366 000 ha under CA, which corresponds to about 56% of all cropland area including pasture, forage crops and arable crops (Baker, 2015). In 2008/09, the area of CA was 166 000 ha. New Zealand is among the first in the world to use and develop the no-till technology. In the beginning in the seventies, pasture renovation without tillage was tried and practiced successfully. Later, also annual crops were seeded with the no-till. However, the majority of the increase in CA area has occurred since 2000.

5.7 Asia

Asian countries have adopted CA in many areas during the past 10–15 years, and since 2008/09, CA area has increased more than fourfold (429.7%), from some 2.6 M ha in 2008/09 to some 13.9 M ha in 2015/16. In 2008/09, CA area was reported in only two countries in the Asia region, but in 2013/14, CA area was reported in 11 countries, and in 2015/16 in 18 countries.

In Central Asia, earlier work on CA practices in Eurasia has been reported by Gan et al. (2008), for Kazakhstan by Suleimenov (2009) and Fileccia (2008), and for Uzbekistan by Nurbekov (2008) and FAO (2009). ICARDA and CIMMYT have also been active in CA research in the Central and West Asia and North Africa (CWANA) region (Pala et al., 2007; Karabayev, 2008; Suleimenov, 2009; Nurbekov, 2008). A faster development of CA can be observed in the last 10 years in Kazakhstan which now has 10.5 M ha under reduced tillage, mostly in the northern drier provinces, and of this some 2.5 M ha (15.6% of crop area) are 'real' CA with permanent no-till and rotation. Kazakhstan is among the top ten countries in the world with the largest crop area under CA systems. No-till adoption has been promoted for some time by CIMMYT and the FAO who introduced no-tillage systems in a CA project from 2002 to 2004. No-till adoption started from 2004 onwards in the north Provinces (North-Kazakhstan, Kostanay and Akmola) where the highest adoption rates have been registered. CA has had an explosive development in recent years as a result of farmers' interest, accumulated research knowledge, facilitating government policies and an active input supply sector (Derpsch and Friedrich, 2009b). Extra incentive is offered to no-till farmers by government which has also supported long-term research work to provide solutions to farmers on issues such as the need to maximize effective winter snowfall through stubble trapping; to increase the generation of biomass through cover crops replacing bare or chemical fallows; to diversify cropping systems; and to improve integrated weed management (Suleimenov and Thomas, 2006; Suleimenov and Akshelov, 2006).

In addition, Uzbekistan, Azerbaijan, Kyrgyzstan and Tajikistan, as well as Laos, Vietnam and Cambodia have made a committed start to promoting rainfed and irrigated CA cropping systems (Nurbekov et al., 2014, 2016; Lienhard et al., 2014) and so have other countries in West Asia such as Iran, Turkey, Lebanon, Syria and Iraq (Loss et al., 2016; Asadi, 2018). Iran and Turkey now report some 150 000 ha and 45 000 ha under CA, respectively. Area under CA in Syria and Iraq has continued to increase because of shortages of fuel (Piggin et al., 2015).

In India, the adoption of no-till practices by farmers has occurred mainly in the wheat-rice double cropping system and initially was adopted primarily for the wheat crop. The main reason for this is the fact that tillage takes too much time resulting in delayed seeding and yield loss of the wheat crop after rice (Hobbs and Gupta, 2003; Hobbs et al., 2008). The Rice–Wheat Consortium for the Indo-Gangetic Plains, an initiative of CGIAR, led by the International Rice Research Institute (IRRI) and CIMMYT, and involves several National Agricultural Research Centres, promoted no-till practices and it is mainly their efforts that have resulted in the massive uptake of no-till wheat in the region (Erenstein et al., 2008). The uptake of the technology was rapid in the north-western states which are relatively better endowed with respect to irrigation, mechanization and where the size of holdings is relatively large (3–4 ha) compared to the eastern region which is less equipped and mechanized and where the average land holding is small (1 ha) (Derpsch and Friedrich, 2009a,b). CA research and extension work now is spread over much of the Indo-Gangetic Plains including the eastern parts in India and in Bangladesh with promising results (see chapter 6, this volume).

In general, in the Indo-Gangetic Plains across India, Pakistan, Nepal and Bangladesh, large adoption of no-till wheat with some 5 M ha is reported, but only modest adoption of permanent no-till systems and full CA (Farooq and Siddique, 2014; Hobbs et al., 2008). The exception appears to be India and Pakistan, where significant adoption (1.5 and 0.6 M ha, respectively) of no-till practices by farmers has occurred in recent years in the rice-wheat double cropping system (Farooq and Siddique, 2014; Kassam et al., 2018), and also in the rainfed upland areas in India for crops such as maize, sorghum, millets, cotton, pigeon pea and chickpea. Recent reports suggest that in the Indo-Gangetic Plains in India, there may be up to 3.5 M ha of CA-based rice-wheat system (Paroda, 2018). Bangladesh has begun to report some CA area with rice-based system, particularly on permanent beds. This is expected to expand because farmers can now access no-till seeding service from service providers with locally produced CA equipment (Haque et al., 2018).

China has been experiencing an equally dynamic development of CA. It began in 1990 with research on no-till and controlled traffic systems (Li et al., 2007), and then the adoption of CA increased during the last few years and the technology has been extended to rice production. A main driver for CA in China has been the need to reduce the danger of dust storms during the Olympic Games in 2008. In the years preceding the games, CA adoption was promoted particularly in the Hebei province, surrounding Beijing, while in Beijing province the sales of ploughs were forbidden. Subsequently, China adopted CA also as a national policy. In 2013/14, some 6.7 M ha were under CA in China and 23 000 ha in DPR Korea were reported. For 2015/16, China reports CA area of some 9 M ha (Li et al., 2016). The introduction of CA in DPR Korea has made it possible to grow two successive crops (rice or maize or soya as summer crop, and wheat or barley as winter crop) within the same year, through direct drilling of the second crop into the stubble of the first. The feasibility of growing potatoes under zero tillage and CA has also been demonstrated in DPR Korea (FAO, 2007) and elsewhere.

5.8 Europe (including Russia and Ukraine)

Since 2008/09, the CA area for annual crops in Europe has changed from 1.6 M ha to 2.0 M ha in 2013/14, and increase of 30%, and to 3.5 M ha in 2015/16, an increase of 127.4%. In 2008/09, CA was reported in 11 countries but in 2013/14, this increased to 15 countries, and in 2015/16 to 29 countries. Since 1999, ECAF and its national association members, comprising many farmers, have been promoting CA systems in Europe, with significant adoption in Spain, Italy, Finland, France, Romania, Poland, United Kingdom and Switzerland. They have also been active in bringing CA to the notice of officials at the European Commission as well as members of the European Parliament. However, progress has been slow in terms of integrating CA principles and practices as part of Common Agricultural Policy (CAP) support to European farmers, although more recently there has been a revival of interest in providing support practices that would improve soil health management. Another positive change in Europe that is facilitating the uptake of CA has been the availability of improved no-till drills and other equipment manufactured in Europe, particularly in the United Kingdom. This has led to farmers and machine companies to organize on-farm events to demonstrate no-till drills and participate in on-farm conferences to discuss their experiences and successes with CA practices which to them seem to be addressing their practical agro-ecological interest in climate-smart regenerative agriculture to build soil health, reduce input costs and raise productivity and profit. One popular event of this nature is the Groundswell No-Till Show and Conference which is held annually in the United Kingdom, where some dozen machine companies demonstrate their CA no-till drills and farmers and CA experts from the United Kingdom and abroad make presentations on different topics related to regenerative and sustainable agriculture based on CA principles and practices (Kassam et al., 2018).

Russia and Ukraine also show significant adoption of CA and they also have active farmer groups promoting CA. In Russia, the area under reduced tillage is believed to be some 15 M ha, but CA according to FAO definition is estimated to be about 5.0 M ha. In Ukraine, CA has reached some 700 000 ha in 2013/14 but an accurate estimate of CA area was not possible in 2015/16.

A description of the status of adoption in some of the countries in Europe is presented below based on Derpsch and Friedrich (2009a,b), Basch et al. (2008), Derpsch et al. (2010), Friedrich et al. (2014) and Kassam et al. (2015, 2018).

No-tillage research in Spain started in 1982. On the clay soils of southern Spain, no-tillage was found to be advantageous in terms of energy consumption and moisture conservation, as compared to both, conventional or minimum tillage techniques (Giráldez and González, 1994).

Spain is the leading country in terms of no-till adoption in Europe. According to AEAC/SV (Spanish Conservation Agriculture Association – Living Soils), no-tillage of annual crops is practiced on 650 000 ha in Spain. Main crops under no-tillage are wheat, barley and much less maize and sunflowers. Besides annual crops grown in the no-tillage system in Spain, many olive plantations and fruit orchards have turned to no-till systems. AEAC/SV reports 893 000 ha of no-tillage being practiced in perennial trees in most cases in combination with cover crops and livestock (generally sheep). Main tree crops in no-tillage systems in combination with cover crops are olives and much less apple, orange and almond plantations. The extent of no-tillage practices in tree crops is not included in the global estimates in Tables 1 and 2. In total, it is reported that CA is applied on about 10% of arable land in Spain, and farmers practicing CA are receiving extra payment (from local, national and EU sources of funds) over and above the CAP-based single farm payment.

In France, long-term experiments with different minimum tillage techniques (including no-tillage) were started by the French National Institute for Agricultural Research (INRA) and the Institute for Cereal and Forage Technology (ITCF) in 1970, mainly with cereals (Boisgontier et al., 1994). The authors concluded that a comprehensive range of technical and economic data are now available in France in relation to where minimum tillage can be developed and how it can be implemented. France is among the more advanced countries in Europe in terms of adoption of CA/No-till farming. The French No-till Farmers Association (APAD) estimates that no-tillage is practiced on about 200 000 ha in this country, corresponding to just over 10% of arable land in France. Some farmers have developed superior no-till systems with green manure cover crops and crop rotation that are working very well. The 2008 International Conference on Sustainable Agriculture, organized by the Institute for Sustainable Agriculture (IAD) under the High Patronage of Mr. Nicolas Sarkozy and the launching of the IAD Charter for Sustainable Agriculture, were expected to show results in terms of greater acceptance of CA practices at all levels and especially at the political level, which is also needed across the whole EU in order to increase farmer acceptance. This notwithstanding, CIRAD has been researching on and promoting CA internationally for many years under the term 'Direct Seeded Mulch-Based Cropping System' (DMC) (Séguy et al., 2006a,b; Seguy et al., 2008).

The adoption of no-tillage technologies was very fast in Finland in the initial years. According to the Finnish Conservation Agriculture Association (FINCA), in less than 10 years no-tillage grew from some 100 ha to 200 000 ha in 2008. In this way, Finland managed to advance to one of Europe's leading no-till countries. The reason for this rapid adoption was that those farmers who believed in the no-till system and made it work communicated their experiences to their peers. The extension service and research organizations as well as agribusiness took interest in this development only later. FINCA has played a major role in spreading no-tillage in Finland. One manufacturer of no-till seeders in Finland took interest in no-tillage very early and claims to have sold almost a thousand no-till seeding machines until 2007, having about 50% of the market share in the country. About ten no-till seeder manufacturers from around the world have been able to place their no-till machines in the Finnish market and four of them are made in Finland. Another interesting fact about no-tillage in Finland is that no-tillage is practiced successfully from the far South of the country up to the Arctic Circle in the North (66° N).

Switzerland has made remarkable progress in terms of research, development and adoption of no-tillage practices. Research performed in Switzerland over more than 10 years has shown equal or better yields under no-tillage in a variety of crop rotations. No-till tends to be more and more accepted in Switzerland. This is because conventional tillage (and also reduced tillage practices as chisel ploughing) exposes the soil to erosion under the topography prevailing in this country. According to Swiss No-Till, CA approach was applied on about 12 500 ha in 2010 and this corresponds to about 3.5% of arable land in this country. In 2013/14, CA area was estimated to be 17 000 ha.

Investigations into no-tillage technologies in Germany started in 1966. Intensive and long-term research has concluded that no-tillage is a viable cultivation system. According to Teebrügge and Böhrensen (1997), no-tillage is a very profitable cultivation system compared to conventional tillage because of the lower machinery costs and lower operating costs. No-tillage decreases the purchase costs, the tractor power requirement, the fuel consumption, the amount of required labour as well as the variable and fixed costs. Since the same crop yields can be achieved by no-tillage compared to plough tillage, on average the profit will be greater with no-tillage systems.

Despite these facts and opportunities and long-term research, adoption of no-till farming in Germany is still very low. Well-informed scientists, farmers and experts with a thorough understanding of no-till farming as practiced in most parts of the world do not coincide so that probably till today there are no more than about 5000 ha of this technology being practiced by farmers in Germany. At the same time, one can recognize that there are outstanding farmers practicing no-tillage in this country like for instance Thomas Sander who farms in Oberwinkel, Saxony and receives many visitors every year. The quality of his no-tillage operation with crop rotations and cover crops earned his farm the Environmental Award of the State of Saxony 2006. With increased fertilizer and fuel prices, erosion problems in some regions and regular droughts in others, interest in no-tillage farming is growing steadily but adoption appears to some 140 000 ha in 2015/16 which is lower than the CA area reported earlier.

Some farmers such as Alfons Bunk from Rottenburg, Suabia have been using continuous no-till for 20 years successfully.

Portugal and Italy, despite showing significant signs of soil degradation and erosion already since antique times (Montgomery, 2007), have still fairly low levels of CA adoption. According to ECAF, Portugal had some 32 000 ha under CA/No-Till system in 2015/16, a modest increase from 28 000 ha in 2008/09. CA in Italy is referred to as *Agricultura Blue* (Pisante, 2007). In 2015/16, CA cropland area covered some 284 000 ha, a considerable increase from 80 000 ha in 2008/09. In Italy, there is also a significant and growing adoption of CA concepts such as no-tillage and cover crops in fruit and olive orchards and regional governments in Italy do subsidize farmers for applying reduced tillage.

Ukraine is a country where estimates on the adoption of no-tillage vary greatly depending on the source of information. Estimates vary from less than 30 000 ha to more than a million ha. Official government statistics on no-tillage state an adoption of 250 000 ha. Unfortunately, CA systems as understood by the authors of this paper (see definition above) have not progressed as much as some people might wish. According to AgroSoyuz (a large cooperative farm in Dnipropetrovsk), there are about 1.1 million ha of Direct Seeding technology being practiced in Ukraine. AgroSoyuz has organized several no-till conferences in Dnipropetrovsk inviting many renowned international speakers and since then understanding has been growing that only low disturbance systems bring additional benefits, justifying the focussing on no-tillage. The estimate of CA in 2013/14 by AgroSoyuz was 700 000 ha.

Compared to other world regions, CA development in Europe has been particularly slow, with some few exceptions, such as for example Finland, Spain, Italy and United Kingdom. There is a number of reasons for this slow adoption in Europe, some of which are the moderate climate which does not cause too many catastrophes urging for action, agricultural policies in the EU including direct payments to farmers and subsidies for certain commodities, which take the pressure off the farmers for extreme cost savings and discourage the adoption of diversified crop rotations. Cheap imports of soya as animal feed has also practically eliminated the market for locally grown legumes, an important component of CA rotations. In addition to this, there are interest groups opposed to the introduction of CA, which results for example in difficulties for a European farmer to buy a good-quality no-till direct seeder with low soil disturbance and high residue handling capacity. Most of the European farmers practicing CA have directly imported CA equipment or have had contact with small import agents. However, the environmental pressure, as well as the effects of climate change in EU are also increasing and the next European CAP soon to be reviewed and reformed seems likely to turn more favourable towards CA.

5.9 Africa

In Africa, innovative participatory approaches are being used to develop supply chains for smallholders to access CA equipment. Similarly, participatory learning approaches such as those based on the principles of FFS and lead-farmer networks are being encouraged to explain the ecological principles underlying CA and to make it attractive for use in local farming.

CA is spreading in eastern and southern Africa, and North Africa, using indigenous and scientific knowledge, and equipment design from Latin America based on the earlier development work promoted by FAO, CIRAD, the ACT Network, ICRAF, CIMMYT, ICRISAT, IITA (Fowler and Rockstrom, 2001; Haggblade and Tembo, 2003; Kaumbutho and Kienzle, 2007; Shetto and Owenya, 2007; Nyende et al., 2007; Baudron et al., 2007; Boshen et al., 2007; SARD, 2007; Erenstein et al., 2008; FAO, 2008; Owenya et al., 2011; Thierfelder et al., 2013, 2018). There is now also collaboration with China, Bangladesh and Australia, and CIMMYT, ICARDA, ICRISAT, ICRAF, CIRAD, ACT, FAO, IFAD, African Development Bank (AfDB) and NGOs. These have all stimulated the trend to have local practices and local equipment, with advantages in maintenance and repair. Farmers in at least 22 African countries are promoting CA (Kenya, Uganda, Tanzania, Rwanda, Sudan, Ethiopia, Swaziland, Lesotho, Malawi, Madagascar, Mozambique, South Africa, Namibia, Zambia, Zimbabwe, Ghana, Burkina Faso, Senegal, Cameroon, Morocco, Tunisia and Algeria). CA has also been incorporated into the regional agricultural policies by NEPAD, and it is recognized as a core element of climate-smart agriculture (Kassam et al., 2017b, 2018).

CA systems help Africa's resource-poor farmers to maintain subsistence with sustainability, so as to meet the challenges of climate change, high-energy costs, environmental degradation and labour shortages. The CA area is still relatively small mainly because of the small land holdings as well as greater attention being paid to the promotion of conventional tillage agriculture, without much success. But there is a developing trend, a CA movement of some 2 million small-scale farmers on the continent. Since 2008/09, CA has spread further (details lacking) in countries such as Kenya, Madagascar, Malawi, Mozambique, Tanzania, Zambia and Zimbabwe. Currently, South Africa is undertaking a national consultation process to facilitate the integration of CA into national agricultural development policy. Similarly, AfDB has announced that it will promote agricultural development in the Guinea Savanna zone of Africa based on CA systems.

In 2008/09, CA was reported in nine countries, but in 2013/14 there were 14 countries with area under CA, and in 2015/16, 17 countries. The total area of CA in Africa in 2015/16 is more than 1.5 M ha, an expansion of some 211% since 2008/09, from 0.48 M ha. From expert and scientific knowledge expressed on building CA-based farming systems at the 1st Africa Congress on Conservation Agriculture in March 2014, and on making climate-smart agriculture real in support of the Malabo Declaration expressed at the 2nd Africa Congress on CA in October 2018, it is clear that a significant research and development effort is being directed in Africa towards transforming conventional agriculture to CA. African Union and FAO have also launched the sustainable agricultural mechanization for Africa (SAMA) with CA as the preferred approach. It is thus likely that the adoption and spread of CA systems in Africa will expand in the coming decades to increase food production with fewer negative effects on the environment and energy costs, and to result in the development of locally adapted technologies consistent with CA principles (Kassam et al., 2017b, 2018; ACT, 2018).

5.10 West Asia and North Africa

Research and practical field demonstrations in the West Asia and North Africa (WANA) winter rainfall Mediterranean region has shown that yields and factor productivities can be improved with CA systems (Kassam et al., 2012; Piggin et al., 2015; Gonzalez-Sanchez et al., 2015; Loss et al., 2016; Bashour et al., 2016; Lalani et al., 2018). Extensive research and development work has been conducted in several countries in the region since the early 1980s (Jat et al., 2014; Farooq and Siddique, 2014) such as in Morocco, Tunisia, Algeria, Syria, Iraq, Lebanon and Jordan, and in Turkey and Iran.

In the WANA region, much of the earlier CA work done in various countries had shown that yields and factor productivities can be improved with no-till systems. Extensive research and development work were conducted in several countries in the WANA region since the early 1980s such as in Morocco (Mrabet, 2007, 2008a,b,c); and more recently in Tunisia (M'Hedhbi et al., 2003; Ben-Hammouda et al., 2007), in Syria, Lebanon and Jordan (Belloume, 2007; Bashour, 2007; Pala et al., 2007; Ghosheh, 2007) and in Turkey (Avci et al., 2007).

In the WANA region, Centero-Martinez et al. (2007), Lahmar and Triomphe (2007) and Pala et al. (2007) endorse the potential benefits that can be harnessed by farmers in the semi-arid Mediterranean environments while highlighting the need for longer-term research including on weed management, crop nutrition and economics of CA systems.

According to Centero-Martinez et al. (2007), the main reasons for adoption of CA are (1) better farm economy (reduction of costs in machinery and fuel and time-saving in the operations that permit the development of other agricultural and non-agricultural complementary activities); (2) flexible technical possibilities for sowing, fertilizer application and weed control; (3) yield increases and greater yield stability; (4) soil protection against water and wind erosion; (5) greater nutrient-efficiency; and (6) better water economy in dryland areas. Also, no-till and cover crops are used between rows of perennial crops such as olives, nuts and grapes. CA can be used for winter crops, and for traditional rotations with legumes, sunflower and canola, and in field crops under irrigation where CA can help optimize irrigation system management to conserve water, energy and soil quality and to increase fertilizer-use efficiency.

Work by ICARDA and CIMMYT has shown benefits of CA especially in terms of increase in crop yields, soil organic matter, water-use efficiency and net revenue. CA also shows the importance of utilizing fallow periods for cropping and of crop diversification, with legumes and cover crops providing improved productivity, soil quality, N-fertilizer-use efficiency and water-use efficiency. CA is perceived as a powerful tool of land management in dry areas according to Lahmar and Triomphe (2007). It allows farmers to improve their productivity and profitability especially in dry areas while conserving and even improving the natural resource base and the environment. However, CA adaptation in drylands faces critical challenges linked to water scarcity and drought hazards, low biomass production and acute competition between conflicting uses including soil cover, animal fodder, cooking/heating fuel, raw material for habitat and so forth. Poverty and vulnerability of many smallholders that rely more on livestock than on grain production are other key factors.

Since 2008/09, the area under CA in WANA region has increased substantially – from 10 000 ha to 103 200 ha in 2013/14, and to 269 300 ha in 2015/16, an increase of some 269%. In 2008/09, only two countries reported the existence of CA area, but in 2013/14 the number increased to six countries, and in 2015/16, it was eight countries.

Morocco, Tunisia and Algeria have shown a modest growth in CA adoption, but the adoption has been enormous in Iran and Syria, increasing in only a few years to 150 000 ha and 30 000 ha, respectively. Iran is the largest adopter in the region followed by Turkey (45 000 ha) and Syria as the second and third largest CA adopters, respectively. Iraq too now has some 15 000 ha of CA, benefitting from the work done by ICARDA in Syria, Iraq and elsewhere (Piggin et al., 2015). The main reason for the rapid uptake is the increased availability of locally produced affordable no-till seeders in Syria, Iran and Turkey, which are also being exported elsewhere in the WANA region, and the efforts of development and promotion activities by organization such as the German International Cooperation (GIZ), ICARDA, FAO and the Arab Centre for the Studies of Arid Zones and Dry Lands (ACSAD) as well as bodies such as INRA in Morocco, American University in Beirut, Aga Khan Foundation in Syria, and the Mediterranean Agro-Systems Innovation Network (RCM) across the WANA Mediterranean region. At the Climate Summit COP 22 in Marrakesh, the Moroccan government proposed the 'Triple A' programme for Africa (Adaptability of African Agriculture to Climate Change) which was accepted. The Moroccan government also set for itself a target of 5 M ha of land under conventional tillage agriculture to be transformed to CA systems over the next 10 years. Thus, the 'Triple A' initiative and the decision taken by Morocco to adopt CA are likely to help African farmers and governments to accelerate the spread of CA across Africa.

International experiences about CA and considerations for its implementation in the Mediterranean region show the potential benefits that can be harnessed by farmers in the semi-arid Mediterranean environments and highlight the need for longer-term research including on weed management, crop nutrition, crop-livestock integration, biomass management and economics of CA systems. Some of the crop-livestock integration issues such as biomass management need to be resolved at the community level because post-harvest crop biomass is in demand by livestock herders and the traditional arrangement between crop farmers and herders is not conducive to CA development (Kassam et al., 2012; Lalani et al., 2018). Besides, unless farmers are engaged through an enabling policy with institutional support and the opportunity to learn CA practices and how to integrate them into crop-livestock production system, rapid uptake of CA is not likely to occur. Examples exist for the technical feasibility of successful crop-livestock integration in such environments. But since the traditions and community structures are country specific, only locally developed procedures for introduction and adoption of CA systems will be successful in the long term.

Work by ICARDA and CIMMYT in the WANA region has shown benefits of CA by the increase in crop yields, soil organic matter, water-use efficiency and net revenue. CA also shows the importance of using cropping and crop diversification with legumes and cover crops, instead of a fallow period, leading to improved productivity, soil quality, N-fertilizer-use efficiency and water-use efficiency. In dry areas, CA allows farmers to improve their productivity and profitability while conserving and even improving the natural resource base and the environment. Yet although CA works better than tillage-based farming, the adaptation in drylands is affected by water scarcity and drought hazard, low biomass production and acute competition between conflicting uses including soil cover, animal fodder, cooking/heating fuel, raw material for habitat and so forth. Many WANA smallholders are poor and rely more on livestock than on grain production for their livelihoods.

6 Conclusion

CA is a new paradigm for farming worldwide. It changes the production system and agricultural land management thinking and mindset. Originally, the adoption of CA was mainly prompted by acute problems faced by farmers, especially wind and water erosion, as for example southern Brazil or the Prairies in North America, or drought in Australia. In all these cases, the farmers' organization generated knowledge that eventually led to mobilizing public, private and civil sector support. More recently, again pressed by erosion and drought problems coming with climate change, exacerbated by increase in cost of energy and production inputs, government support has accelerated the adoption rate of CA in Kazakhstan, China, India and Pakistan but also in some African countries such as Zambia, Zimbabwe, Malawi and Mozambique among others, and this is attracting support from other stakeholders. In Europe too, there has been greater concern shown by the EU towards soil degradation and the need for greater environmental and soil health management in agriculture. Thus, by means of the CAP, Member States of the EU have been able to provide incentives to farmers to adopt soil and water conservation practices that are also climate-smart.

The main reasons for adoption of CA by farmers and societies can be summarized as follows:

1 better farm economy (reduction of production inputs of seeds, fertilizer, pesticides and water, and lower costs in machinery and fuel, and time-saving in the operations that permit the development of other agricultural and non-agricultural complementary activities);
2 delivery of a range of ecosystem services to society and nature and avoided environmental costs, including cleaner water resources, greater carbon sequestration, minimum pollution and soil erosion, greater biodiversity in the agro-ecosystems, and normal functioning of carbon, nutrient and water cycling;
3 equal yields, or yield increases (depending on the starting level of soil degradation) and greater yield stability (as long-term effect) and higher overall seasonal production;
4 greater nutrient-use efficiency and retention;
5 fewer crop protection problems and costs;
6 soil protection against water and wind erosion;
7 better water-use efficiency and retention, and better water economy including in dryland areas;
8 flexible technical possibilities for sowing, fertilizer application and weed control (allowing for more timely operations and adaptation to climate change); and
9 farming based on climate-smart regenerative agriculture.

In reality, farmers have higher profits with CA systems. Otherwise, they would not be practising CA on more than 180 M ha of cropland globally, nor would it be spreading at an annual rate of more than 10 M ha. No-till and cover crops are used between rows of perennial crops such as olives, nuts and grapes or fruit orchards, and in plantation systems. CA can be used for winter crops, and for traditional rotations with legumes, sunflower and canola, and in field crops under irrigation where CA can help optimize irrigation system management to conserve water, energy and soil quality, reduce salinity problems and to make fertilizer use more efficient.

At the landscape level, CA enables several environmental services to be harnessed at a larger scale, particularly C sequestration, cleaner water resources, drastically reduced erosion and run-off, and with this, flooding, as well as enhanced biodiversity. All this does not only bring benefits for the farmer as production resources are saved. It also brings direct economic benefits for society, as treatment costs for drinking water are reduced and damages to infrastructure from flooding or sedimentation are avoided. Overall, CA as an alternative paradigm for sustainable production offers many benefits to producers, the economy, consumers and the environment that cannot be obtained from tillage agriculture. With CA, production intensity becomes a matter of output rather than inputs. So, CA is not only climate-smart, but smart in many other ways.

Globally, the total CA area is still relatively small compared to the total arable areas using tillage. Yet this is changing, and the spread of CA worldwide appears to have been expanding at the rate of 10.5 M ha per annum since 2008/09. It is expected that large areas of agricultural land in Asia, Africa, Europe and Central America will increasingly be transformed by CA in the coming decades as can already be seen in Kazakhstan, India, Pakistan, China, South Africa, Zimbabwe, Zambia, Malawi, Morocco, Spain, Italy, France and the United Kingdom. This is because in the last two decades, the adoption of CA has been important to farmers themselves, to governments, donor agencies, international technical assistance-agencies, NGOs and Foundations and service sectors. In some countries, such as the United States, Canada, Australia, Brazil, Argentina, Paraguay and Uruguay, it appears that CA is being 'mainstreamed' in agricultural development programmes. But only a few countries such as Canada, Switzerland, Italy, Spain, Kazakhstan, China, Zambia, Zimbabwe, Mozambique, Malawi and South Africa *emphasize* its importance. This will change. CA is the future of sustainable agriculture and land use upon which food systems and supply chains will increasingly rely upon to meet national and international food security and environmental service needs, requiring support from all stakeholders in public, private and civil sectors.

The crucial role of the national and international corporate institutions and private business sector is to ensure that CA machinery and equipment, fertilizer and pesticide (against insect pests, weeds and diseases), particularly low-risk herbicides, are available to the farmers through government-assisted programmes, as appropriate. It is in the interest of everyone if the farmers involved in CA adoption were part of a CA-based producer organization.

At the same time, national and international knowledge systems must increasingly align their work in research, education and extension to helping to promote CA systems and practices. Research in particular must help to solve farmer and policy constraints to CA adoption and spread. It would not be out of place to suggest that it would be considered negligent if the stakeholders (including politicians, policy makers, institutional leaders, research scientists, schools, universities and academics, extension agents, private sector) who carry the responsibility of transforming the tillage-based agriculture into CA practices do not earnestly align and support the national and regional agricultural innovation systems towards this goal. In fact, every country in the world must begin to set target for change towards CA and use all available means and processes to set the transformation in motion, thereby securing significant economic, socioeconomic and environmental benefits for the farmers and for the population at large in the world. People and institutions, both public and private sector, everywhere have everything to gain from adopting CA as a basis for sustainable agricultural intensification and ecosystem management. The greater impact that can result from the adoption of CA as a matter of policy and good stewardship is that agriculture development in the future everywhere will become part of the solution of addressing national, regional and global challenges including resource degradation, land and water scarcity, climate change.

CA practices offer a new way of effectively and efficiently managing agricultural environments and the natural resource base for multifunctional services to the society. As full benefits of CA take several years to fully manifest themselves, fostering a dynamic CA sector requires an array of enabling policy and institutional support over a longer-term time horizon. This will allow farmers to take advantage of the future carbon and water markets and support for environmental services currently under discussion internationally. As with climate change, any delayed or missed action now might lead to an Earth where soils are desertified and unproductive and food security for mankind is not any longer secured.

It is perfectly feasible to meet food security needs globally at lower economic and environmental costs through CA systems linked to energy-efficient equipment technologies, precision farming and VRT, other remote-sensing technologies and autonomous machines working alone or in swarms, as well as controlled traffic farming. The transformation to such systems will require effective political will and commitment backed by active support from the farming industry, including the farm machinery sector, which are currently inadequate in most regions.

7 Acronyms

AAPRESID	Argentinean Association of No-till Farmers – Asociación Argentina de Productores en Siembra Directa
ACSAD	Arab Centre for the Studies of Arid Zones and Dry Lands

ACT	African Conservation Tillage Network
AfDB	African Development Bank
ANAPO	Association of Oilcrop and Wheat Producers – Asociación de Productores de Oleaginosas y Trigo
APAD	French No-till Farmers Association – Association pour la Promotion d'une Agriculture Durable
AU-NEPAD	African Union – New Partnership for Africa's Development
AUSID	Uruguayan No-till Farmers Association – Asociación Uruguaya de Siembra Directa
CA	Conservation Agriculture
CAAPAS	American Confederation of Associations for Sustainable Agriculture - Confederación Americana de Asociaciones para la Agricultura Sostenible
CAP	Common Agricultural Policy
CENAPRO	Nacional Center for Productivity – Centro Nacional de Productividad
CGIAR	Consultative Group for International Agricultural Research
CIAT	International Center for Tropical Agriculture – Centro Internacional de Agricultura Tropical
CIRAD	Centre for Cooperation on International Research on Agricultural Development – Centre de coopération internationale en recherche agronomique pour le développement
CIMMYT	International Maize and Wheat Improvement Centre – Centro Internacional de Mejoramiento de Maíz y Trigo
COP	Conference of Parties
CWANA	Central and West Asia and North Africa
DIFD	Department for International Development
DMC	Direct Seeded Mulch-Based Cropping System
EC	European Commission
ECAF	European Conservation Agriculture Federation
EU	European Union
FAO	United Nations Food and Agriculture Organization
FARA	Forum for Agricultural Research in Africa
FEBRAPDP	Brazilian Federation for Direct Seeding into Straw – Federação Brasileira de Plantio Direto na Palha
FFS	Farmer Field Schools
GIZ	German International Cooperation – Deutsche Gesellschaft für Internationale Zusammenarbeit (former GTZ)
GTZ	German Technical Cooperation – Deutsche Gesellschaft für Technische Zusammenarbeit (now GIZ)
GFAR	Global Forum for Agricultural Research

FINCA	Finnish Conservation Agriculture Association
IAD	Institute for Sustainable Agriculture – Institute de l' Agriculture Durable
ICAR	Indian Council for Agricultural Research
ICARDA	International Centre for Agricultural Research in the Dry Areas
ICRAF	World AgroForestry Centre
ICRISAT	International Crops Research Institute for the Semi-Arid Tropics
IFAD	International Fund for Agricultural Development
IITA	International Institute for Tropical Agriculture
INRA	French National Institute for Agricultural Research - Institut national de la recherche agronomiquel
PEAME/EMBRAPA	Brazilian Enterprise for Agricultural and Livestock Research – Empresa Brasileira de Pesquisa Agropecuária
IRRI	International Rice Research Institute
ITCF	Institute for Cereal and Forage Technology - Institut Technique des Céréales et des Fourrages
IWMI	International Water Management Institute
MSSRF	M.S. Swaminathan Research Foundation
NGO	Non-Government Organization
NRI	Natural Resources Institute, Greenwich
RCM	Mediterranean Agro-Systems Innovation Network – Réseau Innovations Agro-Systèmes Méditerranéens
RELACO	Latin American Network for Conservation Tillage/ Agriculture – Red Latino Americana de Labranza/ Agricultura de Conservación
SAMA	Sustainable Agricultural Mechanization for Africa
SDG	Sustainable Development Goal
SEI	Stockholm Environment Institute
TAA	Tropical Agricultural Association
UK	United Kingdom
USA	United States of America
VRT	Variable Rate Technologies
WANA	West Asia and North Africa

8 Where to look for further information

A general source of information on CA is the FAO website (www.fao.org/conservation-agriculture). More information regarding the development of CA systems globally can be found in books and journals, and on websites of national and regional CA organizations.

Books include: Goddard et al. (2006), Baker et al. (2007), FAO (2011, 2016), Jat et al. (2014), Farooq and Siddique (2014), and Chan and Fantle-Lepczyk (2015). Nationally oriented information on CA development is available for several countries including Australia (Crabtree, 2010), Canada (Lindwall and Sonntag, 2010), Brazil (Junior et al., 2012; de Freitas and Landers, 2014), Argentina (Peiretti and Dumanski, 2014), and United States (Lessiter, 2018).

Lessiter (2011) provides a narrative on the 40 legends of the past in no-till farming in North America. The International Soil and Water Conservation Research published a special issue on 'Pioneers in soil conservation and Conservation Agriculture' which provides good information on the adoption of CA in several countries (Dumanski et al., 2014).

9 References

AAPRESID. 2010. Available at: http://www.aapresid.org.ar/institucional_sd.asp.

ACT. 2014. Conservation Agriculture: building entrepreneurship and resilient farming systems. *Book of Condensed Papers of the First Africa Congress on Conservation Agriculture*, 18–21 March 2014, Lusaka, Zambia.

ACT. 2018. *Book of Condensed Papers of the Second Africa Congress on Conservation Agriculture on Making Climate-Smart Agriculture Real in Africa with Conservation Agriculture: Supporting the Malabo Declaration and Agenda 2063*, October 2018, Johannesburg, South Africa.

Amado, T. J. C. and Reinert, D. J. 1997. Plantio direto como instrumento de sustentabilidad ecológica. In: Claverán, A. R. and Rulfo, F. O. (Eds), *Memorias de la IV Reunión Bienal de la Red Latinoamericana de Agricultura Conservacionista*, RELACO, 17–20 November 1997, Michoacán, México. CENAPROS, INIFAP, SAGAR, FAO, pp. 265–79.

Asadi, M. E. 2018. Conservation Agriculture practices in Golestan Province, Iran 2017: turning research into impact. *Agricultural Mechanization* 3(4), 28–32.

Avci, M., Mevveci, K., Akar, T., Ozdemir, B., Yururer, A., Karakurt, E., Surek, D. and Karacam, M. 2007. Turkish experience on dryland agronomy: lessons from the past and the recent experiments. *Journal of Agricultural Research* 45(1), 33–42.

Baig, M. N. and Gamache, P. M. 2009. *The Economic, Agronomic and Environmental Impact of No-Till on the Canadian Prairies*. Alberta Reduced Tillage Linkages, Canada.

Baig, M. N. and Gamache, P. M. 2011. *The Economic, Agronomic and Environmental Impact of No-Till on the Canadian Prairies* (vol. 2). Alberta Reduced Tillage Linkages, Canada.

Baker, C. J. 2015. Personal communications, 25 February 2017.

Baker, C. J., Saxton, K. E., Ritchie, W. R., Chamen, W. C. T., Reicosky, D. C., Ribeiro, M. F. S., Justice, S. E. and Hobbs, P. R. (Eds). 2007. *No-Tillage Seeding in Conservation Agriculture* (2nd edn.). CABI Publishing, Wallingford, UK.

Barbosa dos Anjos, J. 2000. *Implements and Methods for the Preparation of Agricultural Soil. Manual on Integrated Soil Management and Conservation Practices*. FAO Land and Water Bulletin 8. FAO, Rome.

Basch, G., Geraghty, J., Streit, B. and Sturny, W. G. 2008. No-tillage in Europe – state of the art: constraints and perspective. In: Goddard, T., Zoebisch, M. A., Gan, Y. T., Ellis, W.,

Watson, A. and Sombatpanit, S. (Eds), *No-Till Farming Systems*. Special Publication No. 3. World Association of Soil and Water Conservation, Bangkok, pp. 159–68.

Basch, G., Kassam, A., Friedrich, T. F., Santos, F. L., Gubiani, P. I., Calegari, A., Reichert, J. M. and dos Santos, D. R. 2012. Sustainable soil water management systems. In: Lal, R. and Stewart, B. A. (Eds), *Soil Water and Agronomic Productivity, Advances in Soil Science*. CRC Press, Taylor and Francis Group, Boca Raton, FL, pp. 229–89.

Bashour, I. 2007. Impact of conservation agriculture on soil fertility in dry regions. In: *Proceedings International Workshop on Conservation Land Management to Improve the Livelihood of People in Dry Regions*, 7–9 May 2007, Damascus, Syria, pp. 111–19.

Bashour, I., AL-Ouda, A., Kassam, A., Bachour, R., Jouni, K., Hansmann, B. and Estephan, C. 2016. An overview of Conservation Agriculture in the dry Mediterranean environments with a special focus on Syria and Lebanon. *AIMS Agriculture and Food* 1(1), 67–84.

Baudron, F., Mwanza, H. M., Triomphe, B. and Bwalya, M. 2007. *Conservation Agriculture in Zambia: A Case Study of Southern Province*. FAO, Rome.

Belloume, A. 2007. Conservation Agriculture in the Arab region between concept and application. In: Stewart, B., Fares Asfary, A., Belloum, A., Steiner, K. and Friedrich, T. (Eds), *The Proceedings of the International Workshop on Conservation Agriculture for Sustainable Land Management to Improve the Livelihood of People in Dry Areas*, 7–9 May. ACSAD & GTZ, Damascus, Syria, pp. 11–24.

Ben-Hammouda, M., M'Hedhbi, K., Kammassi, M. and Gouili, H. 2007. Direct drilling: an agro-environmental approach to prevent land degradation and sustain production. In: Stewart, B., Fares Asfary, A., Belloum, A., Steiner, K. and Friedrich, T. (Eds), *The Proceedings of the International Workshop on Conservation Agriculture for Sustainable Land Management to Improve the Livelihood of People in Dry Areas*, 7–9 May. ACSAD & GTZ, Damascus, Syria, pp. 37–48.

Benites, J. R. 1997. Labranza de conservación y agricultura sostenible (Conservation tillage and sustainable agriculture). In: Claverán, A. R. and Rulfo, F. O. (Eds), *Memorias de la IV Reunión Bienal de la Red Latinoamericana de Agricultura Conservacionista*, RELACO, 17–20 November 1997, Michoacán, México. CENAPROS, INIFAP, SAGAR, FAO, pp. 57–79.

Benites, J., Vaneph, S. and Bot, A. 2002. Planting concepts and harvesting good results. *LEISA Magazine October* 18(3), 6–9.

Blackshaw, R. E., Harker, K. N., O'Donovan, J. T., Beckie, H. J. and Smith, E. G. 2007. Ongoing development of integrated weed management systems on the Canadian Prairies. *Weed Science* 56(1), 146–50.

Boisgontier, D., Barthelemy, P. and Lescar, L. 1994. Feasability of minimum tillage practices in France. In: *EC-Workshop: Experience with the Applicability of No-Tillage Crop Production in the West European Countries*, 1 June 1994, Giessen, Germany, pp. 81-91.

Borland, T. M. 1974. Which was weed control and tillage? *Agriculture Today (Rhodesia)* 2, 1.

Borland, T. M. 1980. No-tillage. Farmer's Weekly, Parts 1 to 7, 20 February to 2 April 1980, pp. 9 -43.

Boshen, P., Darty, B. A., Dogbe, G. D., Boadi, E. A., Triomphe, B., Daamgard-Larsen, S. and Ashburner, J. 2007. *Conservation Agriculture as Practiced in Ghana*. FAO, Rome.

Centero-Martinez, C., Gabina, D. and Arrue, J. L. 2007. Evaluation of Conservation Agriculture technology in Mediterranean agriculture systems. In: Stewart, B.,

Fares Asfary, A., Belloum, A., Steiner, K. and Friedrich, T. (Eds), *The Proceedings of the International Workshop on Conservation Agriculture for Sustainable Land Management to Improve the Livelihood of People in Dry Areas*, 7-9 May. ACSAD & GTZ, Damascus, Syria, pp. 157–64.
Chan, C. and Fantle-Lepczyk, J. 2015. *Conservation Agriculture in Subsistence Farming: Case Studies from South-Asia and Beyond*. CABI Publishing, Wallingford, 278p.
Crabtree, B. 2004. Strong economics of no-tillage cause widespread adoption in southern Australia. Paper presented at the *First Congress on Conservation Agriculture/No-Till*, Dnipropetrovsk, Ukraine, 18–23 November.
Crabtree, W. L. 2010. Search for sustainability with no-till bill in dryland agriculture. Crabtree Agricultural Consulting, Western Australia, 204p. Available at: www.no-till.com.au.
de Freitas, P. L. and Lander, J. N. 2014. The transformation of agriculture in Brazil through development and adoption of Zero Tillage Conservation Agriculture. *International Soil and Water Conservation Research* 1(2), 35–46.
Derpsch, R. 1997a. Nuevos enfoques (paradigmas) en la producción agrícola. In: Claverán, A. R. and Rulfo, F. O. (Eds), *Memorias de la IV Reunión Bienal de la Red Latinoamericana de Agricultura Conservacionista*, RELACO, 17–20 November 1997, Michoacán, México. CENAPROS, INIFAP, SAGAR, FAO, pp. 327–8.
Derpsch, R. 1997b. Desarrollo y difusión de sistemas sostenibles de producción agrícola en la región oriental de Paraguay. In: Claverán, A. R. and Rulfo, F. O. (Eds), *Memorias de la IV Reunión Bienal de la Red Latinoamericana de Agricultura Conservacionista*, RELACO, 17–20 November 1997, Morelia, Michoacán, México. CENAPROS, INIFAP, SAGAR, FAO, pp. 303–14.
Derpsch, R. 1998. Historical review of no-tillage cultivation of crops. In: *Proceedings of the 1st JIRCAS Seminar on Soybean Research on No-Tillage Culture & Future Research Needs*, 5–6 March 1998, Iguassu Falls, Brazil. Working Report No. 13. JIRCAS, pp. 1–18.
Derpsch, R. 2004. History of crop production, with and without tillage. *Leading Edge* 3, 150–54.
Derpsch, R. and Friedrich, T. 2009a. Global overview of Conservation Agriculture adoption. *Proceedings of the 4th World Congress on Conservation Agriculture*, 4–7 February 2009, New Delhi, India, pp. 429–38.
Derpsch, R. and Friedrich, T. 2009b. Development and current status of no-till adoption in the world. *Proceedings on CD of the 18th Triennial Conference of the International Soil Tillage Research Organization (ISTRO)*, 15–19 June 2009, Izmir, Turkey.
Derpsch, R., Friedrich, T., Kassam, A. and Li, H. 2010. Current status of adoption of no-till farming in the world and some of its main benefits. *International Journal of Agricultural and Biological Engineering* 3(2), 1–25.
Derpsch, R., Duiker, S., Franzluebbers, A., Gall, C., Köller, K. and Reicosky, D. C. 2012. About the necessity of standardizing no-tillage research. *Proceedings of the 19th ISTRO Conference, IV SUCS Meeting Striving for Sustainable High Productivity*, 24–28 September, Montevideo, Uruguay.
Derpsch, R., Franzluebbers, A. J., Duiker, S. W., Reicosky, D. C., Koeller, K., Friedrich, T., Sturny, W. G., Sá, J. C. M. and Weiss, K. 2013. Why do we need to standardize no-tillage research? Letter to the editor. *Soil and Tillage Research* 137, 16–22.
Duiker, S. W. 2017. Planting green – a new cover crop management technique. *Field Crop News*. Available at: https://extension.psu.edu/planting-green-a-new-cover-crop-management-technique.

Duley, F. L. and Fenster, C. R. 1954. *Stubble Mulch Farming Methods for Fallow Areas*. University of Nebraska, Extension Service and USDA, 35p. Revised Version E.C. 54-100 April 1961, 19p.

Dumanski, J., Reicosky, D. C. and Peiretti, R. A. 2014. Pioneers in soil conservation and Conservation Agriculture. *International Soil and Water Conservation Research* 2(1).

Erenstein, O., Sayer, K., Wall, P., Dixon, J. and Hellin, J. 2008. Adapting no-tillage agriculture to the smallholder maize and wheat farmers in the tropics and sub-tropics. In: Goddard, T., Zoebisch, M. A., Gan, Y. T., Ellis, W., Watson, A. and Sombatpanit, S. (Eds), *No-Till Farming Systems*. Special Publication No. 3. World Association of Soil and Water Conservation, Bangkok, pp. 253–77.

FAO. 2007. *Conservation Agriculture in China and the Democratic People's Republic of Korea*. FAO Crops and Grassland Service Working Paper, 23p.

FAO. 2008. *Investing in Sustainable Crop Intensification: the Case for Soil Health*. Report of the International Technical Workshop, FAO, Rome, July. Integrated Crop Management (vol. 6). FAO, Rome.

FAO. 2009. *Conservation Agriculture in Uzbekistan*. Crop and Grassland Service Working Paper 2. FAO, Rome, 38p.

FAO. 2011. *Save and Grow. A Policymakers Guide to the Sustainable Intensification of Smallholder Crop Production*. FAO, Rome. Available at: www.fao.org/ag/save-and-grow/.

FAO. 2012. *Conservation Agriculture in Central Asia: Status, Policy, Institutional Support, and Strategic Framework for its Promotion*. FAO Sub-Regional Office for Central Asia (FAO-SEC), Ankara, Turkey. December 2012.

FAO. 2013. *Policy and Institutional Support for Conservation Agriculture in the Asia-Pacific Region*. Food and Agriculture Organization (FAO) of the United Nations, Regional Office for Asia-Pacific (FAO-RAP). December 2013.

FAO. 2014. What is Conservation Agriculture? FAO CA website. Available at: http://www.fao.org/ag/ca/1a.html.

FAO. 2016. *Save and Grow in Practice: Maize, Rice, Wheat – A Guide to Sustainable Production*. FAO, Rome.

FAO. 2017. *Produce + with – Inputs; Memories of the International Expert Consultation on Conservation Agriculture for Sustainable Development*, 17–21 October 2016, Havana Cuba. FAO, Cuba.

Farooq, M. and Siddique, K. H. M. (Eds). 2014. *Conservation Agriculture*. Springer International, Switzerland.

Faulkner, E. H. 1943. *Plowman's Folly*. University of Oklahoma Press, Norman, OK. Ninth printing, May 1963, 156p.

Fenster, C. R. 1960. *Stubble Mulch Farming with Various Types of Machinery*. University of Nebraska, Extension Service and USDA. Soil Science of America Proceedings (vol. 24), No. 6, November – December 1960, pp. 518–23.

Fileccia, T. 2008. *Conservation Agriculture and Food Security in Kazakhstan*. Working Paper. Plant production and Protection Division, FAO, Rome.

Flower, K., Crabtree, B. and Butler, G. 2008. No-till cropping systems in Australia. In: Goddard, T., Zoebisch, M. A., Gan, Y. T., Ellis, W., Watson, A. and Sombatpanit, S. (Eds), *No-Till Farming Systems*. Special Publication No 3. World Association of Soil and Water Conservation, Bangkopk, pp. 457–67.

Fowler, R. and Rockstorm, J. 2001. Conservation tillage for sustainable agriculture, an agrarian revolution momentum in Africa. *Soil and Tillage and Research* 61, 93–107.

Franco, J. A. and Calatrava, J. 2006. Adoption of soil erosion control practices in Southern Spanish olive groves. *Proceedings of the International Association of Agricultural Economists*, Gold Coast, Australia, 12–18 August, 16p.

Friedrich, T. 1997. Potencial y aspectos de la mecanización agrícola en la Agricultura Conservacionista. In: Claverán , A. R. and Rulfo, F. O. (Eds), *Memorias de la IV Reunión Bienal de la Red Latinoamericana de Agricultura Conservacionista*, RELACO, 17–20 November 1997, Michoacán, México. CENAPROS, INIFAP, SAGAR, FAO, pp. 47–55.

Friedrich, T. 2013. Conservation Agriculture as a means of achieving sustainable intensification of crop production. *Agriculture for Development* 19, 7–11.

Friedrich, T. and Kassam, A. H. 2009. Adoption of conservation agriculture technologies: constraints and opportunities. *Proceedings of the 4th World Congress on Conservation Agriculture*, 4–7 February 2009, New Delhi, India.

Friedrich, T., Kassam, A. H. and Shaxson, F. T. 2009. Conservation Agriculture. In: *Agriculture for Developing Countries. Science and Technology Options Assessment (STOA) Project*. European Technology Assessment Group, Karlsruhe, Germany.

Friedrich,T., Derpsch, R. and Kassam, A. H. 2012. Overview of the global spread of Conservation Agriculture. *Field Actions Science Reports, Special Issue* 6, 1–7.

Friedrich, T., Kassam, A. H. and Corsi, S. 2014. Conservation Agriculture in Europe. In: Jat, R. A., Sahrawat, K. L. and Kassam, A. H. (Eds), *Conservation Agriculture: Global Prospects and Challenges*. CABI Publishing, Wallingford, UK, pp. 127–79.

Fukuoka, M. 1975. *One Straw Revolution*. Rodale Press. English translation of *shizen noho wara ippeon no kakumei*. Hakujusha Co., Tokyo.

Gan, Y. T., Harker, K. N., McConkey, B. and Suleimanov, M. 2008. Moving towards no-till practices in northern Eurasia. In: Goddard, T. M., Zoebisch, A., Gan, Y. T., Ellis, W., Watson, A. and Sombatpanit, S. (Eds), *No-Till Farming Systems*. Special Publication No. 3. World Association of Soil and Water Conservation, Bangkok, pp. 179–95.

Garrity, D., Akinifesi, F. K. and Oluyede, A. 2010. Evergreen agriculture: a robust approach to sustainable food security in Africa. *Food Security* 2, 197–214.

Ghosheh, H. 2007. Application of sustainable agriculture principles in weed management. In: Stewart, B., Fares Asfary, A., Belloum, A., Steiner, K. and Friedrich, T. (Eds), *The Proceedings of the International Workshop on Conservation Agriculture for Sustainable Land Management to Improve the Livelihood of People in Dry Areas*, 7–9 May 2007. ACSAD & GTZ, Damascus, Syria, pp. 209–18.

Giráldez, J. V. and González, P. 1994. No-tillage in clay soils under Mediterranean climate: physical aspects. In: Proceedings of the EC-Workshop: Experience with the Applicability of No-tillage Crop Production in the West European Countries, 1 June 1994, Glessen, Germany, pp. 111–17.

Goddard, T., Zoebisch, M. A., Gan, Y. T., Ellis, W., Watson, A. and Sombatpanit, S. (Eds). 2006. *No-till Farming Systems*. Special Publication No. 3. World Association of Soil and Water Association, Bangkok, Thailand.

Goddard, T., Haugen-Kozyra, K. and Ridge, A. 2009. Conservation Agriculture protocols for green house gas offsets in a working carbon market. Paper presented at the *IV World Congress on Conservation Agriculture*, 3–7 February 2009, New Delhi, India.

Gomez, J. A., Alvaez, S. and Soriano, M. 2009. Development of a soil degradation assessment tool for organic olive groves in southern Spain. *Catena* 79(1), 9–17.

González-Sánchez, E. J., Ordóñez-Fernández, R., Carbonell-Bojollo, R., Veroz-González, O. and Gil-Ribes, J. A. 2012. Meta-analysis on atmospheric carbon capture in Spain through the use of Conservation Agriculture. *Soil and Tillage Research* 122, 52–60.

Gonzalez-Sanchez, E. J., Veroz-Gonzalez, O., Blanco-Roldan, G. L., Marquez-Garsia, F. and Carbonell-Bojollo, R. 2015. A renewed view of conservation agriculture and its evolution over the last decade in Spain. *Soil and Tillage Research* 146, 204–12.

González-Sánchez, E. J., Moreno-Garcia, M., Kassam, A., Holgado-Cabrera, A., Trivino-Tradas, P., Carbonell-Bojollo, R., Pisante, M., Veroz-Gozalez, O. and Basch, G. 2017. *Conservation Agriculture: Making Climate Change Mitigation and Adaptability Real in Europe*. European Conservation Agriculture Federation (ECAF), Brussels, 154p.

González-Sánchez, E. J., Mkomwa, S., Conway, G., Kassam, A., Ordonez-Fernandez, R., Moreno-Garcia, M., Repullo-Ruiberriz, de Torres, M., Gil-Ribes, J., Basch, G., Veroz-Gozalez, O., Trivino-Tradas, P., Holgado-Cabrera, A., Miranda-Fuentes, A. and Carbonell-Bojollo, R. 2018. Making climate change mitigation and adaptability real in Africa with conservation agriculture. *European Conservation Agriculture Federation (ECAF) and African Conservation Tillage Network (ACT)*, 143p.

Gullickson, G. 2018. Planting into green cover crops. *Successful Farming*. Available at: https://www.agriculture.com/crops/soybeans/planting-into-green-cover-crops.

Haggblade, S. and Tembo, G. 2003. Conservation farming in Zambia. EPTD Discussion Paper No. 108. International Food Policy Research Institute, Washington DC.

Haque, E., Bell, R., Kassam, A. and Miz, N. N. 2016. Versatile strip seed drill: a 2-wheel tractor-based option for smallholders to implement conservation agriculture in Asia and Africa. *Environments* 3, 1–13.

Haque, E., Bell, R., Jahiruddin, M., Hossain, M., Rahaman, M., Begaum, M., Hossen, A., Salahin, N., Zahan, T., Hossain, M. M., Hashem, A., Islam, A., Vance, W. H., Hossein, I., Esdale, J. and Kabir, E. 2018. *Manual for Smallholders' Conservation Agriculture in Rice-Based Systems*. Murdoch University, Australia, 108p.

Haugen-Kozyra, K. and Goddard, T. 2009. Conservation Agriculture protocols for greenhouse gas offsets in a working carbon markets. *Proceedings of the 4th World Congress on Conservation Agriculture*, 3–7 February 2009, New Delhi, India.

Hobbs, P. R. 2007. Conservation Agriculture: what is it and why is it important for future sustainable food production? *Journal of Agricultural Science* 145, 127–37.

Hobbs, P. R. and Gupta, R. 2003. Resource-conserving technologies for wheat in the rice-wheat systems. In: Ladha, J. K., Hill, J. E., Duxbury, J. M., Gupta, R. K. and Buresh, R. J. (Eds), *Improving the Productivity and Sustainability of Rice-Wheat Systems – Issues and Impacts*. ASA Special Publication Number 65. ASA-CSSA-SSSA, Madison, WI, pp. 149–72.

Hobbs, P. R., Sayre, K. and Gupta, R. 2008. The role of Conservation Agriculture in sustainable agriculture. *Philosophical Transactions of the Royal Society of London. Series B, Biological Sciences* 363(1491), 543–55. doi:10.1098/rstb.2007.2169.

Howeler, R., Lutaladio, N. B. and Thomas, G. 2013. *Save and Grow Cassava: A Guide to Sustainable Production Intensification*. FAO, Rome, 129p.

Huggins, D. R. and Reganold, J. P. 2008. No-till: the quiet revolution. *Scientific American July 2008*, 70–7.

IPCC. 2014. Climate change 2014: synthesis report. In: Pachauri, R. K. and Meyer, L. A. (Eds), *Contribution of Working Groups I, II and III to the Fifth Assessment Report of the Intergovernmental Panel on Climate Change*. IPCC, Geneva, Switzerland.

Jat, M. L., Gathala, M. K., Ladha, J. K., Saharawat, Y. S., Jat, A. S., Kumar, V., Sharma, S. K., Kumar, V. and Gupta, R. K. 2009a. Evaluation of precision land leveling and double zero-till systems in the rice-wheat rotation: water use, productivity, profitability and soil physical properties. *Soil and Tillage Research* 105, 112–21.

Jat, M. L., Dass, S., Sreelatha, D., Sai Kumar, R., Shekar, J. C. and Chandana, P. 2009b. *Corn Revolution in Andra Pradesh. The Role of Single Cross Hybrids and Zero Tillage Technology*. DMR Technical Bulletin 2009/5. Directorate of Maize Research, Pusa, New Delhi.

Jat, M. L., Singh, R. S., Sidhu, H. S., Singh, U. P., Malik, R. K., Kamboj, B. R., Jat, R. K., Singh, V., Hussain, I., Mazid, M. A., Serchan, D. P., Khan, A., Singh, V. P., Patil, S. G., McDonald, A. and Gupta, R. 2010. *Resource Conserving Technologies in South Asia: Frequently Asked Questions*. Technical Bulletin. CIMMYT, New Delhi.

Jat, R. A., Sahrawat, K. L. and Kassam, A. H. (Eds). 2014. *Conservation Agriculture: Global Prospects and Challenges*. CABI Publishing, Wallingford, UK.

Junior, R. C., de Araujo, A. G. and Llanillo, R. F. 2012. *No-till Agriculture in Southern Brazil: Factors That Facilitated the Evolution of the System and the Development of the Mechanization of Conservation Farming*. FAO and IAPAR.

Karabayev, M. 2008. Improvement of soil and water management in Kazakhstan: Conservation Agriculture for wheat production and crop diversification. *Proceedings of an International Technical Workshop on Investing in Sustainable Crop Intensification: Therapeutic Case for Improving Soil Health*, FAO, Rome, 22–24 July 2008. Integrated Crop Management (vol. 6). FAO, Rome.

Kassam, A. H. and Brammer, H. 2012. Combining sustainable agriculture production with economic and environmental benefits. *The Geographical Journal* 2012, 1–8. doi:10.1111/j.1475-4959.2012.00465.x.

Kassam, A. H., Friedrich, T., Shaxson, T. F. and Pretty, J. 2009. The spread of Conservation Agriculture: justification, sustainability and uptake. *International Journal of Agricultural Sustainability* 7(4), 292–320.

Kassam, A. H., Friedrich, T. and Derpsch, R. 2010. Conservation Agriculture in the 21st century: a paradigm of sustainable agriculture. *European Congress on Conservation Agriculture*, 6–10 October 2010. European Conservation Agriculture Federation (ECAF), Madrid, Spain.

Kassam, A., Friedrich, T., Shaxson, F., Reeves, T. F. and Pretty, T. and de Moraes Sà, J. C. 2011. Production systems for sustainable intensification: integrated productivity with ecosystem services. *Technikfolgenabschatzung – Theorie und Praxis* 2, 39–45.

Kassam, A. H., Friedrich, T., Derpsch, R., Lahmar, R., Mrabet, R., Basch, G., González-Sánchez, E. J. and Serraj, R. 2012. Conservation Agriculture in the dry Mediterranean climate. *Field Crops Research* 132, 7–17.

Kassam, A., Basch, G., Friedrich, T., Shaxson, T. F., Goddard, T., Amado, T., Crabtree, B., Hongwen, L., Mello, I., Pisante, M. and Mkomwa, S. 2013. Sustainable soil management is more than what and how crops are grown. In: Lal, R. and Stewart, R. A. (Eds), *Principles of Soil Management in Agro-ecosystems. 2012. Advances in Soil Science*. CRC Press, Taylor and Francis Group, Boca Raton, FL.

Kassam, A. H., Derpsch, R. and Friedrich, T. 2014a. Global achievements in soil and water conservation: the case for Conservation Agriculture. *International Soil and Water Conservation Research* 2(1), 5–13.

Kassam, A. H., Friedrich, T., Derpsch, R. and Kienzle, J. 2014b. Worldwide adoption of Conservation Agriculture. *6th World Congress on Conservation Agriculture*, 22–27 June 2014, Winnipeg, Canada.

Kassam, A. H., Friedrich, T., Shaxson, F., Bartz, H., Mello, I., Kienzle, J. and Pretty, J. 2014c. The spread of Conservation Agriculture: policy and institutional support for adoption and uptake. *Field Actions Science Reports* 7. Available at: http://factsreports.revues.org/3720.

Kassam, A. H., Friedrich, T., Derpsch, R. and Kienzle, J. 2015. Overview of the worldwide spread of Conservation Agriculture. *Field Actions Science Reports* 8. Available at: https://journals.openedition.org/factsreports/3966.

Kassam, A., Freidrich, T. and Derpsch, R. 2017a. Global spread of Conservation Agriculture: interim update 2015/16. Paper presented at the *7th World Congress on Conservation Agriculture*, 1–4 August 2017, Rosario, Argentina.

Kassam, A. H., Mkomwa, S. and Friedrich, T. (Eds). 2017b. *Conservation Agriculture for Africa*. CABI Publishing, Wallingford, UK.

Kassam, A. H., Friedrich, T. and Derpsch, R. 2018. Global spread of Conservation Agriculture. *International Journal of Environmental Studies*. Available at: https://doi.org/10.1080/00207233.2018.1494927.

Kaumbutho, P. and Kienzle, J. (Eds). 2007. *Conservation Agriculture as Practiced in Kenya: Two Case Studies*. FAO, Rome.

King, A. D. and Holcomb, G. B. 1985. *Conservation Tillage: Things to Consider*. Agricultural Information Bulletin No. 461. USDA, Washington DC, 23p.

Lahmar, R. and Triomphe, B. 2007. Key lessons from international experiences about Conservation Agriculture, and considerations for its implementation in dry areas. *Proceedings of the International Workshop on Conservation Agriculture for Sustainable Land Management to Improve the Livelihood of People in Dry Areas*, 7-9 May. ACSAD & GTZ, Damascus, Syria, pp. 123–41.

Lal, R. 1973. Effects of methods of seedbed preparation and time of planting of maize in western Nigeria. *Experimental Agriculture* 9, 303–13.

Lal, R. 1974a. Soil and water conservation through no-tillage systems. FAO-UNDP International Expert Consultation on the "Use of Improved Technology for Rainfed Areas of Tropical Asia". FAO-UNDP.

Lal, R. 1974b. No-tillage effects on soil conditions and crop growth in western Nigeria. *Plant and Soil* 40, 321–31.

Lal, R. 1975. *Role of Mulching Techniques in Tropical Soil and Water Management*. IITA Tech. Bull. 1, 38p.

Lal, R. 1976a. No-tillage implications to tropical soils. In: Fairbridge, R. K. and Finkl Jr., C. W. (Eds), *Encyclopedia of Earth Science Series*. Dowden, Hutchison, and Ross Inc., Stroudsberg, PA, pp. 616–20.

Lal, R. 1976b. Mulch. In: Fairbridge, R. K. and Finkl Jr., C. W. (Eds), *Encyclopedia of Earth Science Series*. Dowden, Hutchison, and Ross Inc., Stroudsberg, PA, pp. 314–9.

Lal, R. 1976c. No-tillage effects on soil properties under different crops in western Nigeria. *Soil Science Society of America Proceedings* 40, 762–8.

Lal, R. 1983. No-till farming. IITA Technical Bulletin Series, 64p.

Lalani, B., Dorward, P., Holloway, G. and Wauters, E. 2016. Smallholder farmers' motivations for using Conservation Agriculture and the roles of yield, labour and soil fertility in decision making. *Agricultural Systems* 146, 80–90.

Lalani, B., Dorward, P. and Holloway, G. 2017. Farm level economic analysis – is Conservation Agriculture helping the poor? *Ecological Economics* 141, 144–53.

Lalani, B., Aleter, B., Kassam, S. N., Bapoo, A. and Kassam, A. 2018. Potential for conservation agriculture in the dry marginal zone of central Syria: a preliminary assessment. *Sustainability* 10, 1–19.

Landers, J. 2007. *Tropical Crop-Livestock Systems in Conservation Agriculture: the Brazilian Experience*. Integrated Crop Management (vol. 5). FAO, Rome.

Lessiter, F. 2011. 40 legends of the past. *40th Anniversary Issue of No-till Farmer*, November 2011.

Lessiter, F. 2018. *From Maverick to Mainstream: A History of No-Till Farming.* No-Till Farmer, 416p.

Leyva, J. C., Martinez, J. A. F. and Roa, M. C. G. 2007. Analysis of the adoption of soil conservation practices in olive groves: the case of mountainous areas in southern Spain. *Spanish Journal of Agricultural Research* 5(3), 249–58.

Li, H. W., Gao, H. W., Wu, H. D., Li, W. Y., Wang, X. Y. and He, J. 2007. Effects of 15 years of conservation tillage on soil structure and productivity of wheat cultivation in northern China. *Australian Journal of Soil Research* 45, 344–50.

Li, H. W., He, J., Bharucha, Z. P., Lal, R. and Pretty, J. 2016. Improving China's food and environmental security with Conservation Agriculture. *International Journal of Agricultural Sustainability* 3, 1–15.

Lienhard, P., Boulakia, S., Legoupil, J.-C., Gilard, O. and Séguy, L. 2014. Conservation Agriculture in South-East Asia. In: Jat, R. A., Sahrawat, K. L.. and Kassam, A. H. (Eds), *Conservation Agriculture: Global Prospects and Challenges*. CABI Publishing, Wallingford, UK, pp. 180–201.

Lindwall, C. W. and Sonntag, B. (Eds). 2010. *Landscape Transformed: The History of Conservation Tillage and Direct Seeding*. Knowledge Impact in Society, Saskatoon, University of Saskatchewan, Canada.

Llewellyn, R. S., D'Emden, F. and Gobbett, D. 2009. *Adoption of No-Till and Conservation Farming Practices in Australian Grain Growing Regions: Current Status and Trends*. Preliminary report for SA No-till Farmers Association and CAAANZ 26 January 2009.

Loss, S., Haddad, A., Desbiolles, J., Cicek, H., Khalil, Y. and Piggin, C. 2016. *The Practical Implementation of Conservation Agriculture in the Middle East*. International Center for Agricultural Research in the Dry Areas and Australian Center for International Agricultural Research, Aleppo, Syria.

Mannering, J. V. 1979. *Conservation Tillage to Maintain Soil Productivity and Improve Water Quality*. Historical Documents of the Purdue University Cooperative Extension Service, Department of Agricultural communication, 01 January 1979, 5p.

Martinez, J. A. F. 2009. Impacto de la políticaagroambientaleuropea de lucha contra la erosiónsobre la oliviculturaen Andalucía. *Ecología Aplicada* 8(2), 37–45.

MEA. 2005. *Ecosystems and Human Well-Being: Synthesis*. Millennium Ecosystem Assessment. Island Press, Washington, DC.

Mello, I. and van Raij, B. 2006. No-till for sustainable agriculture in Brazil. *Proceedings of the World Association of Soil and Water Conservation*, pp. 49–57.

M'Hedhbi, K., Chouen, S. and Ben-Hammouda, M. 2003. A recent Tunisian experience with direct drilling. In: *Proceedings of the II World Congress on Conservation Agriculture*. Iguassu Falls, Parana, Brazil, pp. 132–5.

Montgomery, D. 2007. *Dirt: the Erosion of Civilizations*. California University Press, Berkeley.

Mrabet, R. 2007. Lasting benefits from no-tillage systems: erosion control and soil carbon sequestration. In: *The Proceedings of the International Workshop on Conservation Agriculture for Sustainable Land Management to Improve the Livelihood of People in Dry Areas*, 7–9 May 2007. ACSAD & GTZ, Damascus, Syria, pp. 72–92.

Mrabet, R. 2008a. Conservation agriculture in Morocco: a research review. In: *The Proceedings of the International Workshop on Conservation Agriculture for Sustainable Land Management to Improve the Livelihood of People in Dry Areas*, 7–9 May 2007. ACSAD & GTZ, Damascus, Syria, pp. 183–208.

Mrabet, R. 2008b. *No-Tillage Systems for Sustainable Dryland Agriculture in Morocco*. Institut National de la Recherche Agronomique (INRA), 153pp.

Mrabet, R. 2008c. No-Till practices in Morocco. In: Goddard, T., Zoebisch, M., Gan, Y. T., Ellis, W., Watson, A. and Sombatpanit, S. (Eds), *No-Till Farming Systems*. World Association of Soil and Water Conservation, Special Publication No. 3. WASWAC, Bangkok, pp. 393–412.

Nocelli, S. 2018. *Update! Evolution of No-Till Adoption in Argentina*. AAPRESID, Argentina.

Nurbekov, A. 2008. *Manual on Conservation Agriculture Practices in Uzbekistan*. Ministry of Agriculture, FAO & ICARDA, Tashkent, Uzbekistan, 40p.

Nurbekov, A., Akramkhanov, A., Lamers, J., Kassam, A., Friedrich, T., Gupta, R., Muminjanov, H., Karabayev, M., Sydyk, D., Turok, J. and Bekenov, M. 2014. Conservation Agriculture in Central Asia. In: Jat, R. A., Sahrawat, K. L. and Kassam, A. H. (Eds), *Conservation Agriculture: Global Prospects and Challenges*. CABI Publishing, Wallingford, UK, pp. 223–47.

Nurbekov, A., Akramkhanov, A., Kassam, A., Sydyk, D., Ziyadaullaev, Z. and Lamers, J. P. A. 2016. Conservation Agriculture for combating land degradation in Central Asia: a synthesis. *AIMS Agriculture and Food* 1(2), 144–56.

Nyende, P., Nyakuni, A., Opio, J. P. and Odogola, W. 2007. *Conservation Agriculture: A Uganda Case Study*. FAO, Rome.

Oldreive, B. 2006. Report on farming god's way. *Farming God's Way*, 26 June 2006. Available at: http://www.farming-gods-way.org/Resources/Reports/FGW%20Old%20reports_files/Page501.htm.

Othman, H., Darus, F. M. and Hashim, Z. 2012. Best management practices for oil palm cultivation on peat: *Mucuna bracteata* as ground cover crop. *Malaysian Oil Palm Board Information Series*, June 2012.

Owenya, M. Z., Mariki, L. W., Kienzle, J., Friedrich, T. and Kassam, A. 2011. Conservation Agriculture (CA) in Tanzania: the case of the Mwangaza B CA farmer field school (FFS), Rhotia Village, Karatu District, Arusha. *International Journal of Agricultural Sustainability* 9(1), 145–52.

Pala, M., Haddad, A. and Piggin, C. 2007. Challenges and opportunities for Conservation Cropping: ICARDA experience in dry areas. In: Stewart, B., Fares Asfary, A., Belloum, A., Steiner, K. and Friedrich, T. (Eds), *The Proceedings of the International Workshop on Conservation Agriculture for Sustainable Land Management to Improve the Livelihood of People in Dry Areas*, 7–9 May 2007. ACSAD & GTZ, Damascus, Syria, pp. 165–82.

Paroda, R. S. 2018. *Strategy Paper for Doubling Farmers' Income*. Publication no. 60. Trust for Advancement of Agricultural Sciences (TAAS), New Delhi, India, 27p. Available at http://www.taas.in/documents/pub60.pdf.

Peiretti, R. and Dumanski, J. 2014. The transformation of agriculture in Argentina through soil conservation. *International Soil and Water Conservation Research* 2(1), 14–20. doi:10.1016/S2095-6339(15)30010-1.

Phillips, R. E. and Phillips, S. H. 1984. *No-Tillage Agriculture: Principles and Practices*. Van Nostrand Reinhold, New York.

Phillips, S. H. and Young, H. M. 1973. *No-Tillage Farming*. Reiman Associates, New York.

Piggin, C., Haddad, A., Khalil, Y., Loss, S. and Pala, M. 2015. Effects of tillage and time of sowing bread wheat, chickpea, barley and lentil grown in rotation in rainfed systems in Syria. *Field Crops Research* 173, 57–67.

Pisante, M. (Ed.). 2007. *Agricoltura Blu. La via italiana dell'agricoltura conservativa: Principi, tecnologie e metodi per una produzione sostenibile*. Edagricole, Bologna, Italy.

Reicosky, D. 2015. Conservation tillage is not Conservation Agriculture. *Journal of Soil and Water Conservation* 70(5), 103A–8A.

RELACO. 1997. Proceedings of the bi-annual meeting of the Latin-American network for conservation tillage. In: Claverán, A. R. and Rulfo, F. O. (Eds), *Memorias de la IV Reunión Bienal de la Red Latinoamericana de Agricultura Conservacionista*, RELACO, 17–20 November 1997, Morelia, Michoacán, México. CENAPROS, INIFAP, SAGAR, FAO, 361p.

Saharawat, Y. S., Singh, B., Malik, R. K., Ladha, J. K., Gathala, M. K., Jat, M. L. and Kumar, V. 2010. Evaluation of alternative tillage and crop establishment methods in a rice-wheat rotation in North Western Indo-Gangetic Plains. *Field Crop Research* 116, 260–7.

SARD. 2007. *SARD (Sustainable Agriculture and Rural Development) and Conservation Agriculture in Africa*. SARD Policy Brief 18. FAO, Rome.

Séguy, L., Bouzinac, S.. and Husson, O. 2006a. Direct seeded tropical soil systems with permanent soil cover: learning from Brazilian experience. In: Uphoff, N., Ball, A. S., Fernandes, E., Herren, H., Husson, O., Laing, M., Palm, C., Pretty, J., Sanchez, P., Sanginga, N. and Thies, J. (Eds), *Biological Approaches to Sustainable Soil Systems*. CRC Press, Taylor and Francis Group, Boca Raton, FL, pp. 323–42.

Séguy, L., Bouzinac, S., Scopel, E. and Ribeiro, F. 2006b. New concepts for sustainable management of cultivated soils through direct seeding mulch based cropping systems: the CIRAD experience, partnership and networks. *2001–2005 Cirad's Contribution to the World Congress on Conservation Agriculture*. Available at: http: //agroecologie.cirad.fr/congres/index.php?lang=en&rub=1.

Séguy, L., Loyer, D., Richard, J. F. and Miller, E. 2008. Sustainable soil management: agro-ecology in Laos and Madagascar. In: Goddard, T., Zoebisch, M. A., Gan, Y. T., Ellis, W., Watson, A. and Sombatpanit, S. (Eds), *No-Till Farming Systems*. Special Publication No. 3. World Association of Soil and Water Conservation, Bangkok, pp. 207–22.

Shaxson, T. F., Kassam, A. H., Friedrich, T., Boddey, B. and Adekunle, A. 2008. Underpinning the benefits of Conservation Agriculture: sustaining the fundamental of soil health and function. *Main Document for the Workshop on Investing in Sustainable Crop Intensification: the Case of Soil Health*, 24–27 July. FAO, Rome.

Shetto, R. and Owenya, M. (Eds). 2007. *Conservation Agriculture as Practiced in Tanzania: Three Case Studies*. FAO, Rome.

Sorrenson, W. J. 1997. *Financial and Economic Implications of No-Tillage and Crop Rotations Compared to Conventional Cropping Systems*. TCI Occasional Paper, Series No. 9. FAO, Rome.

Sorrenson, W. J. and Montoya, L. J. 1984. *Implicaçoes econômicas da erosao do solo e de practices conservacionistas no Paraná, Brasil*. IAPAR, Londrina and GTZ, Eschborn (unpublished).

Sorrenson, W. J. and Montoya, L. J. 1991. Chapter about the economics of tillage practices. In: Derpsch, R., Roth, C. H., Sidiras, N. and Kopke, U. (Eds), *Controle da erosão no Paraná, Brasil: sistemas de cobertura do solo, plantio direto e preparo conservacionista do solo*. GTZ, Eschborn, pp. 165–92.

Suleimenov, M. 2009. From conservation tillage to conservation agriculture. *Proceedings of the International Consultation Conference on "No-till with Soil Cover & Crop Rotation: A Basis for Policy Support to Conservation Agriculture for Sustainable Production Intensification"*, 8–10 July 2009. Astana-Shortandy, Kazakhstan, pp. 56–68.

Suleimenov, M. and Akshalov, K. 2006. Eliminating summer fallow in black soils of Northern Kazakhstan. In: Lal, R., Suleimenov, M., Stewart, B. A., Hansen, D. O. and Doraiswamy, P. (Eds), *Climate Change and Terrestrial Carbon Sequestration in Central Asia*. Taylor and Francis Group, New York, NY, pp. 267–79.

Suleimenov, M. and Thomas, R. 2006. Central Asia: ecosystems and carbon sequestration challenges. In: Lal, R., Suleimenov, M. B., Stewart, A., Hansen, D. O. and Doraiswamy, P. (Eds), *Climate Change and Terrestrial Carbon Sequestration in Central Asia*. Taylor and Francis Group, New York, NY, pp. 165–76.

TAA. 2007. *The Importance of Improving Soil Conditions for Water, Plant Nutrients and Biological Productivity to Sustain Agricultural Growth under Rising Population Pressure in a Changing Climate. Record of the Workshop Organized by the Tropical.* Agriculture Association, University of Newcastle, March 2007.

Teebrügge, F. and Böhrnsen, A. 1997. Crop yields and economic aspects of no-tillage compared to plough tillage: results of long-term soil tillage field experiments in Germany. In: *Proceedings of the EC-Workshop IV. Experience with the Applicability of No-Tillage Crop Production in the West-European Countries*, Boigneville, 12–14 May 1977. Wissenschaftlicher Fachverlag, Giessen, pp. 25–43.

Thierfelder, C., Chisui, J. L., Gama, M., Cheeseman, S., Jere, Z. D., Bunderson, W. T., Eash, N. S. and Rusinamhodzi, L. 2013. Mai-based Conservation Agriculture systems in Malawi: long-term trends in productivity. *Field Crops Research* 142, 47–57.

Thierfelder, C., Baudron, F., Setimela, P., Nyagumbo, I., Mupangwa, W., Mhlanga, B., Lee, N. and Gérard, B. 2018. Complementary practices supporting conservation agriculture in southern Africa. A review. *Agronomy for Sustainable Development* 38(2), 16. doi:10.1007/s13593-018-0492-8.

Tilman, D., Knops, J., Wedin, D., Reich, P., Ritchie, M. and Siemann, E. 1997. The influence of functional diversity and composition on ecosystem processes. *Science* 277(5330), 1300–2.

Veiga, M. da 1997. Plantio direto no Brasil. In: Claverán, A. R. and Rulfo, F. O. (Eds), *Memorias de la IV Reunión Bienal de la Red Latinoamericana de Agricultura Conservacionista*, RELACO, 17–20 November 1997, Michoacán, México. CENAPROS, INIFAP, SAGAR, FAO, pp. 123–37.

World Bank. 2012. *Carbon Sequestration in Agricultural Soils*. Report No. 67395-GLB. World Bank, Washington DC, 85p.

Yuxia, L. and Chi, C. 2007. Canadian experience of Conservation Agriculture and project implementation in China, in APCAEM. In: *The International Seminar on Enhancing the Extension of Conservation Agriculture Techniques in Asia and the Pacific*, Zhengzhou, Henan, China, 24–26 October 2007.

Chapter 3

Conservation Agriculture Systems: soil health and landscape management

Don Reicosky, Soil Scientist Emeritus USDA-ARS and University of Minnesota, USA

1 Introduction

1.1 Global population increase

Food security is critical for the global population, not only for the next generation, but for the 20–30 generations beyond that. While conventional agriculture produces increasing yields, the environmental degradation and consequences of this intensification are not generally addressed. Our planet currently hosts approximately 7.5 billion people, a figure that is projected to increase to nearly 10 billion in 2050. Increasing pressure is being placed on farmers to improve crop yields in order to feed a booming global population. However, increasing those yields has become harder in the past few decades, which have seen escalating degradation of soil health (SH). Declining SH is primarily due to a loss of soil organic matter (SOM), relative to undisturbed soil (30–60%), returned to the atmosphere as carbon dioxide (CO_2) leading to extreme climate events (Foley et al., 2011; Tilman et al., 2011; Hatfield et al., 2017). Agricultural practices need to become more efficient so that every bit of available land is more productive. Farmers need to achieve this capacity improvement at a time when climate change is disrupting weather patterns and ecosystems. The question of how to ensure adequate food supply for everyone often focuses on intensifying the output of agriculture through the common model of large-scale, petrochemical monocrop, intensive farming.

http://dx.doi.org/10.19103/AS.2019.0048.03

1.2 Finite resources

All life develops and exists on natural resources (soil, water, air and sunlight). Nature uses solar power through plant photosynthesis, which generates the carbon fuel for all life. It is also by using these finite resources that farmers and ranchers manage our food supply. Our challenge in agriculture is to use incoming solar energy efficiently, without degrading the soil, water and air resources. Efforts to control land degradation and soil erosion can be traced back over millennia (Pimentel et al., 1995; Sanderman et al., 2017) and there is evidence that ecosystem changes have over the course of history contributed to the demise of several ancient civilizations (Montgomery, 2007a,b; Lal et al., 2017; Diamond, 2005; Foley et al., 2011; Tilman et al., 2011). Can modern society avoid the same fate? The answer will depend on our ability to shift from working against nature to working with it – through, for example, the use of Conservation Agriculture Systems (CAS) principles and better adoption of nature-based solutions (NBS) (WWAP, 2018). To avoid the agricultural tragedies of the past (Pimentel et al., 1995; Montgomery, 2007a,b; Lal et al., 2017), adopting CAS is not only necessary to improve water management outcomes and achieve water security, it is also critical for ensuring the delivery of co-benefits essential to all aspects of sustainable development. Erosion control, increasing organic matter, nitrogen fixation, increasing water infiltration, better soil structure and improving the soil micro flora are all derived from cover crops helping sustain or increase yields through healthier soils. Deep-rooting cover crops enhance the ability of the economic crop to follow those bio-pores deeper (Bolinder et al., 2002), while using diverse annual cropping rotations and cover-crop mixes increases soil organic carbon (SOC), which increases water-holding capacity (WHC) and available nitrogen (Carlson and Stockwell, 2013).

Land degradation by intensive tillage agriculture is causing significant losses in biodiversity and ecosystem services (Van Eerd et al., 2014; Tsiafouli et al., 2015; Wall et al., 2015), such as environmental quality, water purification, food security and bioenergy production (Montanarella et al., 2016). Global agriculture needs to be radically transformed to avoid further environmental and social problems (Foley et al., 2011; Tilman et al., 2011; Sanderman et al., 2017). The environmental problems associated with conventional tillage (CT) agriculture suggest that agroecology – farming that imitates natural ecosystems – is the most promising pathway to sustainable food systems. Agroecology is based on the idea that farms should mimic the structure and function of natural ecosystems where there is no 'waste' and nutrients are recycled indefinitely. Agroecology aims to close nutrient loops – returning all nutrients that come out of the soil, back to the soil. In the case of vegetable farming, for example, this can be achieved through composting vegetable scraps and human and farmyard manure (Diamond, 2005).

1.3 Coping with climate extremes

Global climate change is dramatically increasing the variability of weather conditions worldwide and soil is a critical buffer medium for hydrologic and biogeochemical processes to help mitigate the effects of extreme weather conditions and uncertainty in the availability of water resources (Larson and Pierce, 1994). The present threats of climate change and ozone depletion, through elevated levels of certain atmospheric gases and altered hydrological cycles, necessitate a better understanding of the influence of land management on soil quality (Doran and Safley, 1997). Management systems need to be fine-tuned to balance the need for and priorities of food production with those of a safe and clean environment.

Extreme weather events are becoming more common across the United States and globally. With recent events, we have outgrown 'traditional conservation' practices. Since the Dust Bowl days of the 1930s in the United States, grassed waterways, contour strips, erosion control structures and terraces have all served a useful purpose to protect our soil resources. Improved soil conservation practices are needed in view of more frequent climate extremes. Due to significant benefits to SH, crop productivity and improved resilience are visible on farms practicing the principles of Conservation Agriculture (CA) (Hobbs and Govaerts, 2010). The world relies on healthy soils for food, water and a habitable climate. But soils are rapidly losing their ability to support life due to unsustainable development and conventional intensive agriculture. Is enough being done to address this crisis? Climate change is a food security, humanitarian, economic, public health and global security issue.

Future resilience-building to feed humanity in a world characterized by an increasing population and climate change will require improved water management (Hatfield et al., 2001, 2017; Hobbs and Govaerts, 2010; Lal, 2015b; Kaye and Quemada, 2017). This involves productivity improvements across the full range of agricultural water management options, from purely rain-fed agriculture, to supplementary irrigation with water harvesting, to full irrigation. Other recent reviews of agricultural management practices in response to climate extremes of droughts and floods are provided by Basche et al. (2016a,b), Basche and DeLonge (2017), Basche and Edelson (2017), Basche (2017), Bouma (2015), Branca et al. (2011), Lal (2015b) and Hatfield et al. (2017).

There is growing empirical and scientific evidence of the limitations of tillage-based agriculture. A key alternative is CAS, which seeks to understand the ecological basis of sustainability and to manage soil. As new technology and concepts appear, CAS has evolved sustainably, with additional ecosystem services and environmental benefits (Hobbs and Govaerts, 2010). This form of agricultural production with minimum soil disturbance is one way to assure sustainable production with resource preservation, protection and efficient utilization. Zero-tillage farmers are doing society a huge favor for future generations in developing and utilizing this new technology with cover-crop mixes to protect our finite resources, applying agroecological principles. The objective of this review is to analyse the merits and challenges of CAS concepts with agroecology emphasis on carbon (C) management and SH for enhanced water-use efficiency.

2 Conservation Agriculture Systems (CAS)

2.1 Soil health

Soil 'health' is a metaphor that captures an important parallel between soils and human health; soils that are poorly cared for will have shorter lifespans and require increasing levels of intervention to deliver the functions required. This metaphor serves as a teaching tool that captures an essential parallel between soils and our human health (Shaxson et al., 2008; Pepper, 2013; Oliver and Gregory, 2015). The concept of health as it relates to human health is slowly developing into a movement that helps improve, understand and manage the complex soil system. While this analogy may be useful in understanding broad concepts in the soil, it must be supported by a solid scientific foundation. Healthy soil practices can play a major role in reducing harmful nutrient losses and increasing water and nutritional security. The scientific literature now contains much of the 'SH terminology'. Several recent reviews from a SH perspective include Magdoff (2001), Sherwood and Uphoff (2000), Doran (2002), Kibblewhite et al. (2008), Karlen (2012), Laishram et al. (2012), Moebius-Clune et al. (2016), Al-Kaisi et al. (2017), Mitchell et al. (2017), Bisht et al. (2016), Morrow et al. (2016), Lal (2016), Karlen et al. (2017) and Reicosky (2018a,b) which will be the basis for further discussion.

SH is a function of soil's physical, chemical and biological characteristics, which respond to land management and vary in space and time. The concept of SH dates back to ancient civilizations (Doran et al., 1996). Soil quality is defined as 'the capacity of a soil to function within ecosystem boundaries to sustain biological productivity, maintain environmental quality, and promote plant and animal health' (Doran and Parkin, 1994; Doran et al., 1996). The principles of SH and organic agriculture were first promoted by Sir Albert Howard, founder of organic agriculture and author of *Soil and Health: A Study of Organic Agriculture* (Howard, 1947). Soil quality is comparable to SH in the context of our discussion.

SH is defined as the capacity of soil to function as a vital living system, with ecosystem and land-use boundaries, to sustain plant and animal productivity, maintain or enhance water and air quality and promote plant and animal health (Doran et al., 1996; Doran and Zeiss, 2000; Doran, 2002; Karlen, 2012; NRCS,

2018). The SH movement in the United States has gained momentum in the past few years, with practitioners championing zero-till, diverse rotations and cover-crop usage as methods to improve yields, reduce weed pressure and improve 'resilience' to withstand extreme weather conditions (Carlson and Stockwell, 2013). Much of 'soil' processes and properties are actually biological processes and properties, hence the acceptance of SH concepts. The new conservation movement utilizes the characteristics of the metaphor to employ the simplicity of SH principles and concepts by farmers and consumers.

The importance of the soil–human health nexus has also been recognized ever since the dawn of civilization (Howard, 1947; Magdoff, 2001; Sherwood and Uphoff, 2000; Brevik, 2014; Brevik and Sauer, 2015; Pepper, 2013; Oliver and Gregory, 2015; Wall et al., 2015; Kemper and Lal, 2017). Viewing soil as a living ecosystem transforms our understanding about soil and how to improve it to advance our agricultural production without degrading environmental quality. Science-based SH is slowly evolving in response to information needs for addressing multiple complex problems, including climate change, greenhouse gases and the C cycle, soil erosion and land degradation, crop disease and pests, depletion and pollution of water resources and loss of biodiversity (FAO, 2008a; Eagle et al., 2012; Laishram et al., 2012; Tsiafouli et al., 2015; Stirling et al., 2016). A major goal of the SH movement is to promote a scientific foundation for consumers, farmers and policy-makers to help people understand the need to maintain more carbon in the soil (FAO, 2005; Lal, 2016; Reicosky and Janzen, 2018).

Living soil is full of bacteria, fungi, algae, protozoa, nematodes and many other fragile creatures affected by intensive tillage, reflecting a fundamental shift in care for soils. Soil has long been perceived to harbor the greatest microbial diversity among all ecosystems, and advances in analytical and computational tools suggest that approximately 1 billion bacterial cells, grouped into 1000–1 000 000 species, reside in a single gram of soil. Nature's solar power through plant photosynthesis generates the carbon fuel for all life (Fig. 1). Agriculture producers manage plants to capture solar energy and transform it into biochemical C energy (food) that feeds life on earth. Simple C compounds are transformed into more complex plant biomolecules, which enter the soil through mycorrhizae, root exudates and, upon plant senescence, through litter and root material as part of the C cycle. Not only is carbon the fuel for all life, it is a fundamental biological infrastructure to build all creatures on earth (Fig. 1). Biodiversity with carbon cycling and flow is an element of sustainability and is necessary for SH, harmony and stability in nature (Shaxson, 2006; Lal, 1991; Janzen, 2015; Tsiafouli et al., 2015; Wall et al., 2015; Reicosky and Janzen, 2018).

From understanding that the soil is a living biological system, SH is an intuitive, appealing concept that leads to a clearer understanding and faster

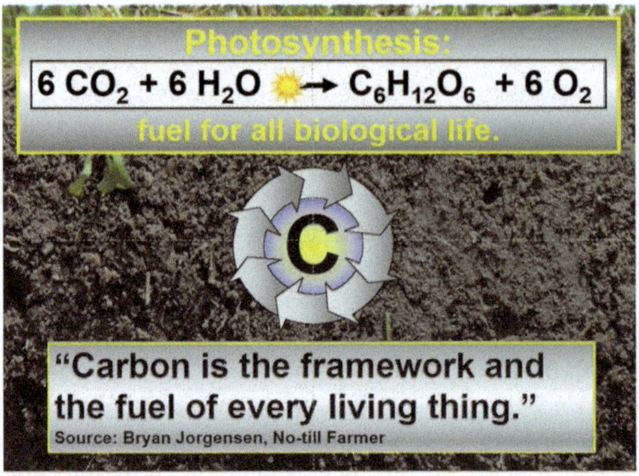

Figure 1 Photosynthesis starts the carbon cycle that supports all life on earth through the capture of carbon and water converting solar energy into biochemical energy and infrastructure required by every living creature.

response from a management perspective. CA and SH system concepts introduced in the United States by the Natural Resources Conservation Service (Moebius-Clune et al., 2016; NRCS, 2018) are closely related. SH, sometimes known as soil quality, includes the first three components (expressed a little differently) of CA and adds a fourth component - to keep living roots in the soil as long as biologically possible. Some promoting the SH movement also encourage incorporation of livestock operations as 'forage harvesters' and 'manure spreaders'. Both CA and SH allow the integration of animals to provide ecosystem services and C transformations in production systems.

The 'SH movement' is shifting the emphasis toward ecological agriculture in the United States. CA with complementary practices and SH systems are providing the same ecosystem services with emphasis on C management. The principles of CA (FAO, 2008b; Hobbs et al., 2008; Kibblewhite et al., 2008; Kassam et al., 2009, 2014; Lehman et al., 2015a,b) and SH (NRCS, 2018; USDA and NRCS, 2013; Moebius-Clune et al., 2016; Lal, 2016; Al-Kaisi and Lal, 2017) are essentially the same, except that SH has a fourth principle that includes keeping a living root system as long as biologically possible. While both CA and SH terminology are still evolving, CA terminology is used more globally while SH terminology is primarily used in the United States and Canada (NRCS, 2018; USDA and NRCS, 2013; Lal, 2015b, 2016; Moebius-Clune et al., 2016).

> 'The health of the soil, plants, animals, people and ecosystems are interdependent, interconnected and indivisible.'
>
> *Source: Rattan Lal*

The terms 'soil health' or 'soil quality' refer to the ability of soil to support and sustain crop growth while maintaining environmental quality (Doran, 2002; Karlen, 2012; Laishram et al., 2012; USDA and NRCS, 2013). Most agricultural soils have the following characteristics:

1. a sufficient, but not excessive supply of nutrients;
2. stable structure (tilth);
3. sufficient depth for rooting and drainage (USDA, 1987; Zhang et al., 2015);
4. free internal drainage;
5. low populations of plant disease and parasitic organisms;
6. high populations of organisms that promote plant growth;
7. low weed pressure;
8. no chemicals that might harm the plant;
9. resistance to being degraded; and
10. resilience following an episode of degradation.

Management intended to improve SH involves practices that enhance the soil's biological, chemical and physical suitability for crop production. SOC strongly impacts soil quality, functionality and health without universal agreement on the interchangeability of SH and soil quality terminology. Lal (2016) suggests soil quality and SH should not be used interchangeably. Soil quality is related to what it does (function), whereas SH treats soil as a living biological entity that affects plant health. Through plant growth, SH is also connected with the health of animals, humans and ecosystems within its domain (Wall et al., 2015).

Measurement of SH requires an index or metric for measurement (Cardoso et al., 2013). Several researchers have reported on the difficulty in identifying an index of the complex interaction of soil biology (Bezdicek, 1996; Seybold et al., 1997; Cardoso et al., 2013; Granatstein and Bezdicek, 1992; Kang et al., 2005; Veum et al., 2014; Lehman et al., 2015a,b). Until the ideal integrated SH index is developed, a list of simple visual indicators is suggested: dark green plant color, vigorous growth, high water infiltration rates, numerous worms and wormholes, deep taproots and bio-pores, darker soil color, strong soil structure, fungal hyphae web networks, sweet soil aroma, slake test versus no soil slaking and soil crusting. Bare, unprotected soil indicates poor health while continuously growing crops and/or crop residue cover indicates good health.

> 'Soil health is like human health; it is hard to define, but you sure know when you don't have it.'
>
> *Source: David Montgomery*

SH is dependent on the maintenance of four major functions - carbon transformations; water and nutrient cycles; soil structure maintenance; and the regulation of pests and diseases (Kibblewhite et al., 2008; Stirling et al., 2016). The integrative interactions among these functions suggest that measurement of individual groups of organisms, processes or soil properties does not suffice to quantify SH (Cardoso et al., 2013). Quantifying the flow of energy and carbon between functions is an essential but non-trivial task for the assessment and management of SH (Janzen, 2015).

2.2 Soil functions

Many soil functions in agricultural ecosystems relate to C management, C cycling and C energy flow through the soil-plant-atmosphere system (Kuzyakov and Cheng, 2004; Janzen, 2015). Sunlight is the ultimate energy source for all agricultural systems, while C is the driver of SH and soil function. Farming starts with the capture of CO_2 and water using solar energy through the process of photosynthesis to provide biochemical energy and food for all living creatures (Fig. 1). The most important process in regenerative farming is photosynthesis, which requires optimization and efficient 'harvesting' of carbon from the atmosphere by green plants using the same resources that farmers and ranchers manage for our food supply.

Soil function can be any service, role or task that soil performs, many of which are vital to society and its environment. There is no single definition of soil function; however, soils provide anchorage for roots, hold water and nutrients and soil management plays a big role in influencing its hydrologic properties (Reicosky et al., 2011; Hatfield et al., 2017). The role of soil management in agricultural ecosystems, especially intensive tillage, in the agricultural water cycle raises questions about sustainability and food security. There are many pressures and threats upon soils, including erosion, decreasing levels of organic matter, point and diffuse contamination, sealing, compaction and declining biodiversity (Tsiafouli et al., 2015). Our understanding of the linkage between soil properties and soil functions and the resultant ecosystem services is incomplete (Nannipieri et al., 2003; Adhikari and Hartemink, 2016; Greiner et al., 2017). The ability of the soil to provide these functions will, however, depend upon the state of the soil properties and the appropriate management (Brady and Weil, 2004).

Soil functions include key life support processes in global ecosystems (Murphy, 2014; Baldock and Skjemstad, 1999; Krull et al., 2004; Weil and Magdoff, 2004). Soil functions are viewed from a 'natural perspective' with minimum soil disturbance and dependent on soil C for optimum results. The primary soil ecosystem functions listed in Table 1 include the soil as a medium for plant growth and nutrient cycling (Tully and Ryals, 2017). In the soil, plant roots

Table 1 Soil functions related to soil health and carbon management

Primary soil functions	Carbon impact
Ecosystem services - food, feed, fibre and fuel	Plants capture CO_2 plus H_2O converting solar energy to biochemical energy in the form of sugars
Base and medium for plant growth and anchorage	C enhances the symbiotic relationship between soil biology and plant root C and exudates
Moderate, control water cycle, flow, infiltration, filter and storage	C enhances soil structure and is the foundation of organic glues that contribute to it. C is the food for macro-organisms creating bio-pores to the soil surface that enhance water infiltration, percolation and soil water storage
Resists run-off and erosion	C in the form of organic residue that protects the soil from raindrop impact and run-off
Exchanges gases with the atmosphere	C enhances soil structure to store CO_2 from the atmosphere and takes O_2 for soil respiration activities
Stores, releases and recycles carbon and nutrients	C decomposition is a continuum from water-soluble sugar exudates to inert humus-releasing nutrients
Source of biodiversity, large and diverse population of soil biota	C is the food and fuel for the biology, releasing and recycling of C and nutrients
Habitat for all soil organisms and gene pool	C enhances the habitat enabling biodiversity and genetic evolution
Support of soil food webs and associated biological activities	C is the food and energy of the soil biology that ranges from micro, meso and macro fauna
Organic humates as catalyst for increasing soil carbon	Recalcitrant C provides stability to soil biology and stimulates further carbon input
Environmental interaction - storage, filtering and transformation	C enhances soil structure for infiltration and additional plant-available water storage in the profile
Decompose biomass-SOM-humus	C used and recycled along a decomposition continuum from water-soluble sugars-fresh plant biomass-SOM-humus
Recycles and supplies nutrients, water and oxygen	Plant biomass supplies most of the nutrients with C enhancing aggregate structure for enhanced water and oxygen flow in and out of the soil
Chemical buffer against changes in pH and solidity	C is being recycled while recalcitrant C in many forms provides chemical stability for the soil system
Remediation of pollutants, toxic materials and heavy metals	C is an energy source for current microbial activity and contributes to recalcitrant carbon in the form of humus
Source of pharmaceuticals - soil chemicals contribute to drugs like antacids to antibiotics	C is available as an energy source for the unique microbiology as predecessors of pharmaceutical drugs
Foundation for physical support of structures and base for infrastructure	Organic C is generally minimized in infrastructure foundation with biological impact reduced

Source: adapted from Murphy (2014), Baldock and Skjemstad (1999), Krull et al. (2004) and Weil and Magdoff (2004).

provide physical anchorage support, for air, water, temperature moderation, protection from toxins and most importantly for nutrients to be recycled for subsequent plants. Soils help regulate and partition water and nutrient flow/supply. Soil affects water supply by storing water for plant use, or by allowing water to drain into the groundwater. Contaminated water may also be purified as it passes through the soil, removing impurities as in wetlands. The soil acts as a 'hydrologic buffer' that stores and regulates run-off and the flow through and over the landscape (USDA, 1987). The presence of SOM and fungi enables the glue formation to bind soil particles to form stable aggregates, cementing and improving critical aggregation functions, for example water infiltration, WHC, aeration and gas exchange (Horne and Sojka, 2002). Maintaining these soil functions requires enhanced C management and minimum soil disturbance.

Soil is nature's premier nutrient cycling and recycling system (Tully and Ryals, 2017). The soil is a 'biochemical or nutrient reactor' that absorbs, releases (i.e. desorbs) and transforms organic and inorganic and biochemical compounds, including toxic metals. Soil buffering is considered to be an important aspect of SH, as it assures reasonable stability in soil pH and influences the amount of chemicals needed to change pH. The increase in the soil cation exchange capacity (CEC) provided by SOM helps retain nutrient cations and conserves nutrients in organic forms slowly released as required nutrients for subsequent plant uptake and growth by the subsequent crop.

Soil functions as habitat for many living organisms, from small mammals and reptiles to tiny insects with many forms that include fungi and microorganisms. The habitat function must be coupled directly with biodiversity to provide a functioning and resilient soil biology system (Brussaard et al., 1997; Brussaard, 2012; Nannipieri et al., 2003). This habitat plus C input and energy flows sustain the biology, diversity, activity and productivity so critical to agricultural ecosystem services. The combination of habitat, biological diversity and energy from the soil C provides potential for reduced disease and insect impact in undisturbed agricultural soils. A less well-known soil function includes the soil as a source of microbial diversity, biotics and antibiotics and pharmaceuticals (Carter et al., 2014). In this case, soil C is necessary as a food source to maintain microbial diversity and activity essential for new medicines, genetic materials and other chemicals.

Biophysical soil functions include nutrient cycling, water dynamics, filtering and buffering, physical stability and support of plant systems and human structures and promotion of biodiversity and habitat (Lal, 1991; Dudley and Alexander, 2017; Hird, 2017). Soil is characterized by a redundancy and resiliency of functions; for example, no relationship has been shown to exist between microbial diversity and decomposition of organic matter. Generally, a reduction in any group of species has little effect on overall processes in soil because other microorganisms can take on its function (Nannipieri et al.,

2003). Many of the benefits of CAS are related to management of the water cycle (Lemordant et al., 2018). Most of the impacts from climate extremes will relate to water, and how it is managed will be at the center of extreme climate adaptation strategies.

Carbon, the fourth most abundant element in the universe, is embodied in many different molecules, compounds and processes to the extent that it is the metaphorical 'backbone' of many organic compounds and the SOM molecules that are the principal currency and building blocks of all life as illustrated in Fig. 2. Carbon is multifaceted and multifunctional as the 'backbone' of the soil-plant-atmosphere system, providing critical connections, integration and energy utilization and flow required for food security. Through the process of photosynthesis, solar energy is transformed into biochemical energy used by all living organisms. This makes carbon a key component of food for everything. Carbon connects components and processes, conducts and facilitates soil organism networking and interaction and facilitates coupled nutrient cycling that fuels and energizes the soil-plant-atmosphere system, increasing WUE. Carbon is a critical part of all living systems, with special emphasis on the ecosystem infrastructure, as a form of energy or fuel for soil organisms, as a form of energy for our food source. Carbon facilitates information and energy transfer throughout the system. Plant C as root exudates and energizes the soil organisms to carry out the appropriate integrated nutrient cycling for

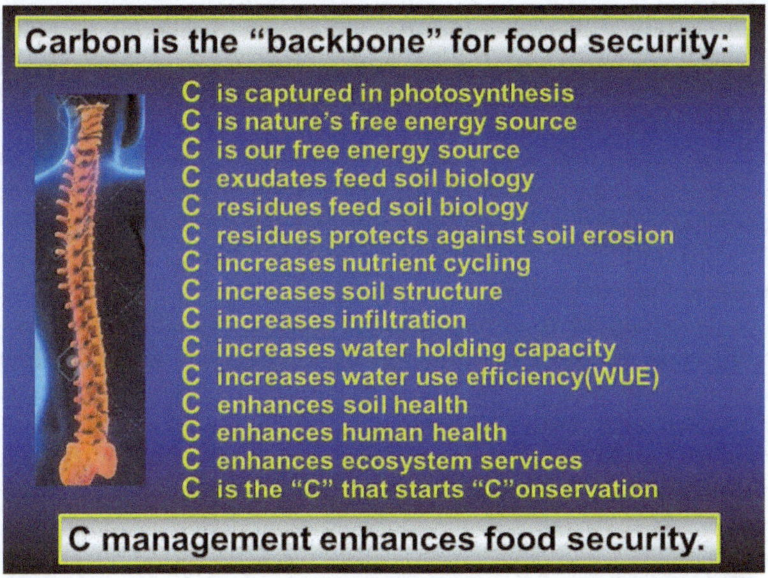

Figure 2 The importance of carbon components and contributions to the 'backbone' of food security.

regenerative agriculture. Within many soil functions, C plays a critical role and must be quantified (Nelson and Sommers, 1996), which is essential for maintaining our food security.

SOM is not a single uniform material but is highly diverse in its chemical and physical properties that influence its capacity to carry out certain functions. An estimate of the capacity for the different soil C fractions to undertake some of the functions of SOM has been discussed by Murphy (2014), Skjemstad et al. (1996, 1998), Baldock and Skjemstad (1999), Krull et al. (2004) and Reicosky and Janzen (2018). The humic and fulvic acids would seem to have lower chemical activity, with the particulate organic C having the highest capacity to maintain the stability of larger soil aggregates. Overall, a balance of the soil C fractions is desirable for all functions. Krull et al. (2004) also identified the importance of the different fractions for different functions. The importance of SOM to carry out certain functions in the soil will vary with soil type. For example, the need for SOM to provide CEC is most important in sandy soils. Baldock and Skjemstad (1999) and Skjemstad (2002) noted that total SOC may not be a good indicator for assessing how well a particular soil function is likely to perform; mainly because the different pools, which make up the bulk of SOC, vary considerably in their physical and chemical properties. At the same time, the need for SOM to provide a food and energy source for the microbial population is needed in all soils, regardless of clay content or texture (Reicosky and Janzen, 2018).

The interrelationship between SOC and soil structure and other physical properties has been extensively reviewed by Tisdall and Oades (1982), Oades (1984) and Carter and Stewart (1996). The addition of SOM not only reduces bulk density and increases WHC, but also effectively increases soil aggregate stability. Angers and Carter (1996) noted that the amount of water-stable aggregates (WSA) was often associated with SOC content, and that particularly labile C was often positively related to macro-aggregate stability (Six et al., 1998). Doran et al. (1996) presented a list of properties affecting soil ecological functions and quality, which included soil bulk density, water infiltration and WHC, total organic C and N, electrical conductivity, pH, plant-available nutrients and measures of microbial biomass and activity.

The integrated effects of the physical, chemical and biological components of soil function as a rooting environment are effectively 'activated', forming a fourth ecohydrologic component reflecting the biological contribution to water management and improvements in WUE (Lin, 2012; Zalewski, 2013; Lemordant et al., 2018). Ecohydrology is an integrative science that focuses on the interaction between hydrology and biota (Brussaard, 2012) to reinforce ecosystem services to reduce anthropogenic impacts. Holistic approaches that manage hydrology and biota aim to achieve sustainability in both ecosystems and human populations. Ecohydrology principles provide the scientific background for regulating the processes and interactions for enhancing water resources, maintaining and restoring biodiversity (Brussaard, 2012), providing ecosystem services for societies and building resilience to climatic and anthropogenic impacts (Lemordant et al., 2018). Ecohydrology promotes the integration of a catchment and its biota into a single entity and the use of ecosystem properties becomes a management tool in which ecohydrology can address fundamental aspects of water resources management (Lin, 2012; Zalewski, 2013).

2.3 Principles of CA

CA and associated carbon management has the potential to mitigate climate and respond to food demands while minimizing environmental damage through improvements in carbon cycling and energy flow (Janzen, 2015; Reicosky and Janzen, 2018). The principles of CAS are universally applicable to agricultural landscapes and land-use practices, which enhances biodiversity and natural biological processes (Lal, 1991, 2015b; Corsi et al., 2012; Kassam et al., 2012, 2013, 2015; Palm et al., 2014; Farooq and Siddique, 2015; Bisht et al., 2016; Al-Kaisi and Lal, 2017). Soil interventions such as mechanical tillage are reduced to an absolute minimum or avoided altogether, and external inputs such as agrochemicals and plant nutrients of mineral or organic origin are applied optimally and in ways and quantities that do not interfere with, or disrupt, biological processes. The zero-tillage system (also referred to as direct seeding, no-tillage, but not conservation tillage (CT1)), according to Eagle et al. (2012), is the only strategy and farming technology to protect and regenerate the soil while enabling seeding for subsequent crops (Friedrich et al., 2012). The zero-till system described here is not to be confused with other forms of CT1, mulch tillage, strip tillage, reduced tillage and so on (Eagle et al., 2012). Rooted historically in conservation (Hobbs, 2007; Farooq and Siddique, 2015; Lal, 2015b), this new CA terminology incorporates continuously improving varieties, technologies and understanding of SH – the outcome of learning from nature over the last 100 years of crop production. The fluid integration of three principles of CA and their complementary supportive SH practices are shown schematically in Fig. 3. The natural interactions and diversity within the CA cropping system contribute to numerous economic and environmental benefits (Lal, 2015b; WWAP, 2018; FAO, 2008b; Corsi et al., 2012; Kassam et al., 2012, 2013, 2015; Palm et al., 2014; Farooq and Siddique, 2015; Bisht et al., 2016; Al-Kaisi and Lal, 2017).

CA is a 'systems-based' approach to managing agroecosystems for improved and sustained productivity, increased profits and food security while regenerating and enhancing the resource base and the environment (FAO, 2008b; Hobbs et al., 2008; Kassam et al., 2009). CAS have been highlighted as a key route toward this system-based goal (Hobbs et al., 2008; Kassam et al.,

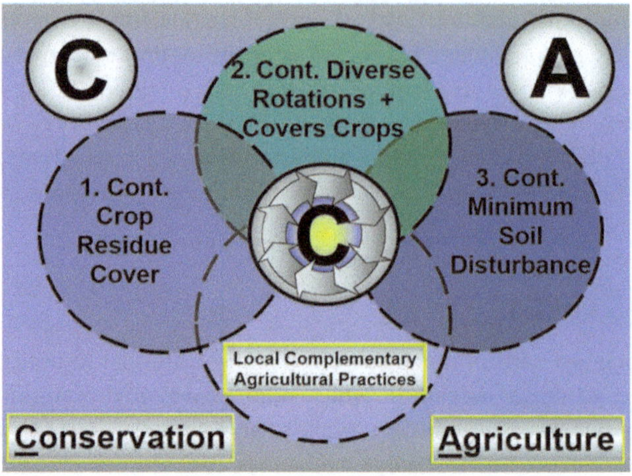

Figure 3 Schematic representation of Conservation Agricultural systems integrating the three primary principles and complementary practices of carbon management.

2009; Erenstein et al., 2012; Kassam and Friedrich, 2012; Friedrich et al., 2012; Corsi et al., 2012; Palm et al., 2014; Pretty and Bharucha, 2014; Lal, 2015b; Reicosky and Janzen, 2018). Recent CA reviews around the globe include Jat et al. (2013), which covers the spread of CA to Brazil (Kassam et al. 2012), Argentina, Canada, Australia, Europe and emerging CA destinations in Asia and Africa. CA is currently practiced on ~180 million ha of cropland globally, and is expanding at around 10 million ha year^{-1} (Kassam et al., 2019). Additional global CAS reviews by Lal (2018) for China; Thierfelder et al. (2018) for South Africa; Farooq and Siddique (2015); and Bisht et al. (2016) for India have been published.

> 'As to methods, there may be a million and then some, but principles are few. The man who grasps principles can successfully select his own method.'
>
> *Source: Ralph Waldo Emerson*

Integrating natural interactions with diversity within CAS cropping contributes to numerous economic and environmental benefits (Liebig et al., 2014; Lal, 2015a; Dudley and Alexander, 2017; Hird, 2017). Soil cover can either consist of live cover crops or dead mulches composed of crop residues that accumulate after previous harvests on the soil surface. Retaining mulch between crops provides better protection against erosion and can also maintain higher soil moisture in dry regions, enrich the soil with organic matter and, if the mulch is sufficiently dense, prevent the regrowth of weeds. Diversified cropping patterns that incorporate at least three plant species, including one legume, are suggested (Chatterjee et al., 2016). To incorporate more diversity, some farmers simultaneously use 10–14 species in cover-crop mixes in the same field (Liebig et al., 2014).

However, CA has not been without its controversies (Sumberg and Thompson, 2012), mainly in sub-Saharan Africa. The positive yield and environmental benefits of CA are not universally agreed (Sumberg and Thompson, 2012; Sumberg et al., 2012; Erenstein et al., 2012; Pittelkow et al., 2015; VandenBygaart, 2016), suggesting additional agronomic and social science research is required to achieve a common understanding of the long-term global impacts of CA. There is a need to develop global consensus on the definition and utilization of CA for food security.

CA is based on the simultaneous use of zero-till techniques, maintaining a soil mulch cover and the use of rotations and cover crops to optimize different aspects of soil and crop health and resilience (Van Eerd et al., 2014). By promoting soil and crop health, CA conserves and regenerates soil, optimizes its productivity and nutrient recycling, reduces input requirements and environmental impact whilst, at the same time, conserving water and improving biodiversity. CAS are characterized by three linked and integrated universal principles.

2.3.1 Permanent organic crop residue soil cover

Crop residue management, zero-till crop production and cover crops are important methods to reduce wind and water erosion, decrease run-off and evaporation and improve SH for a more resilient soil. Basche et al. (2016b), using meta-analysis, found soil water improvements with long-term cover crops in the system. The cover crop increased the water retained in the soil (at water potentials associated with field capacity (FC)) by 10–11% as well as increasing plant-available water (PAW) by 21–22%. Basche et al. (2016b) concluded that a winter rye cover crop, if managed properly, could improve soil water dynamics without impacting cash crop growth and yield.

Crop residue management can add SOM, which improves the quality of the seedbed, increases water infiltration and retention capacity of the soil, buffers the pH and facilitates the availability of nutrients. It stores carbon in the soil, provides nutrients for soil biological activity and plant uptake and captures rainfall on the surface, thus increasing infiltration and soil moisture content. It also provides a protective cover to stop the soil being eroded, reduces evaporation, avoids desiccation from the soil surface and reduces soil temperature extremes. Organic crop residues serve as 'carbon armor', protecting the soil from direct raindrop impact, mitigating soil temperatures and serving as a slow-release energy and nutrient source for the local soil biology.

> 'Almost all advantages of the no-tillage system come from the permanent cover of the soil and only a few from not tilling the soil.'
>
> *Source: Rolf Derpsch*

As reviewed by Turmel et al. (2015), crop residue has been shown to benefit nearly all aspects of SH, including chemical (SOC, pH and CEC), physical (soil structure, run-off, erosion, compaction, soil temperature and water content) and biological properties (biodiversity and biomass) presumably due to increased SOC. The retention of SOC in the system, and potential benefit on SH, is driven by both total crop residue input and decomposition rates (Paustian et al., 1997), which are also influenced by other management factors such as tillage, soil type and climate. As a result, SOC levels may have regional variations that should be considered when determining if crop residue management will be effective in promoting SH (VandenBygaart, 2016).

2.3.2 Diverse crop rotations and cover-crop mixes

Diversity is powerful in natural systems and creates a dynamic mix of species and activities that synergistically enrich CAS productivity. The diversification principle of CA systems is implemented in a continuous diversified cropping rotation (Fig. 3), ideally consisting of at least three different plant species, including one legume (Anderson, 2005; Ostergard et al., 2009; Chatterjee et al., 2016). Associated species and cover-crop mixes can be managed to extend the rotation to a certain extent. Through diverse crop rotations, crop-livestock interactions and the conservation of soil cover, a habitat can be created for a number of species that feed on pests. This will in turn attract more insects, birds and other animals. Thus, rotations and associations of crops and cover crops promote biodiversity and ecological functions (Ostergard et al., 2009; Stavi et al., 2012; Carlson and Stockwell, 2013). Because of the complexity and richness of soil biodiversity, the effects of crop and pasture management are not fully understood. However, the effects on certain functional groups and, hence, specific soil functions, are increasingly being recognized as vital for agricultural productivity and system sustainability.

Diverse crop rotations provide numerous benefits to any cropping system. Rotations with high-residue crops usually increase SOM content when compared with monocultures (Van Eerd et al., 2014) and are critical to reducing the incidence of diseases and pests as well as being credited with improving nutrient use and reducing weeds (Stirling et al., 2016). Cover crops increase the complexity and intensity of crop rotations, effectively contributing to crop rotation benefits (Carlson and Stockwell, 2013). Higher level management may be required to minimize the chance of cover crops adversely affecting other crops in the rotation. Ingram et al. (2014) found that many European farmers are unconvinced as to the economic benefits of practices for managing soil carbon, suggesting incentives are needed, either in the form of subsidies, or evidence of the cost effectiveness of practices. They identified key barriers of CA practices that included perceived scientific uncertainty about the efficacy of practices; lack of real-life 'best practice' examples to show farmers; difficulty in demonstrating the positive effects of soil carbon management practices and economic benefits over a long timescale; and advisors being unable to provide suitable advice due to inadequate information or training.

CA is more about diversity in plant management because of the importance of carbon and carbon cycling. Crop rotation also influences soil carbon dynamics by affecting the proportions of the decomposable organic compounds returned to the soil, leading to potential gains in soil carbon. The guiding CA principles are that nature manages soil better through biodiversity than in isolation. The economic benefits include dramatically reduced fossil fuels and synthetic inputs (fertilizers and pesticides) while increasing SOM and improving soil biology.

Living soil includes a jungle of creatures, seen and unseen, working underfoot, performing countless critical functions (Brussaard, 1999; Wall et al., 2015; Bardgett and van der Putten, 2014). Biodiversity is an element of sustainability and is essential for harmony and stability in nature (WWAP, 2018). Understanding and nurturing this biodiversity is vital to ensure that management practices sustain ecosystem services (Lal, 1991; Brussaard et al., 1997; Brussaard, 2012).

Diverse cover-crop mixes help with drainage management and enhancing evapotranspiration (ET), improving soil and crop production and consuming residual nitrogen and phosphorus (Qi et al., 2011; Stavi et al., 2012; Zhang et al., 2015; Ferchaud and Mary, 2016). Overwintering cover crops are beneficial in reducing surface water flow and extracting residual soil N left over in the crop root zone after harvest, leading to improvements in water quality (Drury et al., 2014; Dabney et al., 2001; Strock et al., 2004). Cereal rye is one of the most widely used cover crops, especially in the Corn Belt of the United States, because of its winter-hardy characteristics and versatility (Singer et al., 2007). Sievers and Cook (2018) evaluated above- and below-ground decomposition and nutrient release of two single winter cover crops – hairy vetch (*Vicia villosa* Roth) and cereal rye (*Secale cereale* L.) – after termination with herbicides over a 16-week period during the cash crop growing season. Hairy vetch, with its lower C:N ratio, decomposed rapidly after termination in the spring, suggesting growers consider delaying termination until close to cash crop planting, to decrease the risk of N leaching loss. Cereal rye residues decompose much slower and may also immobilize N because of its high C:N ratio. Constantin et al. (2010) reported that cover crops increased soil storage of organic N, decreased residual soil NO_3^- concentrations in the root zone by 34–52% and reduced NO_3^- leaching below the root zone by 36–62%. Additionally, use of cover crops may help mitigate climate change through the transformation of CO_2 into SOM (Reicosky and Forcella, 1998; Olson et al., 2014).

Rotations of several species of crops, including legumes as symbiotic sources of plant-fixed atmospheric N, and other usable green manure cover crops, are beneficial for maintaining soil cover at all times, as well as organic residues both at and below the surface. Diverse rotations with high yields can include cover crops, alternate crops, biofuel crops, compound rotations, complex rotations, perennial rotations and stacked rotations (Congreves et al., 2015). Biodiversity that combines different varieties, species and planting times, that spreads diversity throughout the entire year, that captures carbon and reduces erosion is highly beneficial. The nutrient balances in the soil should be maintained from one rotation cycle to the next. Carbon accumulation occurs when there is a legume in the system that fixes more N than is removed in the crop products or otherwise lost from the system. The provision of various qualities of residues, from the most labile to the more-lignified types resistant to decomposition, are important, depending on the plant types. The more-labile/less-lignified forms contribute less to CEC than more-lignified root materials. A wide range of types increases the range of buffering capacities of the soil with regard to pH and nutrient imbalances with respect to plant requirements. Mixed sequences of crops, plus the presence of permanent soil cover, tend to inhibit the build-up of specific weed species that would thrive under monocrop conditions. The placement of organic root residues at different depths occurs according to each crop's characteristics (Mokany et al., 2006; Kell, 2011). The greater the range of plants grown, in mixtures or in sequence, the more varied the biodiversity of organisms above-ground and inhabiting the rooting depth, and the greater the competition that can suppress those detrimental to root function and thus be considered weeds/pests. Crop rotation will further interrupt the infection chain for diseases and may have other pest-repellent and -suppressing characteristics (Stirling et al., 2016).

2.3.3 Continuous minimum mechanical soil disturbance

The third principle of CAS is continuous minimal soil disturbance (Fig. 3), typically achieved using zero-tillage (direct-seeding) management strategies (Eagle et al., 2012). Planting perennial crops such as forages which, by definition, involve no cultivation for the duration of their growth, is another approach. Complementary agricultural practices allow the farmer to adjust operations to economic and site-specific needs. While each of the principles may be considered a separate entity, it is the fluid integration of all three principles and their supportive SH practices that are key to effective CA. The smooth integration and synchrony from a CA systems perspective requires complementary supporting practices for maximum efficiency. The natural interactions and diversity within the cropping system contribute to numerous economic and environmental benefits (Lal, 2015b; Reicosky and Janzen, 2018).

Farmers around the world have relied on intensive tillage for the last 10 000 years (Lal et al., 2007) and since the advent of tractor power, tillage has become much more intensive than before. Although this approach has had some benefits, it also has serious problems, notably the resulting soil erosion and degradation. Tillage destroys SOM through two interrelated processes. Organic matter at depth in the soil is slower to decompose as soil temperature and moisture levels vary more slowly at depth, where oxygen partial pressure can be lower. Plowing brings this organic matter (OM) to the surface and decomposition is speeded up. The second process breaks up the macro-aggregates and exposes the occluded particulate OM (or light fraction) to decomposition. This process has been well described and shown to be true by Denef et al. (2007) and Zotarelli et al. (2007).

Many uncertainties remain in the relationship between tillage, soil carbon and other greenhouse gases (VandenBygaart et al., 2003; Eagle et al., 2012; Luo et al., 2010). Baker et al. (2007) and VandenBygaart (2016) argue that reduced tillage has not been consistently shown to increase SOC. Boddey et al. (2009) and Franzluebbers (2009) argue against the claims made by Baker et al. (2007). Reduced tillage may also affect emissions of nitrous oxide or methane from agricultural cropping (Blanco-Canqui and Lal, 2008). Baker et al. (2007) observed no correlation between tillage systems and differences in the SOC content at soil depths of over 30 cm. Kuhn et al. (2016) showed that the difference in total SOC stocks between NT and CT decreased with soil depth, confirming that the SOC benefits of NT are concentrated to the uppermost topsoil and achieved maximum SOC stocks after about 10 years of NT. Poeplau et al. (2017) suggest caution when measuring C stocks due to the misuse of critical parameters such as bulk density. Pittelkow et al. (2015) conducted a global meta-analysis using 5463 paired yield observations from 610 studies to compare zero-till – the original and central concept of CA – with CT practices across 48 crops and 63 countries. Overall, their variable results showed that zero-till reduces yields and under certain conditions can produce equivalent or greater yields than CT. The widely different research conclusions in the previous results suggest the need for caution and attention to detail in CAS. Brouder and Gomez-Macpherson (2014) found chronic deficits in reporting CA research and therefore proposed a standardized set of 'minimum' data (listed in their Table 7) that should be contained in peer-reviewed papers, and in line with most points raised by Derpsch et al. (2014). The list of quality criteria in meta-analyses suggested by Philibert et al. (2012) also merits consideration.

One possible explanation for the discrepancies and differences in the C storage results in the research presented in the preceding paragraph may be attributed to a 'terminology dilemma' reported by Reicosky (2015). CT1 encompasses a wide range of soil disturbance levels. CT1, minimum tillage, reduced tillage and strip tillage, as examples, are not 'true' zero-tillage systems with continuous minimum soil disturbance required in CAS. Zero-tillage as defined by Eagle et al. (2012) does not include any form of conservation, minimum or reduced tillage as 'true' zero-tillage. The lack of attention to detail on the tillage and planting equipment used in methods and materials of many previous research studies often do not differentiate between CT1, reduced or minimum tillage and zero-tillage and the use of a minimum soil disturbance zero-tillage planter. Following Eagle et al. (2012), this review treats zero-till/direct seeding as a separate activity representing the minimum soil disturbance required to insert the seed. Zero-tillage is not included under the umbrella term 'conservation tillage', which is taken more narrowly to denote any reduced tillage practice (Reicosky, 2015).

Tillage releases CO_2 and mixes soil and crop residue to allow rapid decomposition of SOM (Reicosky and Lindstrom, 1993, 1995; Reicosky, 1997a, 2002; Ellert and Janzen, 1999). Previously decayed carbon, entrapped in soil as CO_2 from previous SOM decomposition, is released as the soil is opened up by the tillage tool. Reicosky and Lindstrom (1993) found the CO_2 emissions were proportional to the volume of soil disturbed using different tillage tools or with the same tool (moldboard plow) at different depths. In this way, tillage is a 'double negative', rapidly releasing C from the soil and contributing to the increase in atmospheric CO_2-increasing greenhouse gases. Tillage intensity, by affecting the amount and distribution of crop residue on and anchored to the soil, and by affecting the size of soil aggregates and their stability, has a large impact on wind and water erosion. Tillage, through the action of soil disturbance and the downward force of gravity, also causes the slow, progressive downslope movement of soil, that is tillage erosion (Lobb et al., 1995).

Olson et al. (2014) found the zero-till system, with cover crops, sequestered the most soil organic C when compared to pre-treatment from the zero-till baseline SOC stock. In addition, the zero-tilled, chisel plowed and moldboard-plowed plots with cover crops all sequestered SOC above the pre-treatment baseline levels of the same tillage treatment. This suggests that SOC stock losses from tillage, water erosion and some disturbance or mixing during zero-till planting, aeration, nitrogen injection in corn years and mineralization were less than the SOC gain from the cover-crop treatment. Bringing together the individual crop benefits into a community whose synergistic effects contribute to subsequent crops is greater than the sum of the individual crop contributions. Presumably, absence of tillage and preservation of the soil structure and root and earthworm bio-pores using zero-till and cover-crop mixes enable water to flow to a greater depth (Ehlers, 1975; Edwards and Lofty, 1982; Edwards et al., 1988, 1990; Brown et al., 2004; Kell, 2011; Kemper et al., 2011).

The three CA principles should be integrated and applied continuously for improved C management and long-term sustainability (Zilverberg et al., 2018). CA derives C accumulation and many of its multiple benefits from synergistic simplicity of minimum soil disturbance (minimizing C and soil loss) and the use of diverse rotations and cover-crop mixes (maximizing soil coverage and net C input) for soil diversity protection and regeneration. Today, CA practitioners aim for greater soil cover because of the additional protective and nutritional benefits of crop residue. Cover crops are critical for capturing off-season C and producing more soil C and surface residue protection, and have the potential to maximize minimum disturbance benefits, resulting in soil C storage.

2.3.4 Complementary supporting practices

CA facilitates good agronomy, such as timely operations, and improves overall land husbandry for rain-fed and irrigated production. Complemented by other good practices, including the use of quality seeds and integrated pest, nutrient, weed and water management and so on, CA is a base for sustainable agricultural production intensification (Shaxson et al., 2008; Stagnari et al., 2009; Kassam et al., 2009; FAO, 2011; Jat et al., 2013). While the implementation of the three principles may be simple in concept, farmers around the world operate in vastly different climates and feature a wide range of cropping and animal production systems. As a result, the complementary practices shown in Fig. 3 are needed to provide tools and flexibility within the management system, as well as meeting the individual needs of every unique farmer while protecting the environment (Thierfelder et al., 2018). Teague et al. (2016) discuss the role of managing livestock within CAS using cover crops as forage and in reducing agriculture's carbon footprint. Part of the reduction in the carbon footprint comes from reduced use of fossil fuels as a result of less intensive tillage and less chemical input, mainly in the form of N fertilizers.

Site-specific adoption of CAS requires locally applied, complementary practices, essential to the integration of system components. Every farmer, farm, soil type and climate will lend itself to different cropping combinations. CA integrates the management of available sun, soil, water, air and biological resources combined with site-specific external inputs to create a system of biological agriculture, with a complementary suite of practices, unique to each individual farm (Corsi et al., 2012; Thierfelder et al., 2018). CA seeks to configure the farm – as a part of the ecosystem – specifically adapted and tailored to local conditions, accounting for the trade-offs associated with technology choice in the short and long term (Zilverberg et al., 2018). Under such a far-sighted, systems-based perspective, the landscape will evolve into a dynamic, site-specific system for each farm and will reflect individual producers' personal decisions about managing farms and their resources (Corsi et al., 2012). Every farm and farmer reflect a unique combination of multiple management decisions. The cascading effects of these supporting practices on the agricultural ecosystem can result in beneficial synergies.

The improved social, economic and environmental efficiencies utilizing CA systems have a direct impact on the consumer. The satisfaction of purchasing quality food produced in an environmentally friendly manner at a reasonable price goes a long way to enhancing our quality of life. CA minimizes impact on infrastructure, rarely creating negative environmental and social externalities. Collaboration with local communities, politicians, industry and NGOs can all help society operate and multiply the social and environmental co-benefits derived from CAS, very closely related to NBS. Utilizing CAS across the entire watershed enables the consumer and the farmer to enjoy recreational activities in clean rivers and lakes while producing pollinator and wildlife habitats. Producing our food using CAS for the whole of society is a responsible investment in our future food security.

2.4 Benefits of CAS

Integrating the principles and complementary supporting practices multiply the social and environmental co-benefits of CAS. The synergistic benefits generated in this operating system provide enhanced ecosystem services, along with food security. CA systems that enhance water management are central to achieving sustainable production because they generate social, economic and environmental co-benefits in the fields of human health and livelihood, food and energy security, sustainable economic growth, decent jobs, ecosystem rehabilitation and maintenance and biodiversity. The substantial value of these co-benefits from NBS decrease the need for further financial expenditure to repair degraded landscapes and water bodies where the economic costs are virtually unknown, but perceived to be very high (WWAP, 2018).

Many of the multiple benefits of CAS are listed in Fig. 4. The large number of individual benefits come from system efficiencies and are a result of synergistic interactions. Of primary interest are the lower input costs in CA from minimizing soil disturbance and the addition of synthetic fertilizers and chemicals. The increased C inputs from diverse rotations and cover-crop mixes provide additional resiliency, which combined with the smaller C footprint of CA, contributes to decreased greenhouse gas emissions (Reicosky and Lindstrom, 1993; Ellert and Janzen, 1999), although much of this is previously decayed C, entrapped in soil as CO_2. Tillage-induced emissions of CO_2 can continue, indicating a prolonged effect on subsequent decomposition of SOM.

Keeping the soil covered and growing living roots is critical to control erosion and improve SH and function in all farming operations. Diverse rotations

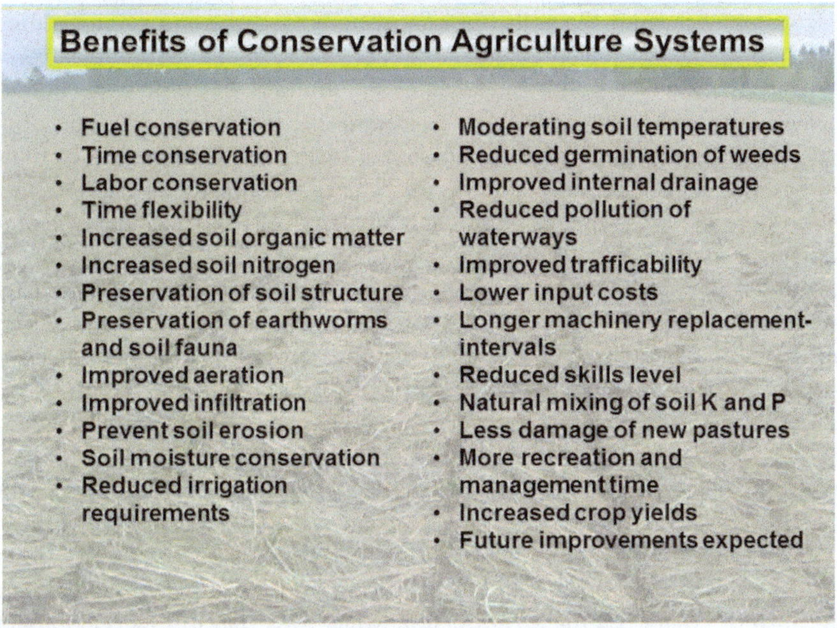

Figure 4 Multiple benefits of Conservation Agriculture systems.

enrich the soil with plant-available nitrogen and organic C, break disease and weed cycles and diminish erosion. Diverse rotations also provide multiple environmental benefits, including reduced synthetic nitrogen use, effective weed control with less herbicide and fewer herbicide carry-over issues. All of these characteristics point to increased system resilience.

3 Plant carbon – impact on hydrology and water-use efficiency (WUE)

CA utilizes the soil profile as a 'giant buffer' for water storage and utilization for optimum WUE. Good C management in CAS enhances WUE at the watershed scale (Lemordant et al., 2018) and suggests the need to better understand the coupling between the hydrologic and C cycles in agroecosystems. It is equally important to understand the benefits and disadvantages in relation to annual versus perennial crops (De Oliveira, et al., 2018). The concept of integration in both CA and SH principles links soil (physical, chemical, biological) properties and ecological properties in performing the required critical ecosystem functions (Karlen, 2012).

From taking a broad view of hydrologic processes and properties involved in managing water in agricultural systems, it is increasingly clear that plant C management is very important. Carbon, the building block of all life, is constantly flowing through the atmosphere to plants via photosynthesis (Fig. 1) and into the soil and animals, then back into the atmosphere as part of the carbon cycle. Carbon is the key element in the >10000 diverse chemical compounds that make up SOM and refers to the non-mineral portion of the soil, including live and dead plant stems, leaves, roots and root exudates and a few other examples listed in Fig. 5 that affect hydrology. Baldock and Nelson (1999) stress that a fundamental function of SOM is the provision of metabolic energy, which drives soil biological processes. In essence, it is the transformation of C by plant, micro- and macro-biological processes that provide energy and results in the establishment of a cycle, that connects above- and below-ground energy transformations. The biomass mulch on the soil surface, particulate organic matter (POM) and SOM are all forms of C important in soil properties. Manure also contributes substantial C. Live and dead microorganisms and earthworms contribute a small portion of C which all eventually degrade, leaving recalcitrant humus and humic acids as part of the C cycle (Janzen, 2015; Reicosky and Janzen, 2018). Morrow et al. (2016) studied long-term tillage impacts and found up to 60% of SOM had been lost from wheat-fallow lands in the US Pacific Northwest. SOM was depleted because of limited biomass production, which was due to low precipitation and fallowing every other year.

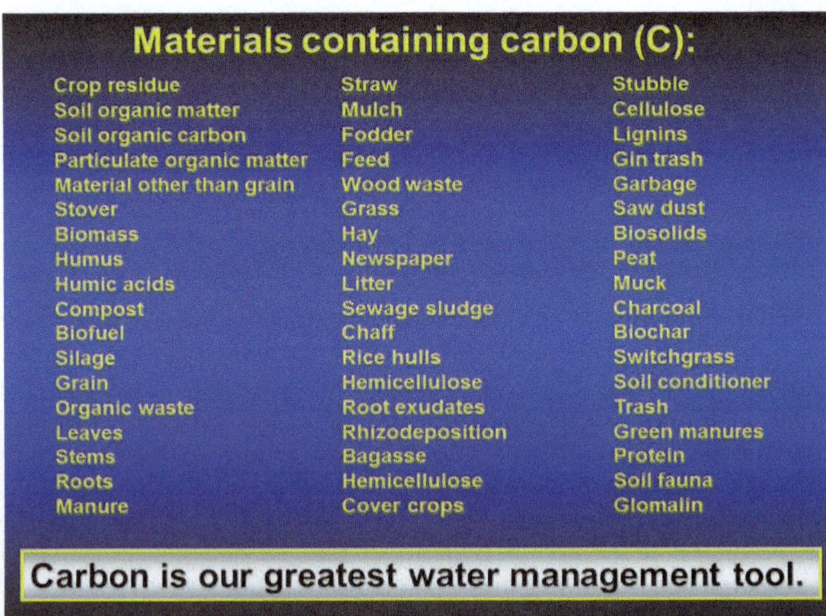

Figure 5 Variety of materials containing carbon useful in soil health and water management.

SOM is a mixture of residual plant material in various stages of decomposition, microbial biomass and all their by-products (Lehmann and Kleber, 2015). SOM is a highly heterogeneous pool that includes numerous diverse carbonaceous compounds that range from easily mineralizable sugars to complex and recalcitrant products of microbial transformations (see Fig. 5), eventually decomposing into humic substances. SOM has been directly and positively related to soil fertility, including water availability and agricultural productivity potential. Matson et al. (1997) describe factors affecting soil C loss from agricultural soils, including climate and soil type (Haddix et al., 2011), tillage intensity and depth, crop rotation decisions, amount of organic inputs, amount and quality of plant residue on the soil surface, soil biological activity, length and time of fallow and erosion. The interchangeable use of terminology suggests that theoretically defined C pools can be equated to SOM fractions. Unfortunately, overlapping terminology is often applied to fractions and pools that are not closely related, leading to confusion. The division between active, slow and passive SOM pools are likely to differ in their hydrologic impact with emphasis on biologically, physically and chemically regulated dynamics. Even though some definitions of SOM exclude fresh plant residues, surface residues can be important in increasing water infiltration and decreasing water evaporation from soil.

3.1 Soil structure – fungal hyphae network

Soil structure is the arrangement of pores and cracks within a matrix of soil particles and OM, with soil particles glued into small clumps, called peds or aggregates. Depending on the composition and conditions in which the peds form (wet and drying cycles, or freezing and thawing cycles, foot traffic, farming etc.), the aggregates may have a specific shape, for example granular (such as gardening soil), blocky, columnar, platy, massive (such as modeling clay) or single-grained (such as beach sand). Structure correlates to the pore space in the soil, which influences microbial habitat, root growth and air and water movement. The solid particles bond and aggregate to create pores and cracks that enable water and gas flow through the soil matrix. The quantity, distribution and arrangement of the pores determine the soil's WHC, infiltration rate, permeability, root and earthworm penetration and soil respiration rate. The continuity of the bio-pores is important for both water and gas flow and storage through and in the soil matrix as well as the synergistic relationship of earthworms and tap roots creating the net result of a larger soil volume from which to extract water and nutrients during stress periods. The preservation of the soil structure and root and earthworm bio-pores using zero-till methods enables water to flow to a greater depth (Ehlers, 1975; Lee, 1995; Kladivko et al., 1997, 2001; Kell, 2011; Kemper et al., 2011; Arnone and Zaller, 2014). Sreelash et al. (2017) estimated that if rain or irrigation are insufficient for ensuring optimal crop growth, then crop growth is highly dependent on soil structure increasing infiltration and available soil water capacity, both dependent on C content.

Soil fungi are important agents of decomposition, pathogenicity and plant health and SH (i.e. nutrient cycling, soil fertility, aggregate stability and SOM turnover). Intensive tillage breaks up fragile fungal hyphae networks and upsets the balance between the fungi and bacteria, impacting soil carbon loss (Bailey et al., 2002). Ecosystems that are tilled have lower fungal activities and stored C than those maintained under native or zero-till systems. Six et al. (2006) showed most tilled agricultural soils are dominated by bacterial activity and zero-tillage systems favored fungal activity and C and N accumulation. They found a quantitative and qualitative increase in SOM with minimum soil disturbance systems favoring a fungal-dominated community. De Vries et al. (2006) suggested that high soil fungal to bacterial (F/B) ratios are indicative of a more sustainable agricultural system. Fungal and bacterial biomass and the F/B ratio were higher in grass than in grass-clover. The F/B ratio decreased with an increasing N application rate, which reflected decreasing fungal activity and increasing bacterial activity. Managing soil disturbance and fertilizer N can impact the soil biology and ecosystem that affects soil C and energy flow. Managing soil fungi to increase SOM requires minimum soil disturbance (Bailey et al., 2002; Six et al., 2000, 2006; Krull et al., 2004). Further, the microbial decay processes, driven by the demand for solar energy, also reconfigures the chemical structure of remaining organic substrates and notably, its affiliation with mineral components of the soil, creating physical arrangements that improve soil physical structure, enhance aeration and infiltration and reduce erodibility. Through these and other mechanisms, the continual cyclical flux of C (solar energy) into and through the soil maintains SH and multiple ecosystem functions (Janzen, 2015; Reicosky and Janzen, 2018).

Fungal contributions to agroecosystems' function are difficult to quantify because of the lack of accurate methods to measure fungal biomass and activity. The benefits and limitations of some of the typical qualification methods are currently being evaluated (Nichols and Wright, 2004). Overall, the diversity of soil organisms is reduced by intensive tillage agricultural practices and mineralization rates that increase the decomposition of SOM, making C the limiting nutrient for the soil fungi. Fungal biomass typically responds positively to zero-till management, for example, Frey et al. (1999) found that fungal hyphae length was 2–2.5 times higher in the zero-till system than conventionally tilled systems. Fungi are favored in the zero-tillage system because hyphal networks can be maintained; fungi can bridge the soil residue interface and utilize spatially separated nutrients, especially C and N, as well as maintain activity, even in dry locations or across air filled pores. The identification of glomalin, a glycoprotein produced by hyphae of arbuscular mycorrhizal fungi, and the correlation between the glomalin and aggregate stability, is proving useful as an indicator of soil quality (Nichols and Wright, 2004).

Healthy soils are characterized by a diverse and active community of organisms that maintain and drive key ecosystem functions, mainly C and nutrient cycling (Tully and Ryals, 2017). The formation of soil aggregates is possible via the synergistic interactions between plant roots, fungal hyphae and microbial extracellular polysaccharides and proteins that help 'glue' soil particles into various-sized aggregates (Six et al., 2004), essential for erosion resistance, microbial habitat and air and water flow.

Experiments looking at the impact of increasing temperature on soil biochemical and physical-chemical processes indicated that soil warming and drying scenarios would decrease SOC via increased carbon loss to the atmosphere, thus adversely impacting SH (Jat et al., 2017). Recent studies by Luo et al. (2017) and Morrow et al. (2017) found that SOC stocks, C input amount and precipitation are the three most dominant variables, which can influence SOC dynamics jointly by the complex interplay of climatic, edaphic and biotic conditions, with implications for SH. Organic C input in colder regions could increase due to increased plant growth at higher temperatures. Maintaining SH in the face of increasing demand for food and pressure on agricultural resources is seen as a major challenge of the twenty-first century (Foley et al., 2011).

3.2 Tillage impacts

Tillage is an 'apocalyptic event' for fungal hyphae, worms and soil insects, causing mortality in addition to all the carbon and water lost. Reductions in residue and SOM associated with long-term tillage restricts food supplies (Zilverberg et al., 2018). Tillage can quickly break apart aggregates, exposing soil C to microbial attack. Tillage also pulverizes the soil surface, creating a crust condition that seals the soil, resulting in more run-off and less effective irrigation. Tillage creates a priming effect for some micro arthropods and microbes (Kuzyakov, 2010) as well as the destruction of the mycorrhizal fungi hyphae network structure. To store more C and N within the soil for subsequent crops, there is need to maximize the fungi:bacteria ratio which means virtually no soil disturbance because of the way any tillage destroys the delicate fungal network. Tillage also destroys soil structure leaving soil susceptible to wind and water erosion (Brown et al., 2003; Reicosky and Lindstrom, 1993; Reicosky, 1997a,b; Reicosky et al., 2011; Hatfield, 2014; Lal, 2015a,b; Reicosky and Janzen, 2018). As a result, much of the best topsoil is washed into rivers and streams, deposited at the bottom of the ocean.

The loss of SOM caused by repeated soil tillage has a number of consequences for associated soil degradation and CO_2 emissions (Reicosky and

Lindstrom, 1993, 1995; Reicosky, 1997a,b; Reicosky and Archer, 2007). Reicosky et al. (1995) reviewed the effects of tillage, crop rotation and time on rates of residual SOM production systems and found with NT there were small gains while most studies with intensive moldboard plow tillage resulted in continued small losses. Jackson et al. (2003) indicated that CO_2 emissions were high the first day after tillage, but respiration declined or remained constant, suggesting that physical soil fracturing processes are responsible for the high flux from the soil surface. Net mineralization and nitrate accumulation increase for several days after tillage, which can be accompanied by higher denitrification rates. Tillage causes immediate changes in microbial community structure, based on phospholipid fatty-acid analysis, but little concomitant change in total microbial biomass. Tillage events also contribute to decreased soil quality by increasing emissions of greenhouse gases, as demonstrated by Reicosky and Lindstrom (1993), Reicosky (1997a,b). Reicosky et al. (2002) evaluated the effects of maize silage removal versus grain removal at two different fertility levels on total C, total N, C:N ratio and tillage-induced CO_2 loss immediately after moldboard plowing. All four treatments had essentially the same SOC content (2.2%) in the 0–20 cm depth and similar short-term CO_2 emissions after 30 years of the same treatments. The results demonstrated the impact of intensive moldboard plow tillage on soil C loss that appeared to over shadow any effect of nitrogen fertility level and silage removal versus grain removal. These negative aspects must be weighed against any 'perceived benefits' of tillage for increasing the health and productivity of crops.

CAS offer a solution that can potentially slow and reverse this damage, and minimize their repetition on newly opened lands. For example, Wortmann et al. (2010) noted slight decreases in microbial biomass, but concluded that one-time tillage of NT can be carried out without measurable differences on yield or soil properties. This inconsistency among findings may reflect, in part, the uniqueness of the ecosystems studied; the effects of sporadic tillage may vary depending on the properties of the site in question, subtle differences in research methods and the complement of other agronomic practices applied. Alternately, Stockfisch et al. (1999) observed that OM stratification and accumulation as a result of long-term (20 years) minimum tillage were completely lost by a single moldboard plow tillage. Cultivating zero-till systems can decrease soil aggregation and accelerate C and N losses so rapidly that years of soil C restoration can be undone within weeks to months, supporting the minimum soil disturbance criteria of Corsi et al. (2012) and Eagle et al. (2012).

A sustainable ecosystem is one with internal mechanisms and resources that maintain productivity, recovers quickly from disturbance (such as tillage) and keeps pest and disease at tolerable levels with only minimal external inputs. Agricultural soils with naturally high amounts of P and K from fertilizers are physically disrupted by tillage, and often vegetated by only one or two plant species in monoculture systems. To decrease pathogenic fungi and enhance the biomass, diversity and function of the mutualistic fungi, one or more of the following management options can be implemented:

- Reduced fertilizer inputs (especially high P fertilizers) will increase the need for mutualistic fungi to scavenge immobile nutrients.
- CAS will minimize disruption of hyphal networks while increasing the number of crops in the rotation.
- Planting interrow crops, or using buffer strips or shelter belts, will increase above-ground diversity and decrease hosts for pathogenic fungi. Planting selected cover crops instead of having fallow periods will maintain the presence of living roots as host for mutualistic fungi.

Microorganisms play an important role in the transformation of SOM and nutrients as 80–90% of the total soil metabolism is due to microbial processes (Brady and Weil, 2004), 1–5% of C and N in soil is stored in living microbial tissue (Duxbury et al., 1989) and microbial biomass in temperate grasslands is estimated to be 1-2 t ha^{-1} (Nannipieri et al., 2003). In fact, a number of soil microbiological parameters (such as microbial biomass C, basal respiration rate) have been suggested as possible indicators of soil quality. Finney et al. (2017) showed the living cover crops almost immediately impact the soil microbial community structure and function. Soil biological activity varied by treatment and was positively correlated with both the size and fungal:bacterial ratio of the microbial community which links cover crops, microbial communities and SH.

Ecosystem services depend on C characterized by its complexity across spatial scales, ranging from its simple atomic structure to its complex molecular structure in SOM and to its role in ecosystem services and biogeochemical cycles. The diverse functions of soil C suggest a key organizing concept for sustainable agriculture while maintaining multiple criteria of environmental, economic and social welfare. Transforming the complex molecules of dead OM into different substances, both labile and resistant, depends on the composition of the substrate. This process leaves transformed materials with differing degrees of resistance to, and thus speed of, subsequent breakdown by biotic process of other soil organisms. Over the long term, this leaves some crop residues less changed than others, providing long-lasting and slowly released remnant reserves of the nutrient and carbonaceous materials. Stagnari et al. (2009) concluded that, when compared to CT agriculture, CA results in 'improved soil structure and stability; increased drainage and WHC; reduced risk of rainfall run-off (see Fig. 6) and reduced pollution of surface waters with pesticides of up to 100% and fertilizers up to 70%; and about one-quarter to one-half lower energy consumption and lower CO_2 emissions'.

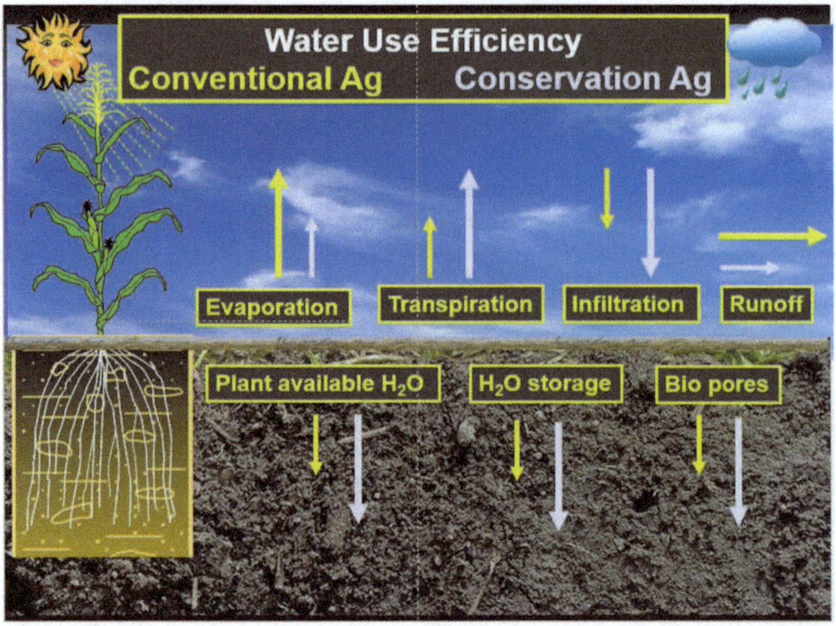

Figure 6 Hydrologic characteristics of Conservation Agriculture (blue arrows) and conventional agriculture (yellow arrows) systems. Arrows are color-coded for each system and represent the relative extent of transpiration, evaporation, run-off, infiltration, soil water storage and plant-available water. Source: adapted from Hillel (1998), Glover et al. (2007) and Basche (2017).

The relationship between plant water use and plant growth is a fundamental facet of agriculture and food production, and has thus received considerable attention (Ali and Talukder, 2008; Hatfield et al., 2001; Stewart and Steiner, 1990; Taylor et al., 1983). Water is required in the photosynthetic equation for C fixation and biochemical energy for all living creatures. Without good carbon management, it is very difficult to enhance WUE at the watershed scale (Lemordant et al., 2018).

Carbon is always in transition in the carbon cycle, from being fixed in photosynthesis to create plant biomass that is returned to the soil and converted to SOM and ultimately to humus and humic acids, yielding CO_2 along the way (Lehmann and Kleber, 2015; Janzen, 2015). Given the importance of soil C cycled or sequestered, and lingering uncertainties as to whether farmers should manage carbon as an energy source for soil biology or as plant biomass, 'carbon armor' for soil protection and preservation and/or climate mitigation provides a conundrum. Stevenson and Cole (1999) stated that the C cycle is the driving force of other cycles, especially the water cycle, which needs enhanced management to protect soil resources from climate extremes (Lemordant et al., 2018). Aeration, WHC and permeability are all favorably affected by

SOM. The frequent addition of easily decomposable organic residues leads to the synthesis of complex organic compounds that bind soil particles into aggregates that maintain a loose, open, granular condition. Water is then better able to infiltrate and percolate downward through the soil. The plant roots need a continual supply of O_2 in order to respire and grow. Large pores permit better exchange of gases between soil and atmosphere.

The mechanisms underlying the positive plant diversity effects on soil C storage are poorly understood. Lange et al. (2015) demonstrated that higher plant diversity increases rhizosphere C inputs into the microbial community, resulting in both increased microbial activity and C storage (Zhu et al., 2016). Increases in soil C were related to the enhanced accumulation of recently fixed C in high-diversity plots, while plant diversity had less pronounced effects on the decomposition rate of existing C. Their study showed elevated C storage at high plant diversity is a direct function of the soil microbial community, indicating that the increase in C storage is limited mainly by the integration of high-residue crops into soil and less by the decomposition of existing soil C.

Active and healthy soil organisms are intrinsic to the provision of life support services, environmental goods and services and food security. If SH, along with water quality and soil biodiversity, is allowed to deteriorate, this impacts the quantity and quality of food produced from the soil, and ultimately quality of life on the planet. Inversion tillage (conventional agriculture) is a major biotic disturbance that seriously disrupts and damages soil biological constituents (Ehlers, 1975; Edwards and Lofty, 1982; Edwards et al., 1988, 1990; Lee, 1995; Clapperton et al., 1997; Brown et al., 2004), whilst also contributing to climate change by rapidly oxidizing soil C and releasing CO_2 into the atmosphere. Inversion tillage most noticeably impacts the larger soil biota, such as earthworms (Brown et al., 2003) and filamentous fungi, slicing them and decreasing their effectiveness.

Earthworms have a large impact on soil structure. In soil, earthworms represent the largest component of the animal biomass and are commonly termed 'ecosystem engineers' (Brussaard et al., 1997; Brussaard, 2012). Blouin et al. (2013) reviewed the contribution of earthworms to ecosystem services through pedogenesis, development of soil structure, water regulation, nutrient cycling, primary production, climate regulation, pollution remediation and cultural services. Earthworms, insects and rodents are the most visible components of the 'living soil' team (Brussaard, 1999), working in tandem with soil microorganisms and fungi to contribute to aeration and nutrient cycling as part of a 'soil factory' team effort. The deep-burrowing earthworm species *Lumbricus terrestris* L. (nightcrawler) can have significant impacts on water infiltration rates in soils (Brussaard, 2012; Wuest, 2001; Chan et al., 2003; Arnone and Zaller, 2014). Many farmers using zero-till practices try to encourage higher nightcrawler populations in their fields (Clapperton et al., 1997; Kladivko et al., 1997; Brown et al., 2003), and a few have questioned whether it would be useful to distribute collected or purchased nightcrawlers throughout their fields. Long-term adoption of CA leads to significant enhancement in earthworm activity, that is over 10 years it was 36% higher under CA compared to CT (Clapperton et al., 1997; Kladivko et al., 1997; Jordan et al., 2000; Brown et al., 2003). CA practices showed positive effects on earthworm populations in drier climates, where earthworm activity was six times higher after 30 years of adoption (Wuest, 2001). Willoughby and Kladivko (2002) reintroduced adult *L. terrestris* into some zero-till fields to determine their survival, population growth and subsequent impact on infiltration rates in the short term. Results from the first 2 years indicate that nightcrawlers may survive and become re-established in many zero-till fields, but the rate of population growth and soil property improvements may not be fast enough to encourage large-scale seeding efforts by farmers. Chan et al. (2003) found a significant reduction in waterlogging after 3 years under zero-tillage compared with CT. The reduction related to a higher density of conducting macropores (140 and 5 m^{-2} for no-tillage and CT, respectively), which was associated with a higher population density of earthworms (240 and 36 m^{-2} for zero-tillage and CT, respectively).

There is limited information on earthworm migration within agricultural production systems. Nuutinen et al. (2011) found earthworm populations were greatest around field margins initially before *L. terrestris* moved into the cultivated field area at 4.6 m $year^{-1}$. Their results demonstrated greater colonization above sub-drain lines in zero-till areas compared with the plowed parts of the field. The field margins were decisive bridgeheads for population establishment and later acted as source areas for colonization of the field. Ruiz et al. (2015) provided a quantitative framework for estimating energy requirements for soil penetration work created by earthworms and plant roots, and delineates intrinsic and external mechanical limits for bioturbation processes. The estimated energy requirements for earthworm bio-pore networks are linked to consumption of SOM, suggesting that earthworm populations are likely to consume a significant amount of ecosystem net primary production to sustain their subterranean activities.

Mehra et al. (2018) reviewed studies on tillage, carbon and climate change interaction, which emphasized that minimum disturbance minimizes soil GHG emissions relative to CT (Tellez-Rio et al., 2015). These zero-tillage practices directly improve SOC, water retention (via increasing infiltration and minimizing run-off), nutrient-use efficiency and soil moisture content (Clapperton, 1999), as well as modifying herbicide efficacy. In addition to many other tangible SH benefits, zero-till contributes to mitigation of climate change through reducing the oxidation rate of SOM and fuel consumption/ha, thus attracting farmers' interest, which is why zero-till farming is becoming a mainstream activity. Zero-till systems may assist in adapting to varying temperature and precipitation conditions (as a result of climate change) which would otherwise downgrade soil's physical, chemical and biological properties in agricultural systems (Brady and Weil, 2004).

Soil macrofauna redistribute organic residues, vertically and horizontally, in the soil profile. This increases the substrate contact surface, loosens the soil (bio-tillage) and improves soil structure through the formation of macropores and aggregates via burrowing activities and interaction with fecal pellets (Berry and Karlen, 1993; Clapperton, et al., 1997). Fauna also produce pores, tunnels and other biological compartments that facilitate the air and water flow, which also stimulates soil microbial activity (Brown et al., 2004; Kuzyakov, 2010). Soils with low faunal activity show more compaction, making it difficult for the plant roots to penetrate (Chen and Weil, 2010). Clearly, soil fauna have a critical role balancing soil processes under terrestrial ecosystems, fragmenting plant residues and establishing relationships with microorganisms at different depth levels. Soil fauna are also good indicators of changes in soil biological health as impacted by climate change and anthropological stresses. However, earthworm invasion of a mixed deciduous forest in Minnesota, USA, decreased SOM to a depth of 50 cm by an estimated 600 kg ha^{-1} $year^{-1}$ (Alban and Berry, 1994).

3.3 Managing agricultural watershed hydrology

The major benefit of CAS on SH is the relationship between SOC and the soil's capacity to infiltrate and hold PAW. At the landscape level, CA enables several environmental services to be harnessed at a larger scale, particularly C storage, cleaner water resources, reduced erosion and run-off and enhanced biodiversity. When the soil can't absorb all the rain in a torrent, the excess water runs off into lakes and streams, taking OM, soil and synthetic chemicals with it. How much rain it receives is important, but of more importance is how well rain is managed and used. Farmers may disagree over the cause of climate change, especially whether it's caused by humans, but it is difficult to dismiss the extreme weather patterns that have developed in recent years (Villarini et al., 2011a,b; Hatfield et al., 2011; Lal, 2013). Extreme rainfall events and flooding have increased during the last century, and this trend is expected to continue, causing more erosion, declining water quality and negative impacts on transportation, agriculture, human health and infrastructure. Current loss and degradation of agricultural soil and water assets will continue to challenge both rain-fed and irrigated agriculture unless innovative conservation methods are implemented. Our water and landscape must be able to withstand these mounting pressures - hence, there is need for global CA, climate-smart agriculture, making farming better suited to cope with impacts of climate extremes that jeopardize our natural resources (Branca et al., 2011; Gattinger et al., 2011; Lal, 2014). Evidence suggests that climate change will bring hotter temperatures, changing rainfall patterns, climate extremes and more frequent natural disasters (Hatfield et al., 2011).

In the US Midwest, up to 50% of the maize ET is lost by soil evaporation (E) during a normal growing season (Peters, 1960). Steele et al. (1995), using non-weighing lysimeters over a 10-year period, measured that evaporation was approximately 50% of the ET and only in a dry year was the evaporation about 40% of the ET. Van Donk et al. (2010) measured soil water content change to a depth of 1.68 m using a neutron probe to characterize the difference between residue covered and bare soils in corn canopies. They found that crop residue decreased evaporation by between 65 and 100 mm in the 2007 growing season and in 2008 it decreased evaporation from 90 to 125 mm. CT under furrow irrigation converted to zero-tillage under overhead irrigation yielded annual water savings of 203 mm (Pryor, 2006). The global water crisis demands agriculture must use water more efficiently. Water saved by minimizing evaporation is used for transpiration and greater yield, which may have even greater social and economic benefits. In addition, with this kind of water conservation, more water would be available for competing needs. CA enhances surface crop residue C management for increased WUE.

Reducing or eliminating tillage can save several cm of irrigation water per year by increasing OM and improving water infiltration rates and reducing evaporation with surface residue (Steiner, 1989). Irrigation research showed that corn and wheat residues reduced evaporation by about 50% (Klocke et al., 2009). Extrapolated over 100 days of a growing season, the water savings totaled 76–127 mm. Crop residues and zero-tillage reduce evaporation and wind effects, with annual irrigation savings of 100–130 mm (Klocke et al., 2009). In addition, long-term zero-till increases infiltration and decreases run-off compared to long-term CT (Zilverberg et al., 2018). Zero-till conserves water by reducing overwinter evaporation and increasing snow trapping. The small amounts of water saved in a typical annual crop growing season, by carbon management from the mulch effect, infiltration/run-off, increase in water storage from increasing SOM +1% and cover-crop deep root access, can be significant. The sum of small amounts of water loss saved due to carbon management plus other carbon synergies all contribute toward food security and ecosystem services.

In San Joaquin Valley's dairy silage production systems, conventional intercrop tillage typically involves a number of tillage passes in the spring between winter wheat or triticale, and corn seeding (Mitchell et al., 2015). Pryor (2006) has shown tillage operations dry the soil before planting to the depth of the tillage layer; typically 8–19 mm of soil water may be lost per tillage pass. In Nebraska, switching from CT to zero-tillage under center-pivot irrigation has been shown to save 80-130 mm of water annually, with added savings from decreased pumping costs (Pryor, 2006).

CAS with continuous crop biomass cover, utilizing dead crop residue, dormant crop biomass and living plant cover, can decrease evaporation and protect the soil surface. The dead crop residue not only protects the soil surface raindrop impact, if sufficiently thick in the form of a residue mat, it can substantially reduce soil evaporation (Steiner, 1989, 1994; Prihar et al., 1996) as well as become a slowly available form of energy for the soil's biology. The residue keeps the soil surface cooler by protecting it from the sun. This cooler, moist surface allows better root development near the surface in the summer, improving standability, and enables roots near the surface to utilize the light rain showers that don't soak into the soil profile.

The management of soil through tillage changes the water storage and evaporation losses. Burns et al. (1971) and Papendick et al. (1973) showed that tillage disturbance of the soil surface increased evaporation compared to untilled areas. Tillage moves moist soil to the surface where evaporation loss may offset increased infiltration rates caused by tillage. Hatfield et al. (2001) found that the total water evaporation fluxes in Iowa were 10–12 mm for a 3-day period following each cultivation operation in the spring. The total evaporation fluxes from zero-till fields were less than 2 mm over the same period. These studies demonstrate the importance of understanding the role of tillage in efficient water use, SH and crop growth, along with any associated soil erosion and environmental degradation summarized in Fig. 7. By adopting continuous zero-till methods and using residue management, more water is available for crop production and the soil becomes more resilient to extreme events, including drought. The retention of SOC in the system, and potential benefit on SH, is driven by both total residue input and decomposition rates (Paustian et al., 1997), which are also influenced by other management factors (e.g. tillage), soil type and climate. As a result, SOC levels may have regional limits to be considered when determining if crop residue management will be effective in promoting SH.

Soil erosion is still a major problem in agricultural production systems. According to Pimentel et al. (1995), about 430 million ha – almost one-third of the global arable land area – has been lost to soil erosion. Efforts to control human-induced land degradation and soil erosion over the last 10000 years have been building on the ruins of past tillage and monoculture concepts (Lal et al., 2007; Reicosky et al., 2011; Montgomery, 2007a,b). Soil erosion on agricultural land is induced by tillage; soil that is loosened by any type of tillage is more easily transported by wind or water, increasing the rate of erosion. Tillage also destroys much of the biological activity in the soil, leaving it lifeless, robbed of its fertility and susceptible to erosion by wind and rain. Tilled soils are vulnerable to raindrop impact, soil crusting, wind, water and tillage

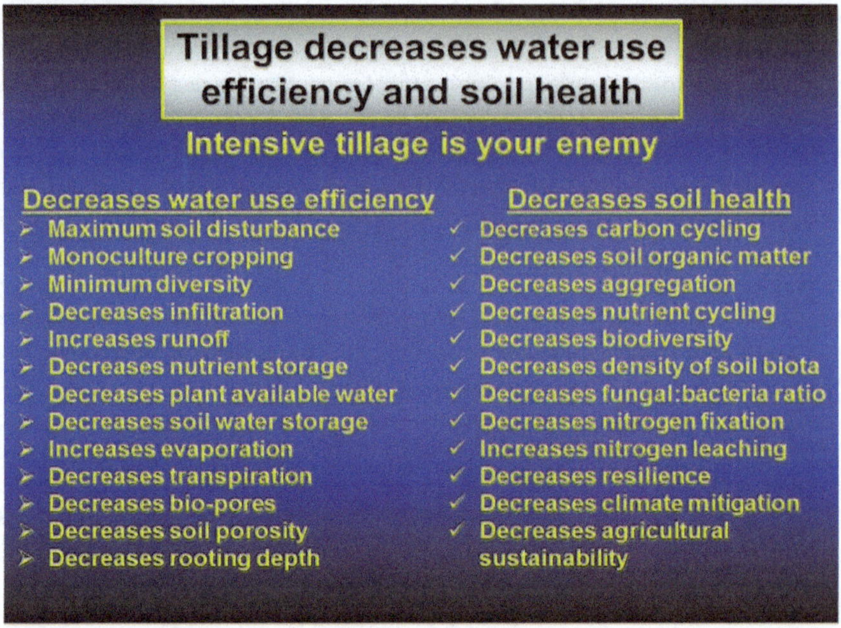

Figure 7 Intensive tillage degrades ecosystems with impacts on water-use efficiency and soil health.

erosion, temperature extremes, water extremes, biology destruction, carbon loss, biodiversity loss, increased run-off, decreased infiltration, increased evaporation, increased leaching and increased pollution (Reicosky et al., 2011). The primary reason for using CA with cover-crop mixes is that this new technology minimizes soil erosion and many associated environmental quality issues. Following Eagle et al. (2012), this work treats zero-till/direct seeding as a separate activity, representing the minimum soil disturbance required to insert the seed. Zero-tillage is not included under the umbrella term 'conservation tillage', taken more narrowly to denote any reduced tillage practice (Reicosky, 2015).

Even with several types of CT1, soil is being lost faster than nature can make it. Seta et al. (1993) measured soil erosion after CT (moldboard plow inversion tillage and disc harrowing), CT1 (practiced in this study as chisel-plow tillage with straight shanks) and zero-till (direct seeding). Conventional plow tillage and disc harrowing had the largest sediment loss, CT1 had intermediate loss and zero-till had the least sediment loss. Sediment lost from CT was 52 times greater than the zero-till treatment, while CT1 was 11 times greater than the zero-till treatment.

Tap-rooted crops such as radishes, turnips and alfalfa can physically generate vertical bio-pores that become 'food paths' for earthworms and soil insects (Chen and Weil, 2010). Intensive tillage destroys these bio-pores and fragments fungal hyphae networks, upsetting the balance between the fungi and bacteria, impacting soil carbon loss (Bailey et al., 2002). Those ecosystems that were tilled had lower fungal activities and lower stored carbon than those that were maintained under native or zero-till systems. Six et al. (2006) indicate that most agricultural soils are dominated by bacterial activity. Briones and Schmidt (2017) showed that disturbing the soil less (e.g. zero-tillage and CA) significantly increased earthworm abundance (mean increase of 137% and 127%, respectively) and biomass (196% and 101%, respectively) compared to soil inverted by conventional plowing.

Basche and DeLonge (2017), using meta-analysis to compare soil hydrology in perennial- and annual-based crop systems, found zero-till and living cover practices significantly increased soil porosity and field water storage. Their findings suggest that continuous living cover practices may be an adaptation strategy to combat rainfall variability amounts and intensity. Presumably, the preservation of the soil structure and root and earthworm bio-pores using zero-till enables water flow to a greater depth (Kell, 2011; Kemper et al., 2011). Diversity is included in extended crop rotations (Poeplau and Don, 2015; Chatterjee et al., 2016; Dudley and Alexander, 2017; Hird, 2017), and cover-crop mixes with cool and warm season species captured the maximum C during favorable temperature and water conditions (Burney et al., 2010). Diversity in the type of cover crops, including legumes and grasses, shallow-rooted and deep-rooted crops (Anderson, 2005; Kell, 2011; Kemper et al., 2011; Kätterer et al., 2011; Paustian et al., 2016), species differing in nutrient uptake and cycling and habitat for wildlife and pollinators, all contribute to synergistic benefits.

In the semi-arid Great Plains, a multi-year experiment found that wheat cover-crop systems (including vetch, pea, clover and triticale cover-crop species) significantly improved WSA compared with wheat–fallow or continuous wheat systems (Blanco-Canqui et al., 2013) in a silt loam soil. At another location, Blanco-Canqui et al. (2011) also found that wheat–sorghum rotations including cover crops increased the mean weight diameter of soil aggregates by 80% in the surface soil and improved water infiltration rates up to three times more than in fields that did not include cover crops. Folorunso et al. (1992) found improved soil strength and water intake (by up to 100%) after 5 years of mixed cover-crop use in orchard and tomato system environments. Gulick et al. (1994) similarly found that just 1–2 years of cover-crop use in sandy loam soils of California increased infiltration rates by more than 140%. Steele et al. (2012) observed that introduction of cover crops into continuous maize systems improved water infiltration and aggregate stability. Sharma et al. (2010) found that green manure crops increased both soil moisture and water infiltration, and ultimately led to greater crop productivity compared to no cover-crop control. Similarly, for intensive rice–wheat cropping in the same region of India, Singh et al. (2007) found that green manure crops increased soil aggregation and infiltration while decreasing bulk density on a loamy sand soil.

3.4 Increase plant-available water (PAW) – sponge effect

The ability of soil to retain water is termed WHC. PAW is measured as the difference between the water content of soil at 'FC' and crop uptake, representing the 'lower limit' of water available. PAW is any water that can be taken up from increased infiltration and redistribution within the bio-pore network from deeper rooted cover crops, allowing for more infiltration and more storage (Bolinder et al., 2002; Williams and Weil, 2004; Chen and Weil, 2010; Kell, 2011; Kemper et al., 2011; Haruna et al., 2017). Increases in SOC can be achieved by adding manure, plant residues or sewage sludge to the soil, increasing WHC (Khaleel et al., 1981; Haynes and Naidu, 1998; Rawls et al., 2003). Khaleel et al. (1981) found that the relative increase in WHC became smaller as the amount of OM from amendments increased. WHC increased at both FC and permanent wilting point with organic waste amendments, but the relative increases varied with soil texture.

The increase in soil carbon content increases PAW (Hudson, 1994; Rawls et al., 2003; Murphy, 2014, 2015; Basche, 2017). SOM high in carbon in the form of 'spongy organic matter' that releases nutrients to crops on decomposition and holds more than its own weight in water provides resilience in both wet and dry periods (Sposito, 2013; Stewart and Peterson, 2015). Carbon provides a critical linkage to water in this symbiotic relationship (Bardgett et al., 2014) and benefits drought-stressed soils (Allison, 1973). While there is some debate about the benefits of SOM acting as a 'sponge' (Minasny and McBratney, 2018), in most situations, more OM is often better with respect to WUE. The challenge is getting more OM stored deeper in the profile rather than at the surface 10–15 cm depth that may provide a practical limit for water storage. Murphy (2015) points out that a change in SOM affects a number of soil properties. The soil properties modified by changing SOM content include aggregate stability, WHC, soil erodibility, soil texture, CEC, nutrient cycling, soil acidity and biological diversity (Brady and Weil, 2004). These changes were not uniform in the soil but isolated near the surface, mainly in the upper 10 cm of the soil profile (Murphy, 2015) and is more critical because the interface between the soil and the atmosphere controls the rates of water infiltration and gas exchange.

Plant roots can rapidly expand deep root cavities at high rates (Dumais and Forterre, 2012), implying macropore creation by biotic processes occurs at rates nearly 20 times faster than that by abiotic processes. Earthworms and plant roots have the capacity to accelerate soil structural restoration by at least an order of magnitude compared to abiotic processes only. Hence the need for enhanced carbon management from diverse cover-crop mixes to provide carbon energy as long as biologically possible.

Diverse cover crops maintain adequate reserves of carbon and deep taproots to create bio-pores for beneficial water flow and storage (Kell, 2011). Without destruction by any type of tillage (Briones and Schmidt, 2017), these bio-pores remain as a 'food tunnel' for earthworms followed by a 'low resistant bio-pore' for the advancing roots of the following crop (Williams and Weil, 2004; Kemper et al., 2011). Bio-pores enable OM (root) penetration, more aeration and deeper water channels for storage and accessibility. Bio-pores provide a 'buffet' enriched with C, N, available phosphate, polysaccharides and other nutrients that stimulate other soil fauna and microbial activity for more nutrient cycling and CO_2 respiration (Tully and Ryals, 2017). Kemper et al. (2011) found 86–100% deeper rooting in a Wisconsin clay soil with 650 mm of rainfall and a large population of nightcrawlers in a long-term (12 years) zero-till field. The number and depth of earthworm pores, the depth, distribution and the type of root structure and the synergistic relationship that was maintained without tillage disrupting the naturally developed bio-pore network provided multiple benefits, minimizing compaction with respect to water storage and management (Williams and Weil, 2004; Chen and Weil, 2010; Kemper et al., 2011). Observations over the years have shown that the productivity and functioning of a soil is directly related to its content of plant-derived OM (Allison, 1973; FAO, 2005; Lal, 2014). The plant depends on the soil and the soil depends on the plant with a strong link between carbon and water.

Plant roots are considered as an integral part of soil biology and therefore the rhizosphere is essential to the function of soil biology maintaining soil diversity and productivity (Foster, 1993; Watt et al., 2006; Gould et al., 2016; Zhu et al., 2016). Fine roots provide biomass and readily available exudates to enhance microbial activity. The roots explore a large volume of soil to variable depths related to the species (Glover et al., 2007; Glover et al., 2010). The root length may be several 1000 km of 'micro tunnels' of carbon and nutrients feeding the soil biology. These living root networks have a very large surface area for root exudates as microbial source of energy as the roots decay with the soil biology releasing CO_2 back to the atmosphere as part of the carbon cycle. Worms are the work horses in the soil and the best for 'natural tillage'. They eat the residue and decompose it, converting it to available nutrients. Fungal hyphae are a very delicate network of 'micro conduits', emanating from the roots, trading water and phosphorus to the plant in exchange for carbon nutrition. The net result is an enlarged volume of soil providing water and phosphorus to the roots, especially under stress conditions. Kong and Six (2010) suggest that root carbon contributes more to overall carbon stabilization than residue carbon, which supports research demonstrating greater retention of root-derived than residue-derived carbon in SOM. The ratio of root-to-residue-derived C (an indicator of relative root contribution) was higher in the microaggregates and silt-and-clay fractions than the carbon POM of low input and conventional systems (Mokany et al., 2006). Balesdent and Balabane (1996) found similar results for maize roots contributing more carbon to the soil.

The various processes and properties affecting the hydrology of an agricultural watershed allude to the complexity and the interactions that affect the overall WUE. Figure 6 is a schematic representing relative quantitative differences in the processes and properties between a CT system and CAS. The size of the arrows represents the relative magnitude for the two different systems. CT systems typically designed around annual crops can experience greater water loss through increased run-off, leaching losses and soil evaporation, to say nothing about off-season wind and water erosion (Hillel, 1998; Glover et al., 2007; Basche, 2017). By contrast, CAS incorporating diverse rotations with perennial crops, pasture grasses and cover-crop mixes ensure nearly continuous land cover and, as a result, can increase crop WUE by simultaneously reducing evaporation and run-off and contributing to soil improvements that create more and deeper water storage (Glover et al., 2007, 2010; Kell, 2011; Basche, 2017). Precipitation is the primary source of freshwater, and soil serves as the reservoir of available water for agriculture, making soil central to ensuring a sustainable path toward food and water security. With recent extreme weather, 'traditional conservation' practices on CT production systems have been outgrown (Hatfield et al., 2017). These soil conservation practices need re-evaluation in light of new climate extremes, combined with improved CA practices.

3.5 New thinking on watershed management at landscape scale

Nature plays a vital role in every part of our lives, from solar energy from the sun to the air we breathe to the water we drink to the food we eat. Working with nature to improve WUE will become imperative in view of the climate extremes being experienced (Villarini et al., 2011a,b; Lal, 2013; Finney and Kaye, 2017; Basche and DeLonge, 2017; Blesh, 2018; King and Blesh, 2018). One way to mimic nature in the function of perennial and native vegetation patterns in temperate agroecosystems is through practices featuring continuous living cover as long as biologically possible or planning prescribed cover-crop mixes to protect the soil with dead or dormant biomass during the winter months (Pimentel et al., 2012; Basche and DeLong, 2017; Zilverberg et al., 2018). There is much to learn from the ecology of natural systems adapted to climatic and environmental limits (Glover et al., 2007, 2010). Unlike exclusively annual cropping systems, there are practices – such as cover crops, agroforestry and various perennial forage, grain or bioenergy crops – that can be managed to ensure canopy cover and living roots in agricultural systems throughout the year within biological limits. Adding perennials and winter annuals to production systems does four things (Glover et al., 2007):

1 Creates value for underused and new resources;
2 Enhances soil, water, wildlife and biodiversity in the landscape;
3 Ensures protection against climate extreme variability from both drought and floods; and
4 Provides an opportunity for economic income through innovative cropping systems implementing winter annuals.

The utilization of perennials and winter annuals helps fill in plant biomass gaps for protecting the soil surface during the entire year (Gesch et al., 2014; Blesh, 2018; King and Blesh, 2018).

Cover crops are being widely considered for enhanced water management at the watershed scale. Cover-crop functions, originally meant to 'cover the soil' to lessen soil erosion and to fix nitrogen, are now being expanded to incorporate plant biodiversity, increased water infiltration, weed and pest control, nitrogen scavenging, decreased compaction, generating vertical bio-pores and being a marketable commodity (Gesch et al., 2014; Finney and Kaye, 2017; Blesh, 2018). The evolution of new terminology requires definition and clarification with expanded functions, perennials and 'continuous living cover' related to agricultural watershed management. Continuous living cover is frequently used in discussions of agricultural water management and quality with enhanced storage and other soil hydrological properties (Basche and DeLonge, 2017). In the northern hemisphere, it is necessary to include the provision 'within biological limits' such as extreme temperature and water availability during the winter months. Depending on the severity of the winter, some cover crops that normally die off in winter may survive. Some winter annual cover crops are terminated by a 'killing frost', while others go dormant during extreme cold and recover when temperature conditions become more favorable. Basche and DeLonge (2017) suggest that continuous living cover may be a potential adaptation strategy to combat extreme rainfall variability by changing soil porosity and FC (Basche, 2017). King and Blesh (2018) evaluated functional diversity as a way of increasing SOC. They found cover-cropped and perennial-cropped rotations, relative to grain-only rotations, increased carbon input by 42% and 23% and SOC concentrations by 6.3% and 12.5%, respectively. King and Blesh (2018) further suggested we increase the 'perenniality' of crop rotations. Functionally diverse perennial- and cover-cropped rotations increase both carbon input and SOC concentrations by exploiting niches in time that would be otherwise unproductive.

Continuous living cover (Anderson, 2005; Jordan and Warner, 2010; Asborjsen et al., 2014; Basche and DeLong, 2017) is an approach to match the function of perennial vegetation in agroecosystems (Glover et al., 2007, 2010; Finney and Kaye, 2017; Basche and Edelson, 2017; Zilverberg et al., 2018; Blesh, 2018; King and Blesh, 2018). While managed agricultural systems cannot directly mimic undisturbed natural systems, practices that 'approximate' continuous living cover, notably maintaining canopy cover and roots in the soil throughout the year within biological limits, present an opportunity to reproduce and/or apply ecological principles in an agronomic setting (Pimentel et al., 2012). Blesh (2018) evaluated functional traits in cover-crop mixtures looking for complementary plant functional traits with emphasis on nitrogen fixation. Nitrogen fixation would be the primary function based on the local needs and soil tests, but would have other multiple ecosystem functions, such as erosion control, weed suppression, carbon input, enhanced WUE and so on. Blesh (2018) found that various cover-crop mixtures had significantly greater multifunctionality than a cereal rye monoculture, reinforcing the need for biodiversity. She found trade-offs among functions, because the highest observed levels of each function were not the same in every case. Blesh (2018) concluded that designing mixtures with complementary plant traits may be particularly effective for increasing multifunctionality and long-term agroecosystem sustainability.

'Perennial crops' means agriculturally produced non-invasive plants that can survive the winter and have a lifecycle of at least 3 years where plants are being grown. 'Cover crops' means grasses, legumes, forbs or other non-invasive herbaceous plants that are either inter-seeded into living agronomic crops or planted on agricultural fields during fallow periods for seasonal cover and conservation purposes (King and Blesh, 2018). In practice, these definitions overlap, since many perennials are grown as cover crops. Alfalfa can be used as an annual cover crop, as can some clover species. Basche and Edelson (2017) suggest improved water resilience with more 'perennially based agriculture' is beneficial for individual farms as well as the broader community and ecosystem health. Asbjornsen et al. (2014) note that perennial grasses, with their deep roots and high WUE, can improve climate resilience. However, in managed agricultural systems, there are challenges combining annual and perennial species (Glover et al., 2007; King and Blesh, 2018).

There are perennial crops, such as fruit and nut trees. However, there are few perennial grain crops. The development of Kernza, a perennial wheat, is making slow progress demonstrating the concept (Culman et al., 2013). Kernza was derived from intermediate wheatgrass (*Thinopyrum intermedium*), which historically has been used as a perennial forage, known for its winter hardiness and disease resistance. De Oliveira et al. (2018) demonstrated the ability of the perennial wheat 'Kernza' in maintaining a relatively high WUE throughout the growing season and its highest ET and net carbon uptake rates (averaging ~370 g C m^{-2} $year^{-1}$ over 4.5 years), particularly when compared to annual counterparts. These findings are important to better understand the coupling between the hydrologic and carbon cycles in these novel agroecosystems as well as to understand the benefits and disadvantages in relation to traditional annual crops.

The concept of perennializing agriculture presents mental challenges because of a long tradition of 'annual cropping' (Basche and Edelson, 2017; Blesh, 2018; King and Blesh, 2018; Zilverberg et al., 2018). With perennials, the soil is not tilled annually, so whatever is in the soil as bacteria, fungi, invertebrates, worms and so forth will continue all year, again, within biological limits. If the soil organisms can stay in place and are not disturbed by intensive tillage, both the crop and soil biology are more efficient using NBS (WWAP, 2018). Even though there are challenges in developing perennial grains, the idea is to 'approximate' the perennial system using diverse combinations of multiple annual winter-hardy crops within the year to have something growing and effectively covering the soil as long as biologically possible. These challenges require more creativity and ingenuity to accomplish these objectives with the biggest challenge overcoming the 'mentality' associated with 'traditional annual cropping'.

CAS enable transition toward 'continuous living cover' in more natural agricultural ecosystems working toward 'continuous cropping' (dead, dormant or alive). Adopting a 'green and growing' approach and developing high-efficiency agriculture systems for as much of the year as possible is the goal. This strategy captures and transforms available sunlight to forms of carbon, for example root exudates, thereby maintaining biological activity and improving the soil structure and aggregate stability. Cover crops can benefit production agriculture, particularly by providing nutrients accumulated or fixed during winter and early spring to the cash crop (King and Blesh, 2018). Continuously growing plants also help absorb and recycle soluble fertilizer nutrients and prevent off-site movement. Figure 8 illustrates three possible scenarios of biomass production over the calendar year in the northern hemisphere. Each of the three seasons will expand as one progresses from north to south toward the equator associated with the increase in temperature. The first scenario is traditional, active growing maize (warm season crop monoculture) for the major annual economic crop (solid line with maize fill) planted in the spring and harvested in autumn. Traditionally, this monoculture system utilizes fall tillage, leaving the soil bare and opened to degradation and erosion over the winter months, essentially wasting any solar energy available during the winter season.

The second scenario is defined by seasons for a second crop inter-seeded in the main economic crop sometime during the summer season. The second crop is a winter-hardy annual commodity crop that can survive low temperatures (dotted line with shaded fill). Managing dead crop residue or dormant crop biomass depends on the temperature tolerance of the winter-hardy crop. With

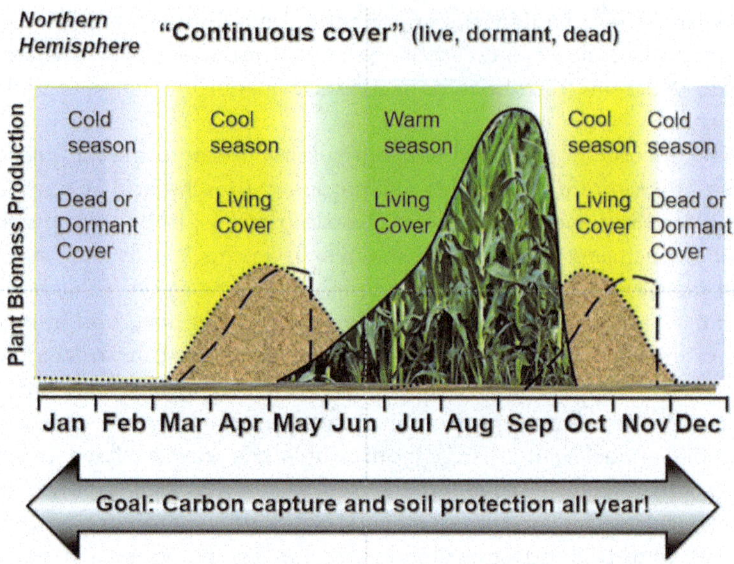

Figure 8 Schematic representation of cover-crop management for soil protection in the northern hemisphere (US northern corn belt) for the full calendar year, segmented into the main growing, cool and cold seasons. Source: Adapted from Heggenstaller et al. (2006).

an extremely cold winter season, there is absorption of some solar energy for photosynthesis within biological limits and the soil is protected with dormant crop biomass and/or snow-covered. The plants re-grow when temperatures become favorable to produce a second economic crop, thereby increasing overall productivity. The crop grain will be harvested, leaving plant residue cover, before the next main economic crop is directly seeded without surface soil disturbance. By leaving crop residue on the surface, this combination of summer annual and a winter annual cover crop with zero-tillage provides a measure of soil protection throughout the entire year.

The third scenario in Fig. 8 allows planting of winter-hardy cover-crop mixes designed to be grown before and after the main crop during the cool season (large dashed lines with no fill). These winter cover crops can tolerate low temperatures and produce biomass for soil protection. Cereal rye is a widely used cover crop in the Corn Belt of the United States, because of its winter-hardy characteristics and versatility (Singer et al., 2007). The risks are the unexpected temperature extremes that may terminate the winter annual cover crop before the cool season ends. Winter cover crops terminated by winterkill leave the dead biomass to protect the soil surface. Winter cover crops planted in the early spring cool season, if they survive any random extreme temperature, can be terminated by a variety of different methods, including herbicide or roller crimping. Depending on the severity of the winter, some cover crops that normally winterkill may survive. Clearly, the range of temperature extremes experienced may shift within any given year (represented by color shading in each season) back or forward in time and cause some unexpected results, depending on the hardiness and composition of the cover-crop mix. These climate seasons generally stay true for the most part; however, factors such as snow cover, soil moisture and plant population all impact a plant's ability to persist up to or beyond their projected growing limits. This is another example where the diversity of the plant species in the cover-crop mix becomes critical for the survival of a few species. Again, this scenario requires minimum soil disturbance.

Johnson et al. (2015) evaluated soybean and a second winter annual crop, field pennycress (*Thlaspi arvense* L.), and proposed them as a means for temporal intensification of agriculture, increasing profitability through oilseed production while providing required protective cover to address soil and water quality issues and weed management. Oilseed yields were variable; however, the combined seed yields of pennycress and soybean were greater than monocrop soybean, illustrating the potential for a double cropping system to increase overall seed yield and reduce early season weed pressure. Another alternative winter oilseed crop being evaluated is camelina (*Camelina sativa* L.) (Gesch et al., 2017). They studied plant growth and yield at different seed depths and found camelina could be planted to a depth of 2 cm before there was a yield decline. Berti et al. (2017) found the environmental impact of double and relay cropping with winter camelina provided lower environmental impacts than maize and soybean in monoculture with increased primary productivity, biodiversity and reduced erosion potential. Because of results like these, more farmers may adapt these systems as a more sustainable production system. These examples suggest progress in environmental protection approaching that of continuous cover to protect soils using winter commodity crops to provide a combination of living, dormant and dead biomass protection on the soil surface. Practices such as crop rotations and/or intercropping, relay cropping, using perennial crops or diverse cover-crop mixes, may also be critical elements of CA because they can extend the photosynthesizing period and capture more free solar energy (Blesh, 2018; King and Blesh, 2018).

The utilization of CAS and SH concepts with diverse cover-crop mixes is an attempt to extend plant C capture in photosynthesis and subsequent transformation into soil protection and as a food source for the soil organisms. Extending living cover as long as biologically possible both into the early spring and into the late autumn with cool season species and developing unique cropping systems and breeding programs to protect soils during the extremely cold 'biological off-season' is critical for cold regions. With CT and annual monocropping, there are substantial periods of soil exposure to climate extremes during the non-cropping season. Cover crops, partly as a result of their carbon input, contribute substantial benefits for physical, chemical and biological soil properties (Basche et al., 2016a,b; Kaye and Quemada, 2017). Cover crops can contribute to agricultural ecosystem resilience, coping with variable rainfall intensity through several mechanisms, including improved soil hydrologic function through enhanced aggregate stability (which improves soil stability and ultimately increases water infiltration and water storage capacity) and with reduced run-off (Blanco-Canqui et al., 2015). Cover crops can restore soil quality and maintain SH through improved carbon management (Reicosky and Forcella, 1998; Franzluebbers, 2005; Blanco-Canqui et al., 2015; Lal, 2015b; Raphael et al., 2016).

Basche and DeLong (2017) suggest using practices that offer 'continuous living cover', notably maintaining canopy cover and roots in the soil throughout the year as long as biologically possible. Exercising this management strategy presents an opportunity to mimic nature and/or apply ecological principles to the agricultural landscape. In seasons with extreme low temperatures, the presence of dead or dormant plant biomass provides adequate cover, protecting the soil from climate extremes with the root system providing stability and energy for the microbiomes until the soil and air temperature increase. More research is needed to quantify 'the biological limits' of actively growing plants to obtain 'continuous living/dormant/dead cover' (Finney and Kaye, 2017; Basche and DeLong, 2017; Blesh, 2018; King and Blesh, 2018) to protect the soil resource all year. These actions in CAS at the watershed scale are beneficial for food security and adaptation to climate change, but also important for farmer resiliency, productivity, viability, environmental quality and food security. Furthermore, by capturing atmospheric CO_2 concentrations and storing carbon in the terrestrial sphere, soils can help mitigate climate change.

4 Summary

Conventional tillage (CT) agriculture and its associated soil and environmental degradation is slowly destroying critically important ecosystem services. Future resilience-building to cope with feeding an increasing population, decreasing resources, environmental degradation and increasing climate extremes, will require changes to improve water management and capture solar energy throughout the year. Agricultural ecosystem services depend on carbon as a biological food source and carbon energy flow through the soil-plant-atmosphere system. There is need for a change in the 'culture' of traditional conventional agriculture toward CA. CA is a management- and knowledge-intensive system approach to farming that is a resilient type of sustainable agriculture to protect our natural resources. CAS, C-based and C-focused, make better use of agricultural resources through the integrated management of available sun, soil, water, air and biological resources, combined with limited external inputs. CAS integrate SOM/C and system concepts based on three key principles: (1) continuous zero- or minimum mechanical soil disturbance; (2) permanent organic mulch cover on the soil surface; and (3) diverse cropping systems with rotations and associations including cover-crop mixes, along with location-specific complementary practices, all important elements of CA. At the core of CA is SH and systems management innovation. Enhanced carbon management enables interactive synergies between the biological, physical and chemical properties and processes, with economic and environmental benefits that are regenerative. Regenerating SH and function is the key to achieving full soil productivity.

Working with nature will be required to enhance water management and improve WUE in the light of present-day climate extremes. The interest in cover crops and carbon input stems from the potential for 'drought-proofing' a landscape by increasing water infiltration and storage. SOM absorbs moisture like a sponge, then releases it and decomposes, enhancing nutrient cycling. With less intensive tillage, there are more environmental benefits accrued with fewer input costs. A few farmers are finding the environmental and economic benefits are going hand-in-hand with this systems approach to food security. By using cover crops, zero-till and diverse crop rotations, some farmers are finding that their soil actually has more available water for their cash crops when those crops need it.

CAS and SH concepts need to be implemented on all producing landscapes, as illustrated in Fig. 9, to maintain the hydrologic control of our landscapes and ensure food security for future generations. The important role of diverse cover crops provides opportunities for carbon management that will make the production systems more resilient and increase WUE. While concepts of SH are still evolving, it is clear that future research on fertilizer use in crop production and cropping systems will need to consider the physical, chemical and biological aspects of soil, in addition to the primary focus of crop yields. Sustainability depends on SH and land capacity to continue yielding both captured carbon in plant products (food) and increased WUE year after year. The ongoing cycling of C, N and other nutrient generation and regeneration of the porous soil architecture – the soil's 'regeneration capacity' – provides soil repair and physical resilience in the face of adverse weather shocks and/or of poor management. Maintaining the vitality of the soil, notably of the number, diversity and activity of the living components that depend on carbon, is a key factor. Cover crops maintain stable soil N and C concentrations, bacterial diversity and bacterial community composition, resulting in a more stable and healthy community, minimizing the feast or famine phenomenon.

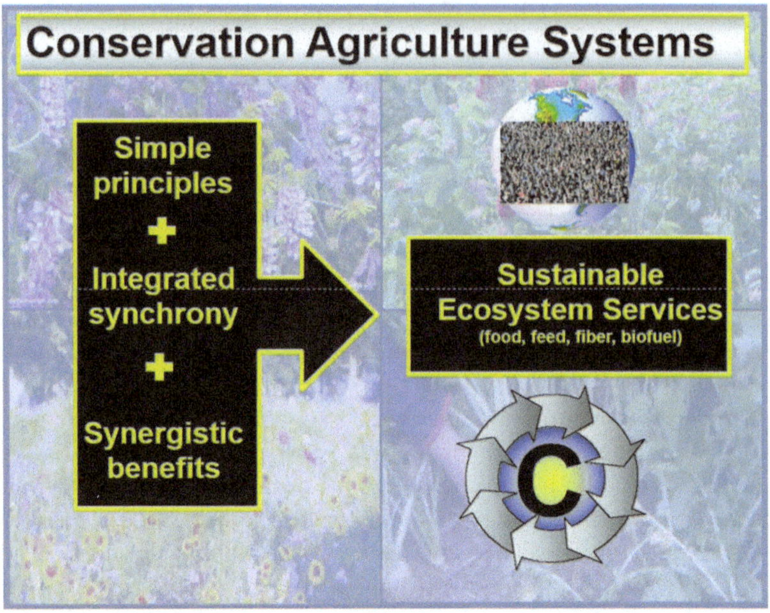

Figure 9 Schematic representation showing how Conservation Agriculture systems utilize simple principles, integration and synergistic benefits to provide carbon management and sustainable ecosystems.

> 'The synergistic simplicity of CAS with minimum soil disturbance (minimizes C and soil loss) and the use of diverse crop rotations and cover-crop mixes (maximizes soil coverage/protection and C input) for soil biodiversity protection and SH regeneration benefits the environment, the farmer and all of society through food security.'

Rapid adoption of innovative solutions for CA require a universally acceptable definition of CAS with clear communication and understanding before real progress can be made. CA provides an opportunity to create a legacy of healthy living soils and healthy farms at the watershed scale, forming the base of our future food security. Although the drive for healthy soils, healthy landscapes and healthy farms must come primarily from the farming community, this must be underpinned by science, supported by both rural and urban sectors, and supported by consumers in the broader society. There must be a strong partnership among all sectors to promote adoption and success of these approaches. Emphasis must be on integrated principles, a scientific foundation and the regeneration of economic and environmental value of CAS to our global society. We can create a world in which nature and people thrive. We owe it to future generations.

The author gratefully acknowledges the constructive criticisms and suggestions provided by two anonymous reviewers.

5 Where to look for further information

5.1 Suggested reading

Barrios, E., Delve, R. J., Bekunda, M., Mowo, J., Agunda, J., Ramisch, J., Trejo, M. T. and Thomas, R. J. 2006. Indicators of soil quality: a South-South development of a methodological guide for linking local and technical knowledge. *Geoderma* 135, 248–59.

Doran, J. W. 2002. Soil health and global sustainability: translating science into practice. *Agriculture, Ecosystems and Environment* 88(2), 119-27. doi:10.1016/S0167-8809(01)00246-8.

Karlen, D. L., Mausbach, M. J., Doran, J. W., Cline, R. G., Harris, R. F. and Schuman, G. E. 1997. Soil quality: a concept, definition, and framework for evaluation. *Soil Science Society of America Journal* 61(1), 4–10. doi:10.2136/sssaj1997.0361599500610 0010001x.

5.2 Recent books on soil health

Al-Kaisi, M. M. and Lowery, B. (Eds). 2017. *Soil Health and Intensification of Agroecosystems*. Academic Press, London, UK, 418pp.

Brevik, E. C. 2014. Soils and human health – an overview. In: Brevik, E. C. and Burgess, L. C. (Eds), *Soils and Human Health* (1st edn.). CRC Press, Boca Raton, FL, pp. 29-56. Chapter 2.

Duiker, S., Myers, J. C. and Blazure, L. C. 2016. *Soil Health in Field and Forage Crop Production*. USDA Natural Resources Conservation Service, Penn State University Extension, Capital Resource Conservation & Development and Clinton County Conservation District, Molly McDonough, USDA NRCS, Renova, PA, p. 27.

Reicosky, D. (Ed.) 2018. *Managing Soil Health for Sustainable Agriculture, Volume 1: Fundamentals*. Burleigh Dodds Science Publishing, Cambridge, UK, 352pp. ISBN: 978 1 78676 188 0. Available at: www.bdspublishing.com.

Reicosky, D. (Ed.) 2018. *Managing Soil Health for Sustainable Agriculture, Volume 2: Monitoring and Management*. Burleigh Dodds Science Publishing, Cambridge, UK, 464pp. ISBN: 978 1 78676 192 7. Available at: www.bdspublishing.com.

Stirling, G., Hayden, H., Pattison, T. and Stirling, M. 2016. *Soil Health, Soil Biology, Soil Borne Diseases and Sustainable Agriculture. A Guide*. CSIRO Publishing, Clayton South VIC, 280pp.

Wall, D. H., Bardgett, R. D., Behan-Pelletier, V., Herrick, J. E., Jones, T. H., Ritz, K., Six, J., Strong, D. R. and van der Putten, W. H. (Eds). 2012. *Soil Ecology and Ecosystem Services* (1st edn.). Oxford University Press, Oxford, UK, pp. 406.

6 References

Adhikari, K. and Hartemink, A. E. 2016. Linking soils to ecosystem services–a global review. *Geoderma* 262, 101–11. doi:10.1016/j.geoderma.2015.08.009.

Alban, D. H. and Berry, E. C. 1994. Effects of earthworm invasion on morphology, carbon, and nitrogen of a forest soil. *Applied Soil Ecology* 1(3), 243-9. doi:10.1016/0929-1393(94)90015-9.

Ali, M. H. and Talukder, M. S. U. 2008. Increasing water productivity in crop production–a synthesis. *Agricultural Water Management* 95(11), 1201–13. doi:10.1016/j.agwat.2008.06.008.

Al-Kaisi, M. M. and Lal, R. 2017. Conservation agriculture systems to mitigate climate variability effects on soil health. In: Al-Kaisi, M. M. and Lowery, B. (Eds), *Soil Health and Intensification of Agroecosystems*. Elsevier Academic Press, London, pp. 79–103.

Al-Kaisi, M. M., Lal, R., Olson, K. R. and Lowery, B. 2017. Fundamentals and functions of soil environment. In: Al-Kaisi, M. M. and Lowery, B. (Eds), *Soil Health and Intensification of Agroecosystems*. Elsevier Academic Press, London, pp. 1–20.

Allison, F. E. 1973. *Soil Organic Matter and Its Role in Crop Production*. Elsevier Scientific Publishing Company, Amsterdam, p. 637.

Anderson, R. L. 2005. Improving sustainability of cropping systems in the Central Great Plains. *Journal of Sustainable Agriculture* 26(1), 97–114. doi:10.1300/J064v26n01_08.

Angers, D. A. and Carter, M. R. 1996. Aggregation and organic matter storage in cool, humid agricultural soils. In: Carter, M. R. and Stewart, B. A. (Eds), *Structure and Organic Matter Storage in Agricultural Soils*. CRC Press, Boca Raton, FL, pp. 193–211.

Arnone III, J. A. and Zaller, J. G. 2014. Earthworm effects on native grassland root system dynamics under natural and increased rainfall. *Frontiers in Plant Science* 5, 152. doi:10.3389/fpls.2014.00152.

Asbjornsen, H., Hernandez-Santana, V., Liebman, M., Bayala, J., Chen, J., Helmers, M., Ong, C. K. and Schulte, L. A. 2014. Targeting perennial vegetation in agricultural landscapes for enhancing ecosystem services. *Renewable Agriculture and Food Systems* 29(2), 101–25. doi:10.1017/S1742170512000385.

Bailey, V. L., Smith, J. L. and Bolton, H. J. 2002. Fungal-to-bacterial ratios in soils investigated for enhanced carbon sequestration. *Soil Biology and Biochemistry* 34(7), 1385–9.

Baker, J. M., Ochsner, T. E., Venterea, R. T. and Griffis, T. J. 2007. Tillage and soil carbon sequestration-what do we really know? *Agriculture, Ecosystems and Environment* 118(1–4), 1–5. doi:10.1016/j.agee.2006.05.014.

Baldock, J. A. and Nelson, P. N. 1999. Soil organic matter. In: Sumner, M. E. (Ed.), *Handbook of Soil Science*. CRC Press, Boca Raton, FL, pp. B25–84.

Baldock, J. A. and Skjemstad, J. O. 1999. Soil organic carbon/soil organic matter. In: Peverill, K. I., Sparrow, L. A. and Reuter, D. J. (Ed.), *Soil Analysis: an Interpretation Manual*. CSIRO Publishing, Collingwood, Vic, pp. 159–70.

Balesdent, J. and Balabane, M. 1996. Major contribution of roots to soil carbon storage inferred from maize cultivated soils. *Soil Biology and Biochemistry* 28(9), 1261–3. doi:10.1016/0038-0717(96)00112-5.

Bardgett, R. D. and van der Putten, W. H. 2014. Belowground biodiversity and ecosystem functioning. *Nature* 515(7528), 505–11. doi:10.1038/nature13855.

Bardgett, R. D., Mommer, L. and De Vries, F. T. 2014. Going underground: root traits as drivers of ecosystem processes. *Trends in Ecology and Evolution* 29(12), 692–9. doi:10.1016/j.tree.2014.10.006.

Basche, A. D. 2017. *Turning Soils into Sponges: How Farmers Can Fight Floods and Droughts*. Union of Concerned Scientists, p. 18. Available at: https://www.ucsusa.org/sites/default/files/attach/2017/08/turning-soils-into-sponges-full-report-august-2017.pdf.

Basche, A. D. and DeLonge, M. 2017. The impact of continuous living cover on soil hydrologic properties: a meta-analysis. *Soil Science Society of America Journal* 81(5), 1179–90. doi:10.2136/sssaj2017.03.0077.

Basche, A. D. and Edelson, O. F. 2017. Improving water resilience with more perennially based agriculture. *Agroecology and Sustainable Food Systems* 41(7), 799–824. doi: 10.1080/21683565.2017.1330795.

Basche, A. D., Archontoulis, S. A., Kaspar, T. K., Jaynes, D. B., Parkin, T. B. and Miguez, F. E. 2016a. Simulating long-term impacts of cover crops and climate change on crop production and environmental outcomes in the Midwestern United States. *Agriculture, Ecosystems and Therapeutic Environment* 218, 95–106. doi:10.1016/j.agee.2015.11.011.

Basche, A. D., Kaspar, T. K., Archontoulis, S. A., Jaynes, D. B., Parkin, T. B., Sauer, T. S. and Miguez, F. E. 2016b. Soil water improvements with the long-term use of a cover crop. *Agricultural Water Management* 172, 40–50. doi:10.1016/j.agwat.2016.04.006.

Berry, E. C. and Karlen, D. L. 1993. Comparison of alternative farming systems. II. Earthworm population density and species diversity. *American Journal of Alternative Agriculture* 8(1), 21–6. doi:10.1017/S0889189300004872.

Berti, M., Johnson, B., Ripplinger, D., Gesch, R. and Aponte, A. 2017. Environmental impact assessment of double- and relay-cropping with winter Camelina in the northern Great Plains, USA. *Agricultural Systems* 156, 1–12. doi:10.1016/j.agsy.2017.05.012.

Bezdicek, D. F. 1996. Development and evaluation of indicators for agroecosystem health. *Agriculture in Concert with the Environment ACE Research Projects Western Region 1991–1995*, 6pp.

Bisht, J. K., Meena, V. S., Mishra, P. K. and Pattanayak, A. (Eds). 2016. *Conservation Agriculture: an Approach to Combat Climate Change in Indian Himalaya*. Springer, p. 559.

Blanco-Canqui, H. and Lal, R. 2008. No-tillage and soil-profile carbon sequestration: an on-farm assessment. *Soil Science Society of America Journal* 72(3), 693–701. doi:10.2136/sssaj2007.0233.

Blanco-Canqui, H., Mikha, M. M., Presley, D. R. and Claassen, M. M. 2011. Addition of cover crops enhances no-till potential for improving soil physical properties. *Soil Science Society of America Journal* 75(4), 1471–82. doi:10.2136/sssaj2010.0430.

Blanco-Canqui, H., Holman, J. D., Schlegel, A. J., Tatarko, J. and Shaver, T. M. 2013. Replacing fallow with cover crops in a semiarid soil: effects on soil properties. *Soil Science Society of America Journal* 77(3), 1026–34. doi:10.2136/sssaj2013.01.0006.

Blanco-Canqui, H., Shaver, T. M., Lindquist, J. L., Shapiro, C. A., Elmore, R. W., Francis, C. A. and Hergert, G. W. 2015. Cover crops and ecosystem services: insights from studies in temperate soils. *Agronomy Journal* 107(6), 2449–74. doi:10.2134/agronj15.0086.

Blesh, J. 2018. Functional traits in cover crop mixtures: biological nitrogen fixation and multifunctionality. *Journal of Applied Ecology* 55(1), 38–48. doi:10.1111/1365-2664.13011.

Blouin, M., Hodson, M. E., Delgado, E. A., Baker, G., Brussaard, L., Butt, K. R., Dai, J., Dendooven, L., Peres, G., Tondoh, J. E., et al. 2013. A review of earthworm impact on soil function and ecosystem services. *European Journal of Soil Science* 64, 161–82. doi:10.1111/ejss.12025.

Boddey, R. M., Jantalia, C. P., Alves, B. J. R. and Urquiaga, S. 2009. Comments on 'no-tillage and soil-profile carbon sequestration: an on-farm assessment'. *Soil Science Society of America Journal* 73(2), 688. doi:10.2136/sssaj2008.0278l.

Bolinder, M. A., Angers, D. A., Bélanger, G., Michaud, R. and Laverdière, M. R. 2002. Root biomass and shoot to root ratios of perennial forage crops in eastern Canada. *Canadian Journal of Plant Science* 82(4), 731–7. doi:10.4141/P01-139.

Bouma, J. 2015. Reaching out from the soil-box in pursuit of soil security. *Soil Science and Plant Nutrition* 61(4), 556–65. doi:10.1080/00380768.2015.1045403.

Brady, N. C. and Weil, R. R. 2004. *Elements of the Nature and Properties of Soils* (2nd edn.). Pearson Education, Incorporated, Upper Saddle River, NJ.

Branca, G., McCarthy, N., Lipper, L. and Jolejole, M. C. 2011. *Climate-Smart Agriculture: A Synthesis of Empirical Evidence of Food Security and Mitigation Benefits from Improved Cropland Management*. Mitigation of Climate Change in Agriculture Series Number 3. Food and Agriculture Organization of the United Nations.

Brevik, E. C. 2014. Soils and human health – an overview. In: Brevik, E. C. and Burgess, L. C. (Eds), *Soils and Human Health* (1st edn.). CRC Press, Boca Raton, FL, pp. 29–56. Chapter 2.

Brevik, E. C. and Sauer, T. J. 2015. The past, present, and future of soils and human health studies. *Soil* 1(1), 35–46. doi:10.5194/soil-1-35-2015.

Briones, M. J. I. and Schmidt, O. 2017. Conventional tillage decreases the abundance and biomass of earthworms and alters their community structure in a global meta-analysis. *Global Change Biology* 23(10), 4396–419. doi:10.1111/gcb.13744.

Brouder, S. M. and Gomez-Macpherson, H. 2014.The impact of conservation agriculture on smallholder agricultural yields: a scoping review of the evidence. *Agriculture, Ecosystems and Environment* 187, 11–32. doi:10.1016/j.agee.2013.08.010.

Brown, G. G., Benito, N. P., Pasini, A., Sautter, K. D., Defguimaraes, M. and Torres, E. 2003. No-tillage greatly increases earthworm populations in Parana State, Brazil. *Pedobiologia* 47, 764–71.

Brown, G. G., Edwards, C. A. and Brussaard, L. 2004. How earthworms affect plant growth: burrowing into the mechanisms. In: Edwards, C. A. (Ed.), *Earthworm Ecology*. CRC Press, Boca Raton, FL, pp. 13–49.

Brussaard, L. 1999. On the mechanisms of interactions between earthworms and plants. *Pedobiologia* 43, 880–5.

Brussaard, L. 2012. Ecosystem services provided by the soil biota. In: Wall, D. H., Bardgett, R. D., Behan-Pelletier, V., Herrick, J. E., Jones, T. H., Ritz, K., Six, J., Strong, D. R. and van der Putten, W. H. (Eds), *Soil Ecology and Ecosystem Services*. Oxford University Press, Oxford, UK, pp. 45–58.

Brussaard, L., Behan-Pelletier, V. M., Bignell, D. E., Brown, V. K., Didden, W., Folgarait, P., Fragoso, C., Freckman, D. W., Gupta, V. V. S. R., Hattori, T., et al. 1997. Biodiversity and ecosystem functioning in soil. *Ambio – A Journal of the Human Environment* 26, 563–70.

Burney, J. A., Davis, S. J. and Lobell, D. B. 2010. Greenhouse gas mitigation by agricultural intensification. *Proceedings of the National Academy of Sciences of the United States of America* 107(26), 12052–7. doi:10.1073/pnas.0914216107.

Burns, R. L., Cook, D. J. and Phillips, R. E. 1971. Influence of no tillage on soil moisture. *Agronomy Journal* 73, 593–6.

Cardoso, E. J. B. N., Vasconcellos, R. L. F., Bini, D., Miyauchi, M. Y. H., dos Santos, C. Ad, Alves, P. R. L., Paula, A. Md., Nakatani, A. S., Pereira, JdM. and Nogueira, M. A. 2013. Soil health: looking for suitable indicators. What should be considered to assess the effects of use and management on soil health? *Scientia Agricola* 70(4), 274–89. doi:10.1590/S0103-90162013000400009.

Carlson, S. and Stockwell, R. 2013. Research priorities for advancing adoption of cover crops in agriculture-intensive regions. *Journal of Agriculture, Food Systems, and Community Development* 3(4), 1–5. doi:10.5304/jafscd.2013.034.017.

Carter, M. R. and Stewart, B. A. 1996. *Structure and Organic Matter Storage in Agricultural Soils*. CRC Press, Boca Raton, FL.

Carter, L. J., Harris, E., Williams, M., Ryan, J. J., Kookana, R. S. and Boxall, A. B. A. 2014. Fate and uptake of pharmaceuticals in soil-plant systems. *Journal of Agricultural and Food Chemistry* 62(4), 816–25. doi:10.1021/jf404282y.

Chan, K. Y., Heenan, D. P., Oates, A. and Munro, K. 2003. Hydraulic properties of an alfisol with pre-existing pan under conservation tillage – effect of earthworms. In: *Proceedings of the 16th ISTRO Conference*, 13–18 July 2003, Brisbane, Australia, pp. 300–4.

Chatterjee, A., Cooper, K., Klaustermeier, A., Awale, R. and Cihacek, L. J. 2016. Does crop species diversity influence soil carbon and nitrogen pools? *Agronomy Journal* 108(1), 427–32. doi:10.2134/agronj2015.0316.

Chen, G. and Weil, R. R. 2010. Penetration of cover crop roots through compacted soils. *Plant and Soil* 331(1–2), 31–43. doi:10.1007/s11104-009-0223-7.

Clapperton, M. J. 1999. Tillage practices, and temperature and moisture interactions affect earthworm populations and species composition. *Pedobiologia* 43, 658–65.

Clapperton, M. J., Miller, J. J., Larney, F. J. and Wayne Lindwall, C. 1997. Earthworm populations as affected by long-term tillage practices in southern, Alberta, Canada. *Soil Biology and Biochemistry* 29(3–4), 631–3. doi:10.1016/S0038-0717(96)00189-7.

Congreves, K. A., Hayes, A., Verhallen, E. A. and Van Eerd, L. L. 2015. Long-term impact of tillage and crop rotation on soil health at four temperate agroecosystems. *Soil and Tillage Research* 152, 17–28. doi:10.1016/j.still.2015.03.012.

Constantin, J., Mary, B., Laurent, F., Aubrion, G., Fontaine, A., Kerveillant, P. and Beaudoin, N. 2010. Effects of catch crops, no till and reduced nitrogen fertilization on nitrogen leaching and balance in three long-term experiments. *Agriculture, Ecosystems and Environment* 135(4), 268–78. doi:10.1016/j.agee.2009.10.005.

Corsi, S., Friedrich, T., Kassam, A., Pisante, M. and Sà, J. de M. 2012. *Soil Organic Carbon Accumulation and Greenhouse Gas Emission Reductions from Conservation Agriculture: A Literature Review*. In: Corsi, S., Friedrich, T., Kassam, A., Pisante, M. and Sà, J. de M. (Eds), *Integrated Crop Management* (vol. 16). Plant Production and Protection Division, Food and Agriculture Organization of the United Nations, Rome, 89pp.

Culman, S. W., Snapp, S. S., Ollenburger, M., Basso, B. and DeHaan, L. R. 2013. Soil and water quality rapidly respond to the perennial grain Kernza wheatgrass. *Agronomy Journal* 105(3), 735–44. doi:10.2134/agronj2012.0273.

Dabney, S. M., Delgado, J. A. and Reeves, D. W. 2001. Using winter cover crops to improve soil and water quality. *Communications in Soil Science and Plant Analysis* 32(7–8), 1221–50. doi:10.1081/CSS-100104110.

Denef, K., Zotarelli, L., Boddey, R. M. and Six, J. 2007. Microaggregate-associated carbon as a diagnostic fraction for management-induced changes in soil organic carbon in two oxisols. *Soil Biology and Biochemistry* 39(5), 1165–72. doi:10.1016/j.soilbio.2006.12.024.

De Oliveira, G., Brunsell, N. A., Sutherlin, C. E., Crews, T. E. and De Haan, L. R. 2018. Energy, water and carbon exchange over a perennial Kernza wheatgrass crop. *Agricultural and Forest Meteorology* 249, 120–37. doi:10.1016/j.agrformet.2017.11.022.

Derpsch, R., Franzluebbers, A. J., Duiker, S. W., Reicosky, D. C., Koeller, K., Sturny, W. G., Sturny, W. G., Sá, J. C. M. and Weiss, K. 2014. Why do we need to standardize no-tillage research? *Soil and Tillage Research* 137, 16–22. doi:10.1016/j.still.2013.10.002.

De Vries, F. T., Hoffland, E., van Eekeren, N., Brussaard, L. and Bloem, J. 2006. Fungal/bacterial ratios in grasslands with contrasting nitrogen management. *Soil Biology and Biochemistry* 38(8), 2092–103. doi:10.1016/j.soilbio.2006.01.008.

Diamond, J. 2005. *Collapse: How Societies Choose to Fail or Succeed* (Revised edn.). Penguin Books, New York, NY.

Doran, J. W. 2002. Soil health and global sustainability: translating science into practice. *Agriculture, Ecosystems and Environment* 88(2), 119–27. doi:10.1016/S0167-8809(01)00246-8.

Doran, J. W. and Parkin, T. B. 1994. Defining and assessing soil quality. In: Doran, J. W., Coleman, D. C., Bezdicek, D. F. and Stewart, B. A. (Eds), *Defining Soil Quality for a Sustainable Environment*. SSSA Special Publication 35. Soil Science Society of America and American Society of Agronomy, Madison, WI, pp. 1–21. doi:10.2136/sssaspecpub35.c1.

Doran, J. W. and Safley, M. 1997. Defining and assessing soil health and sustainable productivity. In: Pankhurst, C. E., Doube, B. M. and Gupta, V. V. S. R. (Eds), *Biological Indicators of Soil Health*. CAB International, Wallingford, UK, pp. 1–28.

Doran, J. W. and Zeiss, M. R. 2000. Soil health and sustainability: managing the biotic component of soil quality. *Applied Soil Ecology* 15(1), 3–11. doi:10.1016/S0929-1393(00)00067-6.

Doran, J. W., Sarantonio, M. and Leibig, M. A. 1996. Soil health and sustainability. *Advances in Agronomy* 56, 1–54. doi:10.1016/S0065-2113(08)60178-9.

Drury, C. F., Tan, C. S., Welacky, T. W., Reynolds, W. D., Zhang, T. Q., Oloya, T. O., McLaughlin, N. B. and Gaynor, J. D. 2014. Reducing nitrate loss in tile drainage water with cover crops and water-table management systems. *Journal of Environmental Quality* 43(2), 587–98. doi:10.2134/jeq2012.0495.

Dudley, N. and Alexander, S. 2017. Agriculture and biodiversity: a review. *Biodiversity* 18(2–3), 45–9. doi:10.1080/14888386.2017.1351892.

Dumais, J. and Forterre, Y. 2012. Vegetable dynamics: the role of water in plant movements. *Annual Review of Fluid Mechanics* 44(1), 453–78. doi:10.1146/annurev-fluid-120710-101200.

Duxbury, J. M., Smith, M. S., Doran, J. W., Jordan, C., Szott, L. and Vance, E. 1989. Soil organic matter as a source and a sink of plant nutrients. In: Coleman, D. C., Oades, J. M. and Uehara, G. (Eds), *Dynamics of Soil Organic Matter in Tropical Ecosystems*. University of Hawaii Press, Honolulu, pp. 33–67.

Eagle, A. J., Henry, L. R., Olander, L. P., Haugen-Kozyra, K., Millar, N. and Robertson, G. P. 2012. *Greenhouse Gas Mitigation Potential of Agricultural Land Management in the United States: A Synthesis of the Literature*. Technical Working Group on Agricultural Greenhouse Gases (T-AGG) Report. Nicholas Institute for Environmental Policy Solutions, Duke University, Durham, NC.

Edwards, C. A. and Lofty, J. R. 1982. The effects of direct drilling and minimal cultivation on earthworm populations. *The Journal of Applied Ecology* 19(3), 723–34. doi:10.2307/2403277.

Edwards, W. M., Norton, L. D. and Redmond, C. E. 1988. Characterizing macropores that affect infiltration into non-tilled soil. *Soil Science Society of America Journal* 52(2), 483–7. doi:10.2136/sssaj1988.03615995005200020033x.

Edwards, W. M., Shipitalo, M. J., Owens, L. B. and Norton, L. D. 1990. Effect of *Lumbricus terrestris* L. burrows on hydrology of continuous no-till corn fields. *Geoderma* 46(1–3), 73–84. doi:10.1016/0016-7061(90)90008-W.

Ehlers, W. 1975. Observations on earthworm channels and infiltration on tilled and untilled loess soil. *Soil Science* 119(3), 242–9. doi:10.1097/00010694-197503000-00010.

Ellert, B. H. and Janzen, H. H. 1999. Short-term influence of tillage on CO_2 fluxes from a semi-arid soil on the Canadian Prairies. *Soil and Tillage Research* 50(1), 21–32. doi:10.1016/S0167-1987(98)00188-3.

Erenstein, O., Sayre, K., Wall, P., Hellin, J. and Dixon, J. 2012. Conservation Agriculture in maize- and wheat-based systems in the (sub)tropics: lessons from adaptation initiatives in South Asia, Mexico, and Southern Africa. *Journal of Sustainable Agriculture* 36(2), 180–206. doi:10.1080/10440046.2011.620230.

FAO. 2005. *The Importance of Soil Organic Matter: Key to Drought-Resistant Soils and Sustained Food and Production*. FAO Soils Bulletin 80. Food and Agriculture Organization, Rome, 95pp.

FAO. 2008a. Investing in sustainable crop intensification: the case for improving soil health. *Integrated Crop Management* (vol. 6). Food and Agriculture Organization, Rome, 149pp. Available at: http://www.fao.org/ag/ca/doc/WORKSHOP-LR.pdf.

FAO. 2008b. *Conservation Agriculture*. 08 July 2008. Food and Agriculture Organization. Available at: http://www.fao.org/ag/ca/index.html.

FAO. 2011. *Save and Grow: A Policymaker's Guide to the Sustainable Intensification of Smallholder Crop Production*. Food and Agriculture Organization, Rome, Italy, pp. 1–37.

Farooq, M. and Siddique, K. H. M. 2015. Conservation Agriculture: Concepts, brief history, and impacts on agricultural systems. In: Farooq, M. and Siddique, K. H. M. (Eds), *Conservation Agriculture*. Springer International Publishing, Switzerland, pp. 3–17. Chapter 1. doi:10.1007/978-3-319-11620-4_1.

Ferchaud, F. and Mary, B. 2016. Drainage and nitrate leaching assessed during 7 years under perennial and annual bioenergy crops. *BioEnergy Research* 9(2), 656–70. doi:10.1007/s12155-015-9710-2.

Finney, D. M. and Kaye, J. P. 2017. Functional diversity in cover crop polycultures increases multifunctionality of an agricultural system. *Journal of Applied Ecology* 54(2), 509–17. doi:10.1111/1365-2664.12765.

Finney, D. M., Buyer, J. S. and Kaye, J. P. 2017. Living cover crops have immediate impact on soil microbial community structure and function. *Journal of Soil and Water Conservation* 72(4), 361–73. doi:10.2489/jswc.72.4.361.

Foley, J. A., Ramankutty, N., Brauman, K. A., Cassidy, E. S., Gerber, J. S., Johnston, M., Mueller, N. D., O'Connell, C., Ray, D. K., West, P. C., et al. 2011. Solutions for a cultivated planet. *Nature* 478(7369), 337–42. doi:10.1038/nature10452.

Folorunso, O. A., Rolston, D. E., Prichard, T. and Loui, D. T. 1992. Soil surface strength and infiltration rate as affected by winter cover crops. *Soil Technology* 5(3), 189–97. doi:10.1016/0933-3630(92)90021-R.

Franzluebbers, A. J. 2005. Soil organic carbon sequestration and agricultural greenhouse gas emissions in the southeastern USA. *Soil and Tillage Research* 83(1), 120–47. doi:10.1016/j.still.2005.02.012.

Franzluebbers, A. J. 2009. Comment on 'no tillage and soil-profile carbon sequestration: an on-farm assessment'. *Soil Science Society of America Journal* 73(2), 686–7. doi:10.2136/sssaj2008.0275l.

Frey, S. D., Elliott, E. T. and Paustian, K. 1999. Bacterial and fungal abundance and biomass in conventional and no-tillage agroecosystems along to climatic gradients. *Soil Biology and Biochemistry* 31(4), 573–85. doi:10.1016/S0038-0717(98)00161-8.

Friedrich, T., Derpsch, R. and Kassam, A. H. 2012. Overview of the global spread of Conservation Agriculture. Field Actions Science Reports, Special Issue 6, pp. 1–7.

Gattinger, A., Jawtusch, J., Muller, A. 2011. No-till agriculture–a climate smart solution? Climate Change and Agriculture Report No. 2. MISEREOR e.V., Aachen, Germany.

Gesch, R. W., Archer, D. W. and Berti, M. T. 2014. Dual cropping winter Camelina with soybean in the Northern Corn Belt. *Agronomy Journal* 106(5), 1735–45. doi:10.2134/agronj14.0215. Available at: http://handle.nal.usda.gov/10113/60081.

Gesch, R. W., Dose, H. L. and Forcella, F. 2017. Camelina growth and yield response to sowing depth and rate in the northern Corn Belt USA. *Industrial Crops and Products* 95, 416–21. doi:10.1016/j.indcrop.2016.10.051.

Glover, J. D., Cox, C. M. and Reganold, J. P. 2007. Future farming: a return to roots? *Scientific American* 297(2), 82–9. doi:10.1038/scientificamerican0807-82.

Glover, J. D., Reganold, J. P., Bell, L. W., Borevitz, J., Brummer, E. C., Buckler, E. S., Cox, C. M., Cox, T. S., Crews, T. E., Culman, S. W., et al. 2010. Increased food and ecosystem security via perennial grains. *Science* 328(5986), 1638–9. doi:10.1126/science.1188761.

Gould, I. J., Quinton, J. N., Weigelt, A., De Deyn, G. B. and Bardgett, R. D. 2016. Plant diversity and root traits benefit physical properties key to soil function in grasslands. *Ecology Letters*, 2016; doi:10.1111/ele.12652.

Granatstein, D. and Bezdicek, D. F. 1992. The need for a soil quality index: local and regional perspectives. *American Journal of Alternative Agriculture* 7(1–2), 12–6. doi:10.1017/S0889189300004380.

Greiner, L., Keller, A., Grêt-Regamey, A. and Papritz, A. 2017. Soil function assessment: review of methods for quantifying the contributions of soils to ecosystem services. *Land Use Policy* 69, 224–37. doi:10.1016/j.landusepol.2017.06.025.

Gulick, S. H., Grimes, D. W., Goldhamer, D. A. and Munk, D. S. 1994. Cover-crop-enhanced water infiltration of a slowly permeable fine sandy loam. *Soil Science Society of America Journal* 58(5), 1539–46. doi:10.2136/sssaj1994.03615995005800050038x.

Haddix, M. L., Plante, A. F., Conant, R. T., Six, J., Steinweg, J. M., Magrini-Bair, K., Drijber, R. A., Morris, S. J. and Paul, E. A. 2011. The role of soil characteristics on temperature sensitivity of soil organic matter. *Soil Science Society of America Journal* 75(1), 56–68. doi:10.2136/sssaj2010.0118.

Haruna, S. I., Anderson, S. H., Nkongolo, N., Reinbott, T., Zaibon, S. 2017. Soil Thermal Properties Influenced by Perennial Biofuel and Cover Crop Management. *Soil Science Society of America Journal* 81: 1147–56. doi:10.2136/sssaj2016.10.0345.

Hatfield, J. L. 2014. Soil degradation, land use, and sustainability. In: Songstad, D. D., Hatfield, J. L. and Tomes, D. T. (Eds), *Convergence of Food Security, Energy Security and Sustainable Agriculture*. Biotechnology in Agriculture and Forestry (vol. 67). Springer, New York, NY, pp. 61–74. doi:10.1007/978-3-642-55262-5_4.

Hatfield, J. L., Sauer, T. J. and Prueger, J. H. 2001. Managing soils to achieve greater water use efficiency: a review. *Agronomy Journal* 93, 271–80.

Hatfield, J. L., Boote, K. J., Kimball, B. A., Ziska, L. H., Izaurralde, R. C., Ort, D., Thomson, A. M. and Wolfe, D. 2011. Climate impacts on agriculture: implications for crop production. *Agronomy Journal* 103(2), 351–70. doi:10.2134/agronj2010.0303.

Hatfield, J. L., Sauer, T. J. and Cruse, R. M. 2017. Soil: the forgotten piece of the water, food, energy nexus. In: Sparks, D. L. (Ed.), *Advances in Agronomy* (vol. 143). Elsevier Academic Press, Burlington, pp. 1–46. ISBN: 978-0-12-812421-5.

Haynes, R. J. and Naidu, R. 1998. Influence of lime, fertilizer and manure applications on soil organic matter content and soil physical conditions: a review. *Nutrient Cycling in Agroecosystems* 51(2), 123–37. doi:10.1023/A:1009738307837.

Heggenstaller, A. H., Menalled, F. D., Liebman, M. and Westerman, P. R. 2006. Seasonal patterns in post-dispersal seed predation of *Abutilon theophrasti* and *Setaria faberi* in three cropping systems. *Journal of Applied Ecology* 43(5), 999–1010. doi:10.1111/j.1365-2664.2006.01198.x.

Hillel, D. 1998. *Environmental Soil Physics: Fundamentals, Applications, and Environmental Considerations*. Academic Press, San Diego, CA, p. 771.

Hird, V. 2017. Farming systems and techniques that promote biodiversity. *Biodiversity* 18(2–3), 71–4. doi:10.1080/14888386.2017.1351395.

Hobbs, P. R. 2007. Conservation agriculture: what is it and why is it important for future sustainable food production? *The Journal of Agricultural Science* 145(2), 127–37. doi:10.1017/S0021859607006892.

Hobbs, P. R. and Govaerts, B. 2010. How Conservation Agriculture can contribute to buffering climate change. In: Reynolds, M. P. (Ed.), *Climate Change and Crop Production*. CABI, Wallingford, Oxfordshire, UK. International Maize and Wheat Improvement Center (CIMMYT), pp. 177–99. Chapter 10.

Hobbs, P. R., Sayre, K. and Gupta, R. 2008. The role of conservation agriculture in sustainable agriculture. *Philosophical Transactions of the Royal Society of London. Series B, Biological Sciences* 363(1491), 543–55. doi:10.1098/rstb.2007.2169.

Horne, D. J. and Sojka, R. E. 2002. Aeration, tillage effects on. In: Lal, R. (Ed.), *Encyclopedia of Soil Science*. Marcel Dekker, Inc., New York, NY, pp. 30–3.

Howard, A. 1947. *The Soil and Health: A Study of Organic Agriculture*. University Press of Kentucky, Lexington, KY.

Hudson, B. D. 1994. Soil organic matter and available water capacity. *Journal of Soil and Water Conservation* 49(2), 189–94.

Ingram, J. C., Wilkie, D., Clements, T., McNab, R. B., Nelson, F., Baur, E. H., Sachedina, H. T., Peterson, D. D. and Foley, C. A. H. 2014. Evidence of Payments for Ecosystem Services as a mechanism for supporting biodiversity conservation and rural livelihoods. *Ecosystem Services* 7, 10–21. doi:10.1016/j.ecoser.2013.12.003.

Jackson, L. E., Calderon, F. J., Steenwerth, K. L., Scow, K. M. and Rolston, D. E. 2003. Responses of soil microbial processes and community structure to tillage events and implications for soil quality. *Geoderma* 114(3–4), 305–17. doi:10.1016/S0016-7061(03)00046-6.

Janzen, H. H. 2015. Beyond carbon sequestration: soil as conduit of solar energy. *European Journal of Soil Science* 66(1), 19–32. doi:10.1111/ejss.12194. Available at: http://onlinelibrary.wiley.com/doi/10.1111/ejss.12194/full.

Jat, R. A., Sahrawat, K. L., Kassam, A. H. and Friedrich, T. 2013. Conservation Agriculture for sustainable and resilient agriculture: global status, prospects and challenges. In: Jat, R. A., Sahrawat, K. L. and Kassam, A. H. (Eds), *Conservation Agriculture: Global Status, Prospects and Challenges*. CABI, Boston, MA, pp. 1–25.

Jat, M. L., Stirling, C. M., Jat, H. S., Tetarwal, J. P., Jat, R. K., Singh, R. and Shirsath, P. B. 2017. Soil processes and wheat cropping under emerging climate change scenarios in South Asia. *Advances in Agronomy* 148, 111–71.

Johnson, G. A., Kantar, M. B., Betts, K. J. and Wyse, D. L. 2015. Field pennycress production and weed control in a double crop system with soybean in Minnesota. *Agronomy Journal* 107(2), 532–40. doi:10.2134/agronj14.0292.

Jordan, N. and Warner, K. D. 2010. Enhancing the multifunctionality of US agriculture. *BioScience* 60(1), 60–6. doi:10.1525/bio.2010.60.1.10.

Jordan, D., Hubbard, V. C., Ponder Jr., F. and Berry, E. C. 2000. The influence of soil compaction and the removal of organic matter on two native earthworms and soil properties in an oak-hickory forest. *Biology and Fertility of Soils* 31(3–4), 323–8. doi:10.1007/s003740050663.

Kang, G. S., Beri, V., Sidhu, B. S. and Rupela, O. P. 2005. A new index to assess soil quality and sustainability of wheat-based cropping systems. *Biology and Fertility of Soils* 41(6), 389–98. doi:10.1007/s00374-005-0857-4.

Karlen, D. L. 2012. Soil health: the concept, its role, and strategies for monitoring. In: Wall, D., Bardgett, R. D., Behan-Pelletier, V., Herrick, J. E., Jones, T. H., Ritz, K., Six, J., Strong, D. R. and van der Putten, W. H. (Eds), *Soil Ecology and Ecological Systems*. Oxford University Press, New York, NY, pp. 331–6. Chapter 5.3.

Karlen, D. L., Goeser, N. J., Veum, K. S. and Yost, M. A. 2017. On-farm soil health evaluations: challenges and opportunities. *Journal of Soil and Water Conservation* 72(2), 26A–31A. doi:10.2489/jswc.72.2.26A.

Kassam, A. and Friedrich, T. 2012. An ecologically sustainable approach to agricultural production intensification: global perspectives and developments. Field Actions Science Reports, Reconciling Poverty Eradication and Protection of the Environment, Special Issue 6. Institut Veolia Environnement, France. Available at: http://factsreports.revues.org/1382.

Kassam, A., Friedrich, T., Shaxson, F. and Pretty, J. 2009. The spread of Conservation Agriculture: justification, sustainability and uptake. *International Journal of Agricultural Sustainability* 7(4), 292–320. doi:10.3763/ijas.2009.0477.

Kassam, A., Mello, I., Bartz, H., Goddard, T., Friedrich, T., Laurent, F. and Uphoff, N. T. 2012. Harnessing ecosystem services in Brazil and Canada. Abstract presented at the *Planet under Pressure Conference*, London, 26–29 March 2012.

Kassam, A., Basch, G., Friedrich, T., Shaxson, F., Goddard, T., Amado, T., Crabtree, B., Hongwen, L., Mello, I., Pisante, M., et al. 2013. Sustainable soil management is more than what and how crops are grown. In: Lal, R. and Stewart, R. A. (Eds), *Principles of Soil Management in Agro-ecosystems*. Advances in Soil Science. CRC Press, Boca Raton, FL, pp. 337–400.

Kassam, A., Derpsch, R. and Friedrich, T. 2014. Global achievements in soil and water conservation: the case of Conservation Agriculture. *International Soil and Water Conservation Research* 2(1), 5–13. doi:10.1016/S2095-6339(15)30009-5.

Kassam, A., Friedrich, T., Derpsch, R. and Kienzle, J. 2015. Overview of the worldwide spread of Conservation Agriculture. Field Actions Science Reports, vol. 8 | 2015, Online since 26 September 2015, connection on 01 December 2017. Available at: http://factsreports.revues.org/3966.

Kassam, A., Friedrich, T. and Derpsch, R. 2019. Global spread of Conservation Agriculture. *International Journal of Environmental Studies* 76(1), 29–51, doi:10.1080/00207233.2018.1494927.

Kätterer, T., Bolinder, M. A., Andrén, O., Kirchmann, H. and Menichetti, L. 2011. Roots contribute more to refractory soil organic matter than aboveground crop residues, as revealed by a long-term field experiment. *Agriculture, Ecosystems and Environment* 141(1–2), 184–92. doi:10.1016/j.agee.2011.02.029.

Kaye, J. P. and Quemada, M. 2017. Using cover crops to mitigate and adapt to climate change. A review. *Agronomy for Sustainable Development* 2017, 37(1). doi:10.1007/s13593-016-0410-X.

Kell, D. B. 2011. Breeding crop plants with deep roots: their role in sustainable carbon, nutrient and water sequestration. *Annals of Botany* 108(3), 407–18. doi:10.1093/aob/mcr175.

Kemper, K. J. and Lal, R. 2017. Pay dirt! human health depends on soil health. *Complementary Therapies in Medicine* 32, A1–2. doi:10.1016/j.ctim.2017.04.005.

Kemper, W. D., Schneider, N. N. and Sinclair, T. R. 2011. No-till can increase earthworm populations and rooting depths. *Journal of Soil and Water Conservation* 66(1), 13A–7A. doi:10.2489/jswc.66.1.13A.

Khaleel, R., Reddy, K. R. and Overcash, M. R. 1981. Changes in soil physical properties due to organic waste applications: a review. *Journal of Environment Quality* 10(2), 133–41. doi:10.2134/jeq1981.00472425001000020002x.

Kibblewhite, M. G., Ritz, K. and Swift, M. J. 2008. Soil health in agricultural systems. *Philosophical Transactions of the Royal Society of London. Series B, Biological Sciences* 363(1492), 685–701: doi:10.1098/rstb.2007.2178.

King, A. E. and Blesh, J. 2018. Crop rotations for increased soil carbon: perenniality as a guiding principle. *Ecological Applications: a Publication of the Ecological Society of America* 28(1), 249–61. doi:10.1002/eap.1648.

Kladivko, E. J. 2001. Tillage systems and soil ecology. *Soil and Tillage Research* 61(1-2), 61–76. doi:10.1016/S0167-1987(01)00179-9.

Kladivko, E. J., Akhouri, N. M. and Weesies, G. 1997. Earthworm populations and species distributions under no-till and conventional tillage in Indiana and Illinois. *Soil Biology and Biochemistry* 29(3–4), 613–5. doi:10.1016/S0038-0717(96)00187-3.

Klocke, N. L., Currie, R. S. and Aiken, R. M. 2009. Soil water evaporation and crop residues. *Transactions of the ASABE* 52(1), 103–10. doi:10.13031/2013.25951.

Kong, A. Y. Y. and Six, J. 2010. Tracing root versus residue carbon into soils from conventional and alternative cropping systems. *Soil Science Society of America Journal* 74, 1201–10.

Krull, E. S., Skjemstad, J. O. and Baldock, J. A. 2004. Functions of soil organic matter and the effects on soil properties. GRDC Project No. CSO 00029. Available at: http://grdc.com.au/uploads/documents/cso00029.

Kuhn, N. J., Hu, Yaxian, Bloemertz, L., He, J., Li, H. and Greenwood, P. 2016. Conservation tillage and sustainable intensification of agriculture: regional vs. global benefit analysis. *Agriculture, Ecosystems and Environment* 216, 155–65. doi:10.1016/j.agee.2015.10.001.

Kuzyakov, Y. 2010. Priming effects: interactions between living and dead organic matter. *Soil Biology and Biochemistry* 42(9), 1363–71. doi:10.1016/j.soilbio.2010.04.003.

Kuzyakov, Y. and Cheng, W. 2004. Photosynthesis controls of CO_2 efflux from maize rhizosphere. *Plant and Soil* 263(1), 85–99. doi:10.1023/B:PLSO.0000047728.61591.fd.

Laishram, J., Saxena, K. G., Maikhuri, R. K. and Rao, K. S. 2012. Soil quality and soil health: a review. *International Journal of Ecology and Environmental Sciences* 38(1), 19–37.

Lal, R. 1991. Soil conservation and biodiversity. In: Hawksworth, D. L. (Ed.), *The Biodiversity of Microorganisms and Invertebrates: Its Role in Sustainable Agriculture*. CAB International, Wallingford, UK, pp. 89-103.

Lal, R. 2013. Intensive agriculture and the soil carbon pool. In: Kang, M. S. and Banga, S. S. (Eds), *Combating Climate Change: an Agricultural Perspective*. Taylor and Francis, Boca Raton, FL, pp. 59–72.

Lal, R. 2014. Climate strategic soil management. *Challenges* 5(1), 43–74. doi:10.3390/challe5010043.

Lal, R. 2015a. Sequestering carbon and increasing productivity by conservation agriculture. *Journal of Soil and Water Conservation* 70(3), 55A–62A. doi:10.2489/jswc.70.3.55A.

Lal, R. 2015b. A system approach to conservation agriculture. *Journal of Soil and Water Conservation* 70(4), 82A–8A. doi:10.2489/jswc.70.4.82A.

Lal, R. 2016. Soil health and carbon management. *Food and Energy Security* 5(4), 212–22. doi:10.1002/fes3.96.

Lal, R. 2018. Intensification of China's agroecosystems by conservation agriculture. *International Soil and Water Conservation Research* 6(1), 1–12. doi:10.1016/j.iswcr.2017.11.001.

Lal, R., Reicosky, D. C. and Hanson, J. D. 2007. Evolution of the plow over 10,000 years and the rationale for no-till farming. *Soil and Tillage Research* 93(1), 1–12. doi:10.1016/j.still.2006.11.004.

Lal, R., Mohtar, R. H., Assi, A. T., Ray, R., Baybil, H. and Jahn, M. 2017. Soil as a basic nexus tool: soils at the center of the food-energy-water nexus. *Current Sustainable/Renewable Energy Reports* 4(3), 117–29. doi:10.1007/s40518-017-0082-4.

Lange, M., Eisenhauer, N., Sierra, C. A., Bessler, H., Engels, C., Griffiths, R. I., Mellado-Va´zquez, P. G., Malik, A. A., Roy, J., Scheu, S., et al. 2015. Plant diversity increases soil microbial activity and soil carbon storage. *Nature Communications* 6, 6707. doi:10.1038/ncomms7707.

Larson, W. E. and Pierce, F. J. 1994. The dynamics of soil quality as a measure of sustainable management. In: Doran, J. W., Coleman, D. C., Bezdicek, D. F. and Stewart, B. A. (Eds), *Defining Soil Quality for a Sustainable Environment*. SSSA Special Publication 35. Soil Science Society of America, Madison, WI, pp. 37–51.

Lee, K. E. 1995. Earthworms and sustainable land use. In: Hendrix, P. F. (Ed.), *Earthworm Ecology and Biogeography in North America*. Lewis Publishers, Boca Raton, FL, pp. 215–34.

Lehman, R. M., Cambardella, C. A., Stott, D. E., Acosta-Martinez, V., Manter, D. K., Buyer, J. S., Maul, J., Smith, J., Collins, H., Halvorson, J., Kremer, R. J., et al. 2015a. Understanding and enhancing soil biological health: the solution for reversing soil degradation. *Sustainability* 7(1), 988–1027. doi:10.3390/su7010988.

Lehman, R. M., Acosta-Martinez, V., Buyer, J. S., Cambardella, C. A., Collins, H. P., Ducey, T. F., Halvorson, J. J., Jin, V. L., Johnson, J. M. F., Kremer, R. J., et al. 2015b. Soil biology for resilient, healthy soil. *Journal of Soil and Water Conservation* 70(1), 12A–8A. doi:10.2489/jswc.70.1.12A.

Lehmann, J. and Kleber, M. 2015. The contentious nature of soil organic matter. *Nature* 528(7580), 60–8. doi:10.1038/nature16069.

Lemordant, L., Gentine, P., Swann, A. S., Cook, B. I. and Scheff, J. 2018. Critical impact of vegetation physiology on the continental hydrologic cycle in response to increasing CO_2. *Proceedings of the National Academy of Sciences of the United States of America* 115(16), 4093–8. doi:10.1073/pnas.1720712115. 29610293.

Liebig, M. A., Archer, D. W. and Tanaka, D. L. 2014. Crop diversity effects on near-surface soil condition under dryland agriculture. *Applied and Environmental Soil Science* 2014, Article ID 703460, 7pp. doi:10.1155/2014/703460.

Lin, H. S. (Ed.). 2012. *Hydropedology: Synergistic Integration of Soil Science and Hydrology*. Elsevier, Waltham, MA.

Lobb, D. A., Kachanoski, R. G. and Miller, M. H. 1995. Tillage translocation and tillage erosion on shoulder slope landscape positions measured using cesium-137 as a tracer. *Canadian Journal of Soil Science* 75(2), 211–8. doi:10.4141/cjss95-029.

Luo, Z., Wang, E. and Sun, O. J. 2010. Can no-tillage stimulate carbon sequestration in agricultural soils? A meta-analysis of paired experiments. *Agriculture, Ecosystems and Environment* 139(1-2), 224-31. doi:10.1016/j.agee.2010.08.006.

Luo, Z., Feng, W., Luo, Yiqi, Baldock, J. and Wang, E. 2017. Soil organic carbon dynamics jointly controlled by climate, carbon inputs, soil properties and soil carbon fractions. *Global Change Biology* 23(10), 4430-9. doi:10.1111/gcb.13767.

Magdoff, F. 2001. Concept, components, and strategies of soil health in agroecosystems. *Journal of Nematology* 33(4), 169-72.

Matson, P. A., Parton, W. J., Power, A. G. and Swift, M. J. 1997. Agricultural intensification and ecosystem properties. *Science* 277(5325), 504-9. doi:10.1126/science.277.5325.504.

Mehra, P., Baker, J., Sojka, R. E., Bolan, N., Desbiolles, J., Kirkham, M. B., Ross, C. and Gupta, R. 2018. A review of tillage practices and their potential to impact the soil carbon dynamics. *Advances in Agronomy*, 185-230. doi:10.1016/bs.agron.2018.03.002.

Minasny, B. and McBratney, A. B. 2018. Limited effect of organic matter on soil available water capacity. *European Journal of Soil Science* 69(1), 39-47. doi:10.1111/ejss.12475.

Mitchell, J. P., Shrestha, A. and Irmak, S. 2015. Trade-offs between winter cover crop production and soil water depletion in the San Joaquin Valley, California. *Journal of Soil and Water Conservation* 70(6), 430-40. doi:10.2489/jswc.70.6.430.

Mitchell, J. P., Shrestha, A., Mathesius, K., Scow, K. M., Southard, R. J., Haney, R. L., Schmidt, R., Munk, D. S. and Horwath, W. R. 2017. Cover cropping and no-tillage improve soil health in an arid irrigated cropping system in California's San Joaquin Valley, USA. *Soil and Tillage Research* 165, 325-35. doi:10.1016/j.still.2016.09.001.

Moebius-Clune, B. N., Moebius-Clune, D. J., Gugino, B. K., Idowu, O. J., Schindelbeck, R. R., van Es, H. M., Thies, J. E., Shayler, H. A., McBride, M. B., Wolfe, D. W., et al. 2016. *Comprehensive Assessment of Soil Health - the Cornell Framework Manual* (3.1 edn.). Cornell University, Geneva.

Mokany, K., Raison, R. J. and Prokushkin, A. S. 2006. Critical analysis of root: shoot ratios in terrestrial biomes. *Global Change Biology* 12, 84-96.

Montanarella, L., Pennock, D. J., McKenzie, N., Badraoui, M., Chude, V., Baptista, I., Mamo, T., Yemefack, M., Singh Aulakh, M., Yagi, K., et al. 2016. World's soils are under threat. *Soil* 2(1), 79-82. doi:10.5194/soil-2-79-2016.

Montgomery, D. R. 2007a. *Dirt: the Erosion of Civilizations*. University of California Press, Berkeley, CA.

Montgomery, D. R. 2007b. Soil erosion and agricultural sustainability. *Proceedings of the National Academy of Sciences of the United States of America* 104(13), 13268-72.

Morrow, J. G., Huggins, D. R., Carpenter-Boggs, L. A. and Reganold, J. P. 2016. Evaluating measures to assess soil health in long-term agroecosystem trials. *Soil Science Society of America Journal* 80(2), 450-62.

Morrow, J. G., Huggins, D. R. and Reganold, J. P. 2017. Climate change predicted to negatively influence surface soil organic matter of dryland cropping systems in the inland Pacific Northwest, USA. *Frontiers in Ecology and Evolution* 5. doi:10.3389/fevo.2017.00010.

Murphy, B. W. 2014. *Soil Organic Matter and Soil Function - Review of the Literature and Underlying Data*. Department of the Environment, Canberra, Australia.

Murphy, B. W. 2015. Impact of soil organic matter on soil properties–a review with emphasis on Australian soils. *Soil Research* 53(6), 605-35. doi:10.1071/SR14246.

Nannipieri, P., Ascher, J., Ceccherini, M. T., Landi, L., Pietramellara, G. and Renella, G. 2003. Microbial diversity and soil functions. *European Journal of Soil Science* 54(4), 655–70. doi:10.1046/j.1351-0754.2003.0556.x.

Nelson, D. W. and Sommers, L. E. 1996. Total carbon, total organic carbon and organic matter. In: Sparks, D. L. (Ed.), *Methods of Soil Analysis Part 3: Chemical Methods*. Soil Science Society of America, Madison, WI, pp. 961–1010.

Nichols, K. A. and Wright, S. F. 2004. Contributions of fungi to soil organic matter in agroecosystems. In: Magdoff, F. and Weil, R. (Eds), *Soil Organic Matter in Sustainable Agriculture*. CRC Press LLC, Boca Raton, FL, pp. 179–98. Chapter 6.

NRCS. 2018. Natural resources conservation services: soil health. Available at: https://www.nrcs.usda.gov/wps/portal/nrcs/main/soils/health/(accessed on 28 September 2018).

Nuutinen, V., Butt, K. R. and Jauhiainen, L. 2011. Field margins and management affect settlement and spread of an introduced dew-worm (*Lumbricus terrestris* L.) population. *Pedobiologia* 54, S167–72. doi:10.1016/j.pedobi.2011.07.010.

Oades, J. M. 1984. Soil organic matter and water-stable aggregates in soils. *Plant and Soil* 76(1–3), 319–37. doi:10.1007/BF02205590.

Oliver, M. A. and Gregory, P. J. 2015. Soil, food security and human health: a review. *European Journal of Soil Science* 66(2), 257–76. doi:10.1111/ejss.12216.

Olson, K., Ebelhar, S. A. and Lang, J. M. 2014. Long-term effects of cover crops on crop yields, soil organic carbon stocks and sequestration. *Open Journal of Soil Science* 4, 284–92.

Ostergard, H., Finckh, M. R., Fontaine, L., Goldringer, I., Hoad, S. P., Kristensen, K., Lammerts van Bueren, E. T., Mascher, F., Munck, L. and Wolfe, M. S. 2009. Time for a shift in crop production: embracing complexity through diversity at all levels. *Journal of the Science of Food and Agriculture* 89(9), 1439–45. doi:10.1002/jsfa.3615.

Palm, C., Blanco-Canqui, H., DeClerck, F., Gatere, L. and Grace, P. 2014. Conservation agriculture and ecosystem services: an overview. *Agriculture, Ecosystems and Environment* 187, 87–105. doi:10.1016/j.agee.2013.10.010.

Papendick, R. I., Lindstrom, M. J. and Cochran, V. L. 1973. Soil mulch effect on seedbed temperature and water during fallow in eastern Washington. *Soil Science Society of America Journal* 37(2), 307–14. doi:10.2136/sssaj1973.03615995003700020039x.

Paustian, K., Andren, O., Janzen, H., Lal, R., Smith, P., Tian, G., Tiessen, H., van Noordwijk, M. and Woomer, P. 1997. Agricultural soil as a C sink to offset CO_2 emissions. *Soil Use and Management* 13, 230–44.

Paustian, K., Lehmann, J., Ogle, S., Reay, D., Robertson, G. P. and Smith, P. 2016. Climate-smart soils. *Nature* 532, 49–57. doi:10.1038/nature17174.

Pepper, I. L. 2013. The soil health-human health nexus. *Critical Reviews in Environmental Science and Technology* 43(24), 2617–52. doi:10.1080/10643389.2012.694330.

Peters, D. B. 1960. Relative magnitude of evaporation and transpiration. *Agronomy Journal* 52(9), 536–8. doi:10.2134/agronj1960.00021962005200090015x.

Philibert, A., Loyce, C. and Makowski, D. 2012. Assessment of the quality of meta-analysis in agronomy. *Agriculture, Ecosystems and Environment* 148(15), 72–82. doi:10.1016/j.agee.2011.12.003.

Pimentel, D., Harvey, C., Resosudarmo, P., Sinclair, K., Kurz, D., McNair, M., Crist, S., Shpritz, L., Fitton, L., Saffouri, R., et al. 1995. Environmental and economic cost of soil erosion and conservation benefits. *Science* 267(5201), 1117–23.

Pimentel, D., Cerasale, D., Stanley, R. C., Perlman, R., Newman, E. M., Brent, L. C., Mullan, A. and Chang, D. T.-I. 2012. Annual vs. perennial grain production. *Agriculture, Ecosystems and Environment* 161, 1–9. doi:10.1016/j.agee.2012.05.025.

Pittelkow, C. M., Liang, X., Linquist, B. A., Van Groeningen, K. J., Lee, J. W., Lundy, M. E., Van Gestel, N., Six, J., Venterea, T. and Van Kessel, C. 2015. Productivity limits and potentials of the principles of conservation agriculture. *Nature* 517, 365–8. doi:10.1038/nature13809. H.

Poeplau, C. and Don, A. 2015. Carbon sequestration in agricultural soils via cultivation of cover crops – a meta-analysis. *Agriculture, Ecosystems and Environment* 200, 33–41.

Poeplau, C., Voss, C. and Don, A. 2017. Soil organic carbon stocks are systematically overestimated by misuse of the parameters bulk density and stone content. *Soil* 3, 61–6. doi:10.5194/soil-2016-78.

Pretty, J. and Bharucha, Z. P. 2014. Sustainable intensification in agricultural systems. *Annals of Botany* 114(8), 1571–96. doi:10.1093/aob/mcu205.

Prihar, S. S., Jalota, S. K. and Steiner, J. L. 1996. Residue management for reducing evaporation in relation to soil type and evaporativity. *Soil Use Management* 12, 150–7.

Pryor, R. 2006. Switching to No-Till Can Save Irrigation water 2006. Univ Nebraska-Lincoln what pre-Ext Pub EC196-3. Available at: www.ianrpubs.unl.edu/epublic/live/ec196/build/ec196-3.pdf.

Qi, Z., Helmers, M. J., Christianson, R. D. and Pederson, C. H. 2011. Nitrate-nitrogen losses through subsurface drainage under various agricultural land covers. *Journal of Environmental Quality* 40(5), 1578–85. doi:10.2134/jeq2011.0151.

Raphael, J. P. A., Calonego, J. C., Milori, D. M. B. P. and Rosolem, C. A. 2016. Soil organic matter in crop rotations under no-till. *Soil and Tillage Research* 155, 45–53. doi:10.1016/j.still.2015.07.020.

Rawls, W. J., Pachepsky, Y. A., Ritchie, J. C., Sobecki, T. M. and Bloodworth, H. 2003. Effect of soil organic carbon on soil water retention. *Geoderma* 116(1–2), 61–76. doi:10.1016/S0016-7061(03)00094-6.

Reicosky, D. C. 1997a. Tillage methods and carbon dioxide loss: fall versus spring tillage. In: Lal, R., Kimble, J. M., Follett, R. F. and Stewart, B. A. (Eds), *Management of Carbon Sequestration in Soil*. CRC Press, Boca Raton, FL, pp. 99–111. Chapter 8.

Reicosky, D. C. 1997b. Tillage-induced CO_2 emission from soil. *Nutrient Cycling in Agro-Ecosystems* 49, 273–85.

Reicosky, D. C. 2002. Long-term effect of moldboard plowing on tillage-induced CO_2 loss. In: Kimble, J. M. and Lal, R. (Eds), *Agricultural Practices and Policies for Carbon Sequestration in Soil*. CRC Press, Boca Raton, FL, pp. 87–97.

Reicosky, D. C. 2015. Conservation tillage is not Conservation Agriculture. *Journal of Soil and Water Conservation* 70(4), 103A–8A.

Reicosky, D. (Ed.). 2018a. *Managing Soil Health for Sustainable Agriculture. Volume 1: Fundamentals*. Burleigh Dodds Science Publishing, Cambridge, UK, 352pp. ISBN: 978 1 78676 188 0.

Reicosky, D. (Ed.) 2018b. *Managing Soil Health for Sustainable Agriculture. Volume 2: Monitoring and Management*. Burleigh Dodds Science Publishing, Cambridge, UK, 464pp. ISBN: 978 1 78676 192 7.

Reicosky, D. C. and Archer, D. W. 2007. Moldboard plow tillage depth and short-term carbon dioxide release. *Soil and Tillage Research* 94, 109–21.

Reicosky, D. C. and Forcella, F. 1998. Cover crop and soil quality interactions in agroecosystems. *Journal of Soil and Water Conservation* 53(3), 224–9.

Reicosky, D. C. and Janzen, H. H. 2018. Conservation Agriculture: maintaining land productivity and health by managing carbon flows. In: Lal, R. and Stewart, B. A. (Eds), *Soil and Climate*. Advances in Soil Science. Taylor and Francis Group, LLC, Boca Raton, FL, pp. 131–61.

Reicosky, D. C. and Lindstrom, M. J. 1993. Fall tillage method: effect on short-term carbon dioxide flux from soil. *Agronomy Journal* 85(6), 1237–43. doi:10.2134/agronj1993.00021962008500060027x.

Reicosky, D. C. and Lindstrom, M. J. 1995. The impact of fall tillage on short-term carbon dioxide flux. In: Lal, R., Kimble J., Levine E., Stewart B. A. (eds), *Soils and Global Change*. Lewis Publishers, Chelsea, MI, pp. 177–87.

Reicosky, D. C., Kemper, W. D., Langdale, G. W., Douglas Jr., C. L. and Rasmussen, P. E. 1995. Soil organic matter changes resulting from tillage and biomass production. *Journal of Soil and Water Conservation* 50(3), 253–61.

Reicosky, D. C., Evans, S. D., Cambardella, C. A., Allmaras, R. R., Wilts, A. R. and Huggins, D. R. 2002. Continuous corn with moldboard tillage: residue and fertility effects on soil carbon. *Journal of Soil and Water Conservation* 57(5), 277–84.

Reicosky, D. C., Sauer, T. J. and Hatfield, J. L. 2011. Challenging balance between productivity and environmental quality: tillage impacts. In: Hatfield, J. L. and Sauer, T. J. (Eds), *Soil Management: Building a Stable Base for Agriculture*. American Society of Agronomy and Soil Science Society of America, pp. 13–38. doi:10.2136/2011.soilmanagement.c2.

Ruiz, S., Or, D. and Schymanski, S. J. 2015. Soil penetration by earthworms and plant roots–mechanical energetics of bioturbation of compacted soils. *PLoS ONE* 10(6), e0128914. doi:10.1371/journal.pone.0128914.

Sanderman, J., Hengl, T. and Fiske, G. J. 2017. Soil carbon debt of 12,000 years of human land use. *Proceedings of the National Academy of Sciences of the United States of America* 114(36), 9575–80. doi:10.1073/pnas.1706103114.

Seta, A. K., Blevins, R. L., Frye, W. W. and Barfield, B. J. 1993. Reducing soil erosion and agricultural chemical losses with conservation tillage. *Journal of Environmental Quality* 22, 661–5.

Seybold, C. A., Mausbach, M. J., Karlen, D. L. and Rogers, H. H. 1997. Quantification of soil quality. In: Lal, R., Kimble, J. M., Follett, R. F. and Stewart, B. A. (Eds), *Soil Processes and the Carbon Cycle*. CRC Press LLC, Boca Raton, FL, pp. 387–404.

Sharma, A. R., Singh, R., Dhyani, S. K. and Dube, R. K. 2010. Moisture conservation and nitrogen recycling through legume mulching in rainfed maize (*Zea mays*)-wheat (*Triticum aestivum*) cropping system. *Nutrient Cycling in Agroecosystems* 87(2), 187–97. doi:10.1007/s10705-009-9327-y.

Shaxson, T. F. 2006. Re-thinking the conservation of carbon, water and soil: a different perspective. *Agronomy for Sustainable Development* 26(1), 9–19. doi:10.1051/agro:2005054.

Shaxson, F., Kassam, A., Friedrich, T., Boddey, B. and Adekunle, A. 2008. Underpinning conservation agriculture's benefits: the roots of soil health and function. *Workshop on Investing in Sustainable Crop Intensification: The Case for Improving Soil Health*, 22–24 July. FAO, Rome, Italy. Available at: http://www.fao.org/AG/CA/doc/SHW_MainDoc_0708.pdf.

Sherwood, S. and Uphoff, N. 2000. Soil Health: Research, practice and policy for a more regenerative agriculture. *Applied Soil Ecology* 15, 85–97.

Sievers, T. and Cook, R. L. 2018. Aboveground and root decomposition of cereal rye and hairy vetch cover crops. *Soil Science Society of America Journal* 82(1), 147–55. doi:10.2136/sssaj2017.05.0139.

Singer, J., Nusser, S. and Alf, C. 2007. Are cover crops being used in the US corn belt? *Journal of Soil and Water Conservation* 62, 353–8.

Singh, G., Jalota, S. K. and Singh, Y. 2007. Manuring and residue management effects on physical properties of a soil under the rice-wheat system in Punjab, India. *Soil and Tillage Research* 94(1), 229–38. doi:10.1016/j.still.2006.07.020.

Six, J., Elliott, E. T., Paustian, K. and Doran, J. W. 1998. Aggregate and soil organic matter accumulation in cultivated and native grassland soils. *Soil Science Society of America Journal* 62, 1367–77.

Six, J., Elliott, E. T. and Paustian, K. 2000. Soil macroaggregate turnover and microaggregate formation: a mechanism for C sequestration under no-tillage agriculture. *Soil Biology and Biochemistry* 32(14), 2099–103. doi:10.1016/S0038-0717(00)00179-6.

Six, J., Bossuyt, H., Degryze, S. and Denef, K. 2004. A history of research on the link between (micro) aggregates, soil biota, and soil organic matter dynamics. *Soil and Tillage Research* 79(1), 7–31. doi:10.1016/j.still.2004.03.008.

Six, J., Frey, S. D., Thiet, R. K. and Batten, K. M. 2006. Bacterial and fungal contributions to carbon sequestration in agroecosystems. *Soil Science Society of America Journal* 70(2), 555–69. doi:10.2136/sssaj2004.0347.

Skjemstad, J. O. 2002. Importance of soil organic matter fractions to crop production, soil structure and soil resilience. Grains Research & Development Corporation Final Report CSO 195.

Skjemstad, J. O., Clarke, P., Taylor, J. A., Oades, J. M. and McClure, S. G. 1996. The chemistry and nature of protected carbon in soil. *Soil Research* 34(2), 251–71. doi:10.1071/SR9960251.

Skjemstad, J. O., Janik, L. J. and Taylor, J. A. 1998. Non-living soil organic matter: what do we know about it? *Australian Journal of Experimental Agriculture* 38(7), 667–80. doi:10.1071/EA97143.

Sposito, G. 2013. Green water and global food security. *Vadose Zone Journal* 12(4). doi:10.2136/vzj2013.02.0041.

Sreelash, K., Buis, S., Sekhar, M., Ruiz, L., Kumar Tomer, S. and Guérif, M. 2017. Estimation of available water capacity components of two-layered soils using crop model inversion: effect of crop type and water regime. *Journal of Hydrology* 546, 166–78.

Stagnari, F., Ramazzotti, S. and Pisante, M. 2009. Conservation Agriculture: a different approach for crop production through sustainable soil and water management: a review. In: Lichtfouse, E. (Ed.), *Organic Farming, Pest Control and Remediation of Soil Pollutants*. Sustainable Agriculture Reviews, Vol. 1. Springer, Dordrecht, The Netherlands, pp. 55–83. doi:10.1007/978-1-4020-9654-9_5.

Stavi, I., Lal, R., Jones, S. and Reeder, R. C. 2012. Implications of cover crops for soil quality and geodiversity in a humid-temperate region in the Mid-western USA. *Land Degradation and Development* 23(4), 322–30. doi:10.1002/ldr.2148.

Steele, D. D., Sajid, A. H. and Prunty, L. D. 1995. New corn evapotranspiration crop curves for Southeastern North Dakota. *Transactions of the ASAE* 39(3), 931–6.

Steele, M. K., Coale, F. J. and Hill, R. L. 2012. Winter annual cover crop impacts on no-till soil physical properties and organic matter. *Soil Science Society of America Journal* 76(6), 2164–73. doi:10.2136/sssaj2012.0008.

Steiner, J. L. 1994. Crop residue effects on water conservation. In: Unger, P. W. (Ed.), *Managing Agricultural Residues*. Lewis Publishers and CRC Press, Chelsea, MI, pp. 41–76.

Steiner, J. L. 1989. Tillage and surface residue effects on evaporation from soils. *Soil Science Society of America Journal* 53(3), 911–6. doi:10.2136/sssaj1989.0361599 5005300030046x.

Stevenson, F. J. and Cole, M. A. 1999. *Cycles of Soils: Carbon, Nitrogen, Phosphorus, Sulfur, Micronutrients* (2nd edn.). John Wiley & Sons, 427pp.

Stewart, B. A. and Peterson, G. A. 2015. Managing green water in dryland agriculture. *Agronomy Journal* 107(4), 1544–53. doi:10.2134/agronj14.0038.

Stewart, B. A. and Steiner, J. L. 1990. Water-use efficiency. In: Singh, R. P., Parr, J. F. and Stewart, B. A. (Eds), *Dryland Agriculture: Strategies for Sustainability*. Advances in Soil Science (vol. 13). Springer Verlag, New York, NY, pp. 151–73.

Stirling, G., Hayden, H., Pattison, T. and Stirling, M. 2016. *Soil Health, Soil Biology, Soilborne Diseases and Sustainable Agriculture. A Guide*. CSIRO Publishing, p. 280. ISBN: 9781486303045.

Stockfisch, N., Forstreuter, T. and Ehlers, W. 1999. Ploughing effects on soil organic matter after 20 years of conservation tillage in Lower Saxony, Germany. *Soil and Tillage Research* 52(1–2), 91–101. doi:10.1016/S0167-1987(99)00063-X.

Strock, J. S., Porter, P. M. and Russelle, M. P. 2004. Cover cropping to reduce nitrate loss through subsurface drainage in the northern UC Corn Belt. *Journal of Environmental Quality* 33(3), 1010–6. doi:10.2134/jeq2004.1010.

Sumberg, J. E. and Thompson, J. (Eds). 2012. *Contested Agronomy: Agricultural Research in a Changing World*. Earthscan Publications, New York, NY.

Sumberg, J., Thompson, J. and Woodhouse, P. 2012. Contested agronomy: agricultural research in a changing world. In: Sumberg, J. and Thompson, J. (Eds), *Contested Agronomy: Agricultural Research in a Changing World*. Routledge, London.

Taylor, H. W., Jordan, W. R. and Sinclair, T. S. (Eds). 1983. *Limitations to Efficient Water Use in Crop Production*. American Society of Agronomy, Madison, WI, 538pp.

Teague, W. R., Apfelbaum, S., Lal, R., Kreuter, U. P., Rowntree, J., Davies, C. A., Conser, R., Rasmussen, M., Hatfield, J., Wang, T., et al. 2016. The role of ruminants in reducing agriculture's carbon footprint in North America. *Journal of Soil and Water Conservation* 71(2), 156–64. doi:10.2489/jswc.71.2.156.

Tellez-Rio, A., García-Marco, S., Navas, M., López-Solanilla, E., Tenorio, J. L. and Vallejo, A. 2015. N_2O and CH_4 emissions from a fallow-wheat rotation with low N input in conservation and conventional tillage under a Mediterranean agroecosystem. *The Science of the Total Environment* 508, 85–94. doi:10.1016/j.scitotenv.2014.11.041.

Thierfelder, C., Baudron, F., Setimela, P., Nyagumbo, I., Mupangwa, W., Mhlanga, B., Lee, N. and Gérard, B. 2018. Complementary practices supporting conservation agriculture in southern Africa. A review. *Agronomy for Sustainable Development* 38, 16.

Tilman, D., Balzer, C., Hill, J. and Befort, B. L. 2011. Global food demand and the sustainable intensification of agriculture. *Proceedings of the National Academy of Sciences of the United States of America* 108(50), 20260–4. doi:10.1073/pnas.1116437108.

Tisdall, J. M. and Oades, J. M. 1982. Organic matter and water-stable aggregates in soils. *Journal of Soil Science* 33(2), 141–63. doi:10.1111/j.1365-2389.1982.tb01755.x.

Tsiafouli, M. A., Thébault, E., Sgardelis, S. P., de Ruiter, P. C., van der Putten, W. H., Birkhofer, K., Hemerik, L., de Vries, F. T., Bardgett, R. D., Brady, M. V., et al. 2015. Intensive agriculture reduces soil biodiversity across Europe. *Global Change Biology* 21(2), 973–85. doi:10.1111/gcb.12752.

Tully, K. and Ryals, R. 2017. Nutrient cycling in agroecosystems: balancing food and environmental objectives. *Agroecology and Sustainable Food Systems* 41(7), 761–98. doi:10.1080/21683565.2017.1336149. *Issue 7*, pp. 761–98.

Turmel, M. S., Speratti, A., Baudron, F., Verhulst, N. and Govaerts, B. 2015. Crop residue management and soil health: a systems analysis. *Agricultural Systems* 134, 6–16. doi:10.1016/j.agsy.2014.05.009.

USDA. 1987. Farm Drainage in the United States. USDA miscellaneous publication no. 1455.

USDA and NRCS. 2013. Soil health key points, February 2013. Available at: https://www.nrcs.usda.gov/Internet/FSE_DOCUMENTS/stelprdb1082147.pdf.

VandenBygaart, A. J. 2016. The myth that no-till can mitigate global climate change. *Agriculture, Ecosystems and Environment* 216, 98–9. doi:10.1016/j.agee.2015.09.013.

VandenBygaart, A. J., Gregorich, E. G. and Angers, D. A. 2003. Influence of agricultural management on soil organic carbon: A compendium and assessment of Canadian studies. *Canadian Journal of Soil Science* 83(4), 363–80. doi:10.4141/S03-009.

Van Donk, S. J., Martin, D. L., Irmak, S., Melvin, S. R., Petersen, J. L. and Davison, D. R. 2010. Crop residue cover effects on evaporation, soil water content, and yield of deficit-irrigated corn in west-central Nebraska. *Transactions of the ASABE* 53(6), 1787–97. doi:10.13031/2013.35805.

Van Eerd, L. L., Congreves, K. A., Hayes, A., Verhallen, A. and Hooker, D. C. 2014. Long-term tillage and crop rotation effects on soil quality, organic carbon, and total nitrogen. *Canadian Journal of Soil Science* 94(3), 303–15. doi:10.4141/cjss2013-093.

Veum, K. S., Goyne, K. W., Kremer, R. J., Miles, R. J. and Sudduth, K. A. 2014. Biological indicators of soil quality and soil organic matter characteristics in an agricultural management continuum. *Biogeochemistry* 17, 81–99.

Villarini, G., Smith, J. A., Baeck, M. L. and Krajewski, W. F. 2011a. Examining flood frequency distributions in the midwest U.S. *Journal of the American Water Resources Association* 47(3), 447–63. doi:10.1111/j.1752-1688.2011.00540.x.

Villarini, G., Smith, J. A., Baeck, M. L., Vitolo, R., Stephenson, D. B. and Krajewski, W. F. 2011b. On the frequency of heavy rainfall for the midwest of the United States. *Journal of Hydrology* 400(1), 103–20.

Wall, D. H., Nielsen, U. N. and Six, J. 2015. Soil biodiversity and human health. *Nature* 528(7580), 69–76. doi:10.1038/nature15744.

Weil, R. R. and Magdoff, F. 2004. Significance of soil organic matter to soil quality and health. In: Magdoff, F. and Weil, R. (Eds), *Soil Organic Matter in Sustainable Agriculture*. CRC Press, Boca Raton, FL, 398pp. Chapter 1. doi:10.1201/9780203496374.ch1.

Williams, S. M. and Weil, R. 2004. Cover crop root channels may alleviate soil compaction effects on soybean crop. *Soil Science Society of America Journal* 68(4), 1403–9.

Willoughby, G. L. and Kladivko, E. J. 2002. Water infiltration rates following reintroduction of Lumbricus terrestris into no-till fields. *Journal of Soil and Water Conservation* 57(2), 82–8.

Wortmann, C. S., Drijber, R. A. and Franti, T. G. 2010. One-time tillage of no-till crop land five years post-tillage. *Agronomy Journal* 102(4), 1302–7. doi:10.2134/agronj2010.0051.

Wuest, S. B. 2001. Earthworm, infiltration, and tillage relationships in a dryland pea–wheat rotation. *Applied Soil Ecology* 18(2), 187–92. doi:10.1016/S0929-1393(01)00145-7.

WWAP (United Nations World Water Assessment Programme)/UN-Water. 2018. *The United Nations World Water Development Report 2018: Nature-Based Solutions for Water*. UNESCO, Paris.

Zalewski, M. 2013. Ecohydrology: process-oriented thinking towards sustainable river basins. *Ecohydrology and Hydrobiology* 13(2), 97–103.

Zhang, T. Q., Tan, C. S., Zheng, Z. M. and Drury, C. F. 2015. Tile drainage phosphorus loss with long-term consistent cropping systems and fertilization. *Journal of Environmental Quality* 44(2), 503–11. doi:10.2134/jeq2014.04.0188.

Zhu, S., Vivanco, J. M. and Manter, D. K. 2016. Nitrogen fertilizer rate affects root exudation, the rhizosphere microbiome and nitrogen-use-efficiency of maize. *Applied Soil Ecology* 107, 324–33.

Zilverberg, C. J., Heimerl, K., Schumacher, T. E., Malo, D. D., Schumacher, J. A. and Johnson, C. 2018. Landscape dependent changes in soil properties due to long-term cultivation and subsequent conversion to native grass agriculture. *CATENA* 160, 282–97.

Zotarelli, L., Alves, B. J. R., Urquiaga, S., Boddey, R. M. and Six, J. 2007. Impact of tillage and crop rotation on light fraction and intra-aggregate soil organic matter in two oxisols. *Soil and Tillage Research* 95(1–2), 196–206. doi:10.1016/j.still.2007.01.002.

Chapter 4

The role of no or minimum mechanical soil disturbance in Conservation Agriculture systems

Theodor Friedrich, Food and Agriculture Organization of the United Nations (FAO), Italy

1 Introduction

Tillage, intentionally changing the mechanical characteristics and structure of the soil, has characterized farming since the very first sedentary agricultural land use. It has become the cultural identity of crop farming to the extent that the term 'arable' land – which means 'land that can be ploughed' – is used to characterize agricultural land that can be used for annual cropping. However, human history has demonstrated that this kind of tillage-based agriculture is the main cause of land degradation worldwide (Montgomery, 2007). In recognition of this, attempts have been made to reduce the impact of tillage on the soil, focusing mainly on soil loss through erosion as the main problem. However, while concepts such as 'conservation tillage' and other soil conservation measures can reduce the amount of soil erosion, soil degradation continues worldwide and the levels are still far greater than those permissible for sustainable land use. Other problems, such as reduced water infiltration capacity, loss of organic matter and therefore soil structure, and loss of soil biodiversity, are not addressed by these conventional soil conservation approaches. Any kind of mechanical soil disturbance and tillage always causes soil degradation in its various forms much faster than the soil can naturally form and recover. Therefore, tillage-based agriculture cannot be considered in any way sustainable (Montgomery, 2007).

http://dx.doi.org/10.19103/AS.2019.0048.04

Conservation Agriculture (CA) aims for minimum or no soil disturbance as the core principle for sustainable intensification of crop production, combined with the two synergistic principles of permanent soil cover and diversity of crops and production systems. The term 'minimum soil disturbance' must not be confused with 'minimum tillage'. In fact, 'minimum soil disturbance' describes a subset of no-till farming systems, including no-till or zero-tillage systems, which causes minimal or no mechanical disturbance to the soil surface and matrix during all cropping operations. Conversely, no-till systems that do not include a dedicated tillage operation, but disturb the entire soil profile or more than 25% of the soil surface with one pass during seeding or planting, are excluded from the concept of minimal soil disturbance (FAO, 2018a).

With our history and habits derived from tillage-based agriculture, it is now a challenge to imagine practical solutions for farming operations that not only exclude tillage, but also any kind of mechanical soil disturbance, as much as possible for each and every farming operation, from land preparation to weed management and harvest. However, while a few crops and operations might still present issues, particularly in mechanized farming, we now have ample experience with techniques and equipment for the various operations in most crops and farming conditions to comply with the principle of minimum soil disturbance.

2 Effects of soil tillage and tillage implements on the soil

2.1 Origins of tillage

When humans shifted from being hunter-gatherers to sedentary agriculture, they started seeding and planting crops to grow and harvest. This involved bringing the seed or seedling into direct contact with the soil. In the beginning, this was achieved by poking a small hole in the soil with a stick and putting the seed in the hole. However, with growing demand and pressure to sow more land, there was a need to open more soil surface and to make sowing more efficient, for example by broadcasting the seed onto an open, loose soil surface and then mixing it with the soil for contact. A very simple tool, a wooden beam pulled by humans or animals, marked the beginning of soil tillage. This tool, which was the precursor of today's ploughs, is still used in many parts of the world and is known as the ard plough (Fig. 1). However, opening the soil surface creates competition from weeds, particularly in moister and cooler climates. In those areas, the ard plough was perfected by adding a board, which enabled the soil to be turned over, burying the weeds and giving the crop a head start. Developments to improve the action of the plough by adding these boards were reported in China several centuries BC, while they were introduced to Europe by the Romans in the first century AD (Thiesing, 2016). Only later, in the late middle ages, was the mouldboard introduced into middle

Figure 1 Wooden ard plough used on steep slope for ploughing in the Andes of Ecuador. Photo: Friedrich.

Europe, and it was only perfected in the eighteenth century to turn the soil better (Eichhorn, 1952). This invention, which at that time was considered a salvation from famine, created the myth of tillage as weed control that persists today. However, ploughing year after year buries weed seeds in the soil and preserves them for long periods, while also bringing a similar amount of weed seeds back to the surface to germinate (Anderson, 2005). This is why weed control is still the biggest issue in farming despite 90% of the world's farming today being tillage-based, and all intensive cropping systems use chemical weed control methods or large amounts of fuel or labour for weeding, despite the supposedly weed-controlling effect of tillage.

2.2 Types of tillage

Tillage is the mechanical alteration of soil structure. It is supposed to have various functions, which are related to the different types of tillage equipment. The main mechanical functions of tillage are:

- Inversion (e.g. with a plough)
- Mixing (e.g. with a chisel or cultivator)
- Loosening (e.g. with a subsoiler or heavy chisel)
- Pulverizing (e.g. with a disk harrow)
- Compacting (e.g. with a packer or surface roller)

These classic tillage operations change the physical condition of the soil, to create the so-called perfect seedbed for example. This should be free of obstacles, level, with a granular structure suitable for the size of the seed to ensure perfect seed-soil contact, and firm enough to allow water capillarity. The mixing action also facilitates rapid decomposition of crop residues and manure to release the mineral nutrients contained in this organic matter (FAO, 2000).

None of these operations actually improve root or plant development, as the structure of the natural, undisturbed soil is destroyed. The diversity of creatures in an undisturbed soil creates a solid and stable soil matrix of different aggregate sizes with connected pores of various sizes to allow water and air circulation, and movement of the soil-inhabiting macrofauna and mesofauna (Nickel, 2011; Castellanos-Navarrete et al., 2012). The vertical continuous macropores or biopores created by earthworms, termites and root channels are of particular importance to the water infiltration capacity of the soil. These continuous pores, which facilitate root growth and water infiltration, are destroyed by soil tillage and replaced by uniform aggregate sizes, organized in horizontal layers (Kassam et al., 2008). Tilled soils are therefore characterized by a homogenous A horizon, which contains most of the roots and nutrients, and a B horizon, which is dead, decomposed mineral material, as is the underlying C horizon. In many cases, there is a hard pan or 'plough pan' between the A and B horizons and, even if there is no compaction layer, there is a marked difference in the bulk density between the A and B horizons which makes crop roots less likely to penetrate into the deeper soil layers. Depending on the texture of the soil, water infiltration capacity in the B horizon may be limited by the absence of biopores. In contrast to this layered structure of a tilled soil, a no-till soil is a relative continuum, with high levels of organic matter at the surface, gradually reducing with depth, and with much deeper root channels and biopores than tilled soils (Kassam et al., 2013).

2.3 Tillage implements and the effect on the soil

As described above, all tillage implements are detrimental to soil health and structure, as they destroy the natural habitat of the soil fauna, the hyphae of soil fungi and soil food webs. They also destroy soil aggregates and expose organic matter, which is the structural element in many soils, to bacterial decomposition that results in carbon dioxide being emitted after each tillage operation (Sá et al., 2012). However, some tillage implements are particularly dangerous, because of their design and function. Obviously, each tool has been designed for a specific purpose, but this chapter does not aim to elaborate on the intended functions of tillage implements. Instead, it will explain the downsides and dangers of the different tillage tools, which are normally not addressed.

The most symbolic implement is probably the mouldboard plough, which turns the soil by about 120°, turning the profile upside down (Feuerlein, 1964). This is obviously catastrophic for the soil-inhabiting fauna, as the edaphic conditions of the habitat change very suddenly. If the cutting edge of the plough share on the mouldboard plough is not constantly sharpened, or is badly adjusted, it can also create a plough pan by smearing the bottom of the furrow (Söhne, 1956). The original purpose of the mouldboard plough, which was to turn the soil upside down as perfectly as possible and create a clean soil surface for seedbed preparation, is also one of its biggest problems, as it leaves a completely clean and exposed soil surface with absolutely no protection against wind or rain.

This is why the disk plough, and disk harrow, have become more popular, especially in tropical countries (Wieneke and Friedrich, 1989), and disk harrows are still considered 'conservation tillage' tools. Disk tools do not turn the soil profile as perfectly as the mouldboard plough, and they do leave some crop residues on the surface. However, they also pulverize and mix the soil more than the mouldboard plough, accelerating internal soil degradation, particularly in tropical climates. Disk tools present yet another problem when compared to mouldboard ploughs and chisel tools: they do not create a vertical force to pull them into the soil, but rely only on weight applied to the disks to enter the soil (Wieneke and Friedrich, 1989). So, although their sharp cutting edge does not tend to create a plough pan by smearing (unlike ploughshares), disk tools create serious compaction problems, as the entire weight of the implement is supported by the cutting edge of the disk, compacting the soil from below. Constant and exclusive use of disk tools causes serious compaction horizons that grow towards the soil surface (FAO, 2000), and these can reach several decimetres in thickness. In time, farmers note that it is more difficult for the disk tool to penetrate the soil, unless additional weight is applied. This compacting effect of the disk harrow is well known, and it is used to compact the foundation layers when building roads, for example.

3 Minimizing soil disturbance in farming

3.1 Land preparation

3.1.1 Initial operations to prepare for CA

Land preparation has traditionally been linked to tillage, but this aspect obviously changes under CA, with the concept of minimum soil disturbance. However, it might be necessary to eliminate severe soil problems, such as plough pans and soil compactions, with a final till when converting a tillage-based system into a permanent no-till system, such as CA. If the compaction is not too severe, it can be broken up by deep-rooted 'biological' subsoilers, such as pigeon peas or radishes. If the compaction is more serious and time is limited, a mechanical subsoiler may be required, but it is important to select the right tool to break up the soil compaction without creating new problems or altering the soil structure more than necessary (FAO, 2000). The dangers of subsoiling include creating a new hard pan by smearing in moist soil or by separation and sedimentation of silt particles in extremely degraded and unstable soils. These dangers are increased with subsoilers that use wide-sweep blades to break up a wider area. Another issue is that large clods of soil can be brought to the surface, particularly by chisels that have a parabolic shape for lower draft requirements. This will require additional levelling passes with a disk harrow to destroy the clods, with the risk of re-compacting the newly loosened soil to an even greater depth. It is preferable to use subsoilers with chisels or shanks angled at 45° to prevent these problems. The 'paraplough®' is a classic subsoiler with this characteristic, and it was originally designed for low-disturbance subsoiling in no-till systems (Wieneke and Friedrich, 1989) (Fig. 2).

Once any severe soil compaction has been rectified, the soil surface might need to be prepared for the intended crop rotation and cropping system, including irrigation. This could involve levelling, or establishing permanent bed and furrow systems, which might require a 'last' full tillage operation, after which the field surface will remain untouched. In the case of permanent bed and furrow systems for furrow irrigation, reshaping the furrows once a year might be the only soil operation to continue after that (Sims et al., 2016).

Figure 2 Subsoiler with shanks bent sidewards for breaking compactions in no-till systems. Photo: Friedrich.

3.1.2 Land preparation in a CA system

Land preparation in a CA system does not usually engage the soil, with the exception of reshaping irrigation furrows, which does not interfere with the crop-rooting environment. The land is prepared to establish conditions for seeding or planting, mainly by managing crop- or residue-mulch covers. In mature CA systems, the land is prepared for the following crop at harvest, ideally in one pass. In general, the cover should be evenly distributed across the field. Other characteristics of the crop or residue cover that facilitate planting and seeding depend on the subsequent crop and the seeding technology (Sims et al., 2016).

In general, a standing and anchored crop or crop residue facilitates seeding, particularly with disk seeders. This can be achieved by using stripper harvesters on grain crops, or by cutting a very high stubble to leave as little loose residue as possible on the ground. In some conditions, the residue or mulch cover is rolled down to the ground: for example, a cover crop or spontaneous vegetation can be rolled down with a knife roller for mechanical management. If the crop is in the right physiological condition, that is after the vegetative growth at the beginning of the generative phase, knife rolling can kill some cover crops without the need for further treatment, and create a surface mulch for planting. The direction of rolling must be established, and the seeder should follow the same direction as the knife roller, or at an angle of 90° to it, depending on the nature of the residue, but never against the rolling direction (Baker et al., 2007).

Loose residues should preferably be left intact and not chopped into pieces in CA systems. This slows decomposition and maintains soil cover for longer throughout the cropping season. This condition is best for disk seeders, while chopped residues are better for chisel-type no-till seeders (Sims et al., 2016).

3.2 Seeding and planting

The objective when seeding and planting in CA systems is the same as in any other farming system: to bring the seed into direct contact with the soil. The advantage of the CA system is that the soil structure, particularly near the surface, is rich in organic matter and naturally has the ideal structure for the seed to germinate. In some cases, seeds can even germinate in dry soils when the capillary water supply is interrupted, but where the deep, air-conducting continuous macropores might provide enough moisture to germinate the seed. The challenge is to penetrate the surface cover and deposit the seed accurately into the undisturbed soil (Baker et al., 2007).

When climatic conditions are moist and seeds are small, this can be achieved by broadcasting the seed under a mulch cover. This operation can be connected to the grain harvest by attaching a seed tube for broadcasting to the rear of the combine header and the broadcast seed is then covered by the straw spread by the combine.

Under normal conditions, however, seeding and planting is carried out in CA with dedicated equipment. Again, keeping soil disturbance to a minimum is of utmost importance, for several reasons. First, soil disturbance in the seeding zone disrupts the continuity of the biopores that carry air and water through the soil. This disturbance also causes soil moisture loss by stimulating evaporation, which can be critical in dry climates. The only exceptions to this are moist and cold climates, where disrupting the seed zone with a narrow strip could be beneficial to dry out and warm up the seed zone. Secondly, soil disturbance, and exposure to light in particular, stimulates germination of the weed seed bank in the soil and leads to weed competition in the crop seed row and germination zone. Thirdly, soil disturbance requires energy and fuel (in mechanized farming systems), which can be avoided (Baker et al., 2007) (Fig. 3).

Equipment is available for minimum disturbance seeding and planting at all levels of mechanization. At the manual level, the jab planter is nothing more than a mechanized planting stick, which has a sharp, pointed beak to penetrate the soil cover into the undisturbed soil, opening a small planting hole and releasing a predetermined amount of seed. Fertilizer can be applied in the same way and in the same operation through a second beak. A metering mechanism provides the predetermined quantity of seed and fertilizer to the beaks with

Figure 3 Different levels of soil disturbance between disk furrow openers (left) and chisel furrow openers (right). Photo: Friedrich.

every cycle of opening and closing. There are various designs of manual no-till planters, but the jab planter is the most popular and sturdy. Other versions resemble a planting stick, or a pointed two-pronged hoe. Less popular are the rolling jab planters, which have problems penetrating some soils and may clog on sticky soils. There are also jab planters with larger diameter feeding tubes and larger beaks for no-till transplanting of seedlings for tobacco or vegetable crops, for example (Baker et al., 2007).

Pull-type no-till seeders and planters use the same functional principals for soil-engaging components, regardless of whether they are animal- or tractor-drawn.

3.2.1 Higher disturbance seeders and planters

The 'minimum soil disturbance' range has been defined by the FAO as disturbance of a maximum of 25% of the soil surface or a slot width not wider than 15 cm, whichever is the lesser, and there are reasons why soil-engaging tools that fall within this range but create some soil disturbance, more than the 'minimum', such as chisels or actively driven rotary strip seeding tools, might be beneficial under certain conditions (FAO, 2018a). Chisels, for example, can penetrate hard-setting soils better, without the requirement for heavy equipment. This makes them ideal for animal-traction seeders and single-axle tractor seeders, but also for soils that have become severely degraded and hard-set under arid conditions, at least in the early years of the transition to CA. Again, the objective here is to open a seed row to the desired depth with as little disturbance as possible, so chisels should be narrow and shaped to reduce soil movement, particularly at higher seeding speeds, to prevent tillage erosion in hilly areas (Fig. 4). Tillage erosion (the downhill movement of soil caused by soil-engaging tools at high speeds) from high-disturbance no-till seeders can exceed the tillage erosion caused by the plough due to their high operating speeds (Papiernik et al., 2007). When used in unchopped residue cover, chisel-type seeders need a cutting disk coulter to prevent the residues or cover crops from wrapping around the chisels.

Rotary harrow strip seeders have been developed for no-till seeding of grain crops with narrow row spacing using a single-axle tractor. The low weight and draft capacity of these tractors make it difficult for them to operate under no-till conditions, having residue covers, using chisel or disk seeders with more than four rows, resulting in low working width and performance. Another reason is the cost: disk openers need to be thin and sharp to cut through the residue and penetrate into the soil without excessive equipment weight. Conversely, they must also be durable and flexible to withstand stones and abrasion. This requires steel of a special quality, which makes these tools too expensive for farmers in some countries, so tools that might disturb the soil more may still

Figure 4 High-disturbance seeding with chisels. Photo: Friedrich.

be useful when starting with CA. However, the terminology surrounding these rotary harrow strip seeders is still confusing: they are derived from the full-width powertiller-cum-seeder, so they are still called 'minimum tillage', or 'strip-till', equipment, although they only open a narrow seed slot without the intention of tillage (Haque et al., 2017).

3.2.2 Minimum disturbance seeders and planters

From a soil disturbance point of view, the best equipment does not leave any visible seed row; it cuts the soil to deposit the seed without altering the natural structure of the soil. This is most commonly achieved with disk tools that can cut through residues and soil, and open a very narrow slot for the seed and fertilizer. There are various designs of disk openers, with different soil disturbance characteristics. Single disk openers are slightly turned out of the line of travel, so they open a seed furrow by pushing soil aside and depositing the seed behind the disk. The better option for reduced soil disturbance is the double disk opener, where the disks are positioned in a V-shape with the pointed side facing the direction of travel. It cuts a slot and positions the seed between the two disks. One disk is ideally slightly bigger than the other, and the axle of that disk should be slightly ahead of the smaller disk. This improves the cutting action with less weight, and reduces soil disturbance even further (Fig. 5). This design also reduces the danger of pushing crop residues into the seed row and

Figure 5 Low-disturbance seeding with disk furrow openers. Photo: Friedrich.

impairing seed-soil contact, an effect commonly called hairpinning (Ashworth et al., 2010).

Offset double disks of different sizes, tilted at an angle of 45° sideways, are an even better design, requiring less weight, reducing the risk of hairpinning, and depositing the seed better onto a completely undisturbed soil. This planter design is particularly useful for heavy soils in wet conditions, where the disks may seal the sidewalls of the seed slot by smearing (Sims et al., 2016).

Another design with very low soil disturbance, which deposits the seed onto completely undisturbed soil, with clear separation of seed and fertilizer and no risk of hairpinning, is the Cross Slot. It has one vertical cutting disk aligned with the direction of travel, which cuts a slot through the residues and soil. There is a winglet on each side of the disk at an adjustable depth below the soil surface, and the seed and fertilizer is deposited under it. Seed and fertilizer are close together, but separated by the slot cut by the cutting disk. This technology is known for its very low-disturbance and high germination rate, even under difficult and dry conditions (Baker et al., 2007).

The star wheel planter is another implement that causes extremely low soil disturbance and has very low energy requirements. The seed is pushed into the soil between the fingers of two star wheels, similar to the manual jab planter. This seeder also handles residues very well (Mitchell, 1993).

The downside of the low-disturbance disk furrow opener is the high cost, which makes its use on very abrasive, sandy soils too expensive due to the high wear on the disks. The Aqua-Till is an interesting option for these conditions, as it uses an ultra-high-pressure water jet to cut through the residues and open the seed slot (Butler, 2011).

3.3 Crop husbandry and weed management

Soil disturbance must also be avoided during crop husbandry operations, such as ridging or earthing-up, which is common in some crops. This is often used for weed control, but ridging is often merely a tradition and is not necessary for physiological crop reasons in most cases. Permanent bed and furrow systems are ideal if the aim is to maintain drainage or irrigation furrows. Other techniques that do not engage the soil are available for some crops, such as potatoes, and will be discussed in the section on harvesting crops with sub-surface yields. However, the most significant crop husbandry operation with regard to soil disturbance is weed management. This aspect is often overlooked when promoting CA, causing CA systems to perform poorly and increasing labour requirements, which is not the case in well-managed CA systems (Edralin et al., 2017). The reason for this misperception is that traditional mechanical weed control is based on soil disturbance, so a no-till system which includes mechanical, soil-engaging weed control, for example by hand hoeing or inter-row cultivation, is not a minimum disturbance system. Residue cover is also incompatible with hoe-weeding or inter-row cultivation, as the residues are raked by the weeding tools (Sims et al., 2012).

Paying insufficient attention to the issue of weed management, particularly when herbicides are unavailable, is a very common mistake and care must also be taken to use weed management techniques that keep soil disturbance to a minimum. It is also the origin of the myth that CA is only possible with herbicides. However, there are other options for weed management that do not mechanically engage the soil (Anderson, 2005).

3.3.1 Biological methods

Biological methods are very effective weed management tools. Some cash crops, such as sunflowers, have an excellent weed control effect, and crop rotations can be designed to include these weed-clearing crops. Cover crops can be used specifically to suppress weeds and the mulch of the cover crop can keep the subsequent cash crop free of weeds. Mulch cover suppresses weeds much more efficiently than attempting to control them with tillage: the mulch cover, combined with minimal soil disturbance, reduces light stimulation and germination of the weed seed bank in the soil (Anderson, 2005). Weeds that spread by roots or stolons, such as grasses, also are very effectively controlled through shading by mulch cover, provided the cover is thick enough and the shading complete. In perennial crops, such as fruit plantations, the soil cover crops can be grazed, which can provide an additional income to the farm and also be beneficial for management of the pests that affect the plantation crop, depending on the type of grazing animals.

The biological herbicides that are currently being developed and may be available on a larger scale in the future are not widely used at present. However, the effect of legume cover crops in rotation or mixed cropping to control parasitic weeds, such as *Striga*, is well known and widely used in practical CA farming (Lee and Thierfelder, 2017).

3.3.2 Mechanical methods

Mechanical weed management methods do not have to engage the soil, and the most important mechanical weed management tool in CA is the knife roller, or crimper roller. It consists of knives placed around the circumference of a cylinder, and it flattens standing vegetation and crimps or crushes it, without cutting. This results in a loss of water and stability, causing the plants to wilt. The plants are not cut at the base, so they will also not easily ratoon from the roots. If the knife roller is used after the plants have shifted from the vegetative growth phase to the reproductive phase, but before the seeds have matured, the plants eventually die without producing fertile seeds, but still provide a soil-shading mulch cover, suppressing future weed growth. This method is used in CA systems in Latin America with black oats as the cover crop, controlling weeds by a method that does not engage the soil for the entire subsequent crop without the need for any herbicide (Baker et al., 2007). Different widths of knife roller can be used for animal traction or tractor use. There is a narrower, manual pushed version for inter-row weeding in a standing row crop.

Depending on the type of accompanying vegetation, often referred to as 'weeds', and the type of crop, the vegetation can be cut and mulched by hand with some kind of blade. The purpose of weed management in CA is not to eliminate weeds completely, as weeds can also serve as soil cover, and the water lost through evaporation from unprotected disturbed soil can exceed the evaporation from a soil covered with dead mulch, and even the transpiration of living plant cover. Therefore, weed management in CA prevents weeds from spreading, propagating and interfering with the crop. In some cases, it can be enough to chop the heads off the weeds when they are flowering, and in other cases, weeds can be suppressed by cutting or rolling until the crop is tall enough to suppress them by itself.

Experience has shown that weed pressure, and therefore the time and cost associated with weed management in CA systems, declines with time if weeds are managed correctly and according to a forward-looking plan, which includes good crop rotation as a key element.

3.3.3 Chemical methods

Chemicals are the most common method of weed management in large-scale mechanized CA farming, as they are in conventional tillage-based systems. However, if combined with a good weed management strategy involving the above-mentioned biological and mechanical methods, herbicide doses and application rates in CA systems tend to be lower than in conventional tillage-based farming systems. This obviously does not apply to no-till systems that do not adhere to the other two principles of CA (Lindwall and Sonntag, 2010).

Chemical herbicides should always be used as part of an integrated weed management programme, according to the FAO definition of integrated pest management:

> Integrated Pest Management (IPM) means the careful consideration of all available pest control techniques and subsequent integration of appropriate measures that discourage the development of pest populations and keep pesticides and other interventions to levels that are economically justified and reduce or minimize risks to human health and the environment. IPM emphasizes the growth of a healthy crop with the least possible disruption to agro-ecosystems and encourages natural pest control mechanisms. (FAO, 2018b)

This also relates to the FAO definition of CA:

> Interventions such as mechanical soil tillage are reduced to an absolute minimum, and the use of external inputs such as agrochemicals and nutrients of mineral or organic origin are applied at an optimum level and in a way and quantity that does not interfere with or disrupt the biological processes. (FAO, 2018a)

One category of herbicides, however, is completely excluded from use in CA systems, and these are the pre-seeding herbicides, which have to be incorporated into the soil. There are also some other pesticides and methods for sterilizing the soil that are not to be used in CA, as an active and diverse soil life is the functional base for CA systems, and the biodiversity in a healthy living soil also helps control pests and diseases naturally.

Soil compaction can occur on large-scale farms that use big, ground-spraying rigs for chemical management of insect pests, as it may not always be possible to time the control measure to respect optimal soil conditions for in-field traffic. In this case, the controlled traffic systems described in Section 3.5.3 can be suitable solutions.

3.4 Harvesting

The harvest operation in CA usually coincides with traditional land preparation. Ideally, harvesting of one crop leaves the field ready for immediate establishment of the subsequent crop, without any further operations.

The issue of minimal soil disturbance is obviously a challenge when the harvested part of the crop is located below the soil surface, such as root and tuber crops. However, some crops that are harvested above the soil surface also present issues when harvesting, particularly in difficult weather conditions, or with wet soils or heavy machinery.

3.4.1 Crops with above-surface yields

Crops with the harvested parts located above the ground should not necessarily create an issue with regard to minimum soil disturbance. Managing the crop residue to prepare the best conditions for the subsequent crop is the main concern when harvesting these crops. However, soil compaction and tyre tracks left by heavy machinery do create a problem for minimum soil disturbance, as repairing this damage requires extensive tillage. Therefore, crops that are harvested with heavy machinery in adverse weather conditions can also pose a problem for CA, as they may disrupt the permanent minimum soil disturbance principle. In some cases, the time pressure for harvest in periods of wet weather might only allow the crop to dry sufficiently to be harvested, but the soil still might not be in a suitable condition to support heavy machinery, resulting in deep tracks in the field.

As a general observation, soil in a mature CA system, that is with more than 5 years permanent no-till and higher levels of organic matter, is more resistant to compaction and track formation than tilled soil. The organic matter layer on top of the soil provides an elastic cover that has a cushioning effect, reducing permanent damage to the soil. The soil itself has a firmer structure and, despite the high pore volume, it 'feels' harder than tilled soil because the matrix is firmly structured. When combined with a better capacity to drain excess water, soils under long-term permanent no-till generally have better trafficability than tilled soils (Baker et al., 2007).

Using flotation tyres and rubber tracks with lower ground pressures on medium to lightweight equipment can help prevent permanent compaction and tracks, even in adverse conditions. However, the supporting capacity of even the best-structured soil can be exceeded by machinery masses at the top end of the grain combine harvesters, sugar cane harvesters and transport bins, and the climatic conditions under which these sometimes have to operate, despite using the best technologies to reduce ground pressure (Sächsische Landesanstalt für Landwirtschaft, 2005). In these cases, the controlled traffic systems described in Section 3.5.3 offer feasible solutions to these problems.

3.4.2 Crops with below-surface yields

Crops with the harvested parts located below the ground, such as root crops or tubers, and also some legumes, such as groundnuts, present a particular challenge for minimum soil disturbance. Establishment and husbandry of these crops is no different from other crops with regard to soil disturbance, but harvesting does create a problem. These crops can be seeded or transplanted perfectly well in no-till soils, and the crops usually develop better under no-till conditions as the soil matrix is firmer, so roots and tubers usually develop more evenly. There is usually no plough pan or compaction horizon in a no-till soil, so the common L-shaped roots are also less of an issue. Many roots, such as cassava, also grow closer to the surface, which facilitates harvest, and the high organic matter soils with stable soil aggregates tend to stick to the roots or tubers less, which not only reduces the uprooting force, but also reduces the dirt content in the harvested product (FAO, 2013).

Soil disturbance cannot be totally prevented when harvesting roots and tubers, but the impact must be minimized, so aggressive 'digger-type' harvesters that excavate the soil with the crop and then separate out the soil with mechanical sieving devices (Fig. 6) should be avoided. These techniques not only destroy soil structure, but also damage the soil aggregates, causing unnecessarily high carbon losses from the soil. Methods that pull the crop out of the soil and only lift and break the soil by passing a horizontal knife underneath the crop at the most, are preferred. These 'pulling-type' harvesters are available for most crops, including groundnuts, although they have mostly been replaced by digger-type harvesters in tillage-based agriculture systems. Sugar beet is one of the few crops where pulling-type harvesters are the norm.

Figure 6 Digger-type root-crop harvester. Photo: Friedrich.

In any case, the damage to the soil's carbon balance by the soil disturbance caused by harvesting should be compensated by long crop rotations with at least 4–5 years between two root crops, and integrating carbon-rich crops, such as cereals, in the rotation (Sá et al., 2012). Cover crops can be established before harvest in some root crops to facilitate a mulch cover immediately after harvest to protect the loose soil from rain, wind and sunlight.

A very special case is the potato, a tuber commonly grown below the soil surface which cannot be harvested by pulling. However, the potato tuber is not actually part of the root, but a sprout of the plant. The only reason that the tubers must be kept underground and out of the light is that they turn green and toxic with sunlight, so materials other than soil could be used to cover potato tubers. The only important factor is that the cover should provide complete protection from sunlight. Cereal straw, particularly rice straw, which decomposes very slowly and shades tubers perfectly throughout the cropping season, is a very good cover. No-till potatoes are therefore grown in rotation with rice, underneath the rice straw from the preceding rice crop in some countries, such as Vietnam, China and DPR Korea (FAO, 2012).

3.5 Compaction management

Soil tillage is often seen as a remedy for soil compaction, or at least to repair the damage caused by soil compaction. This might be true for the type of soil compaction that is created by tillage itself, such as the plough pan, or disk harrow compaction, but compaction from the weight of the machine itself reaches far beyond the capacity of mechanical remedies, such as subsoiling. Unfortunately, research on soil compaction and its effect on yields has only considered the upper soil horizons and little work has been done on subsoil compaction, which affects the water cycle in a watershed by reducing the overall infiltration and drainage capacity of the soil (Sächsische Landesanstalt für Landwirtschaft, 2005). This is why tillage is not really the answer to compaction problems, and no-till and CA cropping systems with minimum soil disturbance offer much more effective and efficient solutions.

3.5.1 Origin of compaction

There are very few soils that can be considered 'self-compacting', mainly due to specific chemical reactions. Soils generally tend to self-compact through a lack of internal structure, which is commonly caused by a lack of organic matter, soil structuring mesofauna and macrofauna and soil fungi. This may result in separation and internal sedimentation of the silt, creating compaction horizons in the soil profile, which can also be triggered by subsoiling in very degraded, unstable soils with inert mineral components, such as silt or sand, which have poor self-binding forces to form aggregates. Most compaction in soils and subsoils, however, originate from tillage and traffic (Sommer, 1985).

3.5.2 Avoiding compaction

This chapter is about minimum soil disturbance and CA systems, so compaction created by tillage and its prevention will not be discussed. The other major source of compaction is traffic of humans, animals or machinery (Söhne, 1953). Even human or animal traffic can cause compaction if repeated constantly over the same piece of land and on degraded soils. The frequency of travel, ground pressure and the total mass supported by the soil affect the level of damage to the soil. Human traffic, for example, can cause compaction by frequency, rather than by mass or ground pressure. Constantly using the same path eliminates plant growth and exposes the soil surface, degrading and compacting the soil in time.

Compaction by animals is mainly characterized by high ground pressure, but the compaction does not penetrate very deeply because of the limited mass. Frequency of passes with animals has the same effect as humans, and it can degrade and compact the soil in the longer term, making remedial action necessary. Compaction by animal traction is less of an issue in CA systems due to the low frequency of field traffic, but it can become an issue in crop rotations that include pasture crops with grazing animals. The pasture land can suffer degradation as well as compaction, depending on soil moisture, animal density and duration of grazing, but this compaction should only be superficial in a good CA system with a healthy crop rotation, and can easily be overcome either by deep-rooting cover crops that can break down the compaction, such as pigeon peas or radishes, or by using chisel furrow openers for fertilizer application instead of disk furrow openers, which cause less soil disturbance.

In reality, integrating crops and livestock provides additional synergistic benefits for the sustainable intensification of agricultural production described by Carvalho et al. (2018):

> Mimicking the nutrient coupling/decoupling processes of natural ecosystems by diversifying plant and animal components of no-till integrated crop-livestock operations is an essential feature of the design of agro-ecological systems that support self-regulating feedbacks and lend resilience while increasing productivity and ecosystem service provision. Focusing on grazing animals as drivers of agroecosystem change, we highlight the benefits of grazed cover crops in rotation with cash crops for primary and secondary production and for soil physical, chemical and biological parameters. However, careful management of grazing intensity is imperative; overgrazing drives soil deterioration, while light to moderate grazing enhances overall system functioning and allows for the generation of emergent properties.

The issue of machinery traffic, particularly large harvesters, is more complex. As a rule, field traffic should be reduced to a minimum; unnecessary traffic should be eliminated completely, and only dedicated field equipment should be allowed on cropland. This refers particularly to the type and design of the drive gear: road traffic vehicles need to run on high tyre pressures and hard tyres to support the weight and reduce tyre wear and rolling resistance, but the opposite is the case for in-field vehicles. Tyres should be soft, with low pressures and as wide as possible to reduce ground pressure and prevent surface deformation, such as tracks and compaction. Rubber tracks are another solution, lowering ground pressure considerably (KTBL, 2011).

Ground pressure under the pressure-creating foot, hoof or agricultural machine extends vertically and horizontally into the soil, with isobars in the shape of 'onion' rings. The wider the footprint, the wider the 'onions', so the rings also reach deeper soil layers than the rings caused by a smaller footprint. This is why compaction by an animal does not penetrate very deeply into the soil, despite the high ground pressure. However, this also means that the pressure isobars from a heavy machine may reach into the subsoil, even though the ground pressure may be so low that it leaves no visible track on the surface. There are fewer structure-building processes in the subsoil, so it compacts with time (Söhne, 1953). This is particularly relevant to tillage-based systems, where even root channels become scarce below the A horizon, and soil life that could provide structure to the subsoil is mostly absent.

3.5.3 Controlled traffic systems

The gradual build-up of subsoil compaction below the reach of subsoilers is still an unrecognized, slow process (Sächsische Landesanstalt für Landwirtschaft, 2005), accompanied by a steady decline in the infiltration capacity of agricultural soils, resulting in more frequent flooding events. Low-pressure tyres or rubber tracks are therefore no longer a safe choice to prevent all compaction by extremely heavy machinery, such as modern combine harvesters or sugar beet harvesters, which can produce wheel loads of over 10 t (KTBL, 2011), especially in moist or wet soil conditions, so controlled traffic systems are the best option in these cases. These systems do not prevent compaction, but reduce the area affected by compaction and keep it off land that is to be used for crop growth (Kerr, 2001) (Fig. 7). Controlled traffic systems are based on permanent tracks that are used as traffic lanes, with crop-growing areas in between. Those permanent traffic lanes can be marked as visible furrows in permanent bed and furrow systems. However, today's technologies of choice are differential GPS-based global positioning systems, which guide the machines on the exact tracks, even without any visibility. It is advisable to combine autosteer systems with the GPS for enhanced accuracy and to prevent driver mistakes. Vehicle

Figure 7 Farm equipment set up for controlled traffic systems. Photo: Friedrich.

and machinery traffic in the field is operated in very compacted lanes, which are even trafficable in difficult weather conditions, without causing any compaction to the cropland. Loss of cropland is low and economically compensated by increased yields from the non-compacted cropping zones (Gaffney and Wilson, 2003).

4 Conclusion

The impact of soil disturbance on soils has been largely ignored throughout human and agricultural history, like the elephant in the room, so soil degradation is still advancing worldwide, despite decades of attempts to promote sustainable farming. Mechanical soil disturbance interferes with a very delicate, still widely unknown habitat, which takes a very long time to develop or recover (Montgomery, 2007). It interferes with habitat biodiversity, both below and above ground, and affects the ecosystem services that a healthy soil can provide, such as controlling insect pests and diseases, and managing micro- and macro-climates and water cycles. This is why minimal mechanical soil disturbance is an essential condition for any kind of sustainable soil and ecosystem management (Kassam et al., 2008, 2013). Agriculture has traditionally been practised with no consideration of the damage caused by soil disturbance. Instead, soil disturbance has become a regular functional element of agriculture and agricultural scientists still believed that soil disturbance was a precondition for successful agriculture until the twentieth century. However, the last 50 years have shown that agriculture is possible without soil disturbance, and this has given rise to the realization that the dream of sustainable agriculture could become a reality. Minimum soil disturbance is a key precondition, which allows the environmental rewards that derive from CA production systems to be reaped. Minimum soil disturbance in this context is much more than just not tilling - it is about avoiding soil disturbance in all its forms at all stages of agricultural production, including disturbance in the form of soil compaction during crop husbandry operations and harvest. Technically, this is possible in most cropping systems today and, with some clever engineering, it might also be perfected for crops where the perfect solution to prevent soil disturbance has not yet been discovered because it is still not considered to be of any importance.

5 Where to look for further information

In order to serve the future development of sustainable agriculture systems, research should focus on no-tillage systems for all research activities, including crop breeding. This is to learn about the characteristics and functions of the components within minimum disturbance no-tillage systems, but also to further develop them and find solutions to problems which might appear. One precondition to this type of research is to define exactly the no-till system under research, as described in Derpsch et al. (2014). Such approach to agricultural research in function of no-till cropping systems is for example used by some research institutes in southern Brazil, such as IAPAR in Paraná or EPAGRI in Santa Catarina.

Some of the organizations working on no-till cropping systems are the Food and Agriculture Organization of the United Nations (FAO) (http://www.fao.org/conservation-agriculture/en/), research institutes of the CGIAR system such as the International Maize and Wheat Research Center (CIMMYT) (https://www.cimmyt.org/), the International Center for Agricultural Research in the Dry Areas (ICARDA) (http://www.icarda.org/), regional organizations and farmers' organizations such as the European Conservation Agriculture Federation (ECAF) (http://www.ecaf.org/), the African Conservation Tillage Network (ACT) (http://www.act-africa.org/), the Argentine Association of No-till Farmers and the Latin American Federation of such associations AAPRESID-CAAPAS (http://www.aapresid.org.ar/blog/tag/caapas/), the Brazilian Federation for Direct Seeding into Straw (FEBRAPDP) (https://febrapdp.org.br/), the Western Australian No Till Farmers Association (WANTFA) (https://www.wantfa.com.au/), the South Australian No Till Farmers Association (SANTFA) (https://www.santfa.com.au/), No-till on the plains (http://www.notill.org/) and many more.

6 References

Anderson, R. L. (2005), 'A multi-tactic approach to manage weed population dynamics in crop rotations', *Agronomy Journal*, 97: 1579–83. doi:10.2134/agronj2005.0194.

Ashworth, M., Desbiolles, J. and Tola, E. K. (2010), *Disc Seeding in Zero-till Farming Systems – A Review of Technology and Paddock Issues*, Western Australian No-Tillage Farmers' Association (WANTFA), Northam, Western Australia, p. 223.

Baker, C. J., Saxton, K. E., Ritchie, W. R., Chamen, W. C. T., Reicosky, D. C., Ribeiro, M. F. S., Justice, S. E. and Hobbs, P. R. (2007), *No-tillage Seeding in Conservation Agriculture* (2nd edn.), CABI and FAO, Wallingford, UK and Rome, Italy, 326p.

Butler, G. (2011), 'Aqua-Till – seeding without soil engagement', *SANTFA – The No-Till Journal*, 8(2): 44–50.

Carvalho, P. C. F., Peterson, C. A., de Albuquerque Nunes, P. A., Posselt Martins, A., de Souzsa Filho, W., Bertolazi, V. T., Kunrath, T. R., de Moraes, A. and Anghinoni, I. (2018), 'Animal production and soil characteristics from integrated crop-livestock systems: Toward sustainable intensification', *Journal of Animal Science*, 96(8): 3513–25. doi:10.1093/jas/sky085.

Castellanos-Navarrete, A., Rodríguez-Aragonés, C., de Goede, R. G. M., Kooistra, M. J., Sayre, K. D., Brussaard, L. and Pulleman, M. M. (2012), 'Earthworm activity and soil structural changes under conservation agriculture in central Mexico', *Soil and Tillage Research*, 123: 61–70.

Derpsch, R., Franzluebbers, A. J., Duiker, S. W., Reicosky, D. C., Koeller, K., Friedrich, T., Sturny, W. G., Sa, J. C. M. and Weiss, K. (2014), 'Why do we need to standardize no-tillage research?', *Soil and Tillage Research*, 137: 16–22.

Edralin, D. A., Sigua, G. C., Reyes, M. R., Mulvaney, M. J. and Andrews, S. S. (2017), 'Conservation agriculture improves yield and reduces weeding activity in sandy soils of Cambodia', *Agronomy for Sustainable Development*, 37: 52. doi:10.1007/s13593-017-0461-7.

Eichhorn, H. (1952), Landwirtschaftliches Lehrbuch Landtechnik (7. Auflage Ausg., Bd. 4), Eugen Ulmer GmbH & Co., Gießen.

FAO (2000), 'Manual on integrated soil management and conservation practices', FAO Land and Water Bulletin 8, FAO, Rome, pp. 37–44.

FAO (2012), 'Vietnam holds National Field Day on minimum tillage on potato production, Hanoi, Vietnam', *FAO News*, 12 January 2012. Available at: http://www.vegetableipmasia.org/News/News47.html (accessed on 13 March 2012).

FAO (2013), *Save and Grow: Cassava. A Guide to Sustainable Production Intensification*, Food and Agriculture Organization, Rome, p. 142.

FAO (2018a), 'What is Conservation Agriculture, Definition of Conservation Agriculture', *FAO*. Available at: www.fao.org/ag/ca (accessed on February 2018).

FAO (2018b), 'AGP Integrated Pest Management', *FAO*. Available at: http://www.fao.org/agriculture/crops/core-themes/theme/pests/ipm/en/ (accessed on July 2018).

Feuerlein, W. (1964), *Geräte zur Bodenbearbeitung*, Eugen Ulmer, Braunschweig.

Gaffney, J. and Wilson, A. (2003), 'The economics of zero tillage and controlled traffic farming for western downs farms', *Proceedings of the 16th ISTRO Conference on Soil Management for Sustainability*, Brisbane, Australia, 13–18 July 2003.

Haque, M. E., Bell, R. W., Islam, A. K. M. S., Sayre, K. D. and Hossain, M. M. (2017), 'An innovative versatile multi-crop planter for crop establishment using two-wheel tractors', *AMA-Agricultural Mechanization in Asia, Africa & Latin America*, 48(3): 34–9.

Kassam, A. H., Friedrich T. and Shaxson F. (Eds) (2008), *Proceedings of an International Technical Workshop on Investing in Sustainable Crop Intensification: The Case for Improving Soil Health*, FAO, Rome, Italy, 22–24 July 2008, Integrated Crop Management, Vol. 6.

Kassam, A., Basch, G., Friedrich, T., Shaxson, F., Goddard, T., Amado, T. J. C., Crabtree, B., Hongwen, L., Mello, I., Pisante, M. and Mkomwa, S. (2013), 'Sustainable soil management is more than what and how crops are grown', in Lal, R. and Stewart, B. A. (Eds), *Principles of Sustainable Soil Management in Agroecosystems*, CRC Press, pp. 337–400. Chapter 14. Print ISBN: 978-1-4665-1346-4, eBook ISBN: 978-1-4665-1347-1.

Kerr, P. (2001), *Controlled Traffic Farming at the Farm Level*, GRDC Research Update, Finley, NSW, Australia.

KTBL (2011), Boden schonen und Kosten senken, KTBL Heft 89, Kuratorium für Technik und Bauwesen in der Landwirtschaft, Darmstadt, Germany. ISBN: 978-3-941583-53-5.

Lee, N., Thierfelder, C. (2017), 'Weed control under conservation agriculture in dryland smallholder farming systems of southern Africa: A review', *Agronomy for Sustainable Development*, 37: 48. doi:10.1007/s13593-017-0453-7.

Lindwall, C. W. and Sonntag, B. (Eds) (2010), *Landscapes Transformed: The History of Conservation Tillage and Direct Seeding*, Knowledge Impact in Society University of Saskatchewan, Saskatoon.

Mitchell, A. (1993), 'Star wheel seed planter', Tinto Industries Limited (Zimbabwe) United States Patent 5205228. Available at: http://www.freepatentsonline.com/5205228.html.

Montgomery, D. (2007), *Dirt: The Erosion of Civilizations*, University California Press, Berkeley, USA.

Nickel, R. (2011), 'Soil biology', *in* Manitoba-North Dakota Zero Tillage Farmers Association (Ed.), *Beyond the Beginning: The Zero Till Evolution* (3rd edn.). www.mandakzerotill.org.

Papiernik, S. K., Lindstrom, M. J., Schumacher, T. E., Schumacher, J. A., Malo, D. D. and Lobb, D. A. (2007), 'Characterization of soil profiles in a landscape affected by long-term tillage', *Soil and Tillage Research*, 93: 335–45.

Sá, J. C. M., Santos, J. B. and Lal, R. (Eds) (2012), *An On-farm Assessment of Carbon Monitoring and Mapping Scaling up in No-till Fields*, Gráfica e Editora Castro e Lima Ltda, Ponta Grossa, Brazil, 58p. ISBN: 978-85-63023-05-6.

Sächsische Landesanstalt für Landwirtschaft (2005), 'Bodendruck und Bodenbelastbarkeit; Abschlussbericht zum Projekt "Risikomanagement Bodengefüge"', Schriftenreihe der Sächsischen Landesanstalt für Landwirtschaft, Heft 15, 10. Jahrgang, 2005, ISSN 1861-5988, Dresden, Germany.

Sims, B. G., Thierfelder, C., Kienzle, J., Friedrich, T., Kassam, A. (2012), 'Development of the conservation agriculture equipment industry in sub-Saharan Africa', *Applied Engineering in Agriculture*, 28(6): 1–11. ISSN: 0883-8542.

Sims, B., Kienzle, J., Mkomwa, S., Friedrich, T. and Kassam, A. (2016), 'Mechanization of smallholder conservation agriculture in Africa: Contributing resilience to precarious systems', in Kassam, A., Mkomwa, S. and Friedrich, T. (Eds), *Conservation Agriculture for Africa: Building Resilient Farming Systems in a Changing Climate*, CABI International, Wallingford, UK and Boston, USA, pp. 183–213.

Söhne, W. (1953), 'Druckverteilung im Boden und Bodenverformung unter Schlepperreifen', *Grundlagen der Landtechnik*, 5/1953: 49–63.

Söhne, W. (1956), 'Einige Grundlagen für eine Landtechnische Bodenmechanik', *Grundlagen der Landtechnik*, 7/1956: 11-27.

Sommer, C. (1985), 'Ursachen und Folgen von Bodenverdichtungen sowie Möglichkeiten zu Ihrer Verminderung', *Landtechnik*, 40(9): 378–84.

Thiesing, M. (2016), 'Kräfte und Kinematik an landtechnischen Pflügen', Dissertation, Agricultural Engineering, University of Kassel, Germany.

Wieneke, F. and Friedrich, Th. (1989), *Agricultural Engineering in the Tropics and Subtropics*, Centaurus Verl., Pfaffenweiler.

Chapter 5

The role and management of soil mulch and cover crops in Conservation Agriculture systems

A. Calegari, Agricultural Research Institute of Paraná State (IAPAR), Brazil; T. Tiecher, Federal University of Rio Grande do Sul (UFRGS), Brazil; E. B. Wutke, Research Agricultural Institute of Campinas (IAC), Brazil; L. B. dos S. Canalli, Agricultural Research Institute of Paraná State (IAPAR), Brazil; R. Bunch, Consultant, Brazil; and D. dos S. Rheinheimer, Federal University of Santa Maria, Brazil

1 Introduction

Conventional agricultural tillage systems, where normally the soil is disturbed by plowing or disk harrowing, or by some other mechanical impact on the uppermost soil layer, generally causes considerable soil damage. In addition, uncovered soil, even without soil tillage, is subject to the impact of rainwater, causing surface sealing and reducing water infiltration in the soil, thus triggering the whole process of water erosion. Mulches, cover crops, and crop residues can mitigate this damage. In Conservation Agriculture (CA), crop residues from cover crops should be maintained on the soil surface as dead or live mulch which, beyond protecting the soil from erosion, can improve

http://dx.doi.org/10.19103/AS.2019.0048.05

chemical, physical and biological attributes of soil (Hargrove, 1991; Scopel et al., 2012; Florentin et al., 2010; Calegari and Costa. 2010; Calegari et al., 2013a,b,c). Research all over the world is showing that soil mulch cover is an efficient strategy for soil protection in different agroecological zones, and in many diversified cropping and farming systems. Across different parts of the world CA systems represent a diversified set of options, offering farmers the possibility of achieving sustainable and intensive crop production under a wide range of agroecological environments, be it under temperate or tropical conditions. They simultaneously allow maintenance or enhancement of soil fertility by protecting and reducing soil erosion, increasing soil organic matter (SOM), and enhancing soil porosity, as well as making efficient use of available natural resources such as light radiation, water and soil nutrients (Scopel et al., 2012).

The presence of soil mulch cover or stubble mulch in CA systems considerably modifies the establishment and performance of crops as well the crop-water balance (Kwaad et al., 1998; Da Silva, 2004). Such cover avoids rain falling on the bare soil and improves the rate of water infiltration and, consequently, surface water runoff is generally reduced in CA systems. Depending on factors such as soil type, the total amount and type of crop residues, slope steepness, the type of crop and its development, between 0% and 85% reductions have been observed in the Brazilian tropics compared with conventional systems (Schick et al., 2000; Mello et al., 2003; Scopel et al., 2004, 2012).

The main advantages of soil mulching to avoid leaving the soil uncovered are to:

(a) protect the soil from the kinetic energy of the rain, avoiding the direct impact of the raindrops on the soil, thus preventing erosion;
(b) keep more moisture in the soil, decreasing water loss through evaporation, and decreasing thermal oscillation in the top layers;
(c) soften channels in the soil profile that will increase the water infiltration into the profile, reducing surface runoff and losses of water, soil and nutrients;
(d) improve soil structure and particle aggregation, aeration and higher porosity, favoring the root system growth of the following crops;
(e) increase nutrient cycling in the soil, reducing nutrient leaching;
(f) provide competition with weeds, reducing the costs with its control;
(g) increase soil organic matter (OM), providing significant improvement to soil chemical, physical and biological properties.

The main aims of using soil mulch cover and cover crops are thus to protect the soil against water erosion, to improve water infiltration and maintain soil moisture (Lal, 1997). In addition such cover promotes improved nutrition for crop roots (Tiecher et al., 2012a,b), suppresses weed growth (Teasdale et al., 2007) and provides shelter and food to the population of soil biota (Blanchart et al., 1997; Balota et al., 2014). Mulch cover also increases natural enemies of pests and diseases/nematodes affecting crops. Crop rotations can also be seen as part of an integrated nutrient management approach (Medeiros et al., 1989; Calegari and Alexander, 1998).

The success of CA systems is therefore directly related to the presence of a protecting mulch of crop residues and its incorporation into the soil, as well as the use of a mix (blend or cocktail) of multifunctional cover crops which fits into an adequate crop rotation system. If mulching rates are too low and cover crops are not introduced, the benefits of CA systems are much reduced or even absent. From a socioeconomic perspective, CA systems have been shown to be generally profitable if their adoption translates into lower production costs, starting with the lower costs in land preparation, coupled with the maintenance of, or an increase in, grain yields (Scopel et al., 2012).

Historical records show that the practice of using cover crops was successfully developed by ancient civilizations (e.g. the Romans, Greeks and Chinese) but that, thanks to intensive modern cultivation techniques relying on external inputs, this practice was almost forgotten in this century. Fortunately, in the last three decades, research studies and the practical experiences of farmers in various parts of the world have allowed this efficient technique to resume its important place in many different agroecological conditions and production systems in different countries, contributing to the maintenance and improvement of soil attributes (physical, chemical and biological). In this chapter, we will discuss the role of soil mulching in CA systems, especially highlighting the use of different cover crops species to promote live or dead soil mulch cover.

2 The main components of an adequate Conservation Agriculture system and general aspects of soil mulching and cover crops

Good no-till (NT) systems require using different species of cover crops, dry matter crop residues and crop rotations as fundamental components in rational and sustainable crop management. NT is now practiced, for example, on more than 5 million ha in Paraná State in Brazil and estimates show that NT covers more than 30 million ha in Brazil as a whole (Casão Júnior et al., 2012; Kassam et al., 2019). Results from the use of CA systems over the years, especially in Southern Brazil, prove that the proper use of cover crops to promote soil mulching, as part of NT systems, are economically feasible as well as ecologically sustainable, providing not only greater crop productivity, but also maintenance and/or recovery of soil fertility. In addition, it reduces the need for nitrogen fertilizers by fixation (using leguminous species), as well as improving soil biology and overall health (Primavesi, 1990).

There are many advantages of using cover crops in local farming systems in different developing countries of Africa, Asia and America, depending on agroecological conditions and the farmer's aims. The selection of the appropriate species, which will fit in a suitable crop rotation, is one of the keys to making the CA system work properly, facilitating the adoption of CA by farmers. Farmers can move toward adopting a full CA system linked to NT practices, with no plowing of soil, the use of cover crops, crop rotation, avoiding burning of residues, maintaining soil cover, according to local conditions and farming infrastructure (Derpsch and Calegari, 1985, 1992; Derpsch et al., 1986, 1991; Calegari, 1991; Hargrove, 1991; Calegari et al, 1998b; Dijkstra, 2002; Pieri et al., 2002; Mielniczuk et al., 2003).

According to Prudencio et al. (2004), the key features of CA systems are:

(a) no plowing, disking or seedbed preparation;
(b) green manure/cover crops integrated into the cropping system;
(c) crop, weed and cover crop residues applied as mulch to protect the soil permanently;
(d) direct seeding or planting;
(e) no burning of crop residues or fallow vegetation;
(f) no uncontrolled grazing;
(g) nutrient cycling through the biomass in and above the soil;
(h) surface application of lime and fertilizers;
(i) specific equipment for seeding and mulch management;
(j) continuous use of cropland;
(k) crop rotations and cover crops used to maximize biological control of pests and diseases.

The use of cover crops in CA systems is applicable to both large-scale farmers in mechanized areas and smallholders-subsistence farmers (Merten and Fernandes, 1998; Ribeiro and Milleo, 2002; Wall and Ekboir, 2002). These cover crops may be grown in the 'off season' between major cash crops to protect soil and build organic matter (OM), biological diversity and soil structure for the benefit of subsequent 'cash' crops. Additionally, cover crops may be sown under permanent crops such as vineyards, orchards and biofuel, annual or perennial crops to provide year-round soil protection, provide a 'catch crop' to capture rainwater and nitrogen, reduce water runoff and maximize infiltration, and enrich the topsoil through increased OM. Careful selection of the cover crop (e.g. shallow rooted species) ensures minimal competition for stores of soil moisture between cash and cover crops; and the selection of nitrogen-fixing crops (grass/legume swards) may help enhance soil fertility and subsequent yields of the main cash crop (Miyasaka, 1984; Calegari et al., 2006a,b; Bolliger et al., 2006; Schomberg et al., 2006; Calegari, 2014). The type/species of cover crops utilized and their in-field management are dependent on region and climate. Strong potential exists to grow perennial crops as animal fodder, hence not only promoting soil health but also providing supplementary feed for animals and more diverse income for farmers.

During their growing stages different crops have distinct effects on soil properties. This is directly linked with biomass production (shoot and root), soil cover capacity, crop residue stability on the soil surface and residues effects on soil properties: soil density, soil aggregates, infiltration rates, porosity etc. Such general functions of cover crops include:

(1) providing additional forage and food and secondary commercial or subsistence products for livestock and humans;
(2) directly adding or conserving N to/from the soil through symbiotic N_2-fixation from the atmosphere;
(3) converting otherwise unused resources, such as sunlight and residual soil moisture, into additional biomass to build up SOM;
(4) capturing and recycling easily leachable nutrients (nitrate, K, Ca and Mg) that would otherwise be lost beyond the rooting zone of commercial crops;
(5) ameliorating soil structure and buffering against compaction by creating additional root channels that differ from those of the main crops and by stimulating soil biological activity through the release of root exudates;
(6) improving the management of acidic soils by releasing various products that can mobilize lime movement through the soil profile, promoting ligand exchange and adding basic cations to the soil;
(7) facilitating weed management by competing against or smothering weeds that would otherwise limit crop growth;
(8) breaking the cycle of certain pests and diseases that could otherwise build-up in continuous monocropping systems.

The function of particular cover crops in terms of building SOM, enhancing nutrient management, alleviating soil compaction and facilitating soil acidity and weed management, must be considered at the cropping system planning stage (Sidiras and Pavan, 1985; Bolliger et al., 2006).

When deciding on a particular CA system, a systematic approach (selection of rotations) is needed to understand and improve the amount and availability of different nutrients to the crops being grown (Calegari et al., 2005; Schomberg et al., 2006). The basic principles of CA are shown in Fig. 1:

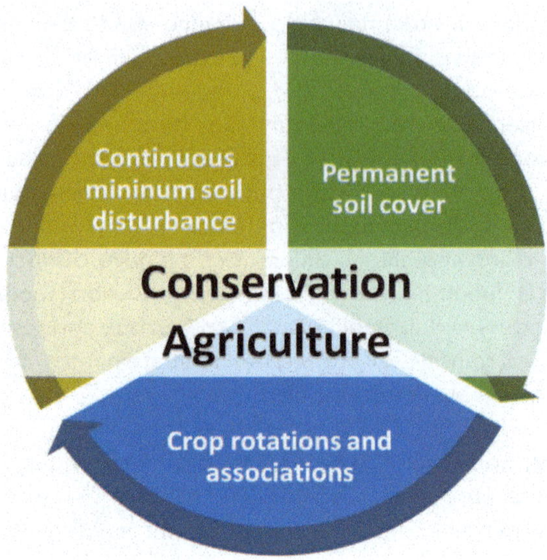

Figure 1 Basic principles of Conservation Agriculture.

- Continuous minimum soil disturbance;
- Permanent soil cover;
- Crop rotations and associations.

All the three principles need to be applied to create a desirable environment for crop growth in a CA system. Different farmers or groups will emphasize a particular theme depending on their current stage of development. They need to ensure that all three elements are taken forward together, with some empahasized more than others depending on existing conditions. Some pre-conditions may be necessary before the process of implementing a CA system starts. As examples, surface runoff needs to be controlled (e.g. by terracing slopes) and chemical activity in the soil needs to be checked, for example, for the presence of aluminum and manganese which can increase acidity and thus inhibit nutrient availability.

To achieve these three principles of CA, appropriate cover crop species must be chosen according to a diagnosis of soil chemical, physical and biological properties. There are many different options and crop rotation systems all over the world, including different species that have been used in Brazil (Calegari, 1985, 1995a,b, 1998a, 2016, 2018a,b; Monegat, 1991), such as:

- Fall/winter species: black oat (*Avena strigosa* Schreb), white oat (*Avena sativa*), radish *(Raphanus sativus)*, vetches (*Vicia sativa and Vicia villosa)*, field pea (*Pisum sativum*), lupin (*Lupinus* spp.), ryegrass (*Lollium multiflorum),* rye (*Secale cereale*), sweet pea *(Lathyrus sativus),* clovers (*Trifolium* spp.), sweet clover (*Melilotus officinalis*), lucerne (*Medicago sativa*), serradella (*Ornithopus sativus),* chickpea (*Cicer arietinum*) and so on.

- Spring/summer species: buck wheat (*Fagopirum esculentum*), pigeon pea (*Cajanus cajan*), velvet bean (*Mucuna* sp.), sunn hemp (*Crotalaria juncea*), *Crotalaria spectabilis, Crotalaria ochroleuca, Crotalaria breviflora*, cowpea (Vigna unguiculata), green gram (*Vigna radiata*), lablab (*Dolichos lablab*), siratro (*Macroptilium atropurpureum*), stylo (*Stylosanthes* spp.), vetch pea *Clitoria ternatea,* jack bean *(Canavalia ensiformis),* brave bean of Ceará *(Canavalia brasiliensis),* pearl millet *(Pennisetum glaucum),* finger millet *(Eleusine coracana), Centrosema* sp., *Desmodium* spp., tropical Kudzu (*Pueraria phaseoloides), Stylosanthes* spp., *Tephrosia* spp., *Calopogonium mucunoides, Neonotonia wightii* and so on.

Beyond the use of individual species, the mixing of cover crops (cocktail/blend) has been studied and recommended in the last few decades (Calegari, 2006a,b, 2016, 2018a,b). These mixes of cover crops include two or more species such as:

- oat + vetch, radish + black oat + vetch;
- buckwheat + radish + pear millet;
- crotalaria + pearl millet or cajanus + pearl millet + crotalaria + buckwheat + finger millet;
- pearl millet + crotalaria sp. + buckwheat.

These and other combinations can be used in annual and perennial cropping systems. Some species can also be oversown (broadcast by tractor or airplane) before harvesting of soybean, maize or bean (around 20–30 days before harvesting). After harvesting, the cover crops achieve a height of 20–30 cm.

To develop an effective NT system including cover crops and crop rotation, it is important to carry out an analysis of each plot in the farm. This needs to consider all possible constraints/challenges, such as climate conditions, soil types, texture, SOM, nutrient status, pest, root disease/nematode population and soil compaction; these must be addressed before planning the different species which will be part of the crop sequence (Calegari, 2000a,b, 2014; Florentin et al., 2001). In general, planting should normally be around two weeks after the cover crops, though, in the case of finger millet or ryegrass, the period should be around 3-4 weeks or ryegrass to avoid alleopathic effects.

The planting of cover plants should achieve a uniform soil cover which can be facilitated using NT planting machinery. The planting machine should have a cutting blade to properly cut the straw and facilitate sowing through double

Table 1 Main cover crop species indicated to be used (in rotation) before soybean, maize and bean

Soybean		Maize		Bean	
Fall/winter species	Spring/summer (early sown)	Fall/winter	Spring/summer (early sown)	Fall/winter species	Spring/summer (early sown)
Oat (black and white)	Pearl millet	Hairy vetch	*Crotalaria* (spectabilis, ochroleuca, sunn hemp)	Oat (black and white)	Pearl millet
Rye + vetch (hairy or common)	Finger millet	White lupin	*Pigeon pea* (giant, dwarf)	Rye	Finger millet
Hairy vetch	Sudan grass	Field pea	*Stylosanthes capitata* × *Stylosanthes macrocephala* (cv. Campo Grande)	Hairy vetch	Sudan grass
Field pea	Buckwheat	Radish + oat	Buckwheat	Field pea	Forage sorghum
Radish + oat	*Stylosanthes capitata* × *Stylosanthes macrocephala* (Campo Grande)	Oat (black and white) + vetch (hairy or common)	Pearl millet	Radish	Buckwheat
Field pea + oat	Buckwheat	Rye + vetch (hairy or common)	Cowpea	Oat + radish + pearl millet	Buckwheat + pearl millet + radish
*Mix - oat + rye + radish and/or vetch	*Mix - Pearl millet + Crotalaria (sunnhemp/ spectabilis/ochroleuca)	*Mix - oat + rye + radish and/or vetch and/or white lupin	*Mix - Pearl millet + Crotalaria (sunn hemp/ spectabilis/ochroleuca)	*Mix - Oat + rye + radish and/or vetch	*Mix - pearl millet + crotalaria (sunnhemp/ spectabilis/ochroleuca)

Source: Calegari, not published.

disk or seed drills. Table 1 shows some alternative combinations in rotation with commercial crops.

Some options for mixing cover crops are listed below:

- Mixes for spring/summer (Calegari, 2018a,b):
  1) Millet (5-8 kg/ha) + buckwheat (15-20 kg/ha) + crotalarias (*spectabilis, breviflora and ochroleuca) (6-8 kg/ha each) *may be mixed or each species individually.
  2) Millet (5-8 kg/ha) + buckwheat (15-20 kg/ha) + crotalaria (spectabilis, breviflora and ochroleuca) (6-8 kg/ha each) + cowpea or dwarf pigeon pea orMucuna (10-15 kg/ha).
  3) Millet (5-8 kg/ha) + buckwheat (15-20 kg/ha) + crotalaria (spectabilis or breviflora or ochroleuca) (6-8 kg/ha) + cowpea (10-15 kg/ha) + sunflower (3-4 kg/ha) + finger millet (4-5 kg/ha) + radish (3 kg/ha).
- Mixes for fall/winter - indicated also for high altitude areas (normally more than 600-700 m) (Calegari, 2016, 2018a,b):
  4) Radish (3-4 kg/ha) + black oat (20-25 kg/ha) + common vetch (15-20 kg/ha).
  5) Black oat (20-25 kg/ha) + white lupine (30-35 kg/ha) + rye (15-20 kg/ha) + radish (3 kg/ha) + vetch - common or hairy (20 kg/ha).
- Other mix options:
  6) Black oat (20-25 kg/ha) + forage pea (30-35 kg/ha) + rye (20-25 kg/ha) + lupine (white and/or blue) (50-70 kg/ha).
  7) Black oat (25-35 kg/ha) + vetch (hairy and/or common) (30-35 kg/ha).
  8) Black oat (25-35 kg/ha) + radish (6-8 kg/ha).
  9) Black oat (25 kg/ha) + forage pea (25-30 kg/ha) + radish (4-5 kg/ha).
  10) Black oat (25 kg/ha) + forage pea (25-30 kg/ha) + radish (4-5 kg/ha) + rye (10-15 kg ha).
  11) Black oat (25 kg/ha) + forage pea (25-30 kg/ha) + radish (4-5 kg/ha) + rye (10-15 kg ha) + vetch (common or hairy) (15-20 kg ha).
  12) Black oat (15 kg/ha) + forage pea (25 kg/ha) + radish (3-4 kg/ha) + vetch (common or hairy) (20-25 kg/ha).
  13) Black oat (15-20 kg/ha) + white lupine (25-30 kg/ha) + radish (5 kg/ha) + vetch (common or hairy) (20-25 kg/ha) + (10-15 kg/ha).Black oat (15-20 kg/ha) + white lupine (25-30 kg/ha) + radish (5 kg/ha) + vetch (common or hairy) (20-25 kg/ha) + rye (10-15 kg/ha).
  14) Black oat (15-20 kg/ha) + white lupine (25-30 kg/ha) + radish (5 kg/ha) + vetch (common or hairy) (20-25 kg/ha) + radish (3 kg/ha) + ryegrass (10-15 kg ha).

15) Black oat (15–20 kg/ha) + white lupine (25–30 kg/ha) + radish (5 kg/ha) + vetch (common or hairy) (20–25 kg/ha) + buckwheat (10–15 kg/ha).

Many cover crop species can increase the abundance of predators, which decrease the population of nematode species. In addition to the cover crops themselves, some biological products also promote greater soil biodiversity with reduced risk of nematodes and root diseases, promoting a healthier, productive soil and thus more sustainable cultivation (Calegari, 2014).

In general, cover crops improve soil physical, chemical and biological properties, increasing chemical and biological activity in the soil, including the ability to recycle different nutrients and promote nitrogen fixation (through legumes) that can be more easily taken up by the following crop. Given their capacity to produce vegetal biomass and roots, with important direct and indirect effects in the soil-water-plant system, cover crops play a fundamental role in a well-planned rotation system with cash and food crops. According to Calegari and Peñalva (1994), the main roles of cover crops are:

1 Soil protection and prevention of soil erosion (by water and wind);
2 Mulch to improve soil cover;
3 Weed suppression;
4 Enhancement of water holding capacity in the soil profile;
5 Limiting temperature fluctuations in soil;
6 Nutrient cycling;
7 Adding nitrogen through biological fixation (legumes);
8 Improvement of soil biology (macro, meso and micro flora and fauna);
9 Strong roots which break up compacted layers;
10 Beneficial physical effects (soil particle aggregation, infiltration, porosity, permeability etc.);
11 Fewer problems with pests and diseases (in both soil and crops);
12 Increased soil organic carbon content;
13 Decreased leaching and losses of nutrients;
14 Promotion of biological activity;
15 Promotion of plant uptake of water and nutrients.

Many different species are recommended, depending on analyses of the area, soil texture, soil compaction, soil biological activity (macro, meso and micro fauna and flora), presence of nematodes or other pests and diseases, soil OM, nutrient status, rainfall regime and temperature, as well how cover crops need to fit in a crop rotation system. The selection of mixes of crops needs to take account of all these factors (Table 2).

Table 2 Some of the major cover crops grown in Brazil

Season/type of crop		Species	Soil and climatic requirements	Days to flowering	DM (t $ha^{-1}yr^{-1}$)	Advantages and limitations
Winter species	Non-legumes	*Avena strigosa* (Schreb.)	S - C; LF-MF	120–160	2–11	AF; WC; decrease soil root diseases (*Fusarium* spp., etc.); FASM
		Lolium multiflorum (L.)	S - C	120–150	2–6	AF; WC
		Raphanus sativus ssp. oleiferus Metzg.	S - L; A-	90–110	3–9	High nutrient recycling capacity; BP; WC; FASM
		Secale cereale (L.)	S - C; LF; A+; Wlog-; DT	100–120	4–8	BP; WC; controls some soil diseases
	Legumes	*Lathyrus sativus* (L.)	S - C; MF	100–120	2.5–4	AF; HF; mech. harvesting difficult; sensitive to aphids and diseases
		Lupinus albus (L.)	S - C; MF; Wlog-	120–140	3.5–5	AF; HF; BNF; BP; sensitive to diseases (*Fusarium* spp., etc.)
		Lupinus angustifolius (L.)	S - C; A+, Wlog-;	120–140	3–6	AF; HF; BNF; BP; sensitive to diseases (*Fusarium* spp., etc.); FASM
		Lupinus luteus (L.)	S - C; LF; A+; Wlog-	130–150	3–4	Recommended for restoring depleted soils (sandy and clay)
		Pisum arvense (L.)	S - C; A-	100–130	2.5–7	AF; FEG; BNF; sensitive to aphids and some diseases
		Vicia sativa (L.)	S - C; HF; A-, Wlog-	120–150	3–5	AF; BNF
		Vicia villosa Roth.	S - C; LF; A+; WL-	140–180	3–5	AF; BNF; WC

(*Continued*)

Table 2 (*Continued*)

Season/type of crop		Species	Soil and climatic requirements	Days to flowering	DM (t $ha^{-1}yr^{-1}$)	Advantages and limitations
Summer species	Non-legumes	*Brachiaria* spp.	S - C; A+	n.a.	>4	AF; BP; high biomass; SOM
		Helianthus annuus (L.)	S - C; A+; LF; DT	70–120	4–8	FEG, high nutrient recycling; WC
		Pannicum maximum (L)	S - C; WD; DT; A+; Wlog-	n.a.	>20	FEG; AF; BP; SOM
		Paspalum notatum Flugge	S; DT; CT	n.a.	3–8	AF; SOM
		Pennisetum glaucum (Schum.)	S; A+; LF; DT	90–120	3.5–21	AF; BP; SOM; WC; FASM
		Setaria italica (L.)	S - C; WD; MF; DT	45–60	2.5–8.5	AF; FEG; FASM; high seed production
		Sorghum bicolor (L.) Moench	S - C; WD; MF; DT	60–110	3.5–18.5	AF; BP; SOM
	Legumes	*Cajanus cajan* (L.) (dwarf variety)	S - L; LF; Wlog-	70–85	2–6.5	AF; NC; high seed production
		Cajanus cajan (L.) Millsp.	S - C; LF; Wlog-	140–180	3–7.5	AF; BP; BNF + nutrient recycling, NC
		Calopogonium mucunoides Desv.	L - C	n.a.	4–10	WC; GC
		Canavalia ensiformis (L.) DC.	S - C; LF; DT	100–120	5–6	WC (allelopathic effects against *Cyperus* spp. & *Cynodon dactylon*)

Crotalaria juncea (L.)	S - C; MF	110–140	3–8.5	BNF; WC; NC; efficient in nutrient cycling
Dolichos lablab (L.)	S - C; LF; A+; DT; WD	75–150	4–13	AF; HF,
Macroptilium atropurpureum (DC.) Urb.	S - C; WD; A+; MF; DT	n.a.	3–6.5	AF; SOM; WC
Mucuna pruriens (L.) DC.	S - C; LF	130–150	2–5	FEG; GC; BNF; NC
M. pruriens (L.) DC. (dwarf varieties)	S - C; LF	80–100	2–4	NC; FASM; rain during harvesting period can damage the seeds
Pueraria phaseoloides (L.)	L; WD; Wlog-; DT	n.a.	3.5–8	AF; GC
Stylosanthes spp.	S - C; A+, LF; DT	n.a.	n.a	AF; BP; SOM
Vigna radiata (L.)	S - C; DT; WL-	60–80	3.5–6.5	AF; HF; high seed production
Vigna unguiculata (L.)	S - C; L/MF; A+; WL-	70–110	2.5–5.7	AF; HF

n.a. = data not available; S = light-textured (sandy) soil; L = medium-textured (loamy) soil; C = heavy-textured (clayey) soil; L/M/H = low/medium/high fertility; WD = well-drained soil; WLog-/+ = intolerant/tolerant of waterlogging; A-/+ = intolerant/tolerant of soil acidity; DT = drought tolerant; AF = animal forage; HF = human food; BNF = high N fixation; GC = produces good cover; WC = weed suppression; BP = biological plowing; SOM = good SOM builder; FASM = facilitates acid soil management; FEG = fast early growth; NC = nematode control.
Source: adapted from Bolliger et al. (2006).

In general, legume crops have a high potential for fixing nitrogen through the symbiosis between roots and bacteria, and also have a large capacity to recycle nitrogen and other nutrients that were leached to deeper soil layers (Tables 3 and 4). It is possible to quantify the amount of a nutrient recycled and/or biologically fixed by legumes, based on the biomass produced and the nutrients contained in leaf tissue, which will be released on the soil surface, with a significant percentage of these nutrients available for the next crop rotated in the cropping system (Table 5). The amount of nutrients recycled or biologically fixed by plants demonstrates the great potential that different cover plants have in leaving nutrients on the upper soil profile layer that can be absorbed by the roots of later crops. In addition, one of the most important contributions cover crops make is their residues, organic carbon compounds that form stable SOM, which will be responsible, directly or indirectly for beneficial chemical, physical and biological processes in the soil-water-plant system. Over many years the results from research and farming practice in the tropical and subtropical conditions of Brazil have shown that cover crops, as part of a crop rotation system, are economically viable as well as ecologically sustainable (Calegari and Pavan, 1995).

3 Soil organic carbon and nutrients

Brazilian research data indicates that the pattern and quality of SOM in NT soils differs from that of plowed soils. Research has shown that the relative amount of free labile (e.g. particulate OM – POM) rather than humified and occluded SOM fractions is higher in NT soils compared to plowed soils; this has important ramifications for soil structure, nutrient cycling and promoting soil microbial biomass (Reicosky et al., 1995; Bolliger et al., 2006; Balota et al., 2014).

A study by Bayer (1996) in the Brazilian State of Rio Grande do Sul found that, after nine years of NT with high-residue addition (14 t of dry matter ha^{-1} yr^{-1}), soil organic carbon (SOC) content increased by 11 t per ha^{-1} in the top 17.5 cm of soil compared to conventional cropping systems, which only yielded additions of 6.5 t ha^{-1} yr^{-1} of residues. In Southern and subtropical Brazil, with average annual temperatures close to or above 20°C and concomitantly high decomposition rates, it is therefore necessary to annually add between 8 and 10 t ha^{-1} yr^{-1} of residue dry mass in order to maintain the SOM stocks under NT. As Mielniczuk et al. (2003) postulates, this is only really feasible if both suitable high-biomass cover crops and main crops are used. The high amounts of biomass needed can only be produced by cover crops such as fodder grasses, which are used as dry season cover crops in the Cerrado (savannah area), as well as by cereal cover crops, such as millet, sorghum, black oat and rye, crucifers such as oilseed radish or legumes such as vetches and lupines (Calegari, 1998b), when grown in combination with main crops able to produce

Table 3 Chemical composition (plant tissue) of different spring/summer species used as cover crop and animal food in the flowering stage

Cover crops	Nutrients (% of dry matter)							ppm		
	N	P	K	Ca	Mg	C	C:N ratio	Cu	Zn	Mn
Crotalaria juncea	2.50	0.19	1.20	2.31	0.47	45.25	18.10	14	44	179
Crotalaria spectabilis	2.17	0.09	1.59	0.49	0.37	50.83	23.42	8	23	126
Cajanus cajan	2.61	0.14	2.61	1.79	0.45	56.30	21.57	7	22	87
Canavalia ensiformis	3.19	0.15	5.62	1.35	0.63	50.15	15.72	9	62	254
Canavalia brasiliensis	2.49	0.13	1.68	0.20	0.16	51.24	20.57	4	14	17
Mucuna pruriens (grey)	2.50	0.15	1.40	1.20	0.27	52.30	21.12	16	28	183
M. pruriens (black)	2.49	0.13	1.40	1.17	0.27	52.15	21.06	14	29	174
M. pruriens (dwarf)	3.10	0.19	4.49	2.14	0.65	50.83	16.39	9	85	179
Vigna radiata	2.09	0.21	4.94	1.48	0.75	52.47	25.10	10	78	127
Vigna unguiculata	2.62	0.20	2.82	0.93	0.28	45.42	17.33	-	-	-
Indigofera sp.	2.17	0.14	1.54	1.20	0.32	40.36	18.60	13	24	53
Calopogonium mucunoides	2.16	0.12	1.56	1.40	0.29	46.73	21.63	9	15	172
Pueraria phaseoloides	3.68	0.29	2.14	1.30	0.41	54.10	14.70	11	27	155
Glycine wightii	2.60	0.23	2.39	0.99	0.35	45.03	17.31	8	32	102
Centrosema pubescens	2.34	0.23	1.19	0.66	0.45	47.60	20.34	10	32	67

Source: data from Calegari et al. (1993) and Calegari (1995b).

Table 4 Chemical composition of fall/winter plant tissue species

Cover crop	Dry matter %								ppm		
	N	P	K	Ca	Mg	C	Protein	C/N ratio	Zn	Cu	Mn
Hairy vetch	3.82	0.30	2.03	0.78	0.27	37.87	23.87	10.05	26	9	61
Common vetch	2.87	0.23	2.88	1.05	0.41	37.1	17.94	12.9	24	9	87
Serradella	1.79	0.14	3.55	1.10	0.45	40.14	11.18	22.43	59	13	97
Radish	2.68	0.17	2.80	1.54	0.76	38.58	16.75	14.45	49	8	84
White lupin	3.20	0.09	2.66	0.46	0.38	47.49	20.0	14.84	57	12	330
Yellow lupin	2.94	0.16	2.50	0.59	0.39	42.25	18.37	14.37	66	14	359
Blue lupin	3.19	0.19	2.29	1.20	0.49	37.83	19.93	11.86	24	13	230
Sweet blue lupin	2.28	0.10	1.75	0.59	0.42	37.87	14.25	16.61	32	16	147
Field pea IAPAR-83	2.09	0.12	`1.50	0.70	0.20	39.77	13.06	19.02	8	22	52
Wheat	0.77	0.06	1.15	0.22	0.10	40.38	4.81	52.71	-	-	-
Sweet pea	2.23	0.10	2.90	0.39	0.19	41.91	13.93	18.79	22	11	52
Black oat	1.93	0.28	2.15	0.43	0.21	39.69	12.06	20.76	11	7	102
White oat	0.81	0.052	2.40	0.24	0.17	38.52	5.06	47.55	9	6	138
Ryegrass	1.34	0.067	2.60	0.41	0.22	59.22	8.37	44.20	23	9	214
Rye	1.22	0.075	1.40	0.18	0.14	44.59	7.62	36.54	15	6	53
Sunflower	1.80	0.15	2.40	1.55	0.62	39.95	11.25	22.19	31	18	96
Corn spurrey	2.13	0.22	3.45	0.52	0.77	41.92	13.31	12.78	44	11	136

Source: data from Calegari et al. (1993).

Table 5 Biomass of different cover crops species and nutrients dry mass

Common name	Scientific name	Cycle	Biomass (t/ha)	Dry mass (t/ha)	N	P	K
					(% dry mass)		
Black oat	*Avena strigosa* Schreb.	Annual	15–40	2-11	0.70-1.68	0.14–0.42	1.08-3.08
Rye	*Secale cereale* L.	Annual	30–35	4-8	0.58–0.66	0.16–0.29	0.75-1.45
Hairy vetch	*Vicia villosa* L.	Annual	20–37	3-5	2.51–4.36	0.25–0.41	2.41-4.26
Common vetch	*Vicia sativa* L.	Annual	20–30	3-5	2.74–3.47	0.27–0.38	2.33-2.56
Field pea	*Pisum sativum* Subesp. Arvense	Annual	15–40	2.5-7	1.77–3.36	0.14–0.41	0.67-3.31
Radish	*Raphanus sativus* L.	Annual	20–65	3-9	0.92-1.37	0.18–0.33	2.02-2.65
White lupin	*Lupinus albus* L.	Annual	30–40	3.5-5	1.22-1.97	0.25–0.29	1.00-1.77
Blue lupin	*Lupinus angustifolius* L.	Annual	25–40	3-6	0.85-2.15	0.12–0.29	1.36-1.49
Black oat + common vetch	-	Annual	15–50	2-10	0.93-1.39	0.15–0.16	1.23-1.47
Common vetch + black oat + white lupin	-	Annual	18–26	3.5-6	1.99-2.99	0.16–0.26	2.86-3.86
field pea + black oat	-	Annual	27–34	4-10	1.62–2.22	0.13–0.18	2.55-3.35
Pearl millet	*Pennisetum americanum* L.	Annual	11–90	3.5-21	0.34–1.46	0.13–0.29	1.05-3.12
Cowpea	*Vigna unguiculata* L.	Annual	20–33	2.5-6	1.67–2.22	0.25–0.50	1.82-2.77
Black Mucuna	*Mucuna pruriens cv. Black*	Annual	12-23	2-5	2.29-2.73	0.11–0.17	1.25-1.55
Grey Mucuna	*Mucuna pruriens cv.grey*	Annual	10-25	2-6	1.56-2.43	0.46–0.57	1.00-1.55

(*Continued*)

Table 5 (*Continued*)

Common name	Scientific name	Cycle	Biomass (t/ha)	Dry mass (t/ha)	N	P	K
						(% dry mass)	
Dwarf Mucuna	*Mucuna pruriens cv.dwarf*	Annual	10-18	2-4	2.85-3.35	0.16-0.23	4.14-4.84
Pigeon pea	*Cajanus cajan* L.	Bi or Tri- Annual	18-45	5-12	2.41-2.85	0.12-0.19	2.40-2.84
Dwarf pigeon pea	*Cajanus cajan* L.	Annual	14- 22	2-6.5	1.02-2.04	0.21-0.28	0.92-1.47
Crotalaria spectabilis	*Crotalaria spectabilis* L.	Annual	12-23	4-7	2.14-2.20	0.07-0.12	1.40-1.78
Jack bean	*Canavalia ensiformis* L.	Annual	14-26	3-7	3.00-3.39	0.12-0.18	5.30-5.94
Ceará Brave bean	*Canavalia brasiliensis* L.	Annual	14-25	3-6.5	2.27-2.71	0.11-0.15	1.58-1.78
Mung bean	*Vigna radiata* L.	Annual	12-22	3-5.5	2.00-2.18	0.15-0.27	4.64-5.24
LabLab	*Dolichos lablab* L.	Annual	14-28	4-7	2.15-2.57	0. 27-0.61	2.14-2.53
Leucaena	*Leucaena leucocephala* L.	Perennial	20-50	10-16	4. 17-4.43	0.17-0.28	1.45-1.94
Indigofera	*Indigofera* sp.	Perennial	13-24	4-7	2.02-2.33	0.09-0.19	1. 45-1.64
Calopogonio	*Calopogonio mucunoides* L.	Perennial	14-23	4-6	2.05-2.28	0.08-0.17	1.43-1.68
Tropical Kudzu	*Pueraria phaseoloides* L.	Perennial	13-25	4-7	3.47-3.88	0.23-0.36	2.06-2.23
Perennial soybean	*Glycine wightii* L.	Perennial	12-23	4-6	2.44-2.85	0.17-0.30	2.24-2.45
Centrosema	*Centrosema pubescens* L.	Perennial	14-27	4.5-6.5	2.21-2.48	0.17-0.29	1.03-1.34
Crotalaria juncea	*Crotalaria juncea* L.	Annual	15-35	2.5-8.5	1.42-1.65	0.19-0.21	0.96-1.38
Sunflower	*Helianthus annuus* L.	Annual	20-46	4-8	1.02-1.12	0.18-0.24	2.50-2.78
Pearl millet + cowpea	-	Annual	19-40	3.5-10	0.61-0.82	0.13-0.17	1.08-1.12

Source: data adapted from Calegari (2000, 2006), Calegari et al. (1993), Florentin et al. (2001) and Bolliger et al. (2006).

high biomass, such as maize (rather than soybean or wheat). In this context, it is essential to consider below-ground biomass, especially as the carbon (C) added in roots can be close to double that of shoots (Mielniczuk et al., 2003). Sisti et al. (2004) found an increased volume of roots well below plowing depth under NT compared to plowed soils.

Sisti et al. (2004) and Amado et al. (2006) have studied the role of N additions in the buildup of SOM under NT in Brazil. They both found that, where rotations with N_2-fixing legumes were included, much more SOM was accumulated, highlighting that there, as well as carbon, cover crops contributed a net external input of N. Sisti et al. (2004) also found that, where a hairy vetch (winter cover crop) was included in a rotation with common oat and wheat in winter and maize or soybean in summer, soil C stocks were increased by approximately 10 t ha^{-1} down to 100 cm depth after 13 years of NT, which represents about 17 t ha^{-1} of soil C higher than observed in the plowed soil. They further postulated that, where net N balance was close to zero over the whole crop rotation, little SOM accumulation was observed (Bolliger et al., 2006). Amado et al. (2006) reported that pigeon pea and mucuna cover crops integrated into NT maize cropping systems had the highest C accumulation rates under NT; they also found that intensive cropping systems, including mixes of black oat with hairy vetch in winter and maize intercropped with cowpea in summer, as well as rotations of oilseed radish and ryegrass, among other crops, effectively increased C accumulation rates in NT compared to more conventional double-crop systems used by many farmers.

According to Prudencio et al. (2004), CA defines soil fertility in ways that go beyond the simple application of fertilizer and manure, focusing on the concept of a living soil. A soil rich in biological activity is able to support plants and animals in a sustained way. Soil fertility is a part of soil, not something we bring in. Soil fertility can be related to the capacity to produce biomass, be it microorganisms or root and shoot biomass (Bolliger et al., 2006). Fertility is about not only the presence of nutrients, as shown in a laboratory test, or in amount of nutrients applied on the soil, but their accessibility by the plant in a timely manner, as well as in the harmonious inter-relationships between soil, water and plants. Limitations to nutrient availability can be biological and physical as well as chemical, and in-situ assessment (diagnosis) needs to be conducted to understand the potential productive capacity of a given soil.

The efficiency of CA has been reported by Calegari et al. (2008) in a 19-year experiment with a very acid Oxisol (Rhodic Hapludox) with a high clay content (72% clay, 14% silt, and 14% sand). They found that, after 19 years of different cropping sequences and types of tillage management, NT management resulted in 6.84 Mg ha^{-1} more organic C when compared to conventional tillage (64.6%) at a soil depth of 0–10 cm, 29.4% more at a depth of 0–20 cm, but equivalent amounts as conventional tillage at a soil depth of 20–40 cm. Greater

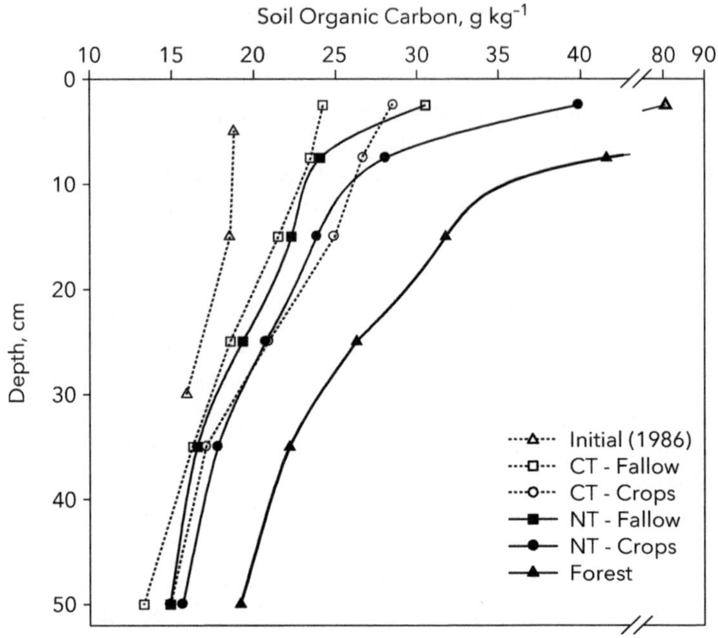

Figure 2 Soil organic carbon (SOC) for different soil management and cropping systems in a Rhodic Hapludox in southwestern region of Paraná State, Brazil. CT – conventional tillage; NT – NT. Source: adapted from Calegari et al. (2008).

amounts of SOC were within the 0–20 cm depth (i.e. the moldboard plow layer). Greater amounts of organic C were generally sequestered when winter cover crops were used in a NT system (Fig. 2). Continuous NT management including the use of winter cover crops resulted in the greatest amount of SOM in the surface soil, and was the only cultivation treatment that approached the undisturbed forest condition. The NT system with winter cover crops has found to store greater amounts of soil organic C, providing a model for sustaining the productivity of Oxisols in tropical and subtropical regions of the world.

In order to find the best crop rotation system for each situation, it is important to know the characteristics of the cover crop species to be used in the rotations, their potential benefits, biomass production capacity, residue production and decomposition rate. According to Leite et al. (2010), the decomposition rate of the vegetative residues determines the rate of nutrient release. The rate of decomposition can vary between different plant materials, especially due to amount and quality of macro and microorganism decomposers, chemical composition of residues (C/N ratio, content of cellulose, lignin and hemicelluloses) and climatic conditions, mainly temperature and humidity (Carvalho et al., 2010; Maluf et al., 2015). Because of high and rapid biomass production with a high C/N ratio, grasses are characterized by slow decomposition which provides high OM input and protects soil against erosion. On the other hand, legumes increase N availability in soil through biological fixation by specific bacteria but, because of low C/N ratio, decompose faster.

When studying plant residue decomposition, many authors (Torres et al., 2005; Kliemann et al., 2006; Acosta et al., 2014) have confirmed that grasses decompose more slowly because of a high C/N ratio. Other authors have compared individual cover crops and combinations of crops (Ziech et al., 2015; Ceretta et al., 2002; Aita and Giacomini, 2003). They have found that combinations of poaceae and fabaceae produce a good amount of biomass with a moderate C/N ratio, with more gradual decomposition when compared to pure fabaceae, whilst also providing soil cover and N fixation for the benefit of the following crops. According to Aita and Giacomini (2003), the decomposition speed and N release from the cover crop residues was inversely proportional to C/N and lignin/N ratios, and directly proportional to concentrations of total N in the biomass and of N and C in the fraction soluble in water. Many authors have reported the important contribution of N input by legumes to C sequestration (Bayer et al., 2000, 2006; Resck et al., 2000; Amado et al., 2001; Aita and Giacomini, 2003; Six et al., 2004; Amado et al., 2006; Calegari et al., 2008).

This research highlights the importance of including fabaceae and poaceae in crop rotations, alternately or in combination. The aim is good biomass production and gradual decomposition, with nutrient recycling compatible with the needs of the following the crop whilst providing effective and lasting soil cover. The improvement of soil fertility and soil structure is a process and depends on various factors affecting macro and microbial activity in the soil, including the quantity and quality of input from crop residues, the timing of crop rotations and soil management systems, climatic conditions (mainly temperature and humidity). Sá et al. (2010) have mapped the evolution of chemical, physical and biological soil attributes in an NT system in Southern Brazil (Table 6). This process of evolution shows the importance of avoiding soil disturbance and maintain crop rotations in order to build up long-term soil quality and improve crop yields with costs reducing over time.

The decomposition and transformation of above- and below-ground plant detritus (litter) is the main process by which SOM is formed (Cotrufo et al., 2013). The amount and nature of organic compounds from crop residues added to the soil surface in a NT system changes SOM concentrations and the total organic C in soil (Canalli, 2009).

Research on a subtropical and tropical region of Brazil by Sá et al. (2015) found that the SOC stock was drastically reduced by a continuous use of plow-based conventional tillage (CT), leading to a depletion of 0.58 (subtropical) and 0.67 (tropical) Mg C ha^{-1} yr^{-1} at a depth of 0–20 cm. In NT systems, the rate of SOC sequestration was 0.59 Mg C ha^{-1} yr^{-1} in the subtropical region and 0.48–1.30 Mg C ha^{-1} yr^{-1} in the tropical region. Any increase in SOC sequestration

Table 6 Evolution scale of the no-tillage system for the Southern region (subtropical), Brazil

Early stage (0–5 anos)	Transition (6–10 anos)	Consolidation (11–20 anos)	Maintenance (>20 anos)
Soil re-aggregation to new soil structure	Increase of the soil density by machinery traffic	Soil density stabilization	Soil density stable
Low straw high decomposition	Start of the straw accumulum on the soil surface	Straw accumulation on the soil surface	High straw accumulation on soil surface
Low SOM content[†]	Início de acúmulo de MOS	Accumulum of C and CTC improvement	High C e N accumulum; Continuous flux of C e N
Low P availability	Increase of the P availability	P accumulum and minor response to P-fertilizer	High P accumulation; Reduction of P-fertilizer
Immobilization of N > Mineralization Restoration of the fauna and soil microbial biomass	N Mineralization ≥ immobilização	N Mineralization > N Immobilization	Continuous N mineralization
Great need of N-mineral to the crops	N-mineral need to the crops	Decrease of mineral N need > water storage and nutrients cycling	Reduction of mineral N application; > water storage and continuous nutrients cycling

Source: adapted from Sá et al. (2010).

in the subsoil of CT was not observed at either of the two sites. In NT, however, SOC accumulation increased with the increase in the input of biomass-C in the tropical region. A high SOC resilience in the tropical NT systems indicates a considerable potential to reverse the process of soil degradation and SOC decline through converting CT to intensive NT systems. The fraction of C input by crop residues converted into SOC stock was ~14.2% in the subtropical region and ~20.5% in the tropical region.

Dieckow et al. (2009) examined the influence of land use change, tillage system and soil texture on organic carbon (C) stocks and on OM composition of tropical and subtropical soils from Brazil in four long-term experiments (11–25 years) based on fine- and coarse-textured soils. The authors concluded that the conversion from CT to NT management is a viable strategy to increase OM content and the proportion of labile C in tropical and subtropical soils. Coarse-textured soils seem to be more responsive (less resistant) than fine-textured soils. The direct relation between increase in C stock and increase in potentially labile moieties (e.g. O-alkyl) and the decrease in more recalcitrant moieties (e.g. aromatics) in NT soils, suggests that spatial inaccessibility by stable aggregates plays a major role in promoting C accumulation in NT soils. Gaiser et al. (2012) highlights the importance of characterizing SOC fractions when evaluating carbon sequestration effects of green manure practices with different plant residues. The mixtures of legumes and grasses, which contain high levels of extractable polyphenols and moderate levels of nitrogen, may be able to increase the content of SOC fractions within relatively short periods, fixing carbon in the soil and supplying additional cation exchange sites to improve cation retention in the soil.

In a clay Oxisol, Sá et al. (2009) reported C stock ranging from 0.64 to 1.30 Mg ha^{-1}, with an average of 0.82 Mg ha^{-1} yr^{-1}. They confirmed that, in the 0–5 cm soil layer, 65%–90% of the SOC was represented by labile C. Other authors have highlighted the importance of labile C in soil C storage and stable forms of C (Bayer et al., 2004; Dieckow et al., 2005; Amado et al., 2006; Calegari et al., 2008; Sá and Lal, 2009).

Briedis et al. (2012) reviewed NT using an Inceptisol soil in Ponta Grossa, Paraná State, Brazil. They evaluated the decomposition of black oat (*Avena strigosa* Schreb.) residues, the relationship between biomass decomposition and SOC content, the particulate organic carbon (POC) content and the soil carbon stratification ratio (SR). The oat dry biomass was 5.02 Mg ha^{-1}. However, after 160 days, only 17.8% of the initial dry biomass was on the soil surface. The SOC in the 0–5 cm soil layer varied from 27.6 (initial) to 30.1 g dm^{-3} (after 160 days). The SR increased from 1.33 to 1.43 in 160 days. There was also an increase in the POC pool in this period, from 8.1 to 10.7 Mg ha^{-1}. The increase in SOC in the 0–5 cm soil layer in 160 days was mainly due to the increase of POC derived from oat residue decomposition. The linear relationship between SOC and POC showed that 21% of SOC was due to the more labile fraction. The authors indicated that the continuous input of residues could be intensified to increase the C pool and sequestration in soils under NT.

Canalli (2009) evaluated crop residue decomposition in a long-term experiment under NT in the same region. They reported that average biomass losses during decomposition of winter crop residues varied as follows: vetch > black oat = radish. In the case of summer crop residues, beans had the highest decomposition rate, followed by soybean, while maize had the lowest decomposition rate. The half-life of crop residues evaluated showed the following sequence: maize (173 days) > soybean (116 days) > bean (87 days) > oat (69 days) = radish (69 days) > vetch (46 days). The high half-life for maize residues (173 days) shows the importance of its inclusion in crop rotations, combining a high input of OM with slow and gradual decomposition. The half-life for vetch residues of 46 days shows that legumes decompose faster than maize. The decomposition rates of crop residues are affected by the residues of the previous crop.

Long-term NT leads to SOM stratification in the soil profile, with the soil respiration (SR) of SOC providing an indicator of soil quality (Sá et al., 2009). The effect of CT and NT on the SR of SOM pools was assessed in an Oxisol in Southern Brazil. In undisturbed NV soil, the SR increased with the increase in soil depth. In contrast, the SR decreased in CT, and the SOM was uniformly distributed in the soil profile. All NT treatments restored the SR and had higher values compared to NF. The SR for SOC ranged from 1.12 to 1.51 for CT-22 compared with 1.64–2.61 SR for NT, surface and subsoil layers, respectively. The SR for POC and nitrogen (PN) were higher than those for stable C and N. However, SR for the biological pools (soil microbial carbon – MBC, soil microbial nitrogen – MBN and BR) was the highest and strongly correlated with the rate of SOC sequestration. The data showed that the SR of SOC is an efficient indicator of C sequestration in long-term NT management. Balota et al. (2014) also reported the benefits of long-term (23 years) winter cover crops in conventional (plow) and NT management on soil microbial quality indicators in an Oxisol at Paraná State, Southern Brazil, concluding that winter cover crop with either tillage is a beneficial practice in enhancing soil microbial quality and soil organic C stocks.

Ferreira et al. (2012) evaluated the C sequestration rate and the minimum amount of crop residues required to maintain the dynamic C equilibrium ($dC/dt = 0$) of two soils (Typic Hapludox), with a medium and clay textural class, in a long-term NT experiment with different rotations: wheat/soybean/black oat + vetch/maize and wheat/maize/black oat + vetch/soybean. The C conversion rate in both soils studied was lower than that reported in other studies in the region, resulting in a greater amount of crop residues left on the soil surface. The estimated minimum amount of crop residues required to maintain the dynamic balance ($dC/dt = 0$) of the system was 6.83 Mg ha^{-1}. A Typic Hapludox medium texture required 0.6 Mg ha^{-1} more crop residue than a Typic Hapludox clay texture to maintain the dynamic balance of the consolidated NT system (in Campos Gerais region, Southern Brazil). Campos et al. (2011) also found that C conservation in NT incorporating crop rotation was higher than in CT with monoculture, and that ratio of soil C stock per unit of C input by crop residues increased with cropping system diversification, regardless of soil tillage.

Soil aggregation and its stability can be influenced by different soil management and cultural practices, which lead to changes in SOM content and biological activity. Soil aggregation can be improved by management practices that reduce soil disturbance, use species with high-biomass production and root density, and with a low decomposition rate (Wei et al., 2006). This improves soil structure and stability and, consequently, promotes greater microbial activity and nutrient cycling (Bronick and Lal, 2005), as well as conserving OM in aggregates which maintains or increases soil carbon stock.

Soil management changes the structure, size and formation of aggregates. NT with crop residues retained on the soil surface stimulates microbial biomass activity, promoting a continuous C and N flow, which affects aggregate structure. In contrast, soil disturbance disrupts soil aggregates and release of C-CO_2 from the soil, decreasing soil C and contributing to an increase in greenhouse gas emissions. This process is particularly affected by climatic conditions such as temperature and humidity (Table 7). Aggregate stability and SOM are closely correlated, and aggregation status and organic C content are attributes that can be used to indicate soil health, since these are directly related to processes such as water infiltration and storage capacity, as well reduction of soil density and erosion (Wuddivira and Camps-Roach, 2007).

Binding agents are essential for the formation and stabilization of aggregates. These may be transitory, composed of polysaccharides, animal and plant residues and root exudates. These are easily decomposed by the soil microbiota, persisting in the soil for hours to a few months. Roots and fungal hyphae may persist in the soil for months to years, and are important in the formation and stabilization of macroaggregates. They can be persistent, represented by decomposed organic compounds which form organomineral complexes that are important in the formation and stabilization of microaggregates (Wei et al., 2006). Macroaggregates are associated with

Table 7 Losses of the soil C content due to continue uses of CT (plow) associated to monocrop

Climate	Latitude	Time[a] (years)	C total loss (%)	C annual loss[b] (%)	Reference
Temperate					
Canada[c]	54° N	50	48	0.96	Campbell and Souster (1992)
EUA[d]	39° N	50	58	1.16	Heinzmann (1985)
Subtropical					
RS, Brazil	30° S	15	30	2.0	Pöttker et al. (1992)
PR, Brazil	25° S	10	35	3.5	Sá et al. (2009)
Tropical					
GO, Brazil	15° S	5	35[e]	7.0	Resck (1992)
MT, Brazil	12° S	5	69[f]	13.8	Seguy et al. (2006)

[a] Canadá e EUA (winter wheat monoculture), Rio Grande do Sul (RS) and Parana (PR) Brazil (winter wheat/summer soybean succession), Goiania (GO) and Mato Grosso (MT) Brazil (soybean monoculture).

[b]
- Canada in 50 years wheat in conv.tillage - losses of 48% of the original soil C content (0.96% per year).
- In Mt-Brazil, in Quartzarenic soil in 5years - losses of C were 69% (13.8% per year).
- In temperate climate with harsh winter (5 to 7 months with snow - Saskatoon - Canada) annual C losses are almost 15 times lower than those occurring in soils under tropical and sandy texture.
- These data are in agreement with the statements of Dr. R. Lal and Terry Logan published in 1995. The colder and less rainfall and clayey soils, the lower the C losses. The warmer and abundant rainfall and sandy soils the greater the C losses.

[c] Saskatchewan.

[d] Missouri.

[e] Latossolo Vermelho (350 g kg^{-1} Clay content).

[f] Neossolo Quartzarênico.

Source: adapted from Sá et al. (2010).

labile organic C and decomposition of OM, underlining the importance of the continuous addition of vegetal residues for the maintenance and stabilization of soil structure. On the other hand, microaggregates are associated with more recalcitrant organic carbon and stable OM.

Canalli (2009) evaluated soil aggregation and C stock in different crop rotations under NT. They reported a significantly high percentage of macroaggregates (19–8 mm) when species with high C:N ratio were included (oat, maize and soybean). They highlighted the importance in the NT system of no soil disturbance and continuous input of crop residues to promote the formation of large, more stable macroaggregates, resulting in an increase of C and N content in the soil. Comparing soil management systems involving crop rotations, Castro Filho et al. (1998) found that the accumulation of vegetal residues on the soil surface improved the state of soil aggregation due to an increase in organic C, especially at a depth of 0–10 cm, and that soil aggregation had a tendency to increase when crop rotations included species with higher C/N ratio (such as corn/maize). The continuous addition of vegetal residues to the soil surface thus plays a major role in the formation and stabilization of soil aggregates, since it activates the soil microbiota which initiate decomposition of plant residues. This results in the release of exudates that act as linking agents between soil primary particles as a first step to formation of soil aggregates. Plant roots and fungal hyphae contribute to this process of formation of stable aggregates in the soil.

The experience of using of cover crops, crop rotation and no soil disturbance has promoted soil recovery and improved productive capacity in countries such as the United States, Australia, Russia and many others (Angers et al., 1997; Baker et al., 2007; Blanco-Canqui et al., 2011; Dimassi et al., 2014). Long-term results in Brazil also prove that cover crops are both economically and ecologically sustainable, including proving higher water storage in the soil profile and preventing water evaporation. They provide not only higher yields of crops such as cotton, soybean, maize, rice, sunflower, sorghum, and wheat, but also conservation, maintenance and/or recovery of soil fertility. In addition, they mean less need for N fertilizers (with the use of leguminous plants), better weed control through the effects of mulch, and a better biological balance in the soil which helps in decreasing pests and diseases. In summary, many studies and farmer experiences in Brazil show higher SOM accumulatation under NT, which means that that farmers can contribute to reducing CO_2 greenhouse gas emissions and reduce the effects of agriculture on climate global change.

4 Effect of cover crops on soil acidity and nutrient availability

The cultivation of different plant species in rotation can alter the dynamics of soil acidity. Changes in soil pH in the rhizosphere can vary up to 1.7 units (Li et al., 2010).

Reasons for this variation include the imbalance in cation-anion absorption which is particularly affected by N sources, efflux of protons induced by P deficiency (Gahoonia and Nielsen, 2004) and exudation of organic acids (Jones, 1998). Soil acidification is intensified by growing legumes because they generally have a higher uptake and exportation of basic cations (Ca, Mg and K) (Li et al., 2013). Moreover, biological N fixation in legumes reduces nitrate uptake, increasing nitrate leaching along with basic cations (Frageria and Nascente, 2014).

Changes in soil pH induced by plants may also alter nutrient availability, especially for those nutrients that are affected by soil pH (e.g. P, Fe, Mn, Cu and Zn). In evaluating cover crop residues, mainly radish and black oat, Miyazawa et al. (1993) found that root organic exudates promote change in nutrients distribution in the soil profile. Moreover, long-term experiments on acidic soils in Brazil have indicated that NT may increase soil pH, KCl-exchangeable Ca and Mg, and Mehlich-1 P, and decreased KCl-exchangeable Al (Calegari, 1995a; Calegari and Pavan, 1995; Sidiras and Pavan, 1985; Tiecher et al., 2018) compared to CT (Machado and Gerzabek, 1993; Muzilli, 1983; Sidiras and Pavan, 1985). Cover crops residues normally contain low-molecular weight organic acids, such as malate, citrate and others which may increase soil pH deeper in the soil, as well as generally increasing Ca and decreasing free Al concentrations in the soil profile. Wang et al. (2008) reported that the acquisition of P was improved by the release of protons by wheat. Rose et al. (2010) also found that the soil pH in the rizosphere and the exudation of carboxylates may partially explain depletion of inorganic P in the soil. The acquisition of P of low lability due to exudation of organic acids has been observed in lupin (*Lupinus* albus L.) by several researchers (Bais et al., 2006; Bayon et al., 2006; Shane et al., 2008; Wang et al., 2008).

The use of cover crops with a high soil–root contact area are particularly important for P and K uptake because they have low mobility in Oxisols due to large amounts of clay minerals and iron oxides (Kaminski et al., 2010; Tiecher et al., 2012b). In the long-term, the lack of any recycling of crop residues in monocropping systems might cause severe K depletion from soil reserves (non-exchangeable K) (Srinivasarao et al., 2014), or K losses by leaching (Alfaro et al., 2006). Different crop rotation management systems can alter soil P availability in the long-term due to increasing exudation of acid enzyme phosphatase, increasing organic P (Chavarría et al., 2016; Cui et al., 2015; Kunze et al., 2011) or by the exudation of organic acids with low-molecular weight that promotes P mobilization by ligand exchange or by occupying P adsorption sites (Bayon et al., 2006; Neumann and Römheld, 1999). Moreover, the mycorrhizal or non-mycorrhizal nature of cover crops has a regulatory effect on enzyme activity linked to organic P mineralization in the soil (Kunze et al., 2011). The use of non-mycorrhizal plants such as lupin and oilseed radish (*Raphanus sativus* L.) increases phosphatase enzyme activity in the soil compared to vetch (*Vicia sativa* spp.) and black oat (*Avena strigosa* Schreb) (mycorrhizal plants) (Dalla Costa and Lovato, 2004).

Results from a long-term experiment established in 1986 in a very clayey Oxisol from Pato Branco, Paraná State, in Southern Brazil, highlight the importance of growing cover crops to increase nutrient availability through crop residue deposition on the soil surface and subsequent decomposition (Calegari, 1995a,b; Calegari et al., 2013b; Tiecher et al., 2012b,a, 2017, 2018). These studies have evaluated the influence of blue lupin, hairy vetch, black oat, oilseed radish, spring wheat and fallow under CT and NT system on soil acidity and nutrient availability. The findings have demonstrated that cover crops have limited influence on soil acidity parameters in the long-term. Although cover plants may not directly affect the exchangeable Ca and Mg contents in the soil, the increase in the OM content resulting from the use of cover crops combined with conservation management systems, such as NT, can alter an important soil chemical property, the cation exchange capacity (CEC).

In highly weathered soils, like most tropical and subtropical soils, there is a very low or negligible amount of clay minerals of the 2:1 type, such as smectites and vermiculites, which have permanent electric charges due to isomorphic substitution (Essington, 2003). In these soils, OM is a key component in the generation of electric charges to retain basic cations, such as Ca, Mg and K, and in increasing soil fertility. After a 23-year study with winter cover crops in an Oxisol from Southern Brazil, there was an average increase of 0.22 $cmol_c\ kg^{-1}$ in the CEC for each increment of one gram in the total organic carbon content (Fig. 3). Other studies involving cover crops and soil management systems in Southern Brazil have also found a similar relationship. In another Oxisol from the

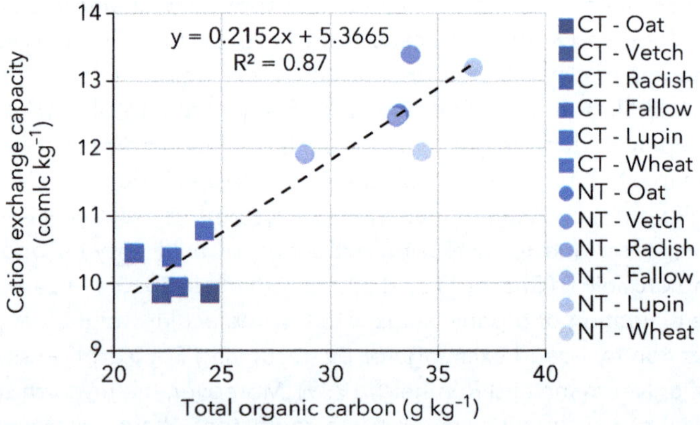

Figure 3 Relationship between total organic carbon content and cation exchange capacity in the 0–5 cm soil layer from a long-term experiment (23-years) with different soil management systems and winter cover crops in Pato Branco, State of Paraná, Southern Brazil. Source: data from Tiecher et al. (2012a,b).

States of Paraná and Santa Catarina in Southern Brazil, Ciotta et al. (2003) and Bayer et al. (2003) found an increase of 0.26 and 0.35 $cmol_c$ kg^{-1} in the CEC for each increment of one gram in the total organic carbon content, respectively.

Among winter cover crops, two cover crops, black oat and blue lupin, have demonstrated particular efficiency in improving the availability of P and K when cultivated under NT (Table 8). Black oat in NT resulted in higher availability of P and K in the topsoil (<10 cm), which may be due to its higher biomass yield (approximately 15 Mg ha^{-1} of dry biomass) which exceeds other winter cover crops (Calegari and Machado, 2013; Tiecher et al., 2017). On the other hand, the higher availability of P in the 0–10 cm soil layer in NT with blue lupin might be due to its capacity for excreting organic acids such as citrate, malate and fumarate, even in conditions of relatively high P availability (Bayon et al., 2006; Li et al., 2010; Shane et al., 2008; Steffens et al., 2010). Blue lupin also promotes the mobilization of P by ligand exchange and/or by dissolving and occupying P adsorption sites (Fink et al., 2016). Blue lupin is able to access inorganic P forms that are inaccessible to other plants, cycling larger amounts of P. Black oat and blue lupin also increase the content of labile organic P (extracted by $NaHCO_3$) in the soil of the 0–5 cm layer in NT, and hairy vetch resulted in greater acid phosphatase enzyme activity among the winter treatments (Tiecher et al., 2012a). However, in a short-term experiment, Teles et al. (2017) observed no influence of cover crops in the more labile P fractions. On the other hand, ryegrass and black oat accumulated more moderately labile P from the 5–10 cm layer than white clover and common vetch.

Growing cover crops in winter reduces soil P and K losses, demonstrating the importance of cover plants in the cycling of these elements, especially under NT. A fallow period in winter resulted in the most negative balance of P and K after 23 years, compared to the other cover plants (Fig. 4). Fallow under NT resulted in K losses while, with winter cover crops, the K budget varies from null (for wheat) to +223 kg ha^{-1} with black oats, demonstrating the great cycling potential of black oats due to its high biomass production compared to other cover crops (Fig. 4). The positive K budget for cover crops in NT may be due to K cycling from deeper soil layers (>40 cm) or due to contributions of non-exchangeable K forms (Li et al., 2016).

The content of C and N as well as the C/N ratio in the tissue of cover plants has been well studied and documented in the literature (Fig. 5a, b, d). It is known that legumes have a lower C/N ratio than grasses (Fig. 5). The lower C/N ratio favors mineralization and accelerates the release of nutrients such as P and K for the subsequent crop. The speed of nutrient release in legumes, such as vetch, is often very rapid, and incompatible with the needs of the following crop. On the other hand, grasses with a high C/N ratio can delay the release of P and K too long. However, a better balance can be achieved between the demand for P and K by commercial crops and the release of these nutrients from crop

Table 8 Summary of the effect of the use of cover crops on the availability of soil nutrients in a long-term experiment from Pato Branco, State of Paraná, Southern Brazil

Soil parameter	Soil depth (cm)	Fallow	Wheat	Radish	Vetch	Lupin	Oat	Reference
Available P (Mehlich-1)	0–5	+	++	++	++	+++	+++	Tiecher et al. (2017)
Available P (Mehlich-1)	0–5	+	+	++	++	+++	++++	Calegari and Machado (2013)
Available P (AER)	0–5	+	++	++	++	+++	+++	Tiecher et al. (2012b)
Available P (AER)	5–10	+	+	+	+	++	+	Tiecher et al. (2012b)
Soluble P	0–5	+	+	+	+	++	++	Tiecher et al. (2012b)
Inorganic P extracted by $NaHCO_3$	0–5	+	+	+	+	++	++	Tiecher et al. (2012b)
Inorganic P extracted by HCl	0–5	+	+	++	+	+++	+++	Tiecher et al. (2012b)
Inorganic P extracted by HCl	5–10	+	++	+	+	++	++	Tiecher et al. (2012b)
Inorganic P extracted by NaOH 0.1 mol L^{-1}	0–40	+	+	+	+	++	+	Tiecher et al. (2012b)
P stored into the SMB	0–40	+	++	++	++	++	++	Tiecher et al. (2012a)
Organic P extracted by $NaHCO_3$	0–5	+	++	++	++	+++	+++	Tiecher et al. (2012a)
Acid phosphatase enzyme activity	0–10	+	++	++	+++	++	++	Tiecher et al. (2012a)
Total organic C	0–10	+	++	++	++	++	++	Tiecher et al. (2012a)
Available K	0–20	+	++	++	++	++	+++	Tiecher et al. (2017)
Available K	0–30	+	++	++	++	++	++	Calegari and Machado (2013)

AER = anion exchange resin. SMB = soil microbial biomass.
The signals + indicate magnitude of effect.

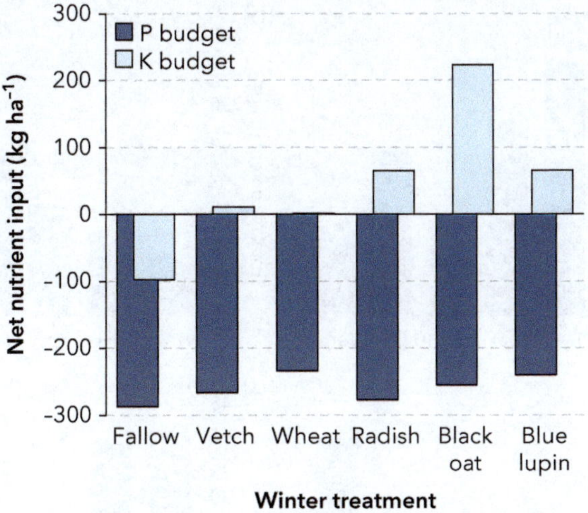

Figure 4 P and K budget after 23 years of different winter cover crops under NT in an Oxisol from Pato Branco, State of Paraná, Southern Brazil. Source: data from Tiecher et al. (2017).

residues when a mix of cover crops such as oats and vetches is used (Giacomini et al., 2003a,b). It is important to note that there are factors other than the C/N ratio that control the release of nutrients from the organic residues, especially P content. The C/P ratio (Fig. 5e) gives an indication of P mineralization, but this can still be affected by the activity of the acidic phosphatase enzymes present in the soil (McGill and Cole, 1981).

Different species of cover plants have a different capacity for absorption and accumulation of P (Fig. 5c). Although plants generally accumulate P in tissue mainly as soluble inorganic P, they may accumulate P in other forms, such as soluble organic P, P contained in lipids and RNA, and this alters the release rate of P. Among the cover crops evaluated by Casali (2012), common vetch and rye had the highest potential to recycle P from the soil. Vetch has the greatest capacity to accumulate P in the tissue (Fig. 5c), mainly in soluble form with a high rate of decomposition. Rye has a lower content of P in the tissue but produces a higher amount of biomass and accumulation of P in soluble inorganic form which is easily released to the soil. Franchini et al. (2000) evaluated the changes in soil fertility after 8 years of different crop rotation systems in two Oxisols from Paraná State in Brazil. They verified that the crop rotation lupine-corn-wheat-soybean (LC) reduced soil pH and the level of exchangeable Ca and Mg, increasing the exchangeable Al and the potential acidity (H+Al), compared to the traditional system wheat-soybean (WS), but no difference was observed for available K. The decrease of Ca content in the LC was ascribed to the soil acidification related to the N fertilization of wheat and

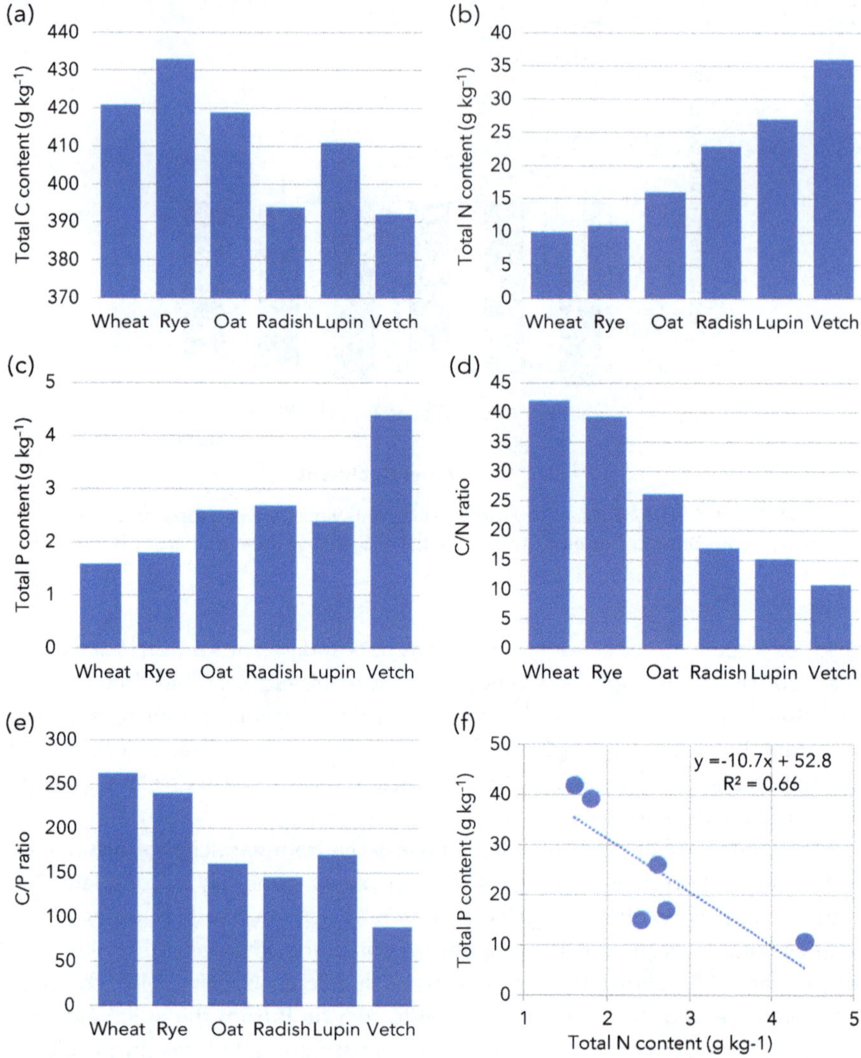

Figure 5 Content of total C (a), N (b) and P (c), C/N (d) and C/P (e) ratios, and the relationship between N and P content in the above-ground biomass of different winter cover crops grown in an Oxisol from the State of Paraná, Southern Brazil. Source: data from Casali (2012).

corn. The maintenance of levels of K and the decrease on the levels of Ca and Mg demonstrate that there was a change in the preference of leaching cations in the LC crop rotation system. The formation of metal-organic complexes with divalent cations was assumed to be the mechanism responsible for the chemical alterations in the soil.

Marcelo et al. (2009) evaluated changes in soil fertility as a result of four summer crop sequences (maize monocrop, soybean monocrop, soybean/maize rotation and rice/bean/cotton rotation) combined with seven winter crops (maize, sunflower, oilseed radish, pearl millet, pigeon pea, grain sorghum and sunn hemp) in an Oxisol from São Paulo State in Brazil. They found that the summer rice/bean/cotton rotation increased soil K levels at 0–10 cm depth when sunn hemp and oilseed radish had previously been grown in the winter, and in the 0–2.5 cm layer for millet. Higher P levels were found at the depths 0–2.5 cm and 0–5 cm, respectively, when sunn hemp and oilseed radish were grown in the winter.

Pavan and Calegari (2003) evaluated different soil management systems in an apple orchard (cv. Gala) in an Oxisol from Paraná State in Brazil, from 1987 to 2000. Compared with the initial soil composition (1986), all the cover crops tested improved soil fertility levels over the 12-year period in the following areas: improved pH, Ca, Mg, K and P and decreased Al. Black oat mulch and Pinus needles had the best effects on the soil. Both contributed to the decrease of pH and Al and increased Ca. The effects are possibly associated with the combination of organic binders with metal cations. These changes clearly show that such soil cover can be an effective way to manage and improve acid soils, Ca deficiency and Al toxicity. Besides increasing P and K, the use of mulch also increased the average fruit size and yield. The results suggest the value of keeping soil covered with mulch in orchards.

Rosolem and Calonego (2013) evaluated P and K after three years of different crop rotations: triticale (X *Triticosecale*) and sunflower (*Helianthus annuus*) cropped in autumn–winter; pearl millet, forage sorghum, and sunn hemp grown in the spring, and soybean in the summer. Results demonstrated that nutrient losses occurred in all of the management systems. P losses varied between 15% and 33%, and K losses varied between 7% and 21% compared with the original levels in the soil profile plus the amount of fertilizer applied. However, among the crop rotations, the use of high K-accumulating plants such as triticale-maintained K in the soil–plant system, avoiding losses by leaching.

The effect of residual straw and N fertilization in a maize crop combined with subsequent cover crops in an NT system was studied in an Ultisol from São Paulo State in Brazil. Acidity was reduced from high to medium, base saturation (BS) was maintained at mean values and there was a 16% increase in OM. Macro and micronutrients increased with these green manure crops (Table 9) (Esteves et al., 2014; Wutke et al., 2013).

An experiment was carried out to study the effects of three different crop rotation cycles on a Haplorthox soil. These were: fallow (spontaneous managed vegetation) and leguminous green manure species (oats, sunnhemp/*Crotalaria juncea* L., pigeon pea/*Cajanus cajan* (L.) Millsp., and black velvet beans/*Mucuna*

Table 9 Accumulation of macro and micronutrients in the shoot of green manure species in full bloom. Mococa, SP, 2012.

Green manure[a]	N	P	K	Ca	Mg	S	B	Cu	Fe	Mn	Zn
	g m^{-2}						mg m^{-2}				
CB	10.2c	0.52c	4.3e	5.6b	1.17b	0.45d	6.24b	3.19c	77.88cd	10.07d	6.29e
CJ	14.4b	0.92b	6.3c	5.1b	1.54a	0.83ab	7.94a	3.83b	158.94 b	22.46b	11.16ab
CS	8.1c	0.50d	5.0de	4.1c	0.84c	0.50d	8.81a	2.89cd	48.86 e	8.94d	8.82cd
CE	14.5b	0.84b	8.7ab	8.4a	1.33b	0.72bc	8.28a	2.54d	107.42 c	20.67b	7.58de
CC	18.6a	1.17a	9.1a	5.7b	1.58a	0.89 a	8.83a	4.99a	350.14 a	31.44a	12.42a
MD	5.1d	0.35e	2.1f	3.2d	0.53d	0.29 e	3.41c	1.56e	75.23 d	13.52c	4.72f
PG	5.2d	0.65c	7.9bc	2.4d	1.26b	0.63 c	0.37d	3.11cd	158.17 b	19.43b	10.34bc
dms 5%	2.0	0.13	1.0	0.8	0.20	0.10	1.06	0.58	31.13	3.36	1.40
CV%	8.0	7.8	6.8	7.0	7.2	7.1	7.2	8.0	9.6	8.0	6.8

[a] CB: *Crotalaria breviflora*; CJ: *Crotalaria juncea*; CS: *Crotalaria spectabilis*; CE: *Canavalia ensiformis*; CC: *Cajanus cajan*; MD: Mucuna deeringiana; PG: *Pennisetum glaucum.*

Means followed by same letters in the column are not different by the Duncan test, $p < 0.05$.

aterrima), as well as common bean 'IAC-Carioca'/short cycle corn (*Zea mays* L.). Pigeon pea and corn crops were not viable options but the inclusion of black velvet bean, sunn hemp and corn in rotation with common bean and corn is feasible. These crops produced the greatest amounts of green mass, provided adequate soil cover, increased the grain yield of irrigated common beans in relation to fallow treatment, maintained soil OM, reduced soil acidity, increased the base saturation index (BS%) and enhanced water infiltration rate in relation to fallow (Tables 10 and 11). The results indicated that bean yields and infiltration of soil water were related. Effective bean root depth (80% of the total) was around 0.35–0.40 m deep (Wutke et al., 1998, 2000; Grego et al., 2006).

Tanaka et al. (2009) assessed the impact of soybean cultivation in restoring a Nitosol soil after many years of sugarcane cultivation. A CT system with soybean 'BRS-48' and a NT system with soybean 'M-Soy 7908 RR' were evaluated. The results of chemical analysis of both soil samples are presented in Table 10. After

Table 10 Average chemical characteristics at 0–20 depth of a typic Haplorthox soil cultivated with common bean 'IAC-Carioca' after three cycles of crop rotations with corn followed by fallow (spontaneous managed vegetation) and green manure crops. Ribeirão Preto, State of São Paulo, Brazil

Previous crop	MO[a] (g dm^{-3})	pH	P (mg dm^{-3})	K	Ca	Mg	H+Al	CEC	BS (%)
				$mmol_c\ dm^{-3}$					
Fallow	31.8	5.0	36.7	2.4	30.1	9.0	43.3	71.9	49
Corn	31.1	5.1	38.3	2.4	32.5	9.3	43.7	75.1	50
Black oat	31.9	4.9	36.6	2.4	32.7	8.6	49.4	79.2	47
Sunnhemp	30.3	5.0	45.1	2.9	36.0	9.4	45.3	81.0	51
Pigeon pea	30.3	4.8	34.5	2.5	29.5	7.5	48.2	66.8	45
Black velvet bean	33.7	5.0	37.4	2.5	35.6	9.3	46.4	80.1	50

[a] MO: OM; CEC: cation exchange capacity; BS: base saturation index.

Table 11 Average green biomass yield of green manure crops, average yield of common beans and corn grains, and water infiltration rate (WIR) in a typic Haplorthox soil during three cycles of crop rotations. Ribeirão Preto, State of São Paulo, Brazil

Previous crop	Cover crop biomass (t/ha)	Common bean grains (t/ha)	Maize grains (t/ha)
Fallow	5.6	1.79 bc	6.18 a
Corn	23.7	1.83 abc	6.37 a
Black oat	14.7	1.58 c	6.67 a
Sunnhemp	20.9	2.04 ab	6.69 a
Pigeon pea	8.3	1.67 c	7.08 a
Black velvet bean	19.3	2.06 a	6.86 a
CV%	-	20.4	13.5

Means followed by same letters in the column are not different by the Tukey test, $p < 0.05$.

Table 12 Chemical characteristics of an Nitisol cultivated with soybean in two sugarcane renewal areas under CT and NT systems in Mogi-Mirim, São Paulo State, 2009

	OM		pH ($CaCl_2$)		BS		H + Al		Base sum	
	CT	NT	CT	NT	CT	NT	CT	NT	CT	NT
Soil depth (cm)	g dm^{-3}				%		mmol$_c$ dm^{-3}			
0-7	39	35	5.5	5.2	69	56	26.8	43.1	60.0	55.6
7-20	39	32	5.6	5.2	70	56	26.2	41.1	61.4	52.7
20-40	35	29	5.6	5.4	70	60	23.6	34.6	56.9	53.4
Average	*37.5*	*31.7*	*5.6*	*5.2*	*69.7*	*57.2*	*25.5*	*39.6*	*59.5*	*53.9*
	CTC		**P**		**K**		**Ca**		**Mg**	
	mmol$_c$ dm^{-3}		**mg dm^{-3}**		**mmol$_c$ dm^{-3}**					
0-7	86.8	98.8	24	26	1.3	3.3	48	42	10	11
7-20	87.7	93.9	18	15	1.0	2.5	50	41	10	10
20-40	80.8	88.1	11	11	0.6	1.5	49	42	8	10
Average	*85.1*	*93.6*	*18*	*17*	*1.0*	*2.4*	*49*	*41*	*9.4*	*10*
	B		**Cu**		**Fe**		**Mn**		**Zn**	
					mg dm^{-3}					
0-7	0.26	0.64	4.9	8.5	13	17	35	55	3.5	3.5
7-20	0.25	0.49	5.1	7.4	12	15	33	49	3.1	2.2
20-40	0.20	0.43	5.1	6.7	12	12	24	43	2.4	2.0
Average	**0.23**	**0.52**	**5.0**	**7.5**	**12**	**15**	**31**	**49**	**3.0**	**2.5**

soybean cultivation OM contents were high indicating high nutritional reserves after mineralization; pH values were not altered; P contents in the topsoil were higher; K content was superior in NT system in relation to the CT and Ca and Mg contents were also high (Table 12). The soybean grain yield under NT (1.199 kg ha^{-1}) was double the yield under CT (0.599 kg ha^{-1}).

Wutke et al. (2008) reviewed cultivation of the cover crop species, sunn hemp *(Crotalaria juncea*) 'IAC-1' and pigeon pea (*Cajanus cajan*) 'IAC-Fava Larga'. These were grown as an intercrop between rows of citrus (tangor). Dry biomass values of 3.7 and 8.0 t per hectare were obtained with sunn hemp and pigeon pea. The yield of tangor crop was increased by cultivating green manures from the cover crops. Yields of about 79.8 and 107.3 kg per plant and 1.9 and 2.6 fruit boxes per plant were obtained with interrow cultivation with sunn hemp and pigeon pea (Table 13).

All these studies demonstrate that cover crops play an important role in improving soil fertility, especially when soil is not disturbed in NT system. This occurs most likely because they

1 stimulate soil microbiological activity (Mbuthia et al., 2015);
2 increase nutrient cycling and soil OM content (Conceição et al., 2013);

Table 13 Yield, number of fruit boxes per plant and fruit mass of a tangor Murcott crop with interrow cultivation with green manure species in a NT system, without and with utilization of 150 kg of nitrogen per hectare at the first application of the crop nitrogen fertilization. Moji-Mirim, State of São Paulo, Brazil

Green manure species*	Yield per plant			Fruit boxes per plant			Fruit mass		
	No N	With N	Average	No N	With N	Average	No N	With N	Average
		kg $plant^{-1}$			Number			Grams	
Urochloa brizantha	79.8 Bb	101.0 Aa	90.4 a	1.9 a	2.5 a	2.2 a	266.5	284.0	275.3 a
Crotalaria juncea	98.4 Aa	86.9 Aab	95.0 a	2.4 a	2.1 a	2.3 a	262.8	250.3	256.5 a
Cajanus cajan	107.3Aa	82.8 Bb	95.0 a	2.6 a	2.0 a	2.3 a	274.3	260.5	267.4 a
Average	95.1 A	90.2 A	-	2.3 A	2.2 A	-	267.8A	264.9A	

Yield: CV% Green manure species 12.6%; CV% N 7.4%.
Fruit mass: CV% Green manure species 4.6%; CV% N: 6.2%.
Means followed by small letters in the column and capital letters in the line are not different by the Duncan test, $p<0.05$.

3 improve nutrient acquisition from the soil in situ or from added fertilizers (Teles et al., 2017);
4 improve soil physical properties and alleviate soil compaction stress, allowing the search of nutrients in deeper layers of the soil (Calonego et al., 2017).

The use of cover crops is an efficient and practical alternative to optimize the use of soil nutrients, contributing to a more sustainable intensification of agricultural systems, increasing productivity while reducing fertilizer consumption.

5 Effect of cover crops on soil physical properties

Generally, when a soil is covered by mulch, the soil surface remains protected. This helps to protect the soil from physical impacts and effects of rain, avoiding sealing and crusting of the soil surface, increasing water infiltration rates and reducing runoff. Surface infiltration depends on a number of factors including aggregation and stability, pore continuity and stability (including biopores), the existence of cracks, hydraulic conductivity and the soil surface condition (rugosity). The organic residues added in an NT system help to increase soil life (biology), and indirectly soil fauna activities, which will enhance soil porosity, soil particle aggregation and water infiltration rate (Roth, 1985; Calegari et al., 1995, 2006, 2011, 2013). The improved pore space is a consequence of the activities of earthworms and other macroorganisms and channels left in the soil by decayed plant roots.

In southern Brazil, studies shown that rainwater infiltration increased from 20 mm/h under CT to 45 mm/h under no-tillage, including cover crops and crop rotation (Calegari et al., 1998a). Improvement in water infiltration increases groundwater recharge. Increased water infiltration combined with a higher OM content increases soil water storage (Gassen and Gassen, 1996; Tormena et al., 2004).

SOM influences the physical conditions of a soil in several ways. SOM enhance the soil aggregates stability and pores through the bonding or adhesion of organic acids or other compounds, such as bacterial waste products, organic gels, fungal hyphae (polysaccharides) as well as worm secretions and casts. When mixed with mineral soil materials OM significantly increases moisture holding capacity (Tormena et al., 2004). Working with a clayey Oxisol in Southern Brazil with a 5% slope, Ruedell (1999) found black oat soil mulch cover in an NT system reduced runoff and soil losses while improving water infiltration (Table 14).

After 19 years of different cropping sequences on a Rhodic Hapludox soil in Southern Brazil, Calegari et al. (2008, 2011) were able to measure the total amount of organic residues (winter crop, maize and soybean residues) added to the soil. The winter crop was the main factor in the amount of biomass introduced to the soils. Soil bulk densities are shown in Table 15. They were

Table 14 Crop residue effects on runoff, water infiltration and soil losses

Amount of residues (t/ha)	Runoff (%)	Water infiltration (%)	Soil losses (t/ha)
0.00	45.3	54.7	13.7
0.55	24.3	74.7	1.6
1.10	0.5	99.5	0.3
2.20	0.1	99.9	0.0
4.41	0.0	100.0	0.0

Source: adapted from Ruedell (1999).

Table 15 Bulk density in clayed Oxisol of Brazil affected by soil management and crops system

	Soil layer (cm)				
	0–10	10–20	20–30	30–40	40–60
Soil management			(kg dm^{-3})		
NT	1.25aA	1.19bB	1.11bC	1.06aD	1.00aE
CT	1.24aB	1.28aA	1.17aC	1.08aD	1.02aE
*Forest	0.80	0.90	1.00	1.00	1.00

* The Forest area is located neat to the Experimental plots. So, it´s not a plot, and it can´t be statiscally analyzed and compared wit the experimental plots. Just to have an idea when compared with disturbed (CT) and undisturbed soil (NT).
Source: data from Calegari et al. (2008).

broadly similar between CT and NT treatments. In the NT system the proportion of larger aggregates (> 2 mm) was higher than in CT, for both topsoil layers (0–10 and 10–20 cm), suggesting the NT system was characterized by coarse aggregates (Table 16). Winter crop treatments show a lower proportion of larger aggregates (>2 mm) and a higher proportion of small aggregates (<0.25 mm) in fallow compared to other treatments (Calegari et al., 2008). In the two soil depths studied (0–10 and 10–20 cm), the three soil aggregation parameters (mean weight diameter-MWD, geometric mean diameter-GMD, and aggregate stability-AS) were enhanced under the NT system.

Castro Filho et al. (2002) also found improvement in physical properties in an Oxisol, including higher AS% in NT when compared with CT. NT resulted in higher macroporosity compared to CT. These differences were significant at 0–10, 20–30 and 40–60 cm soil depth. The higher values for microporosity were found on NT at the upper layers (0–10 and 10–20 cm). There was significantly higher microporosity with oat and the fallow treatments.

6 Effect of cover crops on soil biological properties

The use of plants that increase mycorrhiza populations must also be considered in rotation systems. Soil vesicular arbuscular mycorrhiza (VAMP) increase

Table 16 Aggregate size class distributions in clayed Oxisol in Brazil affected by soil management and crops system

		Aggregate size (mm)				
Soil layer (cm)	Soil management	> 2.00	2.00–1.00	1.00–0.50	0.50–0.25	< 0.25
				(g/100g)		
0–10	NT	39.97 aA	6.65 bB	5.28 bB	3.45 bB	4.92 bB
	Conventional	22.39 bA	12.30 aA	11.38 aA	7.23 aA	9.29 aA
	Forest[a]	28.37	8.78	8.93	5.24	5.81
10–20	NT	36.08 aB	8.40 bA	7.37 bA	4.34 bA	4.85 bB
	Conventional	23.76 bA	11.73 aA	11.57 aA	6.59 aB	8.63 aA
	Forest[a]	31.05	9.05	7.77	5.47	6.07
Winter treatment						
	Fallow	27.31 b	10.18ns	9.36ns	5.82ns	8.91 a
	Vetch	31.83 a	9.88	9.03	5.38	6.08 b
	Lupin	31.31 a	9.58	8.59	5.18	6.47 b
	Radish	31.26 a	9.21	8.76	5.27	6.92 b
	Oat	31.14 a	9.98	8.74	5.23	6.41 b
	Wheat	30.45 a	9.80	8.93	5.53	6.74 b

Means followed by the same lower case letters in the same column comparing soil management for each depth into the same aggregate size class, and also winter treatment into each aggregate size class, do not differ at the 5% level of probability by the *F*-test in the analysis of variance. Means followed by the same upper-case letters in the same column comparing soil layer into each soil management into the same aggregate size class do not differ at the 5% level of probability by the *F*-test in the analysis of variance.
[a] Forest soil is not included in statistical analysis.
Source: data from Calegari et al. (2008).

the root surface area, improving nutrient and water absorption, and protect the roots against nematodes and other root diseases (Fig. 6) (Colozzi et al., 1994; Balota et al., 2014). Mycorrhiza can promote release of some otherwise unnutrients such as P in clay soils (Balota et al., 2014). Some crops like apple, potatoes, citrus, banana, cassava, coffee, among others, are particularly dependent on mycorrhiza infection to optimize nutrition. Studies carried out with the Santa Fé System show that Brachiaria intercropped with maize can suppress *Rhizoctonia solani* and *Fusarium solani* f. sp. phaseoli or serve as a physical barrier to the dissemination of *Sclerotinia sclerotiorum* (*white mold*) (Table 18).

Studies show higher mycorrhizal populations in NT systems, where the soil is not disturbed and crop residues remain on the soil surface, when compared with CT. Some cover crop species have a low capacity for encouraging root

High dependency	Low dependency	Non-hosts
Peas, Beans, and other legumes, vetch, crotalarias, cowpea, Flax, Oat, Sunflowers Maize and other warm season cereals, Coffee, banana, Brachiarias	Wheat and other cereals	Cruciferae (rape seed, radish, Mustard and other brassicas) Lupins

Figure 6 The relationship between some crop species and VAMF. Source: Dr. Jill Claperton, USA (pers. comm.).

Table 17 Colonization of soybean, maize and cotton roots by mycorrhizal fungi in NT and CT system

Crop rotation with wheat	No-tillage (%)	Conventional tillage (%)
Soybean	36	25
Corn	31	23
Cotton	41	21
Soybean/corn	35	20
Cotton/corn	34	30
Soybean/cotton	37	23

Source: data from Colozzi-Filho et al. (1994, 2001).

Table 18 Crop rotation and Brachiaria effects on microbiological activity (Santa Fé System) and bean soil pathogens

Rotation	Microbiological activity	*Fusarium solani* (ppg)	*Rhizoctonia solani* (ppg)	*Sclerotinia sclerotiorum* (escler. m^{-3})
Rice	0.46	1.120	83	0
Maize	0.55	2.720	42	0
Brachiaria ruziziensis	0.45	1.560	28	0
Brachiaria brizantha	0.50	1.340	24	0
Soybean	0.29	3.160	32	3

Source: adapted from Costa and Rava (2003).

colonization by mycorrhiza. Examples include lupin, radish, rape and wheat. Other species are significant promoters of mycorrhiza root populations. These species include sunn hemp, other species of crotalarias (spectabilis, ochroleuca, breviflora), millet, mucunas, cowpea, pigeon pea, sunflower, buckwheat, stylosanthes, vetches, rye, pearl millet, finger millet, oats and clovers in general. Mixtures including legumes are excellent hosts for mycorrhizal fungi (Birch, 1958; Landers et al., 1994; Séguy et al., 1996, 2006), (Table 17). Integrated crop-livestock systems, also involving agroforestry, are becoming more common in different areas of Brazil mainly in Savannah (Cerrado) region. In these systems *Brachiaria* sp. is intercropped with maize and, after the maize is harvested, the pasture is used to feed livestock. Soybean is then planted in the dessicated grass, sometimes integrated with strips of trees (Broch et al., 1997; Fontaneli et al., 2000; Kluthcouski et al., 2000; Kluthcouski and Aidar, 2003; Resck et al., 2000; Aidar et al., 2003). The value of integrating agroforestry practices into NT systems is discussed in Bunch (1985).

Key factors involved in the increase in the population of P-solubilizing microorganisms, ammonitors, nitrifying and mycorrhizal fungi are lower temperature fluctuations and lower maximum temperatures which are found when mulches are used in NT systems. Roots can provide the mechanical structure for the initial formation of macroaggregates by enclosing particles and producing cementing agents (root exudates), which will stimulate microbial activity (Jastrow and Miller, 1998). Fungi then form on macroaggregates (Beare et al., 1993; Drijber et al., 2000; Frey et al., 1999, 2000), producing a higher ratio of glucosamine (fungus derivatives) than muramic acids (derived from bacteria) (Frey et al., 1999; Guggenberger et al., 1999). The preferential stabilization of fungi on bacterial biomass may lead to more efficient cycling of carbon and nitrogen (Beare et al., 1992; Frey et al., 1999).

More heavily-fertilized monocultures in CT systems can provide favorable growth conditions for pests and diseases (Magdoff and van Es, 2009). In contrast, growing a diverse range of cover crops in NT systems promotes greater biological diversity, including beneficial insects and other organisms which promote natural biological control of pests and diseases (Phelan et al. 1995); Blouin et al., 2005; Altieri et al., 2007; Calegari, 2014). Intercropping can also help to attract beneficial insects whilst dissuading predators who are not attracted to the second crop, as in the example of soybeans or peanuts intercropped with maize in reducing pests such as termites (Roland Bunch, pers. comm., 2018). NT systems also promote earthworm numbers. Earthworms have the ability to form galleries that raise the water infiltration rate and nutrients into the soil profile, and their excrements are small aggregates of soil mixed with nutrient rich organic wastes that are left on the soil surface (coprolites). They act as 'biological rippers' by promoting soil structuring and deep incorporation of 'pools' of organic material, facilitating aeration, water penetration and plant roots in depth. Results from 15 years of NT in Paraná in Brazil found more than 600 thousand earthworms per ha, while CT only achieved 7% of this total (Gassen and Gassen, 1996).

7 Effect of cover crops on soil nematode control

Nematodes are a concern for agriculture in countries such as Brazil, due to their wide distribution in different agricultural regions and their capacity to damage crops such as soybean, cotton, bean, coffee, and sugarcane (De Waele and Elsen, 2007; Machado, 2015). In Brazil, according to the Brazilian Nematological Society, nematodes cause losses of about R$ 35 billion (around US$ 9.5 billion) though, in soybean alone this damage can reach R$ 16 billion (more than US$4 billion) (Machado, 2015). The main nematode species in Brazilian agriculture are root-knot nematodes, *Meloidogyne incognita* and *M. javanica*, the root-lesion nematode, *Pratylenchus brachyurus*, the soybean cyst nematode, *Heterodera glycines*, and the reniform nematode, *Rotylenchulus reniformis* (Fig. 7) (Machado, 2015).

The main management tools in nematode control are the use of resistant crop varieties, the biological and/or chemical nematicides, and the use of cover crops in a crop rotation system which disrupt nematode life cycles. The use of

Figure 7 Root symptoms caused by *Meloidogyne* sp., *Pratylenchus brachyurus*, *Heterodera glycines* and *Rotylenchulus reniformis* (in clockwise direction). Photo: Santino Aleandro da Silva.

resistant cultivars is an ideal method but there are few varieties which combine adequate resistance with adequate yields (Machado, 2016). Nematicides have a variable efficacy depending on factors such as soil condition, time of application etc.

Crop rotations, including specific cover crop species, can be very effective in managing nematode populations (Calegari and Machado, 2012). The species *Crotalaria*, for example, is a poor host to many nematode species, including *Meloidogyne* spp., *R. reniformis, Radopholus similis, Belonolaimus longicaudatus,* and *Heterodera glycines* (Wang et al., 2002). Crotalarias acts as a 'trap plant', allowing nematode penetration until the adult phase (Wang et al., 2002; Silva et al., 1989). Crotalarias also produce secondary metabolites that have nematicide potential, such as pyrrolizidine alkaloid or monocrotaline (Rich and Rahi, 1995) and can provide a niche for antagonistic flora and fauna (Kloepper et al., 1991). Silveira and Rava (2004) incorporated 13 Mg ha^{-1} of aerial dry mass of *C. spectabilis* and observed that the population of *Pratylenchus* spp. in common bean was reduced by more than 99%.

Machado et al. (2007) found that *C. spectabilis* and *C. breviflora*, can reduce population densities of *P. brachyurus*. Inomoto et al. (2006) reported that dwarf pigeon pea (*Cajanus cajan*), C. *breviflora*, *C. spectabilis*, and black mucuna (*Mucuna pruriens*) reduced the population of M. *javanica* whilst dwarf pigeon pea, C. *breviflora* and *C. spectabilis* reduced population of P. *brachyurus*. In another study, Machado et al. (2014) characterized the host response of cover crops to four species of root-knot nematodes, *Meloidogyne incognita*, *M. javanica*, *M. paranaensis,* and *M. enterolobii*. Results showed that, among the 30 different species tested, at least four showed resistance to all of the root-knot nematodes tested: *Clitoria ternatea*, resistant to the four root-knot species, and *Crotalaria juncea, C. spectabilis,* and *C. breviflora*, resistant to *M. incognita, M. javanica* and *M. paranaensis*. *Lupinus albus, Mucuna pruriens, M. deeringiana*, *M. aterrima*, and *M. pruriens* 'Utili' were also resistant to *M. paranaensis*, an important pathogen for coffee plantations, at Paraná State, Brazil. *Mucuna pruriens* and *M. deeringiana*, *Cajanus cajan* 'Fava Larga', 'Dwarf Iapar 43' and 'BRS Mandarim' were also resistant to *M. enterolobii*, a minor root-knot species with importance to guava plantations in Brazil.

Guinea grass (*Panicum maximum*) and brachiaria grasses (*Brachiaria* spp.) are useful in areas infested by root-knot nematodes, because they are non-hosts or poor hosts of *M. incognita* and M. *javanica* (Brito and Ferraz, 1987; Dias-Arieira et al., 2003) and *R. reniformis* (Asmus and Cargnin, 2005). In relation to *P. brachyurus*, Inomoto et al. (2007a) characterized the host status of five species and one hybrid of brachiaria grasses [*Brachiaria decumbens*, B. *brizantha*, B. *humidicola*, B. *dictyoneura*, B. *ruziziensis*, and Mulato grass (B. *ruziziensis* x B. *brizantha*)], and two cultivars of P. *maximum* ('Mombaça' and 'Tanzânia'). *Brachiaria dictyoneura* proved to be a poor host of P. *brachyurus*. Inomoto et al. (2007b) evaluated the effect of ten species or hybrids used as cover crops on the population of the lesion-nematode, *P. brachyurus,* and found that *Sorghum bicolor* 'IPA 7301011' and 'BRS 800' were the most susceptible hosts for P. *brachyurus*, while *Amaranthus cruentus* 'BRS Alegria' and *Chenopodium quinoa* 'BRS Piabiru' were the most resistant.

Asmus et al. (2005) evaluated some plant species (*Amaranthus cruentus, Chenopodium quinoa, Eleusine coracana, Hibiscus cannabinus, Avena strigosa, Pennisetum glaucum, Raphanus sativus* var. *Oleiferus, Sorghum bicolor, Helianthus annuus,* and *Eragrostis tef*) used as autumn or winter cover crops in NT systems to *M. incognita* and found that all species, except *A. cruentus*, were susceptible to this nematode. In relation to white oat (*Avena sativa*), another winter cover crop widely cultivated in the Brazilian south region, Machado et al. (2015) characterized the host response of oat genotypes to *Meloidogyne incognita*, *M. javanica*, *M. paranaensis*, *M. enterolobii*, and *P. brachyurus* under greenhouse conditions. Results revealed variation in the susceptibility of cultivars to nematodes tested, except for *M. enterolobii*, in which all cultivars tested were highly resistant. The cultivar IPR Afrodite was the most resistant to root-knot species whereas all cultivars tested were considered susceptible to *P. brachyurus*.

Asmus et al. (2008) evaluated cover crops under greenhouse conditions and their effect on reniform nematode populations (*R. reniformis*) . *Raphanus sativus,* Mulato grass – *Brachiaria ruziziensis* x *B. brizantha, Sorghum bicolor, Eragrostis tef, Setaria italica, Avena byzantina, A. strigosa, Pennisetum glaucum* and *Eleusine coracana* were classified as poor hosts to *R. reniformis*, while *Amaranthus cruentus* and *Chenopodium quinoa*, as good hosts. In the field, Mulato grass, *R. sativus* and *S. bicolor* reduced nematode populations.

If well planned and well managed, integrated crop-livestock production systems also allow a reduction in the population densities of phytonematodes present in the soil. In this context, Silva et al. (2015) accessed the host status of *Cynodon* spp. and *Pennisetum* spp. to *P. brachyurus*, *M. incognita*, *M. javanica*, and *M. paranaensis*, under greenhouse conditions, and found that 'Napier Paraíso', 'Flora Kirk', and 'Jiggs' were classified as resistant to *P. brachyurus*, 'Napier Roxo', 'Napier Anão', 'Napier Paraíso', 'Napier Cameron', 'Coast cross', and 'Hemarthria IAPAR 37', as resistant to *M. javanica*; 'Napier Roxo', 'Napier Cameron', 'Coast cross', and 'Hemarthria IAPAR 37', as resistant to *M. paranaensis*, and all genotypes evaluated were classified as resistant to *M. incognita*.

In relation to the effect of crop rotations on nematode population densities in the soil, under field conditions, Mendes et al. (2013) observed lower population densities of *P. brachyurus* in soybean cropped after *C. spectabilis, C. ochroleuca, C. juncea, C. spectabilis* + pearl millet 'ADR 300' and mechanical fallow, compared with fallow, chemical fallow, corn 'GNZ 2005', pearl millet 'ADR 300', *C. spectabilis* + *B. ruziziensis, B. ruziziensis,* and *B. brizantha* 'Marandu'.

Machado et al. (2007) have demonstrated that *C. juncea* hosts *P. brachyurus*, Mendes et al. (2013) attributed the lower population densities of this nematode observed in their field experiment to the faster development of this plant, compared with *C. spectabilis* and *C. ochroleuca*, generating higher biomass production and compensating for the higher multiplication of *P. brachyurus* in *C. juncea*. In this study, soybean yield was higher after *C. spectabilis* + pearl millet 'ADR 300', pearl millet 'ADR 300', and *C. spectabilis*, when compared with the crop of another plants.

The influence of soil management with cover crops and green manures on soil populations of *P. brachyurus* was studied by Debiasi et al. (2011), in a naturally infested field in Mato Grosso State, Brazil. Population densities were lower in the plots where *C. spectabilis* and *C. ochroleuca* were cropped and in the mechanical fallow; by contrast, the use of *B. ruziziensis* and corn increased the soil nematode populations (Fig. 8). Higher yields were obtained in the mechanical fallow and in the *B. brizantha* 'Marandu', corn 'BRS 1010' and chemical fallow treatments (Fig. 7). Debiasi et al. (2011) suggested the use of *C. spectabilis* or *C. ochroleuca* are better options to reduce *P. brachyurus* populations in the mid-season of soybean crops.

An important point about the use of cover crops aiming the nematode control is the effect of the biomass production on the composition of the soil OM. Higher amount of OM in the soil protects plants against the attack of nematodes. Almeida (2008) reported that the main effect of the OM for the nematode control is the increase of the population of natural enemies to nematodes in the soil, such as nematophagous fungi and bacteria, predator nematodes and protozoa. The release of toxic compounds to nematodes, such as butyric acid and volatile fat acids, also reduces nematode soil populations (Nazareno, 2009). Giebel (1974) reported that phenolic compounds produced by some plant species are involved in the resistance to phytopathogens, because

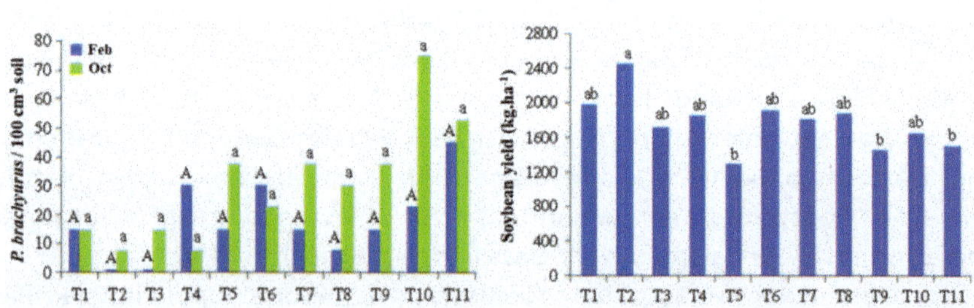

Figure 8 Number of *Pratylenchus brachyurus* in the soil in relation to different crops in the mid-season of soybean and soybean yields (T1 = *Crotalaria spectabilis*; T2 = mechanical fallow; T3 = *C. ochroleuca*; T4 = *C. juncea*; T5 = chemical fallow; T6 = pearl millet 'ADR 7010'; T7 = *Brachiaria ruziziensis*; T8 = fallow; T9 = corn 'BRA 1010'; T10 = corn 'BRS 1010' + *B. ruziziensis*; T11 = *B. brizantha* 'Marandu'). Source: adapted from Debiasi et al. (2011).

these compounds stimulate the synthesis of indoleacetic acid oxidase (IAA) that favor the auxin degradation, compromising the cell formation induced by root-knot nematodes. Increases in the phenol concentrations of were observed in tomato roots cropped in substrates with organic residues, which resulted in lower penetration of *M. incognita* juveniles (Alam et al., 1980; Alves et al., 2007).

In Fig. 9, adapted from Debiasi et al. (2016), it is possible to observe the effect of the soil OM content and of the previous crop grown in a naturally infested area with *P. brachyurus* in the Mato Grosso State, Brazil, on soybean grain yield. A higher amount of SOM was related to lower nematode population densities and higher soybean grain yields. Machado et al. (2016) also analyzed nematode community composition in a long-term experiment in South Brazil, with different winter cover crop species under NT and CT systems. This experiment is part of the OSCAR Project – Optimising Subsidiary Crop Applications in Rotations (https://www.wur.nl/en/show/oscar.htm) that aims to develop improved conservation tillage systems, based on cover crops. Cropping pattern, tillage system and cover crops affected the composition of nematode fauna. In the CT, there was a predominance of semi-endoparasitic nematode species, followed by the omnivores. In the NT system, there was a predominance of fungivores and omnivores nematode species. The authors concluded that crop rotation including cover crops is an effective way to control soil nematode communities.

In summary, the host reaction of cover crops to nematodes and their effect on nematode soil populations is variable according to different plant species and the different nematodes (Table 19). Choosing an alternative crop to manage a single nematode species in an infested field is a complicated task. This can be more complicated when we need to select a crop resistant to two or more nematode species. There is a risk of controlling one but, at the same time, favoring the development of another nematode species also present in the field. However,

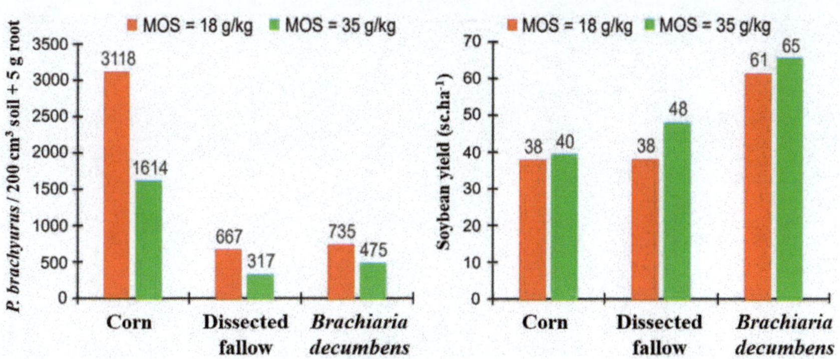

Figure 9 *Pratylenchus brachyurus* population density and soybean yield in relation to the OM content (MOS) and to crop used previously in a naturally infested field in Lucas do Rio Verde, Mato Grosso State, Brazil. Source: adapted from Debiasi et al. (2016).

Table 19 Host reaction of cover crops to the main nematodes for Brazilian agriculture (R = resistant; MR = moderately resistant; S = susceptible). Information compiled from previous published papers

Plant species	*Meloidogyne incognita*	*Meloidogyne javanica*	*Heterodera glycines*	*Rotylenchulus reniformis*	*Pratylenchus brachyurus*
Crotalaria spectabilis	R	R	R	R	R
Crotalaria breviflora	R	R	R	R	R
Crotalaria ochroleuca	-	-	MR	-	MR
Crotalaria mucronata	R	R	-	R	MR
Crotalaria paulina	R	R	R	S	MR
Crotalaria juncea	S	R	R	MR	S
Mucuna pruriens Black	R	MR	R	R	S
Mucuna pruriens Grey	R	MR	R	-	S
Mucuna pruriens Dwarf	MR	MR	R	-	S
Pigeon pea Dwarf 'lapar 43'	-	R	-	S	R
Pigeon pea 'Fava Larga'	S	S	MR	S	MR
Dolichos lablab	S	S	MR	-	S
Ricinus communis	R	R	R	R/S	R
Sorghum bicolor 'IPA 7301011'	MR	S	S	R	S
Sorghum bicolor 'BR 700'	-	S	S	R	S
Sorghum bicolor 'BRS 701'	-	S	S	R	S
Sorghum bicolor 'BRS 800'	MR	-	S	R	S
Sorghum bicolor '8311'	S	-	-	-	-
Pennisetum glaucum 'BN 2'	-	R	R	R	R
Pennisetum glaucum 'BRS 1501'	MR	S	R	R	MR
Pennisetum glaucum 'ADR 300'	-	-	R	R	R
Pennisetum glaucum 'ADR 500'	-	-	R	R	MR
Pennisetum glaucum 'ADR 7010'	S	R	R	-	R

Plant species	*Meloidogyne incognita*	*Meloidogyne javanica*	*Heterodera glycines*	*Rotylenchulus reniformis*	*Pratylenchus brachyurus*
Eleusine indica	MR	–	R	R	–
Eragrostis tef	S	S	–	R	S
Hibiscus cannabinus	S	S	–	MR	MR
Avena strigosa 'Comum'	MR	S	R	R	R
Avena byzantina 'São Carlos'	–	–	–	R	S
Panicum miliaceum	MR	–	–	–	MR
Helianthus annuus 'IAC Uruguai'	S	S	R	R	MR
Helianthus annuus 'Embrapa 122'	–	–	R	–	MR
Helianthus annuus 'Hélio 358'	–	–	R	–	R
Brachiaria brizantha	R	R	R	R	S
Brachiaria decumbens	R	R	R	R	S
Brachiaria dictyoneura	–	R	R	R	MR
Brachiaria humidicola	–	R	R	R	S
Brachiaria ruziziensis	–	R	R	R	S
Brachiaria 'Mulato'	–	R	R	R	S
Raphanus sativus 'Siletina'	S	S	R	R	MR
Amaranthus cruentus 'Alegria'	R	S	–	S	R
Chenopodium quinoa 'BRS Piabiru'	S	–	–	S	R

the right selection of crops can improve OM content in the soil, promote crop growth and manage nematode populations with minimal yield losses.

8 Effect of cover crops on weed control

Weeds compete with crops for water, nutrients and light, and can promote nematode population growth. Common species include *Cynodon dactylon* (Bermuda grass), *Cyperus rotundus*, *Galinsoga parviflora*, *Bidens pilosa*, *Argemone mexicana* (Mexican poppy). One of the major tools in integrated weed management (IWM) in Brazilian NT systems is the use of cover crops. Different cover crops species are important in weed management mainly because they can compete against weeds during their development and, at the end of their cycle, their mulch can suppress weed emergence; they can also have allelopathic effects (Almeida et al., 1984; Almeida, 1988; Kliewer et al., 1998; Skóra Neto, 1998; Tardin et al., 1998; Petersen et al., 1999; Thiesen et al., 2000; Teasdale et al., 2007). Several winter and summer cover crops have been shown to suppress weeds through their fast growth (Calegari et al., 1993). Favero et al. (2001), reducing weed biomass by between 22% and 96%. Using appropriated cover crop species in a rotation, Skóra Neto and Campos (2004) noted a weed population reduction of 93% after three years.

In the Brazilian Savannah (Cerrado) Vasconcelos and Landers (1993) studied planting grain crops into permanent cover crops, including maize into siratro (*Macroptilium atropurpureum* L. urb.) which removed the need for herbicide. Fernandes et al. (1999) observed that *Crotalaria breviflora*, *Crotalaria spectabilis* and pigeon pea plots reduced densities of weeds. Skóra Neto (1993) also noted that pigeon pea intercropped with maize decreased weed infestation at and after harvest; this research was later corroborated by Severino and Christoffoleti (2004) who found that sunn hemp and pigeon pea were effective as smother crops against numerous weed species.

Kliewer et al. (1998), working on clay soils in Paraguay, showed the benefits of cover crops and crop rotation in decreasing weed populations. They evaluated the effects of different winter species residues on soybean (next crop) in an NT system (Table 20). Crop residues of species such as oats (black and white), wheat and rye were effective in weed control (shadow and allelopathic effects), emphasizing the importance in a crop rotation of including species able to control weeds (Table 21).

Table 20 Effects of winter species residue and fallow on weed dry mass (NT). Iguazu Colony, Cetapar, Paraguay, 1996/97

Treatments	Dry mass (kg ha^{-1})
Winter fallow	7.390 A
Radish	4.258 B
White lupin	3.735 BC
Field pea cv. Iapar-78	2.263 CD
Triticale	1.818 D
Sunflower	1.653 DE
Rye	760 DE
White oat cv. IAC-7	723 DE
Wheat	498 DE
Black oat cv. Iapar-61	93 E

LSD 5% = 1.810 kg/ha; F (treatment) = 12.98.
Source: adapted from Kliewer et al. (1998).

Table 21 Allelopathic and suppress effects of cover crops on different weeds

Cover crop specie/residue with allelopathic and suppress effects	Weeds controlled
Crotalaria juncea (Sunnhemp)	*Brachiaria* sp., *Bidens pilosa*, *Galinsoga parviflora*, *Cenchrus echinatus*, *etc.*
Avena strigosa (black oat)	*Brachiaria plantaginea*, *other grass species*
Mucuna pruriens: Mucunas (black and grey), *Crotalaria juncea*, *Canavalia ensiformis* (jack bean)	*Cyperus rotundus*
Mucunas (black and grey)	*Bidens pilosa*, *Galinsoga parviflora*, *Cenchrus echinatus*
Secale cereale (rye) and Mucuna	*Brachiaria plantaginea* can be controlled by many months
Vicia sativa (Common vetch)	Sowed before corn promote efficient control of *Brachiaria plantaginea*
Lollium multiflorum (ryegrass)	*Sida rhombifolia*
Wheat straw	*Cassia tora*
Tagetes patula	*Ipomoea* sp., *Amaranthus* sp., *Desmodium purpureum*, *Momordica charantia*, *Euphorbia heterophylla*

Source: Lorenzi (1984), Almeida and Rodrigues (1985) and Calegari et al. (1993).

Adegas (1998) described a study of an IWM program on 58 farms in Paraná in Brazil, observing that, after three years, weed control costs decreased on average by 35% with reductions in herbicide use of 25%. Ruedell (1995) reviewed an IWM program in South Brazil, Rio Grande do Sul state and found a reduction of 42% in weed control costs.In southern Paraguay, the cassava/jack bean system eliminates two of the three weeding operations normally needed to grow cassava (i.e. about 65% of the total weeding labor). The maize/mucuna system can reduce weeding labor by up to 95%.

In Africa, a particularly noxious parasitic weed is striga (*Striga hermonthica*). Striga can significantly reduce yields of maize, sorghum and millet in areas with depleted soils. Since striga only grows in depleted soils, any green manure/cover crop that produces an above-average amount of biomass will rid a field of this weed, often in three to four years. Controlled experiments in Kenya have shown that the use of legumes can reduce striga counts by 68% in just two years. Another serious weed, Imperata grass (*Imperata cylindrica*, often called 'speargrass' in Africa and 'cogon grass' in parts of Asia) also grows best in poor soils and therefore becomes less problematic as soils are restored. Three or four years of green manure/cover crop use can almost completely eliminate it. Farmers using these green manure/cover to fertilize their fields significantly *reduce* their total costs, including labor.

9 Effect of cover crops on grain yield

The use of cover crops has been found to increase cash crop yields (Schomberg et al., 2006; Calegari, 2014, 2016, 2018a,b; Calegari et al., 2012, 2013a,b,c; Wutke et al., 2014).

Calegari and Alexander (1998), Figure 10 reviewed 12 years of using cover crops and rotations in an NT system in Southern Brazil in a Rhodic Hapludox with 73% of clay. They found NT achieved improved maize yields in most cases (Table 10). Yields were not improved when wheat was uses given N immobilization in wheat straw. This could be corrected by adding N. NT legume treatments (lupin and vetch) showed that, in maize crops, it is possible to obtain the equivalent of more than 90 kg N ha^{-1} when compared with CT. Over time, the NT system promoted better soil conditions for C and P that consequently lead to P and N becoming more available to maize crops, leading to yield improvements.

Calegari et al. (2012) reviewed the effects of a 23 year trial with cover crops and crop rotation in an NT system on maize yields (Table 22). The maize grain harvested after white lupin + oat, common vetch, hairy vetch + black oat, black oat and blue lupin at NT were greater than a comparable CT system. When all winter treatments were compared, the highest maize yield was obtained after NT common vetch (9.48 Mg ha^{-1}). An example of improved cash crop yields in NT systems can be seen in an NT area in Rio Grande do Sul State in South Brazil, in a clay Oxisol (Fig. 11) (Calegari, 2018b). Mixing five different species (radish, rye, oat, vetch and white lupin) achieved the highest yields in the following maize crop.

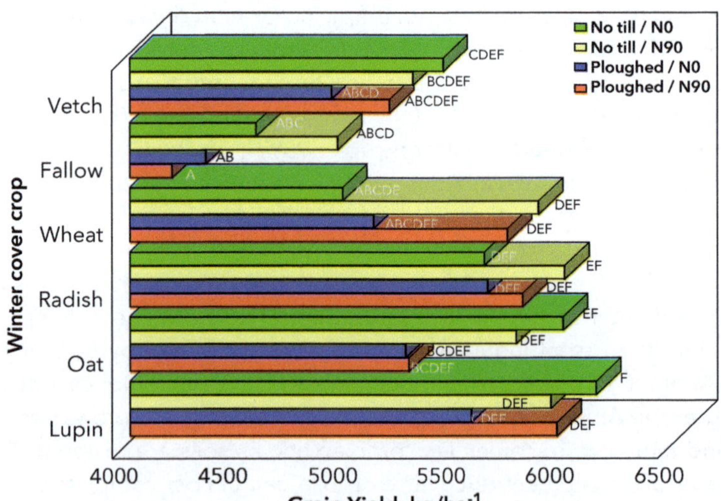

Figure 10 Effect of different soil management systems and winter cover corps on corn grain yield. Source: adapted from Calegari and Alexander (1998).

Table 22 Maize grain yield (Mg ha^{-1}) after winter treatments. IAPAR – Experimental Station, Pato Branco, Paraná State, Brazil (2009)

	Soil management system	
Winter treatments	No-tillage	Conventional tillage
White Lupin + black oat	8.888 ab A	7.166 a B
Hairy vetch	8.814 ab A	7.926 a A
Common vetch	9.481 a A	8.074 a B
Wheat	8.055 bc A	8.111 a A
Ryegrass	8.555 abc A	8.037 a A
Hairy vetch + black oat	9.092 ab A	7.055 a B
Rye	7.999 bc A	7.536 a A
Radish	7.462 c A	7.222 a A
Black oat	8.444 abc A	7.000 a B
Blue lupin	9.111 ab A	7.500 a B
Fallow	8.333 abc A	8.036 a A
Average[a]	8.567 A	7.606 B

[a] Means followed by the same small letter down rows or capital letter across columns are not significantly different using the Tukey test at $p < 0.05$.
Source: adapted from Calegari et al. (2012).

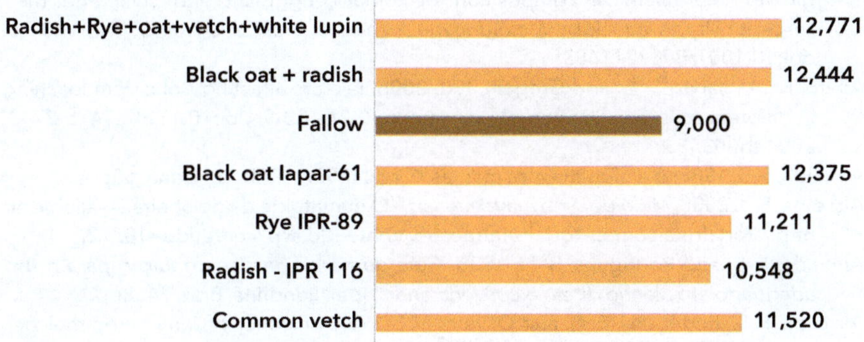

Figure 11 Effect of winter crops on maize grain yield (Mg ha^{-1}), Elias Abel Barboza Farmer, Não-me-Toque RS, Brazil – 2017. Source: Report of Living Soil Project, Renovagro, Uberaba, MG, Brazil, Calegari, (2017) – CAWC – Argentina, 2017.

10 Conclusion

Generally, in tropical and subtropical regions, where land is intensively cultivated, OM decomposition processes are accelerated, causing a severe decrease in the productive potential of the soil. This process can be reversed using cover crops in rotation, particularly when used as mixes (cocktails/ blends). Residues/mulches promote soil and water conservation, maintain and/or recover soil fertility, increase soil OM, stimulate biological activity in the soil, and improve crop yields. They reduce reliance on fertilizers, promote biodiversity and reduce the incidence of pests and diseases. In doing so they promote the long-term sustainability of agriculture.

11 References

Acosta, J. D. A., Amado, T. J. C., da Silva, L. S., Santi, A. and Weber, M. A. 2014. Decomposição da fitomassa de plantas de cobertura e liberação de nitrogênio em função da quantidade de resíduos aportada ao solo sob sistema plantio direto. *Ciência Rural* 44(5), 801–9. doi:10.1590/S0103-84782014005000002.

Adegas, F. S. 1998. Manejo integrado de plantas daninhas em plantio direto no Paraná. In: *Seminário nacional sobre manejo e controle de plantas daninhas em plantio direto, 1, Palestras*. Aldeia Norte, Passo Fundo, RS, Brazil, pp. 17–26.

Aidar, H., Rodrigues, J. A. S. and Kluthcouski, J. 2003. Uso da integração lavoura-pecuária para produção de forragem na entressafra. Em: Kluthcouski, J., Stone, L. F., Aidar, H. (Eds), *Integração Lavoura-Pecuária* (1a. ed.). Embrapa Arroz e Feijão, Santo Antonio de Goiás, pp. 225–62. Cap. 8.

Aita, C. and Giacomini, S. J. 2003. Decomposição e liberação de nitrogênio de resíduos culturais de plantas de cobertura de solo solteiras e consorciadas. *Revista Brasileira de Ciência do Solo* 27(4), 601–12. doi:10.1590/S0100-06832003000400004.

Alam, M. M., Ahmad, M. and Khan, A. M. 1980. Effect of organic amendments on the growth and chemical composition of tomato, eggplant and chilli and their susceptibility to attack by *Meloidogyne incognita*. *Plant and Soil* 57(2–3), 231–6. doi:10.1007/BF02211683.

Alfaro, M. A., Jarvis, S. C. and Gregory, P. J. 2006. Factors affecting potassium leaching in different soils. *Soil Use and Management* 20(2), 182–9. doi:10.1111/j.1475-2743.2004.tb00355.x.

Almeida, F. S. 1988. A alelopatia e as plantas. Circular 53. IAPAR, Londrina, 60pp.

Almeida, E. J. 2008. *Meloidogyne mayaguensis* – O nematoide da goiabeira. Available at: http://todafruta.com.br/todafruta/mostra_conteudo.asp?conteudo=18272.

Almeida, F. S. and Rodrigues, B. N. 1985. *Guia de herbicidas; contribuição para o uso adequado em plantio direto e convencional*. Iapar, Londrina, Brazil, 468pp.

Almeida, F. S., Rodrigues, B. N. and Oliveira, V. F. 1984. Influence of winter crop mulches on weed infestation in maize. In: *Proceedings of the 3rd EWRS Symposium on Weed Problems in the Mediterranean Area*, Lisbon, pp. 351–8.

Altieri, A. M., Ponti, L. and Nicholls, C. I. 2007. Melhorando o manejo de pragas através da saúde do solo: direcionando uma estratégia de manejo do habitat solo. In: *Controle biológico de pragas através do manejo de agroecossistemas*. Ministério do Desenvolvimento Agrário, Secretaria da Agricultura Familiar, Brasília, DF, pp. 17–31.

Alves, F. R., Freitas, L. G. D., Martinelli, P. R. P., Meira, R. M. S. A., Ferraz, S., Demuner, A. J., Borges, E. E. D. L. and Jesus Júnior, W. C. 2007. Efeitos de diferentes níveis de matéria orgânica no solo e de inóculo sobre a interação planta-*Meloidogyne* spp. e a produção massal de *Pasteuria penetrans*. *Summa Phytopathologica* 33(4), 397–401. doi:10.1590/S0100-54052007000400013.

Amado, T. J. C., Bayer, C., Eltz, F. L. F. and Brum, A. C. R. 2001. Potencial de culturas de cobertura em acumular carbono e nitrogênio no solo no plantio direto e a melhoria da qualidade ambiental. *Revista Brasileira de Ciência do Solo* 25(1), 189-97. doi:10.1590/S0100-06832001000100020.

Amado, T. J. C., Bayer, C., Conceição, P. C., Spagnollo, E., Campos, B. H. C. and Veiga, M. 2006. Potential of carbon accumulation in no-till soils with intensive use and cover crops in Southern Brazil. *Journal of Environmental Quality* 35, 1599-607.

Angers, D. A., Bolinder, M. A., Carter, M. R., Gregorich, E. G., Drury, C. F., Liang, B. C., Voroney, R. P., Simard, R. R., Donald, R. G., Beyaert, R. P. and Martel, J. 1997. Impact of tillage practices on organic carbon and nitrogen storage in cool, humid soils of eastern Canada. *Soil and Tillage Research* 41(3-4), 191-201. doi:10.1016/S0167-1987(96)01100-2.

Asmus, G. L. and Cargnin, R. A. 2005. Reação de culturas de cobertura a *Rotylenchulus reniformis*. *Nematologia Brasileira* 29, 136.

Asmus, G. L., Inomoto, M. M., Sazaki, C. S. S. and Ferraz, M. A. 2005. Reação de algumas culturas de cobertura utilizadas no sistema plantio direto a *Meloidogyne incognita*. *Nematologia Brasileira* 29, 47-52.

Asmus, G. L., Inomoto, M. M. and Cargnin, R. A. 2008. Cover crops for reniform nematode suppression in cotton: greenhouse and field evaluations. *Tropical Plant Pathology* 33(2), 85-9. doi:10.1590/S1982-56762008000200001.

Bais, H. P., Weir, T. L., Perry, L. G., Gilroy, S. and Vivanco, J. M. 2006. The role of root exudates in rhizosphere interactions with plants and other organisms. *Annual Review of Plant Biology* 57, 233-66. doi:10.1146/annurev.arplant.57.032905.105159.

Baker, J. M., Ochsner, T. E., Venterea, R. T. and Griffis, T. J. 2007. Tillage and soil carbon sequestration - what do we really know? *Agriculture, Ecosystems and Environment* 118(1-4), 1-5. doi:10.1016/j.agee.2006.05.014.

Balota, E. L., Calegari, A., Nakatani, A. S. and Coyne, M. S. 2014. Benefits of winter cover crops and no-tillage for microbial parameters in a Brazilian Oxisol: a long-term study. *Agriculture, Ecosystems and Environment* 197, 31-40. doi:10.1016/j.agee.2014.07.010.

Bayer, C. 1996. Dinâmica da matéria orgânica em sistemas de manejo de solos. PhD thesis. Universidade Federal do Rio Grande do Sul, Porto Alegre, Brazil, 240pp.

Bayer, C., Martin-Neto, L., Mielniczukc, J. and Ceretta, C. A. 2000. Effect of no-till cropping systems on soil organic matter in a sandy clay loam Acrisol from Southern Brazil monitored by electron spin resonance and nuclear magnetic resonance. *Soil and Tillage Research* 53(2), 95-104. doi:10.1016/S0167-1987(99)00088-4.

Bayer, C., Spagnollo, E., Wildner, LdP., Ernani, P. R. and Alburqueque, J. A. 2003. Incremento de carbono e nitrogênio num latossolo pelo uso de plantas estivais para cobertura do solo. *Ciência Rural* 33(3), 469-75. doi:10.1590/S0103-84782003000300012.

Bayer, C., Martin-Neto, L., Mielniczuk, J. and Pavinato, A. 2004. Armazenamento de carbono em frações lábeis da matéria orgânica de um Latossolo Vermelho sob plantio direto. *Pesquisa Agropecuária Brasileira* 39(7), 677-83. doi:10.1590/S0100-204X2004000700009.

Bayer, C., Martin-Neto, L., Mielniczuk, J., Pavinato, A. and Dieckow, J. 2006. Carbon sequestration in two Brazilian Cerrado soils under no-till. *Soil and Tillage Research* 86(2), 237-45. doi:10.1016/j.still.2005.02.023.

Bayon, R. C. L. Le, Weisskopf, L., Martinoia, E., Jansa, J., Frossard, E., Keller, F., Föllmi, K. B. and Gobat, J. -M. 2006. Soil phosphorus uptake by continuously cropped *Lupinus albus*: a new microcosm design. *Plant and Soil* 283(1-2), 309-21. doi:10.1007/s11104-006-0021-4.

Beare, M. H., Parmelee, R. W., Hendrix, P. F., Cheng, W., Coleman, D. C. and Crossley Jr., D. A. 1992. Microbial and faunal interactions and effects on litter nitrogen and decomposition in agroecosystems. *Ecological Monographs* 62(4), 569–91. doi:10.2307/2937317.

Beare, M. H., Coleman, D. C., Pohlad, B. R. and Wright, D. H. 1993. Residue placement and fungicide effects on fungal communities in conventional and no-tillage soils. *Soil Science Society of America Journal* 57(2), 392–9. doi:10.2136/sssaj1993.03615995005700020018x.

Blanchart, E., Lavelle, P., Braudeau, E., Le Bissonnais, Y. and Valentin, C. 1997. Regulation of soil structure by geophagous earthworm activities in humid savannas of Côte d'Ivoire. *Soil Biology and Biochemistry* 29(3–4), 431–9. doi:10.1016/S0038-0717(96)00042-9.

Blouin, M., Zuily-Fodil, Y., Pham-Thi, A. T., Laffray, D., Reversat, G., Pando, A., Tondoh, J. and Lavelle, P. 2005. Belowground organism activities affect plant aboveground phenotype, inducing plant tolerance to parasites. *Ecology Letters* 8(2), 202–8. doi:10.1111/j.1461-0248.2004.00711.x.

Birch, H. F. 1958. The effect of soil drying on humus decomposition and nitrogen availability. *Plant and Soil* 10(1), 9–31. doi:10.1007/BF01343734.

Blanco-Canqui, H., Schlegel, A. J. and Heer, W. F. 2011. Soil-profile distribution of carbon and associated properties in no-till along a precipitation gradient in the central Great Plains. *Agriculture, Ecosystems and Environment* 144(1), 107–16. doi:10.1016/j.agee.2011.07.004.

Bolliger, A., Magid, J., Amado, J. C. T., Skóra Neto, F., Ribeiro, MdFdS., Calegari, A., Ralisch, R. and de Neergaard, A. 2006. Taking stock of the Brazilian "Zero-Till Revolution". A review of landmark research and farmers' practice. *Advances in Agronomy* 91, 47–110. doi:10.1016/S0065-2113(06)91002-5.

Briedis, C., Sá, J. C. D. M., De-Carli, R. S., Antunes, E. A. P., Simon, L., Romko, M. L., Elias, L. S. and Ferreira, A. D. O. 2012. Particulate soil organic carbon and stratification ratio increases in response to crop residue decomposition under no-till. *Revista Brasileira de Ciência do Solo* 36(5), 1483–90. doi:10.1590/S0100-06832012000500012.

Brito, J. A. and Ferraz, S. 1987. Antagonismo de *Brachiaria decumbens* e *Panicum maximum* cv. Guiné a *Meloidogyne javanica*. *Nematologia Brasileira* 11, 270–85.

Broch, D. L., Pitol, C. and Borges, E. P. 1997. *Integração Agricultura-Pecuária: Plantio direto da soja sobre pastagem na integração agropecuária*. Informativo Técnico 01/97. Fundação MS para Pesquisa e Difusão de Tecnologias Agropecuárias, Maracaju-MS, p. 24.

Bronick, C. J. and Lal, R. 2005. Soil structure and management: a review. *Geoderma* 124(1–2), 3–22.

Bunch, R. 1985. Two ears of corn: a guide to people-centered agricultural improvement (No. 630 B881t Ej. 1 025044). World Neighbors.

Calegari, A. 1985. Tremoço no sudoeste do Paraná. In: *Anais do II Reunião Brasileira de Tremoço - Lupinus* sp. Documentos 10. Passo Fundo - EMBRAPA - CNPT, 99pp.

Calegari, A. 1991. Effects of winter cover crops on corn yield in Paraná, Brazil. In: *Cover Crops for Clean Water*. Proceedings of the International Conference West Tennessee Experiment Station, 9–11 April, Jackson, Tennessee, USA. Soil and Water Conservation Society, Ankeny, IA, pp. 97–8.

Calegari, A. 1995a. The effects of tillage and cover crops on some chemical properties of an Oxisol in South western Paraná, Brazil. Dissertation thesis. University of Aberdeen, Department of Plant and Soil Science, Aberdeen, Scotland, UK, 81p.

Calegari, A. 1995b. Leguminosas para adubação verde de verão no Paraná. Circular 80. Instituto Agronômico do Paraná (IAPAR), Londrina, PR, Brazil.

Calegari, A. 1998a. Espécies para cobertura de solo. In: *Plantio Direto. Pequena propriedade sustentável*. Circular 101. Instituto Agronômico do Paraná (IAPAR), Londrina, PR, Brazil, pp. 65–94.

Calegari, A. 1998b. Culturas de cobertura para plantio direto em solos argilosos. In: *Resumos. II Conferência Anual de Plantio direto*. Pato Branco, Paraná, Brazil.

Calegari, A. 2000a. Coberturas verdes em sistemas intensivos de produção. In: *Nitrogênio na sustentabilidade de sistemas intensivos de produção agropecuária Workshop Proceedings, Dourados, MS, Brazil*. 13–15 junho. Documentos Embrapa Agropecuária Oeste (26), pp. 141–53.

Calegari, A. 2000b. Adubação verde e rotação de culturas. In: *Feijão: Tecnologia de Produção*. Circular 135. Instituto Agronômico do Paraná (IAPAR), Londrina, PR, Brazil, pp. 29–34.

Calegari, A. 2006a. Plantas de cobertura. In: Casao-Junior, R., Siqueira, R., Mehta, Y. R. and Passini, J. J. (Eds), *Sistema plantio direto com qualidade*. Instituto Agronômico do Paraná (IAPAR), Londrina, pp. 55–73.

Calegari, A. 2006b. Sequestro de carbono, atributos físicos e químicos em diferentes sistemas de manejo em um latossolo argiloso do sul do Brasil. Tese (Doutorado em Agronomia). Universidade Estadual de Londrina, Londrina, PR, 191f.

Calegari, A. 2014. Perspectivas e estratégias para a sustentabilidade e o aumento da biodiversidade dos sistemas agrícolas com o uso de adubos verdes. In: Lima Filho, O. F., Ambrosano, E. J., Rossi, F. and Carlos, J. A. D. (Eds), *Adubação verde e plantas de cobertura no Brasil: fundamentos e prática*. Embrapa, Brasília, DF, 1:1, pp. 21–36.

Calegari, A. 2016. *Plantas de cobertura. Manual Técnico* (3a. ed.). Penergetic, Uberaba, MG, 24p.

Calegari, A. 2017. Bioactive Production Systems – action and equilibrium of soil attributes trough adequate use of cover crops and crop rotation, Summaries. In: *7th World Congress on Conservation Agriculture (7WCCA)*, 1-4 August, Rosario-Santa Fe – Argentina. Published by AAPRESID – CAAPAS.

Calegari, A. 2018a. Cover crops. In: *Penergetic the Natural Biotechnology. With the Impulse of Nature. For intelligent Agriculture*. Penergetic International AG, Romanshorn, Switzerland, pp. 7–33. Available at: www.penergetic.com.

Calegari, A. 2018b. Advantages in coffee planting when using cover crops and penergetic products. In: *Penergetic the Natural Biotechnology. With the Impulse of Nature. For intelligent Agriculture*. Penergetic International AG, Romanshorn, Switzerland, pp. 7–33. Available at: www.penergetic.com.

Calegari, A. and Alexander, I. 1998. The effects of tillage and cover crops on some chemical properties of an oxisol and summer crop yields in southwestern Paraná, Brazil. *Advances in GeoEcology* 31, 1239–46.

Calegari, A. and Costa, A. 2010. Manutenção da cobertura melhora atributos do solo. In: *Visão agrícola*. USP, Esalq, ano 6, jul/dez. 2009 (vol. 9). Piracicaba, SP, pp. 13–6.

Calegari, A. and Machado, A. C. Z. 2012. *Milheto e Crotalária: Dupla de sucesso no controle dos fitonematoides*. Revista Campo e Negócios, Uberlândia, MG, pp. 26–9, 01 out. 2012.

Calegari, A. and Machado, A. C. Z. 2013. *Efeito das crotalárias no controle de nematoides*. Revista Campo e Negócios, Uberlândia - MG, pp. 64-5, 30 jul. 2013.

Calegari, A. and Pavan, M. A. 1995. Efeitos da rotação de milho com adubos verdes de inverno na agregação do solo. *Arquivo de Biologia e Tecnologia* 38, 45–53.

Calegari, A. and Peñalva, M. 1994. *Abonos verdes em el Sur del Uruguay*. MGAP/GTZ, Montevideo, Uruguay, 151pp.

Calegari, A., Mondardo, A., Bulisani, E. A., Wildner, L. do P., Costa, M. B. B., Alcântara, P. B., Miyasaka, S. and Amado, T. J. C. 1993. *Adubação verde no Sul do Brasil* (2a. ed.). AS-PTA, Rio de Janeiro, RJ, 346p.

Calegari, A., Ferro, M., Grzesiuk, F. and Júnior, L. J. 1995. Plantio direto, rotação de culturas. Cooperativa dos Cafeicultores de Maringá – COCAMAR, 64p.

Calegari, A., Darolt, M. R. and Ferro, M. 1998a. Towards sustainable agriculture with a no-tillage system. *Advances in GeoEcology* 31, 1205–9.

Calegari, A., Heckler, J. C., Santos, H. P., Pitol, C., Fernandes, F. M., Hernani, L. C. and Gaudêncio, C. A. 1998b. Culturas, Sucessões e Rotações. In: *Sistema Plantio Direto: O produtor pergunta, a Embrapa responde*. Embrapa-CPAO, pp. 59–80.

Calegari, A., Ashburner, J. and Fowler, R. 2005. *Conservation Agriculture in Africa*. FAO Regional Office for Africa, Accra, Ghana, 91pp. ISBN: 9988-627-04.

Calegari, A., Castro Filho, C. D., Tavares Filho, J., Ralisch, R. and Guimarães, M. D. F. 2006. Melhoria da agregação do solo através do sistema plantio direto. *Semina: Ciências Agrárias* 27(2), 147–58. doi:10.5433/1679-0359.2006v27n2p147.

Calegari, A., Hargrove, W. L., Rheinheimer, D. D. S., Ralisch, R., Tessier, D., de Tourdonnet, S. and de Fatima Guimarães, M. 2008. Impact of long-term no-tillage and cropping system management on soil organic carbon in an Oxisol: a model for sustainability. *Agronomy Journal* 100(4), 1013–9. doi:10.2134/agronj2007.0121.

Calegari, A., Rheinheimer, D. S., Tourdonnet, S., Tessier, D., Hargrove, W. L., Ralisch, R., Guimarães, M. F. and Tavares Filho, J. 2011. Soil physical properties affected by soil management and crop rotation in a long term experiment in Southern Brazil. Proceedings of the 5th World Congress of Conservation Agriculture. Integrating 3rd Farming Systems design Conference. Resilient food systems for a changing World; Published on DVD; http://www.wcaa2011.org; Theme 1: More efficient management practice for CA; 26–29 September 2011, Brisbane, Australia, pp. 360–1.

Calegari, A., Costa, A., Rheinheimer, D., Tiecher, T. and Casali, C. A. 2012. Conservation agriculture including cover crops and crop rotation can improve maize yield. In: *The 3rd International Conference on Conservation Agriculture in Southeast Asia*. Conservation Agriculture and Sustainable Upland livelihoods. Innovations for, with and by farmers to adapt to local and global changes. Proceedings. Ha Noi, Vietnam, 10–15 December 2012. Printed by: Steps Media Co, Ltd. Hanoi Vietnam. CIRAD, NOMAFSI, University of Queensland, pp. 253–5. ISBNCIRAD: 978-2-87614-687-7. EAN CIRAD: 9782876146877.

Calegari, A., Araújo, A. G., Costa, A., Lanillo, R. F., Casão Junior, R. and Rheinheimer, D. S. 2013a. Conservation Agriculture in Brazil. In: Jat, R. A., Sahrawat, K. L. and Kassam, A. H. (Eds), *Conservation Agriculture- Global Prospects and Challenges*. CAB International, pp. 54–88.

Calegari, A., Tiecher, T., Hargrove, W. L., Ralisch, R., Tessier, D., de Tourdonnet, S., Guimarães, M. D. F. and Dos Santos, D. R. 2013b. Long-term effect of different soil management systems and winter crops on soil acidity and vertical distribution of nutrients in a Brazilian Oxisol. *Soil and Tillage Research* 133, 32–9. doi:10.1016/j.still.2013.05.009.

Calegari, A., Tourdonnet, S., Tessier, D., Rheinheimer, D. S., Ralisch, R., Hargrove, W., Guimarães, M. F. and Filho, J. T. 2013c. Influence of soil management and crop rotation on physical properties in a long-term experiment in Paraná, Brazil. *Communications in Soil Science and Plant Analysis* 44(13), 2019–31. doi:10.1080/00103624.2013.790406.

Calonego, J. C., Raphael, J. P. A., Rigon, J. P. G., de Oliveira Neto, L. and Rosolem, C. A. 2017. Soil compaction management and soybean yields with cover crops under no-till and occasional chiseling. *European Journal of Agronomy* 85, 31–7. doi:10.1016/j.eja.2017.02.001.

Campbell, C. A. and Souster, W. 1982. Loss of organic matter and potentially mineralizable nitrogen from Saskatchewan soils due to cropping. *Canadian Journal of Soil Science* 62(4), 651–6.

Campos, B. C. de, Amado, T. J. C., Bayer, C., Nicoloso, R., da, S. and Fiorin, J. E. 2011. Carbon stock and its compartments in a subtropical Oxisol under long-term tillage and crop rotation systems. *Revista Brasileira de Ciência do Solo* 35, 805–17.

Canalli, L. B. S. 2009. Decomposição de resíduos culturais e sua contribuição nos macroagregados e na fração lábil da matéria orgânica do solo no sistema plantio direto. Tese (Doutorado) – Curso de Agronomia. Universidade Federal do Paraná, Curitiba, 101f.

Carvalho, A., Dantas, R., Coelho, M. C., Lima, W. M., Souza, J., Fonseca, O. P. and Guimarães Júnior, R. 2010. *Teores de hemiceluloses, celulose e lignina em plantas de cobertura com potencial para sistema plantio direto no Cerrado*. Embrapa Cerrados, pp. 1–14.

Casali, C. A. 2012. Sistemas de culturas sob diferentes manejos por longa duração alteram as formas de fósforo do solo? Tese (Doutorado em Ciência do Solo). Universidade Federal de Santa Maria – UFSM, Santa Maria, 149f.

Casão Junior, R., Araújo, A. G. and Llanillo, R. 2012. *No-Till Agriculture in Southern Brazil: Factors that Facilitated the Evolution of the System and the Development of the Mechanization of Conservation Farming* (The Food and Agriculture Organization of the United Nations and Instituto Agronômico do Paraná). FAO and Agricultural Research Institute of Paraná State (IAPAR), Supported by FAPEAGRO, 81p. ISBN 978-92-5-107102-1.

Castro Filho, C., Muzilli, O. and Podanoschi, A. L. 1998. Estabilidade dos agregados e sua relação com o teor de carbono orgânico num Latossolo Roxo distrófico, em função de sistemas de plantio, rotações de culturas e métodos de preparo das amostras. *Revista Brasileira de Ciência do Solo* 22(3), 527–38. doi:10.1590/S0100-06831998000300019.

Castro Filho, C., Lourenço, A., Guimarães, M. de F. and Fonseca, I. C. B. 2002. Aggregate stability under different soil management systems in a red latosol of Paraná, Brazil. *Soil and Tillage Research* 65, 45–51.

Ceretta, C. A., Basso, C. J., Herbes, M. G., Poletto, N. and Silveira, M. J. 2002. Produção e decomposição de fitomassa de plantas invernais de cobertura de solo e milho, sob diferentes manejos da adubação nitrogenada. *Ciência Rural* 32(1), 49–54. doi:10.1590/S0103-84782002000100009.

Chavarría, D. N., Verdenelli, R. A., Serri, D. L., Restovich, S. B., Andriulo, A. E., Meriles, J. M. and Vargas-Gil, S. 2016. Effect of cover crops on microbial community structure and related enzyme activities and macronutrient availability. *European Journal of Soil Biology* 76, 74–82. doi:10.1016/j.ejsobi.2016.07.002.

Ciotta, M. N., Bayer, C., Mara, S., Fontoura, V., Ernani, P. R. and Albuquerque, J. A. 2003. Soil organic matter and cation exchange capacity increase in a low activity clay soil under no-tillage system. *Ciência Rural* 33336, 1161-4.

Colozzi-Filho, A., Siqueira, J. O., Saggin Júnior, O. J., Guimarães, P. T. G. and Oliveira, E. 1994. Efetividade de diferentes fungos micorrízicos arbusculares na formação de mudas, crescimento pós transplante e produção do cafeeiro. *Pesquisa Agropecuária Brasileira, Brasília* 29(9), 1397–406.

Colozzi-Filho, A., Andrade, D. S. and Balota, E. L. 2001. Atividade microbiana em solos cultivados em sistema plantio direto. *Informe Agropecuário (Belo Horizonte), Belo Horizonte* 22(208), 84–91.

Conceição, P. C., Dieckow, J. and Bayer, C. 2013. Combined role of no-tillage and cropping systems in soil carbon stocks and stabilization. *Soil and Tillage Research* 129, 40–7. doi:10.1016/j.still.2013.01.006.

Costa, J. L., da, S. and Rava, C. A. 2003. Influência da braquiária no manejo de doenças do feijoeiro com origem no solo. In: Kluthouski, J., Stone, L. F. and Aidar, H. (Eds), *Integração lavoura-pecuária*. Embrapa Arroz e Feijão, Santo Antônio de Goiás, pp. 523–33.

Cotrufo, M. F., Wallenstein, M. D., Boot, C. M., Denef, K. and Paul, E. 2013. The Microbial Efficiency-Matrix Stabilization (MEMS) framework integrates plant litter decomposition with soil organic matter stabilization: do labile plant inputs form stable soil organic matter? *Global Change Biology* 19(4), 988–95. doi:10.1111/gcb.12113.

Cui, H., Zhou, Y., Gu, Z., Zhu, H., Fu, S. and Yao, Q. 2015. The combined effects of cover crops and symbiotic microbes on phosphatase gene and organic phosphorus hydrolysis in subtropical orchard soils. *Soil Biology and Biochemistry* 82, 119–26. doi:10.1016/j.soilbio.2015.01.003.

Dalla Costa, M. and Lovato, P. E. 2004. Fosfatases na dinâmica do fósforo do solo sob culturas de cobertura com espécies micorrízicas e não micorrízicas. *Pesquisa Agropecuária Brasileira* 39(6), 603–5. doi:10.1590/S0100-204X2004000600013.

Da Silva, F. A. M. 2004. Parametrização e modelagem do balanço hídrico em sistema de plantio direto no cerrado brasileiro. Ph.D. Dissertation of Campinas State University, Campinas, 218pp.

Debiasi, H., Moraes, M. T., Franchini, J. C., Dias, W. P., Silva, J. F. V. and Ribas, L. N. 2011. Manejo do solo para controle cultural do nematoide das lesões radiculares na entressafra da soja. In: *XXXIII Congresso Brasileiro de Ciência do Solo, Uberlândia, MG*. Available at: htpp://ainfo.cnptia.embrapa.br/ digital/bitstream/item/48290/1/debiasi.manejo.PDF.

Debiasi, H., Franchini, J. C., Dias, W. P., Ramos, J. R., Balbinoti, J. R., Ponce, G. and Ribas, L. N. 2016. Qualidade do solo e manejo de nematoides em sistemas de produção de soja. FERTBIO, Goiânia, GO. Available at: www.eventosolos.org.br/fertbio2016/palestrantes/palestras/Henrique % Debiasi.pdf.

Derpsch, R. and Calegari, A. 1985. Guia de adubação de inverno. Documentos 9. IAPAR, Londrina, Paraná, Brazil, 96p.

Derpsch, R. and Calegari, A. 1992. Plantas para adubação verde de inverno (2ª. ed.). Circular 73. Instituto Agronômico do Paraná (IAPAR), Londrina, PR, Brazil, 78pp.

Derpsch, R., Sidiras, N. and Roth, C. H. 1986. Results of studies made from 1977 to 1984. To control erosion by cover crops and zero tillage techniques in Paraná, Brazil. *Soil and Tillage Research* 8, 253–63. doi:10.1016/0167-1987(86)90338-7.

Derpsch, R., Roth, C. H., Sidiras, N. and Köpke, U. 1991. *Controle da erosão no Paraná, Brasil: Sistemas de cobertura do solo, plantio direto e preparo conservacionista do solo*. GTZ/IAPAR, Eschborn, 272p.

De Waele, D. and Elsen, A. 2007. Challenges in tropical plant nematology. *Annual Review of Phytopathology* 45, 457–85. doi:10.1146/annurev.phyto.45.062806.094438.

Dias-Arieira, C. R., Ferraz, S., Freitas, L. G. and Mizobutsi, E. H. 2003. Avaliação de gramíneas forrageiras para controle de *Meloidogyne incognita* e *M. javanica* (Nematoda). *Acta Scientiarum* 25(2), 473–7. doi:10.4025/actasciagron.v25i2.2163.

Dieckow, J., Mielniczuk, J., Knicker, H., Bayer, C., Dick, D. P. and Kogel-Knabner, I. 2005. Composition of organic matter in a subtropical Acrisol as influenced by land use, cropping and N fertilization, assessed by CPMAS ^{13}C NMRespectroscopy. *European Journal of Soil Science* 56, 705–15.

Dieckow, J., Bayer, C., Conceição, P. C., Zanatta, J. A., Martin-Neto, L., Milori, D. B. M., Salton, J. C., Macedo, M. M., Mielniczuk, J. and Hernani, L. C. 2009. Land use, tillage, texture and organic matter stock and composition in tropical and subtropical Brazilian soils. *European Journal of Soil Science* 60(2), 240–9. doi:10.1111/j.1365-2389.2008.01101.x.

Dijkstra, F. 2002. Conservation tillage development at the ABC Cooperatives in Paraná, Brazil. In: van Santen, E. (Ed.), *Making Conservation Tillage Conventional: Building a Future on 25 Years of Research*. Proceedings of the 25th Annual Southern Conservation Tillage Conference for Sustainable Agriculture held 24–26 June 2002 in Auburn. Alabama Agricultural Experimental Station and Auburn University, AL, pp. 12–18.

Dimassi, B., Mary, B., Wylleman, R., Labreuche, J., Couture, D., Piraux, F. and Cohan, J. J. P. 2014. Long-term effect of contrasted tillage and crop management on soil carbon dynamics during 41 years. *Agriculture, Ecosystems and Environment* 188, 134–46. doi:10.1016/j.agee.2014.02.014.

Drijber, R. A., Doran, J. W., Parkhurst, A. M. and Lyon, D. J. 2000. Changes in soil microbial community structure with tillage under long-term wheat-fallow management. *Soil Biology and Biochemistry* 32(10), 1419–30. doi:10.1016/S0038-0717(00)00060-2.

Essington, M. E. 2003. *Soil and Water Chemistry: an Integrative Approach*. CRC Press, Boca Raton, FL.

Esteves, J. A. F., Wutke, E. B., Zambrosi, F. C. B. and Gallo, P. B. 2014. Fertilidade do solo em função de sucessão de culturas em plantio direto recém-implementado / Soil fertility due to the crop sequences in no-till newly implanted system. In: *Fertbio. Anais...Araxá: Sociedade Brasileira de Ciência do Solo*, p. 1.

Favero, C., Jucksch, I., Alvarenga, R. C. and Costa, L. Md. 2001. Modificações na população de plantas espontâneas na presença de adubos verdes. *Pesquisa Agropecuaria Brasileira* 36(11), 1355–62. doi:10.1590/S0100-204X2001001100005.

Fernandes, M. F., Barreto, A. C. and Emidio-Filho, J. 1999. Fitomassa de adubos verdes e controle de plantas daninhas em diferentes densidades populacionais de leguminosas. *Pesquisa Agropecuária Brasileira* 34(9), 1593–600. doi:10.1590/S0100-204X1999000900009.

Ferreira, A. D. O., Sá, J. C. D. M., Harms, M. G., Miara, S., Briedis, C., Quadros Netto, C., Santos, J. B. and Canalli, L. B. 2012. Carbon balance and crop residue management in dynamic equilibrium under a no-till system in Campos Gerais. *Revista Brasileira de Ciência do Solo* 36(5), 1583–90. doi:10.1590/S0100-06832012000500022.

Fink, J. R., Inda, A. V., Tiecher, T. and Barrón, V. 2016. Iron oxides and organic matter on soil phosphorus availability. *Ciência e Agrotecnologia* 40(4), 369–79. doi:10.1590/1413-70542016404023016.

Florentin, M. A., Peñalva, M., Calegari, A. and Derpsch, R. 2001. Abonos verdes y rotación de cultivos en siembra directa Pequeñas propriedades. Proyecto Conservación de Suelos MAG-GTZ. Editado por Artes Gráficas, Asunción Paraguay, San Lorenzo, 83p.

Florentin, M. A., Peñalva, M., Calegari, A. and Derpsch, R. 2010. *Green Manure/Cover Crops and Crop Rotation in Conservation Agriculture on Small Farmers* (vol. 12). FAO, Rome, 97p. ISBN 978-92-5-106839-7 ISSN 1020-4555.

Fontaneli, R. S., Ambrosi, I., Santos, H. P. D., Ignaczak, J. C. and Zoldan, S. M. 2000. Análise econômica de sistemas de produção de grãos com pastagens de inverno, em sistema plantio direto. *Pesquisa Agropecuária Brasileira* 35(11), 2129–37. doi:10.1590/S0100-204X2000001100002.

Frageria, N. K. and Nascente, A. S. 2014. Management of soil acidity of South American soils for sustainable crop production. *Advances in Agronomy* 128, 221–75. doi:10.1016/B978-0-12-802139-2.00006-8.

Franchini, J. C., Borkert, C. M., Ferreira, M. M. and Gaudêncio, C. A. 2000. Alterações na fertilidade do solo em sistemas de rotação de culturas em semeadura direta. *Revista Brasileira de Ciência do Solo* 24(2), 459–67. doi:10.1590/S0100-06832000000200022.

Frey, S. D., Elliott, E. T. and Paustian, K. 1999. Bacterial and fungal abundance and biomass in conventional and no-tillage agroecosystems along two climatic gradients. *Soil Biology and Biochemistry* 31(4), 573–85. doi:10.1016/S0038-0717(98)00161-8.

Frey, S. D., Elliott, E. T., Paustian, K. and Peterson, G. A. 2000. Fungal translocation as a mechanism for soil nitrogen inputs to surface residue decomposition in no-tillage agroecosystem. *Soil Biology and Biochemistry* 32(5), 689–98. doi:10.1016/S0038-0717(99)00205-9.

Gahoonia, T. S. and Nielsen, N. E. 2004. Root traits as tools for creating phosphorus efficient crop varieties. *Plant and Soil* 260(1/2), 47–57. doi:10.1023/B:PLSO.0000030168.53340.bc.

Gaiser, T., Stahr, K., Bernard, M. and Kang, B. T. 2012. Changes in soil organic carbon fractions in a tropical Acrisol as influenced by the addition of different residue materials. *Agroforestry Systems* 86(2), 185–95. doi:10.1007/s10457-011-9417-0.

Gassen, D. N. and Gassen, F. R. 1996. *Plantio direto. O caminho do futuro*. Passo Fundo, Aldeia Sul, 207pp.

Giacomini, S. J., Aita, C., Vendruscolo, E. R. O., Cubilla, M., Nicoloso, R. S. and Fries, M. R. 2003a. Matéria seca, relação C/N e acúmulo de nitrogênio, fósforo e potássio em misturas de plantas de cobertura de solo. *Revista Brasileira de Ciência do Solo* 27(2), 325–34. doi:10.1590/S0100-06832003000200012.

Giacomini, S. J., Aita, C., Hübner, A. P., Lunkes, A., Guidini, E. and Amaral, E. B. do. 2003b. Liberação de fósforo e potássio durante a decomposição de resíduos culturais em plantio direto. *Pesquisa Agropecuária Brasileira* 38(9), 1097–104. doi:10.1590/S0100-204X2003000900011.

Giebel, J. 1974. Biochemical mechanism of plant resistance to nematodes: a review. *Journal of Nematology* 6, 157–84.

Grego, C. R., Wutke, E. B., Vieira, S. R. and Pereira, J. C. V. N. A. 2006. Variabilidade espacial da infiltração da água na cultura do feijoeiro irrigado em sucessão de culturas. In: *Reunião Sul-Brasileira De Feijão*, Londrina. Resumos Expandidos.

Guggenberger, G., Elliott, E. T., Frey, S. D., Six, J. and Paustian, K. 1999. Microbial contributions to the aggregation of a cultivated grassland soil amended with starch. *Soil Biology and Biochemistry* 31(3), 407–19. doi:10.1016/S0038-0717(98)00143-6.

Hargrove, W. L. 1991. *Cover Crops for Clean Water*. Soil and Water Conservation Society, Ankeny, Iowa, p. 198.

Heinzmann, F. X. 1985. Resíduos culturais de inverno e assimilação de nitrogênio por culturas de verão. *Pesquisa Agropecuaria Brasileira* 20, 1021–30.

Inomoto, M. M., Motta, L. C. C., Beluti, D. B. and Machado, A. C. Z. 2006. Reação de seis adubos verdes a *Meloidogyne javanica* e *Pratylenchus brachyurus*. *Nematologia Brasileira* 30, 39–44.

Inomoto, M. M., Motta, L. C. C., Machado, A. C. Z. and Sazaki, C. S. S. 2007a. Reação de dez coberturas vegetais a *Pratylenchus brachyurus*. *Nematologia Brasileira* 30, 151–7.

Inomoto, M. M., Machado, A. C. Z. and Antedomênico, S. R. 2007b. Reação de *Brachiaria* spp. e *Panicum maximum* a *Pratylenchus brachyurus*. *Fitopatologia Brasileira* 32(4), 341–4. doi:10.1590/S0100-41582007000400009.

Jastrow, J. D. and Miller, R. M. 1998. Soil aggregate stabilization and carbon sequestration: feedbacks through organominearl associations. In: Lal, R., Kimble, J. M., Follett, R. F. and Stewart, B. A. (Eds), *Soil Processes and the Carbon*. CRC Press, Boca Raton, FL, pp. 207–23.

Jones, D. L. 1998. Organic acids in the rhizosphere – a critical review. *Plant and Soil* 205(1), 25–44. doi:10.1023/A:1004356007312.

Kaminski, J., Moterle, D. F., Rheinheimer, D., dos, S., Gatiboni, L. C. and Brunetto, G. 2010. Potassium availability in a hapludalf soil under long term fertilization. *Revista Brasileira de Ciência do Solo* 34, 49–57.

Kassam, A., Friedrich, T. and Derpsch, R. 2019. Global spread of Conservation Agriculture. *International Journal of Environmental Studies* 76(1), 29–51. doi:10.1080/00207233.2018.1494927.

Kliemann, H. J., Braz, A. J. P. B. and Silveira, P. M. 2006. Taxas de decomposição de resíduos de espécies de cobertura em Latossolo Vermelho Distróférrico. *Pesquisa Agropecuária Tropical* 36(1), 21–8.

Kliewer, I. J., Casaccia, F. and Vallejos, F. 1998. Viabilidade da redução do uso de herbicidas e custos no controle de plantas daninhas nas culturas de trigo e soja no sistema de plantio direto, através do emprego de adubos verdes de curto período. In: *Seminário Nacional sobre Manejo e Controle de plantas Daninhas em Plantio Direto*, I, Passo Fundo, RS, Resumo de Palestras. Ed. Aldeia Norte, Passo Fundo, RS, Brazil, pp. 120–3.

Kloepper, J. W., Rodríguez-Kábana, R., Mcinroy, J. A. and Collins, D. J. 1991. Analysis of populations and physiological characterization of microorganisms in rhizospheres of plants with antagonistic properties to phytopathogenic nematodes. *Plant and Soil* 136(1), 95–102. doi:10.1007/BF02465224.

Kluthcouski, J. and Aidar, H. 2003. Uso da integração lavoura-pecuária na recuperação de pastagens degradadas. In: Kluthcouski, J., Stone, L. F. and Aidar, H. (Eds), *Integração lavoura-pecuária* (1ª. ed.). Embrapa Arroz e Feijão, Santo Antonio de Goiás, pp. 185–223. cap. 6.

Kluthcouski, J., Cobucci, T., Aidar, H., Yoko-yama, L. P., de Oliveira, I. P. da, S., Costa, J. G., da Silva, L., et al. 2000. Sistema Santa Fé -Tecnologia Embrapa: integração lavoura-pecuária pelo consórcio de culturas anuais com forrageiras, em áreas de lavoura, nos sistemas direto e convencional. Circular Técnica 38. Embrapa Arroz e Feijão, Santo Antonio de Goiás, GO, 28p.

Kunze, A., Costa, M. D., Epping, J., Loffaguen, J. C., Schuh, R. and Lovato, P. E. 2011. Phosphatase activity in sandy soil influenced by mycorrhizal and non-mycorrhizal cover crops. *Revista Brasileira de Ciência do Solo* 35(3), 705–11. doi:10.1590/S0100-06832011000300005.

Kwaad, F. J. P. M., van der Zijp, M. and van Dijk, P. M. 1998. Soil conservation and maize cropping systems on sloping loess soils in the Netherlands. *Soil and Tillage Research* 46(1–2), 13–21. doi:10.1016/S0167-1987(98)80103-7.

Lal, R. and T.J. Logan 1995. Agricultural activities and carbon emissions from soils of the tropics. In: R. Lal, J. Kimble, E. Levine and B. A. Stewart (Eds) Advances in Soil Science, Lewis Publishers, Chelsea, MI: 293–308

Lal, R. 1997. Residue management, conservation tillage and restoration for mitigating greenhouse effect by CO_2-enrichment. *Soil and Tillage Research* 43(1–2), 81–107. doi:10.1016/S0167-1987(97)00036-6.

Landers, J. N., Teixeira, S. M. and Milhomen, A. 1994. Possíveis impactos da técnica de Plantio Direto sobre a sustentabilidade da produção de grãos na região dos cerrados. In: *Congresso Brasileiro de economia e sociologia*, 32, Brasilia. Desafio do Estado diante de uma agricultura era transformação. Anais. 2, pp. 799–820.

Leite, L. F. C., Freitas, R. D. C. A., Sagrilo, E., Silva Galvão, S. R. 2010. Decomposição e liberação de nutrientes de resíduos vegetais depositados sobre Latossolo Amarelo no Cerrado Maranhense. Revista Ciência Agronômica. 41(1): 29–35.

Li, H., Shen, J., Zhang, F., Marschner, P., Cawthray, G. and Rengel, Z. 2010. Phosphorus uptake and rhizosphere properties of intercropped and monocropped maize, faba bean, and white lupin in acidic soil. *Biology and Fertility of Soils* 46(2), 79–91. doi:10.1007/s00374-009-0411-x.

Li, S., Wang, Z. and Stewart, B. A. 2013. Responses of crop plants to ammonium and nitrate N. *Advances in Agronomy* 118, 205–397. doi:10.1016/B978-0-12-405942-9.00005-0.

Li, T., Wang, H., Zhou, Z., Chen, X. and Zhou, J. 2016. A new grading system for plant-available potassium using exhaustive cropping techniques combined with chemical analyses of soils. *Scientific Reports* 6, 37327. doi:10.1038/srep37327.

Lorenzi, H. 1984. *Considerações sobre plantas daninhas no plantio direto*. São Paulo, Escola Superior de Agricultura 'Luiz de Queirós,' Piracicaba. Plantio direto no Brasil. Fundação Cargill, Campinas, São Paulo, pp. 13–46.

Machado, A. C. Z. 2015. Praga custa R$ 35 bilhões ao agronegócio brasileiro. *Revista Agrícola*, pp. 12–13, nov.

Machado, A. C. Z. 2016. Genética a favor da soja: mais resistência aos nematoides. *Revista Campo e Negócios*, pp. 20–3, ago.

Machado, P. L. O. and Gerzabek, M. H. 1993. Tillage and crop rotation interactions on humic substances of a Typic Haplorthox from southern Brazil. *Soil and Tillage Research* 26(3), 227–36. doi:10.1016/0167-1987(93)90046-R.

Machado, A. C. Z., Motta, L. C. C., Siqueira, K. M. S., Ferraz, L. C. C. B. and Inomoto, M. M. 2007. Host status of green manures for two isolates of *Pratylenchus brachyurus* in Brazil. *Nematology* 9, 799–805.

Machado, A. C. Z., Mattei, D., Silva, S. A., Dorigo, O. F. and Ito, D. S. 2014. Host status of green manures to four species of root-knot nematodes in Brazil. In: *VI International Congress of Nematogy*, Cape Town, South Africa. Available at: journals.fcla.edu/jon/article/view/83290/80155.

Machado, A. C. Z., Silva, S. A., Dorigo, O. F., Riede, C. R. and Garbuglio, D. D. 2015. Phenotypic variability and response of Brazilian oat genotypes to different species of root-knot and root-lesion nematodes. *European Journal of Plant Pathology* 141(1), 111-7. doi:10.1007/s10658-014-0529-z.

Machado, A. C. Z., Silva, S. A. and Dorigo, O. F. 2016. Nematode Community analyses under no-till and conventional tillage with different crop rotations in a long-term experiment in South Brazil. In: *Joint Meeting SON/ONTA, 55th Annual meeting of the Society of Nematologists and 48th Annual Meeting of the Organization of Nematologists of Tropical America*, Montréal, Canada, p. 130.

Magdoff, F. and Van Es, H. 2009. *Building Soils for Better Crops: Sustainable Soil Management* (3rd ed.). Sare, Washington DC, 353p.

Maluf, H. J. G. M., Soares, E. M. B., Silva, I. R., Neves, J. C. L. and Silva, L. O. G. 2015. Decomposição de resíduos de culturas e mineralização de nutrientes em solo com diferentes texturas. *Revista Brasileira de Ciência do Solo* 39(6), 1681–9.

Marcelo, A. V., Corá, J. E., Fernandes, C., Martins, M. D. R. and Jorge, R. F. 2009. Crop sequences in no-tillage system: effects on soil fertility and soybean, maize and rice yield. *Revista Brasileira de Ciência do Solo* 33(2), 417–28. doi:10.1590/S0100-06832009000200019.

Mbuthia, L. W., Acosta-Martínez, V., DeBruyn, J., Schaeffer, S., Tyler, D., Odoi, E., Mpheshea, M., Walker, F. and Eash, N. 2015. Long term tillage, cover crop, and fertilization effects on microbial community structure, activity: implications for soil quality. *Soil Biology and Biochemistry* 89, 24–34. doi:10.1016/j.soilbio.2015.06.016.

McGill, W. B. and Cole, C. V. 1981. Comparative aspects of cycling of organic C, N, S and P through soil organic matter. *Geoderma* 26(4), 267–86. doi:10.1016/0016-7061(81)90024-0.

Medeiros, G. B., Calegari, A. and Gaudêncio, C. 1989. Rotação de culturas. In: *Manual técnico do sub-programa de manejo e conservação do solo*. Secretaria da Agricultura e do Abastecimento, Curitiba, PR, Brazil, pp. 189–95.

Mello, E. L., Bertol, I., Zaparolli, A. L. V. and Carrafa, M. R. 2003. Perdas de solo e água em diferentes sistemas de manejo de um Nitossolo Haplico submetido a chuva simulada. *Revista Brasileira de Ciência Do Solo* 27(5), 901–9. doi:10.1590/S0100-06832003000500015.

Mendes, F. L., Araújo, K., Debiasi, H., Franchini, J. C., Dias, W. P., Ramos Junior, E. U. and Silva, J. F. V. 2013. Alternativas culturais para o manejo do nematoide das lesões radiculares durante a entressafra da soja no Mato Grosso. In: *VIII Jornada Acadêmica da Embrapa Soja*. Documentos 339, pp. 95–101.

Merten, G. H. and Fernandes, F. F. 1998. Manejo de solo de baixa aptidão. In: *Plantio Direto. Pequena propriedade sustentável*. Circular 101. Instituto Agronômico do Paraná (IAPAR), Londrina, PR, Brazil, pp. 46–64.

Mielniczuk, J., Bayer, C., Vezzani, F. M., Lovato, T., Fernandes, F. F. and Debarba, L. 2003. Manejo de solo e culturas e sua relação com os estoques de carbono e nitrogênio do solo. In: Curi, N., Marques, J. J., Guilherme, L. R. G., Lima, J. M., Lopes, A. S., Alvarez, V., et al. (Eds), *Tópicos em ciência do solo* (vol. 3). Sociedade Brasileira de Ciência do Solo, Viçosa, MG, pp. 209–48.

Miyasaka, S. 1984. Histórico do estudo de adubação verde, leguminosas viáveis e suas características / History of green manure studies; viable leguminous species and its characteristics. In: *Encontro Nacional Sobre Adubação Verde, 1. Rio de Janeiro.*

Adubação verde no Brasil: trabalhos apresentados. Fundação Cargill, Campinas, pp. 64–123.

Miyazawa, M., Pavan, M. A. and Calegari, A. 1993. Efeito de material vegetal na acidez do solo. *Revista Brasileira de Ciência do Solo* 17, 411–6.

Monegat, C. 1991. *Plantas de cobertura do solo. Características e manejo em pequenas propriedades* (2ª. ed.). Ed. do Autor, Chapecó, SC, Brasil, 337p.

Muzilli, O. 1983. Influência do sistema de plantio direto, comparado ao convencional, sobre a fertilidade da camada arável do solo. *Revista Brasileira de Ciência do Solo* 7, 95–102.

Nazareno, G. G. 2009. Utilização de matéria orgânica no controle de nematoide das galhas em alface sob cultivo protegido. Dissertação de Mestrado. Universidade de Brasília, 59pp. Available at: repositorio.unb.br/bitstream/10482/4092/1/2009_GlenioGomesNazareno.pdf.

Neumann, G. and Römheld, V. 1999. Root excretion of carboxylic acids and protons in phosphorus-deficient plants. *Plant and Soil* 211(1), 121–30. doi:10.1023/A:1004380832118.

Pavan, M. A. and Calegari, A. 2003. Soil cover management for improving apple productivity under minimum disturbance. In: *II World Congress on Conservation Agriculture*. Extended Summary vol. II. FEBRAPDP, FAO, CAAPAS, CIRAD. Foz do Iguaçu, Brazil, 11–15 August, pp. 183–5.

Petersen, P., Tardin, J. M. and Marochi, F. 1999. *Desenvolvimento participativo de sistemas de plantio direto sem herbicidas voltados para a agricultura familiar: A experiência da região Centro-Sul do Paraná*. Assessoria e Serviços a Projetos em Agricultura Alternativa (AS-PTA), Rio de Janeiro.

Phelan, P. L., Mason, J. F. and Stinner, B. R. 1995. Soil fertility management and host preference by European corn borer, *Ostrinia nubilalis*, on *Zea mays*: a comparison of organic and conventional chemical farming. *Agriculture, Ecosystems and Environment* 56(1), 1–8. doi:10.1016/0167-8809(95)00640-0.

Pieri, C., Evers, G., Landers, J., O'Connell, P. and Terry, E. 2002. *Zero Tillage Farming for Sustainable Rural Development*. Agriculture and Rural Development. Working Paper. Rural Development Department, World Bank, Washington DC.

Pöttker, D. 1977. Efeito do tipo de solo, tempo de cultivo e da calagem sobre a mineralização da matéria orgânica em solos do Rio Grande do Sul. Tese (Mestrado em Ciência do Solo)-Universidade Federal do Rio Grande do Sul, Porto Alegre, 128f.

Primavesi, A. M. 1990. *agricultura em regiões tropicais: manejo ecológico do solo*. Nobel, São Paulo, 549p.

Prudencio, Y. C., Kaumbutho, P. G., Calegari, A. and Bwalya, M. 2004. Conservation agriculture for sustainable land management, food security and improved livelihood in Africa. Extension & Training Manual. Summary. Publication of the Soil Fertility Initiative (World Bank/AFR/ESSD), and the Africa Conservation Tillage Initiative (ACT).

Reicosky, D. C., Kemper, W. D., Langdale, G. W., Douglas, C. L. and Rasmussen, P. E. 1995. Soil organic matter changes resulting from tillage and biomass production. *Soil and Water Conservation* 50, 253–61.

Resck, D. V. S. 1992. Dinâmica da matéria orgânica e seus efeitos nas propriedades do solo. In: *Encontro Nacional De Rotacao De Culturas, 2., Campo Mourão. Ata...* Associação dos Engenheiros Agrônomos de Campo Mourão, Campo Mourão, pp. 117–38.

Resck, D. V. S., Vasconcellos, C. A., Vilela, L. and Macedo, M. C. M. 2000. Impact of conversion of Brazilian cerrados to cropland and pastureland on soil carbon pool and dynamics. In: Lal, R., Kimble, J. M. and Stewart, B. A. (Eds), *Global Climate Change and Tropical Ecosystems*. Advances in Soil Science. CRC Press, Boca Raton, FL, pp. 169–96.

Ribeiro, M. F. S. and Milleo, R. D. S. 2002. *Referências em Plantio Direto para Agricultura Familiar do Centro-Sul do Paraná*. Instituto Agronômico do Paraná (IAPAR), Londrina, PR, Brasil.

Rich, J. R. and Rahi, G. S. 1995. Suppression of *Meloidogyne javanica* and *M. incognita* on tomato with ground seed of castor, Crotalaria, hairy indigo and wheat. *Nematropica* 25, 159–64.

Rose, T. J., Hardiputra, B. and Rengel, Z. 2010. Wheat, canola and grain legume access to soil phosphorus fractions differs in soils with contrasting phosphorus dynamics. *Plant and Soil* 326(1–2), 159–70. doi:10.1007/s11104-009-9990-4.

Rosolem, C. A. and Calonego, J. C. 2013. Phosphorus and potassium budget in the soil-plant system in crop rotations under no-till. *Soil and Tillage Research* 126, 127–33. doi:10.1016/j.still.2012.08.003.

Roth, C. H. 1985. Infiltrability of Latossolo-roxo soils in North Parana, Brazil, in field trials for erosion control under different tillage systems. Thesis. Institut für Bodenwissenschaften, Universität Göttingen, Germany.

Ruedell, J. 1995. *Plantio direto na região de Cruz Alta*. FUNDACEP/BASF, Cruz Alta, p. 134.

Ruedell, J. 1999. Plantio direto na região de Cruz Alta. *Cruz Alta: Pesquisa Agropecuária Brasileira, Brasília* 34(4), 519–26.

Sá, J. C. M. and Lal, R. 2009. Stratification ratio of soil organic matter pools as an indicator of carbon sequestration in a tillage chronosequence on a Brazilian Oxisol. *Soil and Tillage Research* 103(1), 46–56. doi:10.1016/j.still.2008.09.003.

Sá, J. CdM., Cerri, C. C., Lal, R., Dick, W. A., Piccolo, M. C. and Feigl, B. E. 2009. Soil organic carbon and fertility interactions affected by a tillage chronosequence in a Brazilian Oxisol. *Soil and Tillage Research* 104(1), 56–64. doi:10.1016/j.still.2008.11.007.

Sá, J. C. M., Séguy, L., Sá, M. F. M., Ferreira, A. O., Briedis, C., Santos, J. B. and Canalli, L. 2010. Gestão da matéria orgânica e da fertilidade do solo visando sistemas sustentáveis de produção. In: *Boas práticas de manejo do solo*, pp. 1–38.

Sá, J. C. M., Séguy, L., Tivet, F., Lal, R., Bouzinac, S., Borszowskei, P. R., Briedis, C., dos Santos, J. B., Hartman, D. C., Bertoloni, C. G., Rosa, J. and Friedrich, T. 2015. Carbon depletion by plowing and its restoration by no-till systems. *Land Degradation and Development* 26, 531–43.

Schick, J., Bertol, I., Batistela, O. and Balbinot Junior, A. A. 2000. Erosão hidrica em cambissolo humico aluminico submetido a diferentes sistemas de preparo e cultivo do solo: I. Perdas de solo e agua. *Revista Brasileira de Ciência Do Solo* 24(2), 427–36. doi:10.1590/S0100-06832000000200019.

Schomberg, H. H., Endale, D. M., Calegari, A., Peixoto, R., Miyazawa, M. and Cabrera, M. L. 2006. Influence of cover crop on potential nitrogen availability to succeeding crops in a Southern Piedmont Soil. *Biology and Fertility of Soils* 42(4), 299–307. doi:10.1007/s00374-005-0027-8.

Scopel, E., Macena, F., Corbeels, M., Affholder, F. and Maraux, F. 2004. Modelling crop residue mulching effects on water use and production of maize under semi-arid and humid tropical conditions. *Agronomie* 24, 1–13. doi:10.1051/agro:2004029.

Scopel, E., Triomphe, B., Affholder, F., Macena da Silva, F. A., Corbeels, M., Valadares Xavier, J. H., Lahmar, R., Recous, S., Bernoux, M., Blanchart, E., Carvalho Mendes, I. and Tourdonnet, S. de. 2012. Conservation agriculture cropping systems in temperate and tropical conditions, performances and impacts. A review. INRA and Springer-Verlag, France. *Agronomy for Sustainable Development*. doi:10.1007/s13593-012-0106-9.

Séguy, L., Bouzinac, S., Trentini, A. and Cortez, N. A. 1996. L'agriculture brésilienne des fronts pionniers. *Agriculture et Développement* 12(décembre), 2-61.

Séguy, L., Bouzinac, S. and Husson, O. 2006. Direct-seeded tropical soil systems with permanent soil cover: learning from Brazilian experience. In: Uphoff, N., et al. (Eds), *Biological Approaches to Sustainable Soil Systems*. Taylor and Francis Group, CRC Press Publishing, Boca Raton, FL, pp. 323-42.

Severino, F. J. and Christoffoleti, P. J. 2004. Supressão de plantas daninhas por adubos verdes e herbicidas seletivos. *Scientia Agricola* 61(1), 21-6. doi:10.1590/S0103-90162004000100004.

Shane, M. W., Lambers, H., Cawthray, G. R., Kuhn, A. J. and Schurr, U. 2008. Impact of phosphorus mineral source (Al-P or Fe-P) and pH on cluster-root formation and carboxylate exudation in *Lupinus albus* L. *Plant and Soil* 304(1-2), 169-78. doi:10.1007/s11104-007-9535-7.

Sidiras, N. and Pavan, M. A. 1985. Influência do sistema de manejo do solo no seu nível de fertilidade. *Revista Brasileira de Ciência do Solo* 9, 249-54.

Silva, G. S., Ferraz, S. and Santos, J. M. 1989. Atração, penetração e desenvolvimento de larvas de *Meloidogyne javanica* em raízes de *Crotalaria* spp. *Nematologia Brasileira* 13, 151-63.

Silva, S. A., Zeni, F., Amaro, P. M. and Machado, A. C. Z. 2015. Host reaction of forage grasses to root-knot and lesion nematodes. *Nematoda* 2(1), e01215. doi:10.4322/nematoda.01215.

Silveira, P. M. and Rava, C. A. 2004. Utilização de crotalária no controle de nematoides da raiz do feijoeiro. Comunicado Técnico 74. EMBRAPA Arroz e Feijão, Santo Antonio de Goiás, 2p.

Sisti, C. P. J., dos Santos, H. P., Kochhann, R., Alves, B. J. R., Urquiaga, S. and Boddey, R. M. 2004. Change in carbon and nitrogen stocks in soil under 13 years of conventional or zero tillage in southern Brazil. *Soil and Tillage Research* 76(1), 39-58. doi:10.1016/j.still.2003.08.007.

Six, J., Bossuyt, H., Degryze, S. and Denef, K. 2004. A history of research on the link between (micro)aggregates, soil biota, and soil organic matter dynamics. *Soil and Tillage Research* 79(1), 7-31. doi:10.1016/j.still.2004.03.008.

Skóra Neto, F. 1993. Controle de plantas daninhas através de coberturas verdes consorciadas com milho. *Pesquisa Agropecuária Brasileira* 28, 1165-71.

Skóra Neto, F. 1998. Manejo de plantas daninhas. In: *Plantio Direto. Pequena propriedade sustentável*. Circular 101. IAPAR, Londrina, PR, Brazil, pp. 128-58.

Skóra Neto, F. and Campos, A. C. 2004. Alteração populacional da flora infestante pelo manejo pós-colheita e ocupação de curtos períodos de pousio com coberturas verdes. *Boletim Informativo da Sociedade Brasileira da Ciência das Plantas Daninhas* 10(Suppl.), 135.

Srinivasarao, C., Kundu, S., Ramachandrappa, B. K., Reddy, S., Lal, R., Venkateswarlu, B., Sahrawat, K. L. and Naik, R. P. 2014. Potassium release characteristics, potassium balance, and fingermillet (*Eleusine coracana* G.) yield sustainability in a 27-year long experiment on an Alfisol in the semi-arid tropical India. *Plant and Soil* 374(1-2), 315-30. doi:10.1007/s11104-013-1877-8.

Steffens, D., Leppin, T., Luschin-Ebengreuth, N., Min Yang, Z. and Schubert, S. 2010. Organic soil phosphorus considerably contributes to plant nutrition but is neglected by routine soil-testing methods. *Journal of Plant Nutrition and Soil Science* 173(5), 765–71. doi:10.1002/jpln.201000079.

Tanaka, R. T., Wutke, E. B., Pires, R. C. M., Medina, P. F. and Cardoso, F. P. 2009. Monitoramento da nutrição, da produtividade de soja, da fertilidade e da compactação do solo em áreas de reforma da cana-de-açúcar. *Revista Alcoolbrás* 124, 42–51.

Tardin, J. M., Cordeiro, A., Marochi, F., Marques, A. C. and Petersen, P. 1998. *Experimentação participativa no desenvolvimento de sistemas de plantio direto sem herbicidas.* Assessoria e Serviços a Projetos em Agricultura Altemativa (AS-PTA), Rio de Janeiro.

Teasdale, J. R., Brandsaeter, L. O., Calegari, A. and Skora Neto, F. 2007. Cover crops and weed management. In: Upadhyaya, M. K., Blackshaw, R. E. and Reading, U. K. (Eds), *Non-Chemical Weed Management: Principles, Concepts and Technology*, pp. 49–64.

Teles, A. P. B., Rodrigues, M., Herrera, W. F. B., Soltangheisi, A., Sartor, L. R., Withers, P. J. A. and Pavinato, P. S. 2017. Do cover crops change the lability of phosphorus in a clayey subtropical soil under different phosphate fertilizers? *Soil Use and Management* 33(1), 34–44. doi:10.1111/sum.12327.

Thiesen, G., Vidal, R. A. and Fleck, N. G. 2000. Reduction of *Brachiaria* plantaginea infestation on soybean crop with black oat residues. *Pesquisa Agropecuária Brasileira* 35(4), 753–6. doi:10.1590/S0100-204X2000000400011.

Tiecher, T., dos Santos, D. R. and Calegari, A. 2012a. Soil organic phosphorus forms under different soil management systems and winter crops, in a long term experiment. *Soil and Tillage Research* 124, 57–67. doi:10.1016/j.still.2012.05.001.

Tiecher, T., Rheinheimer, D. S., Kaminski, J. and Calegari, A. 2012b. Forms of inorganic phosphorus in soil under different long term soil tillage systems and winter crops. *Revista Brasileira de Ciência do Solo* 36(1), 271–82. doi:10.1590/S0100-06832012000100028.

Tiecher, T., Calegari, A., Caner, L. and Rheinheimer, D. S. 2017. Soil fertility and nutrient budget after 23-years of different soil tillage systems and winter cover crops in a subtropical Oxisol. *Geoderma* 308, 78–85. doi:10.1016/j.geoderma.2017.08.028.

Tiecher, T., Calegari, A., Caner, L. and Rheinheimer, D. D. S. 2018. Soil fertility and nutrient budget after 23-years of different soil tillage systems and winter cover crops in a subtropical Oxisol. *Geoderma* 308, 78–85. doi:10.1016/j.geoderma.2017.08.028.

Tormena, C. A., Friedrich, R., Pintro, J. C., Costa, A. C. S. and Fidalski, J. 2004. Propriedades físicas e taxa de estratificação de carbono orgânico num Latossolo Vermelho após dez anos sob dois sistemas de manejo. *Revista Brasileira de Ciência do Solo* 28(6), 1023–31. doi:10.1590/S0100-06832004000600011.

Torres, J. L. R., Pereira, M. G., Andrioli, I., Polidoro, J. C. and Fabian, A. J. 2005. Decomposição e liberação de nitrogênio de resíduos culturais de plantas de cobertura em um solo de Cerrado. *Revista Brasileira de Ciência do Solo* 29(4), 609–18. doi:10.1590/S0100-06832005000400013.

Vasconcelos, H. and Landers, J. N. 1993. Projeto Morrinhos: relato de uma experiência de Manejo de resíduos no cerrado. In: Arantes, N. E. and de M de Souza, P. I. (Eds), *Cultura de soja nos cerrados*. Potafos, Piracicaba, SP, Brazil, pp. 523–35.

Wall, P. C. and Ekboir, J. 2002. *Conservation Agriculture for Small Farmers: Challenges and Possibilities*. ASA, CSCA, SSSA Meeting, Indianapolis, USA, 10–14 November.

Wang, K. H., Sipes, B. S. and Schmitt, D. P. 2002. Crotalaria as a cover crop for nematode management: a review. *Nematropica* 32, 35–57.

Wang, X., Tang, C., Guppy, C. N. and Sale, P. W. G. 2008. Phosphorus acquisition characteristics of cotton (*Gossypium hirsutum* L.), wheat (*Triticum aestivum* L.) and white lupin (*Lupinus albus* L.) under P deficient conditions. *Plant and Soil* 312(1–2), 117–28. doi:10.1007/s11104-008-9589-1.

Wei, C., Gao, M., Shao, J., Xie, D. and Pan, G. 2006. Soil aggregate and its response to land management practices. *China Particuology* 4(5), 211–9. doi:10.1016/S1672-2515(07)60263-2.

Wuddivira, M. N. and Camps-Roach, G. 2007. Effects of organic matter and calcium on soil structural stability. *European Journal of Soil Science* 58(3), 722–7. doi:10.1111/j.1365-2389.2006.00861.x.

Wutke, E. B., Fancelli, A. L., Pereira, J. C. V. N. A. and Ambrosano, G. M. B. 1998. Rendimento do feijoeiro irrigado em rotação com culturas graníferas e adubos verdes. *Bragantia* 57(2), 325–38. doi:10.1590/S0006-87051998000200014.

Wutke, E. B., Arruda, F. B., Fancelli, A. L., Pereira, J. C. V. N. A., Sakai, E., Fujiwara, M. and Ambrosano, G. M. B. 2000. Propriedades do solo e sistema radicular do feijoeiro irrigado em rotação de culturas. *Revista Brasileira de Ciência do Solo* 24(3), 621–33. doi:10.1590/S0100-06832000000300015.

Wutke, E. B., Pio, R. M. and Negri, J. D. 2008. Impacto do primeiro ano de cultivo intercalar de adubos verdes e do nitrogênio na produtividade de plantas de tangor Murcott. In: *Congresso Brasileiro de Fruticultura*, 54., 2008, Vitória. Anais... Vitória.

Wutke, E. B., Esteves, J. A. F., Gallo, P. B., Zambrosi, F. C. B., de Maria, I. C. and Pires, R. C. M. 2013. Produtividade de Massa e Extração de Nutrientes por Adubos Verdes cultivados na Safrinha em Sistema de Semeadura Direta. In: *Congresso Brasileiro de Ciência do Solo*, 34., 2013, Florianópolis. Anais... Florianópolis: Epagri.

Wutke, E. B., Calegari, A. and Wildner, L. P. 2014. Espécies de adubos verdes e plantas de cobertura e recomendações para seu uso. In: Lima Filho, O. F., Ambrosano, E. J., Rossi, F. and Carlos, J. A. D. (Eds), *Adubação verde e plantas de cobertura no Brasil: fundamentos e prática* (vol. 1). Embrapa, Brasília, DF, pp. 59–167.

Ziech, A. R. D., Conceição, P. C., Luchese, A. V., Balin, N. M., Candiotto, G. and Garmus, T. G. 2015. Proteção do solo por plantas de cobertura de ciclo hibernal na região Sul do Brasil. *Pesquisa Agropecuária Brasileira* 50(5), 374–82. doi:10.1590/S0100-204X2015000500004.

Chapter 6

The role of crop and cropping system management in Conservation Agriculture systems

Peter Hobbs, Cornell University, USA; Christian Thierfelder, International Maize and Wheat Research Center (CIMMYT), Zimbabwe; Patrick Wall, Independent Consultant – Sustainable Agricultural Systems, Mexico; Raj Gupta, Center for Advancement of Sustainable Agriculture, India; and Ademir Calegari, Agricultural Research Institute of Paraná State (IAPAR), Brazil

1 Introduction

The world faces a daunting challenge to feed a growing population now and in the future. This is made even more challenging by the need to intensify production on land already being used for agriculture. Expanding agricultural area is also limited since intensification must be done by maintaining ecosystem services and minimizing environmental impacts so future generations also have the ability to feed the world. Climate change influenced by greenhouse gas (GHG) emissions, declining freshwater resources, future fossil fuel price increases, dietary changes, the need to produce more nutritious food, competition for agricultural land (urbanization and commercial use, and biofuel production) and the fact that known sources of productivity growth are already being used by farmers are factors that will make this challenge even more difficult (Hobbs, 2008).

There is a need to develop land management systems that result in more efficient use of natural resources. These sustainable land management (SLM) practices should also provide farmers with the means (equipment, knowledge etc.) and incentives (credit, markets, prices, payment for eco-system services etc.) to experiment with them and share with their neighbours so adoption is

http://dx.doi.org/10.19103/AS.2019.0048.06

accelerated. Conservation Agriculture (CA: minimal soil disturbance, permanent soil cover and rotations) offers hope for responsibly intensifying agriculture and meeting the challenge mentioned above but also requires farmers to use improved varieties; integrated pest, disease and weed management; integrated nutrient and water management; and measures to reduce storage losses (Hobbs, 2007; Hobbs et al., 2008). The latest data indicates that 'in 2015–16, CA was practised on about 180 Mha of cropland, corresponding to about 12.5% of the total global cropland' (Kassam et al., 2018). This paper also indicates that this is a 69% increase since 2008/09.

One of the key resources this CA management system seeks to attain is soil health improvement. Improving and maintaining soil health is the basis for increasing productivity on any farm. 'Soil health is defined as the continued capacity of the soil to function as a vital living ecosystem that sustains plants, animals and humans. This definition speaks to the importance of managing soils so they are sustainable for future generations' (USDA, 2012). Soil properties include not just the chemical properties most commonly measured in soil laboratories for farmers but also the physical and equally important biological properties that sustain the living organisms which play a major role in soil health (Moebius-Clune et al., 2016).

This chapter uses three agricultural case studies from three different geographical areas to highlight the challenges encountered in increasing crop productivity, some of which are common to all three regions, while others are specific to the climatic, edaphic, economic and social situations found in a particular region. In all three regions CA has been promoted to farmers in the past few decades as one way to responsibly intensify food production. The case studies explore the drivers of CA adoption in the region and the adaptations that have been made to adapt CA to the particular local conditions. The first two case studies end with a list of agronomic research requirements for the particular region, while the third explores what a Brazilian type of experience with green manure cover crops means for CA systems worldwide.

2 Rice-wheat systems of the Indo-Gangetic Plains of South Asia

2.1 Introduction

The rice-wheat (R-W) system in the Indo-Gangetic Plains (IGP) of South Asia is critical to maintaining food security for this most populous region of the world. It is estimated to cover 13.5 million ha (Hobbs and Gupta, 2004; Ladha et al., 2000), stretching from the Indus plains of Pakistan in the west to the floodplains of Bangladesh in the east, the Nepal *tarai* in the north and the states of Punjab, Haryana, Uttar Pradesh, Bihar and West Bengal in India to the south. Per GIS data obtained in 2001, it is estimated that 450 million people make up the population of this floodplain where R-W is grown, with many of them being resource-poor subsistence farmers (Landscan, 2001). This floodplain has fertile alluvial soils which made the Green Revolution (GR) a great success in Asia starting in the late 1960s. GR technology was based on shorter-statured, earlier maturing, disease- and pest-resistant, fertilizer- and water-responsive varieties of rice and wheat. Rice was grown in the wet monsoon summer season and wheat in the cooler more temperate winters. Productivity of both crops increased dramatically with the introduction of GR technology that has allowed the four countries in this area to maintain food security for the past 50 years. Other crops grown in this complex cropping system include sugarcane, maize, potato, legumes (including fodder crops), oilseeds and vegetables.

Long-term soil fertility experiments undertaken by various experiment stations have shown declining factor productivity over time suggesting that soil fertility and other issues are reducing the ability to maintain high yields over time in this area, and thus causing future food insecurity (Ladha et al., 2003). The region has a dry climate in the western areas and higher monsoon rains in the east. Irrigation from the melting glaciers of the Himalayas in the north replenishes groundwater and surface water (dammed reservoirs) and allows intensification of the R-W system production. However, with climate change and declining water tables the sustainability of the water supply from glaciers to farmers in the IGP is under threat, particularly during the dry winter season; a threat that will become a major issue in the future. Climate change also affects the length of the growing season due to more erratic timing of the monsoon season in rainfed areas and higher temperatures (especially the minimum ones) during the grain-filling period of winter crops.

Mechanization in the form of tractor power, but also combine harvesters, has increased over time, especially in the western regions of the IGP and also in the eastern areas like Bangladesh where land preparation and threshing of rice, wheat and maize involved two- and four-wheel tractors as power sources, and low-lift pumps and tube wells for irrigation. However, many resource-poor farmers still rely on manual and animal power for growing their crops especially in the eastern IGP. The following is a description of the issues faced by sustainable intensification in this region, in addition to the technologies and methods used to overcome these problems.

2.2 The contrasting methods of R-W production

The first issue relates to the way rice and wheat are grown in this region. Rice is mostly planted by puddling the soil (i.e. ploughing while wet), raising seedlings in a seedbed and then uprooting and transplanting the seedlings into the puddled field in the wet monsoon summer season, although a winter rice crop (locally called *boro*) is grown in some warmer areas in the east with irrigation using a double rice crop system. Puddling and transplanting seedlings has been a traditional way to grow rice since the GR because it reduces water infiltration and provides ponded water conditions that help control weeds. There are also chemical effects on soil pH that tend to be 7.0 under anaerobic, flooded conditions which makes soil nutrients more available to plants (Ponnamperuma, 1984). However, the physical properties of the soil are destroyed by puddling, and biological activity is changed or affected (Zhou et al., 2014). Wheat, on the other hand, is grown in an aerobic soil in the cool winter season following rice harvest. In order for the farmers to prepare a good seedbed for wheat, especially on heavier textured soils, they have to plough many times with either mechanical or animal traction (Byerlee et al., 1989). This takes time and also dries out the soil resulting in late plantation of wheat, resulting in a loss of yield potential (Hobbs et al., 2017). In some instances, the soil needs to be irrigated before wheat is planted, further delaying planting. This delay in wheat planting further exacerbates the problem of higher temperatures during grain filling, leading to shrivelled grain and lower yields. One of the effects of climate change in this region is increasing spring temperatures (when the wheat grain is filling) that further reduces crop yield.

2.3 The use of no-till in wheat fields

The delay in wheat planting was one of the major reasons for introducing no-till wheat to farmers in this region in the late 1980s. Several requirements need to be met before the implementation of this technology, including farmer participation in the testing of the technology, as well as the involvement of local equipment manufacturers and innovative farmers in the design of suitable equipment. This needed to be within the price range of farmers through adapting local, existing seed drills by adding suitable coulters to get the seed into the ground without tillage (Hobbs et al., 2017; Hobbs and Gupta, 2004). Once the above requirements are met, farmers overcome their reservations about no-till technology and become the technologies' best extension agents. A question still remains: How would resource-poor farmers who do not own tractors benefit from this technology? The answer is that resource-poor farmers are able to utilize the existing practice (used for ploughing) of hiring farmers with tractors to sow their land. Initially these service providers were concerned that introduction of no-till would affect their income from ploughing, but they soon learnt they could still use their tractors to handle the locally produced no-till drills to plant other farmers' fields on time.

Today, no-till wheat is an accepted practice in this region, although, because statistics are not collected on this practice, it is hard to provide a figure on how much wheat or other crops are planted this way. Paroda (2018) has indicated 'that the area under CA in rice-wheat cropping system in the Indo-Gangetic plains, is around 3.5 Mha only, whereas scope exists for almost 10.0 Mha'. In this paper, the 3.5 Mha is for the Indian IGP and not the total for the IGP in South Asia, that has been estimated to have 13.5 Mha of R-W (Hobbs and Gupta, 2003). Paroda (2018) also says 'the CA innovation also has vast scope under rainfed farming covering around 55% of the total 141.0 Mha cultivable area in India'.

Raj Gupta (*pers. comm.*) says 'the promotion of rotovators in NW India as a result of Government subsidies are resulting in deterioration of soil physical properties (soil compaction and plough pans) that result in more lodging, waterlogging and yellowing of wheat and even promoting the spread of *Phalaris minor* again in wheat. This situation is resulting in farmers returning to use of ZT in wheat', although data on the use of zero till (ZT) still needs to be collected. Interestingly, the Indian government is not subsidizing the price of diesel or gasoline, which may help to promote CA further.

Raj Gupta (*pers. comm.*) also mentions 'that after wheat more than 30% of rice is seeded in un-puddled conditions as a result of new herbicides that can control weeds in direct seeded rice (DSR)'. Direct seeded rice (DSR) saves nearly 25% water compared to the puddling and transplanting system, reduces labour and production costs and allows farmers to plant rice independent of monsoon rains and make better use of rainwater (Hobbs et al., 2017). Lower GHG emissions in Indian cereal systems were due to the use of more precise nitrogen use rates and frequency of application, tillage and residue management and water management. Data obtained from R-W systems show that changes in tillage and crop establishment through CA practices reduces overall GHG emissions (Sapkota et al., 2015, 2017, 2018).

Initially the no-till drills developed used fixed tine openers obtained from New Zealand (Hobbs and Gupta, 2004) that worked well in hand-harvested rice fields. Recently, as more farmers began using locally made combine harvesters, the issue of loose straw clogging the drills required a solution. Local manufacturers in association with agronomists and farmers came to the rescue by producing openers and modifications to drills that would work well under clogging conditions that arise after harvesting. Initially, farmers burnt the straw creating severe air quality problems; later on, some used locally made equipment to pick up the loose straw for animal feed and other uses, some used drills that combined picking up loose straw and seeding the crop in the cleaned soil before placing the loose straw back on the soil as surface mulch ('Happy' and Turbo Seeders), while others used drills with disc openers that could cut through the straw to place the seed in the soil. This has led to the development of a strong local industry that both manufactures equipment and even exports the drills and other equipment overseas (National Agro Industries, 2018). This mechanization movement results in timelier planting, higher yields, more efficient use of water, less diesel use and lower GHG emissions, and provides farmers with an alternative to burning their residues and thus improving air quality. By using no-till in wheat, farmers can also save one irrigation cycle which was previously needed when the seedbed was too dry after too much tillage. It was also found that water infiltration was faster in no-till than with tilled soil, and water moved across the field faster when irrigated, resulting in more efficient water use, an important constraint in this region. The Indian government is now providing subsidies for machinery that promotes *in situ* management of crop residues mentioned above to stop rice residue burning and reduce air pollution in NW India.

Other crops, such as maize, pulses and oilseeds that are also grown in this area, can be successfully raised using the same no-till drill after local manufacturers adapted the drill with appropriate seed metering systems for these crops. This enables farmers to diversify their cropping systems except in the areas where only rice can be grown in the monsoon season, and thus provide them options for rotating their crops, the third important practice in CA systems.

One unintended consequence, and a reason that no-till wheat was accepted by farmers, first in the western areas of the IGP but later on in the east, was that it helped control a major weed in wheat, *Phalaris minor* (Malik and Singh, 1995). This grassy weed is small seeded and survives in the soil during the rice phase and does not germinate until the temperatures drop in the fall. Populations as high as 3500 weeds/m^2 were measured in farmer fields that essentially had to be harvested for fodder and not for wheat, even though the quality of this weed was poor as a fodder crop. Tillage essentially brings this seed to the surface whereas no-till keeps it buried; data shows that populations of this weed could be halved by just practising no-till (Malik and Singh, 1993). The combination of no-till, some new herbicides and earlier planting, so the wheat could germinate before the *Phalaris* and then competitively shade out the weed, resulted in reducing this weed problem to manageable levels. This is an excellent example of where no-till actually results in less weeds than with tillage. However, integrated weed control is still needed for rice and other crops as weed dynamics and species change over time.

2.4 The rice phase and use of no-till within the system

In order for farmers to get the full benefits of no-till they need to also grow rice without puddling or directly seeding it instead of transplanting. This is a challenge because ponded water that results from puddling soils helps control weeds. One solution was to do some tillage to uproot and kill weeds before planting rice with the no-till drill. DSR yields are the same or higher compared to transplanted rice if the weeds are controlled, while this technique reduces tillage, labour, seed, fertilizer, water and fuel, resulting in higher net returns (Singh et al., 2013). Allowing the weeds to germinate and then ploughing them under before seeding is the definition of a stale seed bed and is a traditional practice for weed control. Herbicides can also be used to control the weeds that germinate in the wet season. It is important to note that more sedges and some broadleaf weeds emerge when soils are not puddled. Mulching with the previous crop residue, a form of permanent soil cover (one of the pillars of CA), can also help to control weeds. Some reports also indicate that wheat straw has some allelopathic properties that help to control weeds (Steinsiek et al., 1982; Alsaadawi, 2001). Rotations using other crops that allow weeds to be controlled are also important. Maize can be substituted for rice on better drained soils and allows inter-cultivation and use of other herbicides for weed control. Fodder crops can be used where the weeds are controlled when the crop is cut for fodder. Essentially there is a need to use an integrated weed management system to overcome this problem which may even include the use of herbicide-tolerant rice. This is a future research issue that needs financial and administrative support from everyone rather than opposition that will sap the enthusiasm of farmers and other stakeholders involved in food production.

2.5 Future directions for CA in the IGP's R-W systems of South Asia

1 There is a continued need to improve the equipment needed for minimum soil disturbance planting (e.g. no-till, zero-till, strip-till etc.) in South Asia by the various local manufacturers of such equipment. This includes seed and fertilizer drills, sprayer equipment, harvesters and reliable seed storage equipment. Improved equipment is needed for both four-wheel and two-wheel tractors since the latter are popular and better suited to the small plot sizes in eastern areas of the IGP.

2 The development of a cadre of service providers, well trained and funded to provide an array of farm services for farmers of all types would help promote CA in this region and allow any farmer to adopt these responsible intensification practices.

3 Further research with strong farmer participation is needed for the rice phase and other crops grown in the R-W system. Emphasis on promoting the switch from puddled transplanted rice to direct seeded, no-till rice, non-puddled transplanting of rice and other crops is important to overcome water and labour regional constraints and allow the benefits of no-till in wheat to be seen in the entire system.

4 Research and testing of various integrated weed management systems in DSR and other crops, including use of mulches, development of crops competitive with weeds, crop rotations, herbicides and maybe best of all herbicide-resistant crops. Training on proper use of herbicides with the correct sprayers and safe use of herbicide is also required, especially for service providers if they will provide this service.

5 Variety development for all crops in the system that are better adapted to no-till but also better for earlier planting would help improve system productivity and also allow the inclusion of a third crop like mung bean in the system. This is already being researched in wheat where varieties with weak vernalization requirement like HD2967 and CSW18 (Indian varieties) show promise with October planting without sacrificing yield while still maturing in March/April, resulting in avoidance of terminal heat stress (Hobbs et al., 2017). In some cases there is also the possibility of adding a short duration legume, such as mung bean, to the system after wheat and before rice using no-till to reduce turnaround time between wheat and the legume (Hobbs et al., 2017). There may also be differences in performance of rice bred under puddled transplanted conditions compared to direct seeding that needs evaluation in terms of yield, weed competition and disease tolerance.
6 Policies are needed to promote the expansion of responsible intensification practices like CA and also support the development of a network of service providers that enable farmers to adopt these practices.
7 It would be useful if the statistical data collection used in these countries could include data on adoption of CA and improved practices in the R-W system and other systems in South Asia.

3 Rainfed maize systems in East and southern Africa

3.1 Introduction

Increased climate variability may have drastic effects on agricultural production, food and nutrition security in southern Africa (Wheeler and von Braun, 2013). Rainfall predictions show an increasingly erratic pattern, a delayed start, increased intensity and, above all, a 2.6°C increase in temperature by the year 2050 (Cairns et al., 2013). Temperature increase and erratic rainfall are expected to be the most critical factors affecting major food crops and leading to predicted yield reductions of between 15% and 30% if no appropriate adaptation measures are taken (Burke and Lobell, 2010; Lobell et al., 2008).

Soils in southern Africa are inherently poor in fertility due to either strong weathering or the granitic parent material from which they originate (Nyamapfene, 1991). There has been a declining trend in soil fertility in the whole region and a significant soil organic matter decline due to unsustainable land use practices and limited biomass input (Elwell and Stocking, 1988; Zingore et al., 2005). In addition, soils have become increasingly compacted, more acidic and are lacking in essential macro and minor nutrients (N, P, Zn, Bo, Mo among others), and are prone to soil erosion (Elwell, 1989; Thierfelder et al., 2012). In the past, farmers would practice shifting cultivation systems to replenish their soil fertility. However, leaving land to fallow for an extended period of time and opening up virgin land is not an option any more due to escalating population increase and limited carrying capacity of the land.

Declining soil fertility, increasing soil degradation and the need to adapt to a changing climate were the primary reasons to start experimenting with new and more sustainable cropping systems in southern Africa. CA has been tested in southern Africa since the 1970s and some early innovations were pioneered in Zimbabwe and South Africa (Wall et al., 2014). However, most of these earlier experiences were on large commercial farms where tractor-drawn seeding equipment could be used and where fuel and cost efficiency were the primary drivers, especially during the time when both countries were under economic embargos.

Increased promotion of CA systems based on minimum soil disturbance, crop residue retention and crop rotations on smallholder farms only started after 1982/83 in Zimbabwe and later Zambia, and was intensified after 2000 (Oldrieve, 1989; Haggblade and Tembo, 2003). This section summarizes some of the experiences with CA cropping systems in southern Africa. We start by explaining the main seeding systems tested in southern Africa, and then describe the benefits and challenges associated with other components of the system. We will end with a summary and outlook on how CA systems should be developed and adapted in the future.

3.2 Manual and animal traction seeding systems

From the start, the promotion of CA systems to smallholder farmers has been based on either manual or animal traction seeding. Most smallholders do not own tractors and, in rare cases where tractors are available, appropriate seeding equipment for tractors is unavailable in the region. Early attempts therefore focused mostly on how to make CA systems function using hand hoe systems or animal traction (Johansen et al., 2012). In Zimbabwe and later in Zambia, a system similar to the *zai* pit system of the Sahel (Fatondji et al., 2012) was developed in 1982 by a local commercial farmer, Brian Oldrieve, to make use of the principles of precision planting, timely operations, water harvesting and concentration of nutrients (Oldrieve, 1993). Basin planting or conservation farming, as it was later called, is based on digging small planting holes during the dry season which are then fertilized, limed and planted with maize or other crops after the first effective rains without having to wait for the land to be prepared (Sims et al., 2012; ZCATF, 2009). Indeed, a major benefit of basin planting is that fields are already prepared before the season starts and farmers do not have to wait until the soil has softened to till the soil. Making use of the first effective rains gives a significant yield advantage and one of the primary reasons why the basin system has greater yields than the conventional systems. Furthermore, it provides smallholder farmers who do not own or have access to animal draft power the possibility of producing a crop using a precision agriculture system. Other CA principles (residue retention and rotations) were promoted alongside the basin system, though not practised as effectively by farmers due to local constraints.

With more than 20 years of experience there is now sufficient evidence to confirm that basin planting leads to significant yield benefits, mostly for farmers without access to animal traction and in years of erratic rainfall. However, digging basins during the dry season is labour intensive and has led to the partial rejection of the system after some years despite visible and measurable benefits (Thierfelder et al., 2016a). In addition, basins are primarily developed for station crops (maize, sorghum, cotton) and are less suitable with row crops (soybean, cowpea, groundnuts) which require closer plant spacing (Bunderson et al., 2017).

The other manual planting system, promoted mostly in Malawi, Eastern Zambia and parts of Mozambique, was based on a planting stick (dibble stick), sometimes in the advanced form of a Jab planter, where seed and fertilizer are applied simultaneously (Thierfelder et al., 2016a). Dibble sticks are fast and effective and do not require previous land preparation or digging, which has made them more attractive to farmers (Sims et al., 2012). For planting, the seed is dropped into a hole made to the required depth with the dibble stick; additional holes for fertilization can be made. Seeding with a dibble stick results in minimal soil disturbance and usually leads to excellent germination and crop stand. However, the possibility of manure application is limited in this system.

Experiences on CA with animal traction seeding in southern Africa have focused mainly on seeding into a narrow trench opened with a modified chisel plough or ripper or using an animal traction direct seeder imported from Brazil. Both systems have been extensively tested and lead to yield benefits over time (Thierfelder et al., 2015a). Ripper attachments can also be easily mounted on a plough beam with minimal costs (US$25) which makes them more attractive to smallholders. Rip lines are created at the onset of the cropping season at the required row spacing, while seeding and fertilization is done by hand. For animal traction owners this is an easy way to plant under CA as has been evidenced by an increase in farmers using rip-line seeding systems in Zambia and Zimbabwe (Grabowski et al., 2014).

The use of small two-wheel tractors for planting under CA is still in its infancy and, although these small tractors may provide excellent opportunities for mechanization, there is a need for further testing, adaptation and creation of demand (Baudron et al., 2015a).

3.3 Crop residues – benefits, conflicts and trade-offs

Crop residues are generally considered to be essential for successful CA cropping systems. They protect the soil from the impact of heavy rainfall, reduce evaporation and markedly reduce soil erosion and run-off through greater infiltration and impedance to water flow across the soil surface. Consequently, crop residue retention leads to increased soil moisture, thus making the system more resilient against drought stress (Thierfelder and Wall, 2009). Residues encourage soil life and enhance biological activity (De la Cruz-Barrón et al., 2017). Although they may be associated with the carry over of some pest and diseases (e.g. stalk borers and the spores of some fungal leaf diseases may carry over from one season to the next on cereal crop residues), they also provide additional benefits in the form of biological pest control. Crop residues lead to increase in predatory spider populations and provide shelter for ants (Kaluzi et al., 2017). However, due to limited soil moisture in southern Africa and a relatively short growing period such residues are a scarce resource (Mupangwa and Thierfelder, 2014) intensively used by smallholder farmers, especially for animal feed (Wall, 2009; Valbuena et al., 2012). Crop residues are also used for fodder, building material, fuel and grazing during the dry season, and thus are in very high demand. Leaving crop residues on the soil surface under such circumstances is very challenging and has been one of the main reasons why farmers did not make use of the full benefits of CA (Mupangwa and Thierfelder, 2014). Alternative strategies to retain crop residues have been explored which range from temporarily removing the residues and applying them again, the use of grass and leaf litter as available replacements, growing living crop residues in the form of intercrops or applying leaves and branches from intercropped shrubs (Thierfelder et al., 2015b). However, none of these options have so far been fully satisfactory. Community agreements to improve local by-laws against free grazing systems may need to be changed to allow CA farmers to keep their residues (Wall, 2009).

3.4 Rotations and other diversification options

Traditionally, farmers in southern Africa use some form of diversification (Giller, 2001), although rotations and intercropping strategies are often not strategic, vary in space and time and are highly dependent on available markets for rotational crops or alternative benefits (Thierfelder and Wall, 2010). In addition, in areas of limited land availability, such as in Malawi, farmers dedicate most land to the staple maize crop which is key to their food security, with little land given to leguminous or cash crops. The dominance of maize in the farming system is one reason why soils are increasingly degraded and the nutritional status of some smallholder farmers is compromised.

Nevertheless, some rotations have been tested under CA in southern Africa. In the higher rainfall areas of Zimbabwe and Zambia, CA systems with maize-soybean and maize-cowpea rotations have been successfully extended (Thierfelder et al., 2012). In central Malawi and eastern Zambia, maize-groundnut rotations have been widely promoted (Bunderson et al., 2017) and in southern Zambia, the use of maize-cotton-sunn hemp rotations were common, although the drop in cotton prices in recent years has made the rotation unviable for smallholders (Thierfelder et al., 2013). For farmers with limited landholdings, intercropping has been an acceptable way to diversify. Farmers across southern Africa traditionally intercrop maize with pumpkins. There has been a push since the late 1990s to also incorporate grain legumes as intercrops. One of the most successful examples is the maize-pigeonpea system as well as the groundnut-pigeonpea double legume systems (Smith et al., 2016; Snapp et al., 2002, 2003). Due to the slow growth of pigeonpea in the intercropping systems, there is little competition at the onset and once the pigeonpea matures, the companion crop is already harvested. Other intercropping systems including green manures and grain legumes are currently being researched with the aim of increasing ground cover, improving soil fertility, suppressing weeds, increasing fodder for livestock and improving the nutrition of farm families (Mhlanga et al., 2015b, 2016). A range of crops have been tested (e.g. lablab, velvet bean, jack bean, cowpea etc.) and even tree-based systems with *Gliricidia sepium* using chop and drop strategies are being promoted by several NGOs (Thierfelder et al., 2018; Lewis et al., 2011).

3.5 Weeds and their management

Weeds and their control have been a major deterrent for smallholder farmers to adopting CA systems in southern Africa (Muoni et al., 2013; Mashingaidze et al., 2012). The primary reason of tillage is to remove the weeds and to prepare a clean seedbed (Wall, 2007; Chauhan et al., 2012). If ploughing is stopped, there have to be alternative practices to manage weeds. As most of the weed control on smallholder farms in southern Africa is manual with hand hoes and/or with animal traction cultivators, this increase in manual labour requirements is a disincentive to farmers who turn away from CA systems in the first year(s) of practice if no alternatives are presented. The use of herbicides, at least for the first years of CA promotion, has been one strategy that successfully led to increased adoption in Malawi and could be one of the entry points to make CA attractive to farmers as it addresses one of their most critical constraints (Ngwira et al., 2013; Baudron et al., 2015b). CA systems using herbicides have also been considered more economical than systems using manual farm labour (Muoni et al., 2013) and lead to gradually decreasing weed populations (Muoni et al., 2014). However, the accessibility and affordability of herbicides for smallholders, as well as environmental concerns, have been much debated in recent years (Lee and Thierfelder, 2017). Research from southern Africa confirmed that rotations with competitive green manures as well as intercrops and increased ground cover are alternative strategies to reduce weed pressure and to make CA systems more attractive to smallholders (Mhlanga et al., 2015a, 2016).

3.6 Combining CA systems with complimentary practices

CA systems as developed elsewhere need to be adapted to local contexts as farmers do not apply similar amounts of fertilizers, often use unimproved varieties, have limited access to and rarely use chemicals for crop protection, and have limited access and use of appropriate scale machinery (Wall, 2007; Thierfelder et al., 2018). In addition, limited biomass regrowth and unsystematic use of rotations necessitate complimentary practices to support the functioning of CA systems under southern African conditions. Increased fertilization, which is currently well below other regions, can be achieved through various means including additional use of animal or green manures and compost. To increase groundcover, leaf litter and grass can be applied in addition to leguminous intercrops or shrubs in 'chop and drop' systems to generate more on-site mulch. A range of new and improved varieties that are both adapted to climate stress and are more nutritious have been bred in the last decade and should be incorporated into CA systems (Thierfelder et al., 2016b). Crop chemicals including herbicides should be considered at least in the first years of conversion from a tillage-based agriculture to a no-tillage system to reduce weed pressure and labour requirements which particularly affect women and young people. Extending the three principles of CA without complimentary agriculture practices will not enhance soil quality and productivity over time (Thierfelder et al., 2018; Powlson et al., 2016).

3.7 Future outlook for successful CA extension in southern Africa

The pressures of declining soil fertility and the urgent need to adapt to climate change in southern Africa require farmers to change their current extractive land use systems to more sustainable agriculture practices (Wall et al., 2014). Conservation Agriculture, as it has been extended to smallholders since the 1990s, can serve this purpose and can help to adapt to emerging challenges. Some of the entry points for CA in Africa have been summarized by Baudron et al. (2015b). However new avenues need to be explored to make the systems more attractive to smallholders:

1 Most of southern Africa has a unimodal cropping season, lasting from November to April, which limits biomass regrowth and diversification options. The use of intercrops, green manures and tree cuttings can reduce this problem to provide fodder, ground cover and improve soil fertility.

2 Cropping systems have to be profitable from the start to make CA more attractive to cash constrained farmers. Rotation and intercropping systems that give additional cash benefits to smallholders can improve the entry points.
3 Integrated weed control strategies are critical in CA systems and farmers will not accept increases in manual labour requirements. The introduction of improved weed control strategies, including herbicides, has proved to be very effective in the promotion of CA to farmers (e.g. in Malawi).
4 Limited moisture is one of the biggest impediments to successful crop production in southern Africa. Water-saving and water-harvesting CA systems will therefore give immediate benefits. However, they should not increase the labour burden on farmers (e.g. to dig basins) as this will make CA systems unattractive. Farmers who do not adopt CA practices prepare their lands by ploughing it at the monsoon onset either manually or using animals. Both of these traditional systems are tedious, time consuming and will delay the planting of the crop. Planting with a dibble stick into basins prepared before the monsoon rains and kept permanent after formation or into rip-line seeding systems helps to reduce drudgery and time and allows the crop to be planted soon after the start of the rains on time.
5 Small machinery for smallholders (e.g. the use of direct seeders with two-wheel tractors) is increasingly popular in the region but can only be profitable if it provides additional benefits and business opportunities (e.g. through shelling, threshing, pumping of water, transport etc.) for small entrepreneurs. A model that includes service providers for some farm operations is considered much more viable than farmers owning their own equipment.
6 Functional markets for both inputs and outputs and enabling policies for 'climate-smart' agriculture and SLM are both necessary pre-conditions for successful CA scaling and adoption as markets will serve as push and pull for the adoption of more diversified cropping systems. Enabling policies and enforcing local by-laws on grazing will further encourage farmers to adopt CA in southern Africa.

4 The importance of green manure cover crops in CA systems in Brazil

4.1 Introduction

Rapid expansion of soybean and wheat cultivation in southern Brazil in the 1960s, involving intensive tillage twice each year, resulted in extensive soil erosion. Initial attempts to control this erosion, based on ever bigger contour bunds, were largely unsuccessful, given the high intensity rainfall in the region.

The first reported farm application of zero-tillage systems in Brazil was in 1972 on the Rhenania farm of Herbert Bartz near Londrina in the southern state of Paraná, Brazil. Bartz was looking for a solution to the rampant soil erosion on his farm and travelled to Britain and the United States looking for solutions. He came back armed with new ideas and a no-till seed drill. Coincidentally, 2 years previously a German-funded project had been initiated at the Parana Agricultural Research Institute (IAPAR) focusing on no-tillage agricultural systems. Rolf Derpsch, the German agronomist who headed up this project, was a strong advocate of no-tillage, also of the importance of green manure cover crops (gmcc) in no-till systems and was able to share ideas and experiences with Bartz. Soon two other farmers in the eastern part of Parana State, Manoel Henrique Pereira and Franke Dijkstra, also began to try no-till systems, again as a means to overcome soil erosion on their farms.

From these beginnings began the exciting development of CA systems in Brazil that at recent estimates (2012) now cover over 31 million ha, or some 44% of all agricultural land in the country (Aquastat, 2018). Most of this area is in two of the major farming systems described by Dixon et al. (2001) – the intensive mixed cropping system of the south and south-east of the country, and the extensive cropping system of the central Cerrados – although more recently there are also considerable areas of CA in the dryland mixed system of the north-east of the country. Much has been written about the expansion of CA in Brazil, largely driven by commercial farmers, but here we will concentrate on one interesting aspect of Brazilian CA systems that has not been widely replicated elsewhere – the inclusion of green manure/cover crops (gmcc) as an important component of CA systems.

Probably as a result of the importance of erosion control as one of the main factors driving the adoption of CA, Brazilian CA farmers attach great importance to residue retention and ground cover. Generally gmcc are used by farmers to enhance residue cover using available time, water and radiation that are insufficient for a cash crop (Derpsch et al., 1986; Calegari et al., 2007, 2014) or because they increase total system profitability, generally by benefitting the main cash crop while replacing a marginally profitable crop. Cover crops have been a part of Brazilian agriculture for almost 100 years (Calegari et al., 1993, 2014) and numerous books and farmer manuals on gmcc in the different environments of Brazil have been published (e.g. Wütke et al., 2008, 2014). Bolliger (2006) and Calegari (2009, 2010) and have summarized the principal benefits of gmcc in Brazil as:

1 forage production (and occasionally food, especially for small farmers),
2 increasing available nitrogen (N) for the succeeding crop through symbiotic N fixation,
3 converting otherwise unused sunlight and soil moisture resources into additional biomass,

4 helping reduce weed populations,
5 breaking cycles of certain pests and diseases,
6 capturing and recycling easily leachable nutrients (NO_3^- , K, S, B) that would otherwise be lost beyond the rooting zone,
7 ameliorating soil structure and creating additional root channels, especially through dense soil layers,
8 stimulating soil biological activity through the release of root exudates,
9 helping the amelioration of soil acidity,
10 generally increasing soil physical, chemical and biological fertility.

The relative benefits of different gmcc to the succeeding crop and the cropping system in general are defined by the relative importance of these different processes. However, early work also showed that it was virtually impossible to predict effects of a certain gmcc species on the succeeding crops, but rather that these needed to be defined through field testing (Derpsch et al., 1986). For instance, although a yield benefit to the succeeding wheat crop could be expected from a previous legume gmcc such as lupins (*Lupinus albus*) or vetch (*Vicia sativa*), similar benefits from a previous oilseed radish (*Raphanus sativus*) crop were more difficult to explain. Subsequent work has, however, elucidated some of the less obvious benefits of the various gmcc species such as the liberation of root exudates by lupins.

4.2 Green manure cover crops in Brazil

4.2.1 Southern Brazil

While southern, subtropical Brazil has rain all year round, winter rainfall decreases in the south-east region as one moves north of the Tropic of Capricorn. This rainfall pattern, and annual rainfall generally in excess of 1000 mm, allows for two crops per year without irrigation. In the 1980s and early 1990s, the most common production system in this region, the most intensive agricultural region of the country, was a double-crop wheat-soybean system, with wheat grown in the winter and soybeans during the main summer season. However, reduced subsidies on wheat production in the late 1980s, combined with the advent of a new disease – wheat blast (*Magnaporthe oryzae*) – to which there were no resistant varieties, combined to reduce the profitability of the wheat crop. Farmers found that replacing wheat with a gmcc increased total profitability through reduced costs and increased soybean yields. Initially the most common gmcc was black oats (*Avena strigosa*), but later other species such as vetch (*Vicia sativa* and/or *V. villosa*), oilseed radish (*Raphanus sativus*), ryegrass (*Lolium multiflorum*), rye (*Secale cereale*) and white or blue lupins (*Lupinus albus* and *L. angustifolius*), and more recently clovers (*Trifolium* spp.), triticale (×*Triticosecale*) and buckwheat (*Fagopirum esculentum*) have been used (Calegari et al., 2014). After years of research studies and validation in different agro-ecological zones, as well as in different cropping systems, many farmers now grow mixtures of gmcc species. Decisions on the species to mix are based on field fertility and the succeeding crop(s), and include mixtures such as oats and vetch; oats, vetch and oilseed radish; oats, vetch and ryegrass; sunn hemp, buckwheat and pearl millet; and oilseed radish, buckwheat and pearl millet.

4.2.2 The 'Cerrados'

The central 'Cerrados' region of Brazil has high rainfall, generally well in excess of 1000 mm/year, but a marked winter dry season of 4–5 months that limits the cropping system to one major crop per year in the absence of irrigation. With irrigation two crops per year can be grown. However, the relatively high temperatures, coupled with the high rainfall, lead to rapid breakdown of plant matter, and farmers managing CA systems are worried by the lack of ground cover for erosion control. Therefore it is common for farmers to sow fast-growing species immediately after the harvest of the main soybean or maize crop and so use the remaining moisture in the profile to produce ground cover. In annual cropping systems a short-season sorghum (*Sorghum bicolor*) or pearl millet (*Pennisetum glaucum*) were the most common second crops (Landers, 1994), the latter sown as a gmcc, but today crotalaria species (C. *juncea, C. spectabilis, C. ochroleuca, C. breviflora*), pigeon pea (*Cajanus cajan*), cowpea (*Vigna unguiculata*), buckwheat (*Fagopirum esculentum*), finger millet (*Eleusine coracana*) and stylo (*Stylosanthes* spp.) are also common, either alone, or more often in mixtures with pearl millet or sudan grass (*Sorghum sudanense*) (Calegari et al., 2014). Pearl millet was initially used as a gmcc because of its rapid growth, but its popularity has increased after it was found that it reduces populations of three important soil nematodes, as well as the fact that in only 60–90 days the pearl millet roots may reach over 2 m depth, enabling the gmcc to capture nutrients, especially nitrogen, that may otherwise be leached beyond the reach of the roots of commercial crops.

Some innovative farmers in the 'Cerrados' now plant millet at the beginning, rather than at the end, of the rainy season. The millet is then desiccated with glyphosate 45–80 days later and soybeans sown directly into the millet residues. Instead of using herbicide the pearl millet can also be managed with a knife roller when approximately 50% of the plants are flowering. The millet sprouts are knife-rolled again 25–35 days later when the shoots reach an height of 50–60 cm. The second rolling controls more than 95% of the millet plants, with the remainder controlled by the normal herbicide applications.

A third option being used by a growing number of farmers is to seed pearl millet, finger millet (*Eleusine coracana*) or some crotalaria species (C.

spectabilis or C. ochroleuca) at the end of the crop season, and let them grow out on the residual moisture. If there is enough moisture they produce some seed that germinates with the first rains the following season. This early growth is left for 20–30 days, desiccated with herbicide and sown with soybean into the mulch.

Under irrigated conditions, or in the wetter areas that receive more than 1500 mm rainfall per year, larger farmers with livestock often rotate 3 or 4 years of pasture, generally with brachiaria (*B. decumbens, B. humidicola, B. brizantha* etc.) with 3 or 4 years of annual cropping, producing two commercial crops per year and including gmcc in the crop phase of the system (Calegari et al., 2014). With these systems Séguy et al. (2001) showed that total annual biomass production (both above and below ground) has increased from 4 t/ha to 8 t/ha, with a single annual crop, to nearly 30 t/ha in the most efficient zero-till systems.

It is estimated that there are at least 11–12 million ha of gmcc grown every year in Brazil, covering over 35% of the area of CA and nearly 20% of the total cropped area in the country. However, there is relatively little use of gmcc on small farms, despite the potential benefits of gmcc as low-cost options to improve weed control and crop nutrition using internal farm inputs rather than externally purchased inputs. Florentin et al. (2011) has listed various options for including gmcc in smallholder farming systems in Paraguay, including gmcc in rotation with annual crops and in association with both annual and perennial crops. However, it appears that the combined effects of small landholdings and the need for subsistence, risk aversion, lack of knowledge, high cost and relative unavailability of gmcc seed result in low levels of adoption.

4.3 Handling green manure cover crops

Because gmcc are not harvested for grain (if they are harvested for grain they are defined as grain crops rather than gmcc), mechanization has been relatively simple. Crop zero-tillage seeders are adapted, if necessary, for different seed sizes and spacing. Gmcc are generally selected for their resistance to, or tolerance of, pests and diseases, and thus require, at most, infrequent pesticide sprays. Termination of the crop cycle of the gmcc depends on the species, the amount of dry mass and the timing of seeding of the next crop. In Brazil, gmcc may be left standing (and the next crop seeded directly into the dry gmcc which may fall with time or be knocked over during the seeding operation), rolled with a knife roller – an implement developed by Brazilian farmers to manage gmcc, or sprayed with glyphosate or paraquat. The gmcc is terminated, either by rolling or by herbicide, before the seed is set so as to limit future germination of the gmcc seed and therefore avoid problems of competition with commercial crops.

4.4 Use of green manure cover crops in other parts of the world

How does gmcc use in Brazil compare with other regions? Crop residue cover and the use of cover crops have received considerably less attention in the United States than reduced tillage (Clark, 2007). Although the 2012 USDA Census of Agriculture reported 10.3 million acres of gmcc in the United States, it appears that much of this is in tilled farming systems. However, the area of gmcc is growing rapidly, as evidenced by the results of the National Cover Crop Survey (CTIC, 2017). Respondents to the survey reported 280 000 ha of gmcc in 2016, mostly (80%) in field crop (commodity crop) systems, of which approximately half was in no-till systems. Farmers reported a 25% increase in area of gmcc in 2016 compared to 2015, and a huge 60% increase compared to 2014, attesting to the rapid growth of gmcc area in the United States. However, considerable research is still required to 'broaden the further appeal and adoption of cover crops' (Duiker and Thomason, 2014).

In Argentina, Albertengo et al. (2014) report that including legumes such as hairy vetch as a gmcc to increase biological nitrogen fixation is becoming a common practice, but we are unable to ascertain areas of gmcc in the CA systems of Argentina. Rochecouste and Crabtree (2014) state that the use of cover crops in Australia is not common, except in those areas with reliable rainfall in the off-season. In Europe the use of cover crops 'is viewed as an unnecessary cost by the majority of farmers and advisers', although recently interest in gmcc has increased in Germany (Friedrich et al., 2014). Although there has been some research on gmcc and efforts to facilitate their adoption in other continents, there are few reports of adoption. Crop rotation, in general, appears to have received less attention than direct seeding and residue retention in the development of CA systems on small farms in Asia and Africa. This may be due to the fact that crop rotations are already in place in the wetter regions where two crops per year are allowed by the rainfall pattern, whereas in the drier regions, where only one crop per year is possible, farmers prioritize staple crops and the development of markets for new crops is difficult. However, in India there is some adoption of *Sesbania* spp. as a gmcc between the winter and summer crops. This practice, known locally as brown manuring, involves desiccating the young (approximately 60-day) *Sesbania* with herbicide before sowing the summer rice crop (Nawaz et al., 2016). In Tanzania, Mariki (2004) has developed a system through farmer-participatory research for the use of *Lablab purpureus* as a gmcc. However, given the price of the lablab grain in Kenya, farmers tend to harvest the grain rather than use it as a gmcc. There are still benefits from the biomass produced but much of the nitrogen accumulated by the crop through biological nitrogen fixation is exported in the grain. Bunch (2012) reports that gmcc options have been identified for seven countries in sub-Saharan Africa, but extensive adoption has not yet taken place.

4.5 Conclusions – opportunities for green manure cover crops in CA systems

Why have gmcc been so widely adopted in Brazil and not elsewhere? Are gmcc potential components of all farming systems, and what limits their potential? We suggest that the different benefits of gmcc species in different situations are so numerous that it is impossible to generalize. First, adoption of gmcc on small farms (defined as farms of 50 ha or less) in Brazil has been considerably lower than adoption on large farms, for the reasons stated earlier. These limitations of risk aversion, lack of knowledge of gmcc, problems of cost and accessibility of seed and information on gmcc are likely to be even more important for the considerably smaller farmers of Asia, Africa and other regions of Latin America, but also depend to a large degree on similar institutional factors that currently inhibit the spread of CA in these regions. Once strong research has identified beneficial, profitable and sustainable gmcc options, focus on overcoming institutional limitations to expanding adoption through multi-stakeholder innovation platforms (Ekboir, 2003) will be needed to enable widespread smallholder gmcc utilization.

Secondly, as in the Australian situation, gmcc generally fit better in situations where there is more soil moisture available during the year than that required by one crop. In areas with less than 400–500 mm of precipitation per year, that is most of the semi-arid regions of the world, it is unlikely that a gmcc and a cash crop can be grown in the same year.

Above 500 mm precipitation there may be options in many situations to use the soil moisture in excess of the needs of the main crop to produce ground cover or other cropping system services such as N fixation, pest, diseases and nematode reduction and so on. However, there are few examples of widespread farmer adoption of gmcc outside the Southern Cone of South America, even in wetter areas. We hypothesize that the history of research on gmcc in Brazil, and the intensive research on gmcc in no-tillage CA systems since the start of the adoption of CA in Brazil, as well as the demonstrated importance of ground cover for soil erosion control and soil fertility and quality enhancement, have maintained gmcc in the forefront of options for effective CA systems among farmers, extension services and research. Research on gmcc in other parts of the world where CA is expanding has not received the same emphasis as it has in Brazil. The conviction that gmcc can usefully be part of CA systems is common in Brazil among researchers and extension personnel, as well as among farmers, whereas we have not found a similar level of conviction among practitioners elsewhere. We believe that gmcc do have much to offer CA systems worldwide, but an intensive research effort to identify the best species and mixtures of species, the cropping system services that they provide and the selection of lines with enhanced adaptation and system-enhancing characteristics (as opposed to the common practice of simply testing one, or at most a few, selections of lines of a particular species) are necessary to be able to identify efficient strategies for gmcc incorporation into more intensive, productive and sustainable systems.

The relative lack of research has resulted in few exciting options for the use of gmcc in different farming systems, lack of information for farmers on the benefits of gmcc and therefore little interest in adoption. As there is little demand, seed is not available and/or very expensive, equipment such as knife rollers are not commonly available (or even known), and farmers, especially smallholder farmers, are reticent to take the risk of incorporating gmcc into their farming systems. There are researchers, projects and programmes in many places trying to change this situation but we argue that a much greater and intensive effort is needed.

Finally, policy can be an important impediment to change in agricultural practices. In a recent article in *Scientific American*, David Montgomery (2018) stated that 'every farmer I interviewed said that such practices (Conservation Agriculture including gmcc) would see more rapid adoption if we simply stopped subsidizing conventional practices that degrade the soil'. Subsidies that prioritize production as opposed to economic and environmental efficiency of farming systems are likely to reduce the intensity of research, extension and adoption of gmcc, and thus likely to reduce the sustainability of farming systems.

5 Summary

Conservation Agriculture has been adopted successfully in all three of the selected geographical areas (Kassam et al., 2018, Table 1; Paroda, 2018) as a management system to sustainably intensify cultivation in the existing agricultural land and help maintain food security in these regions. However, there are differences by region in the drivers for adopting this management system. In South Asia, issues of late planting of wheat, herbicide resistance of the grassy weed (*Phalaris minor*) and a spike in diesel costs for tractor power in the 1980s encouraged some innovative farmers to experiment with no-till wheat after rice once suitable equipment was available from local manufacturers to enable farmers to see for themselves the benefits including many that were recognized after no-till was tried. Once no-till wheat was cultivated successfully, the farmers found the confidence to experiment with CA in the rice phase, prompted by labour and water shortages. In southern Africa, the drivers were related to degradation of soil fertility, climate change and the need to make better use of rainfall. There is also a need in this region to identify manual and animal-drawn systems since tractors were not available. Chemicals for weed control were also not available and issues of weed control and labour had to be considered. In Brazil, no-till was a farmer-driven management system to help combat serious soil erosion problems. Brazil was also one of the pioneer countries that experimented and adopted no-till farming. The agronomy in this South American section of this chapter looks at the importance of green manure cover crops to help sustain yields over time and manage the weed, soil fertility and soil health issues.

Each of the three regions have common challenges suggesting that ways to handle these challenges in one eco-region could be useful for providing solutions in another. For example, a common theme in all these regions is the issue of biological, physical and chemical soil health degradation as a result of intensive tillage and reliance on only chemical fertilizer for providing nutrients to plants. Adopting CA management systems helps remediate these soil health problems but also includes improved efficiency of water resources whether rainfed or irrigated. Lack of soil cover, an essential component for successful CA and improving soil health (e.g. it provides nutrients for promoting biological soil health), is a constraint in southern Africa (residues are needed for feeding household animals) and in some parts of Brazil, leading to soil organic matter depletion which reduces soil health. Use of cover crops as outlined in the Brazil eco-region is a possible solution for southern Africa, especially if the cover crop can fit the cropping system and, in addition to soil cover, provide a product that could be used for fodder. In South Asia, the issue is too much residue after a rice crop that is dealt with by burning the residue. This reduces the available feed for the biological component of the soil and leads to a decline in soil organic matter. In addition, it creates severe air pollution at the time of rice harvest. Developing machinery that can plant into the rice stubble is a logical way to overcome this constraint, and exchanges of prototypes and designs between regions may help develop better machinery faster.

Other constraints that are common in all three regions include issues of timely planting for optimizing productivity; reducing fossil fuel so that GHG emissions can be reduced; reducing labor needs for land preparation, planting and harvest since labor costs are high and availability declining in all three regions; overcoming soil physical problems of drainage and compaction and plough pans; controlling growth of weeds in the region's cropping systems; and optimizing and improving the efficiency of applied nutrients. The use of CA management systems in these three regions has been shown in this chapter to help resolve many of these constraints, and sharing and describing how farmers and other stakeholders in these and other regions augers well for tackling the challenge of feeding the world now and in the future.

6 References

The following are an extensive set of references for each of the three geographic areas. The following website also has an electronic bibliography of many journals, books, chapters, videos, PowerPoint presentations, photos and more for scientists and practitioners interested in finding more information on this important agricultural management system. Under the 'Research' menu item there is information on how to use this database but also how to join the Cornell Conservation Agriculture Network. The link to this site is: http://soilhealth.org/.

References for the Indo-Gangetic Plains section

Alsaadawi, S. (2001). Allelopathic influence of decomposing wheat residues in agroecosystems. *Journal of Crop Production*, 4(1): 185–96. doi:10.1300/J144v04n02_05.

Byerlee, D., Heisey, P. and Hobbs, P. R. (1989). Diagnosing research priorities for small farmers: Experiences from on-farm research in Pakistan. *Quarterly Journal of International Agriculture*, 28(3): 254.

Hobbs, P. R. (2007). Conservation agriculture: What is it and why is it important for future sustainable food production? *Journal of Agricultural Science, Cambridge*, 145: 127–38.

Hobbs, P. R. and Gupta, R. K. (2003). Resource conserving technologies for wheat in rice-wheat systems. In: J. K. Ladha, J. Hill, R. K. Gupta, J. Duxbury and R. J. Buresh (Eds), *Improving the Productivity and Sustainability of Rice-Wheat Systems: Issues and Impact*. ASA Spec. Publ. 65. ASA, Madison, WI, USA, pp. 149–71. Chapter 7.

Hobbs, P. R. and Gupta, R. K. (2004). Problems and challenges of no-till farming for the rice-wheat systems of the Indo-Gangetic Plains in South Asia. In: R. Lal, P. Hobbs, N. Uphoff and D. O. Hansen (Eds), *Sustainable Agriculture and the Rice-Wheat System*. Ohio State University, Columbus, Ohio, USA, pp. 101–19. Chapter 6.

Hobbs, P. R., Sayre, K. D. and Gupta, R. K. (2008). The role of conservation agriculture in sustainable agriculture. *Philosophical Transactions of Royal Society B (UK)*, 363: 543–55.

Hobbs, P., Gupta, R., Jat, R. K. and Malik, R. K. (2017). Conservation agriculture in the Indo-Gangetic Plains of India: Past, present and future. *Experimental Agriculture*, 51: 1–19. doi:10.1017/S0014479717000424.

Kassam, A., Friedrich, T. and Derpsch, R. 2018. Global spread of Conservation Agriculture. *International Journal of Environmental Studies*, 23p. Available at: https://www.tandfonline.com/doi/full/10.1080/00207233.2018.1494927.

Ladha J. K., Fischer, K. S., Hossain, M., Hobbs, P. R. and Hardy, B. (2000). Improving the productivity of rice-wheat systems of the Indo-Gangetic Plains: A synthesis of NARS-IRRI partnership research. IRRI Discussion Paper No. 40. IRRI, Los Baños, Philippines.

Ladha, J. K., Dawe, D., Pathak, H., Padre, A. T., Yadav, R. L., Bijay-Singh, Yadvinder-Singh, Singh, Y., Singh, P., Kundu, A. L., Sakal, R., Ram, N., Regmi, A. P., Gami, S. K., Bhandari, A. L., Amin, R., Yadav, C. R., Bhattarai, E. M., Gupta, R. K. and Hobbs, P. R. (2003). How extensive are yield declines in long-term rice-wheat experiments in Asia? *Field Crops Research*, 80: 223–34.

Landscan. (2001). *Geographic Information Science and Technology*. Oak Ridge National Laboratory, Tennessee, USA. Available at: https://landscan.ornl.gov/ (accessed July 2018 but used 2001 for population data).

Malik, R. K. and Singh, S. (1993). Evolving strategies for herbicide use in wheat: Resistance and integrated weed management. In: *Volume 1. Proceedings of an International*

Symposium of the Indian Society of Weed Science, 18–20 November 1993, Hisar, India, pp. 225–38.

Malik, R. K. and Singh, S. (1995). Littleseed canarygrass (*Phalaris minor* Retz.) resistance to isoproturon in India. *Weed Technology*, 9: 419–25.

Moebius-Clune, B. N., Moebius-Clune, D. J., Gugino, B. K., Idowu, O. J., Schindelbeck, R. R., Ristow, A. J., van Es, H. M., Thies, J. E., Shayler, H. A., McBride, M. B., Kurtz, K. S. M., Wolfe, D. W. and Abawi, G. S. (2016). *Comprehensive Assessment of Soil Health – The Cornell Framework, Edition 3.2*. Cornell University, Geneva, NY. Retrieved 22 June 2018 from https://soilhealth.cals.cornell.edu/training-manual/.

National Agro Industries. 2018. Retrieved 22 June 2018 from http://www.nationalagro.com/.

Paroda, R. S. (2018). Strategy paper for doubling farmers' income. Publication no. 60. Trust for Advancement of Agricultural Sciences (TAAS), New Delhi, India, 27p. Available at: http://www.taas.in/documents/pub60.pdf

Ponnamperuma, F. N. (1984). Effects of flooding on soils. In: T. T. Kozlowski (Ed.), *Flooding and Plant Growth*. Academic Press, London, pp. 10–42. Chapter 2.

Sapkota, T. B., Jat, M. L., Shankar, V., Singh, L. K., Rai, M., Grewal, M. S. and Stirling C. M. (2015). Tillage, residue and nitrogen management effects on methane and nitrous oxide emission from rice-wheat system of Indian Northwest Indo-Gangetic Plains. *Journal of Integrative Environmental Sciences*, 12(1): 31–46. doi:10.1080/1943815X.2015.1110181.

Sapkota, T. B., Shankar, V., Rai, M., Jat, M. L., Stirling, C. M., Singh, L. K., Jat, H. S. and Grewel, M. S. (2017). Reducing global warming potential through sustainable intensification of *Basmati* rice-wheat systems in India. *Sustainability*, 9(6): 1044. doi:10.3390/su9061044.

Sapkota, T. B., Aryal, J. P., Khatra-Chetri, A., Shirsath, P. B., Arumugam, P. and Stirling, C. M. (2018). Identifying high-yield low emission pathways for cereal production in South Asia. *Mitigation and Adaptation Strategies for Global Change*, 23(4): 621–41. doi:10.1007/s11027-017-9752-1.

Singh, U. P., Singh, Y., Sutaliya, J. M., Singh, N. K., Kumar, R. and Singh, R. P. (2013). Direct seeded rice: A resource conserving technology for enhancing productivity and profitability. Extended summaries: ARRW Golden Jubilee International Symposium, 2–5 March 2013, Cuttack, India.

Steinsiek, J. W., Oliver, L. R. and Collins, F. C. (1982). Allelopathic potential of wheat (*Triticum aestivum)* straw on selected weed species. *Weed Science*, 30(5): 495–7.

USDA. Natural Resources Conservation Services: Soil Health. (2012). Retrieved 22 June 2018 from http://www.nrcs.usda.gov/wps/portal/nrcs/main/soils/health/.

Zhou, W., Lv, T.-F., Chen, Y., Westby, A. P. and Ren, W.-J. (2014). Soil physiochemical and biological properties of paddy-upland rotation: A review. *The Scientific World Journal*, 2014, Article ID 856352, 8p. doi:10.1155/2014/856352.

References for the southern Africa section

Baudron, F., Sims, B., Justice, S., Kahan, D. G., Rose, R., Mkomwa, S., Kaumbutho, P., Sariah, J., Nazare, R. and Moges, G. (2015a). Re-examining appropriate mechanization in Eastern and Southern Africa: Two-wheel tractors, conservation agriculture, and private sector involvement. *Food Security*, 7(4): 889–904.

Baudron, F., Thierfelder, C., Nyagumbo, I. and Gérard, B. (2015b). Where to target conservation agriculture for African smallholders? How to overcome challenges associated with its implementation? Experience from eastern and southern Africa. *Environments*, 2(3): 338–57.

Bunderson, W. T., Jere, Z. D., Thierfelder, C., Gama, M., Mwale, B. M., Ng'oma, S. W. D., Museka, R., Paul, J. M., Mbale, B., Mkandawire, O. and Tembo, P. (2017). Implementing the principles of conservation agriculture in Malawi: Crop yields and factors affecting adoption. In: A. Kassam, S. Mkomwa and T. Friedrich (Eds), *Conservation Agriculture for Africa: Building Resilient Farming Systems in a Changing Climate*. CABI Publishing, Wallingford, UK.

Burke, M. and Lobell, D. (Eds). (2010). Climate effects on food security: An overview. *Climate Change and Food Security*. Springer, 13-30

Cairns, J. E., Hellin, J., Sonder, K., Araus, J. L., MacRobert, J. F., Thierfelder, C. and Prasanna, B. M. (2013). Adapting maize production to climate change in sub-Saharan Africa. *Food Security*, 5(3): 345–60.

Chauhan, B. S., Singh, R. G. and Mahajan, G. (2012). Ecology and management of weeds under conservation agriculture: A review. *Crop Protection*, 38: 57–65.

De la Cruz-Barrón, M., Cruz-Mendoza, A., Navarro-Noya, Y. E., Ruiz-Valdiviezo, V. M., Ortíz-Gutiérrez, D., Ramírez-Villanueva, D. A., Luna-Guido, M., Thierfelder, C., Wall, P. C., Verhulst, N., Govaerts, B. and Dendooven, L. (2017). The bacterial community structure and dynamics of carbon and nitrogen when maize (*Zea mays* L.) and its neutral detergent fibre were added to soil from Zimbabwe with contrasting management practices. *Microbial Ecology*, 73(1): 135–52.

Elwell, H. A. (1989). *Soil Erosion. Post Production Activities and Marketing of Food Crops in Small-Scale Farming Areas of Zimbabwe*. Department of Agriculture Research and Extension Service (AGRITEX), Harare, Zimbabwe.

Elwell, H. A. and Stocking, M. (1988). Loss of soil nutrients by sheet erosion is a major hidden farming cost. *Zimbabwe Science News*, 22: 7-8.

Fatondji, D., Bationo, A., Tabo, R., Jones, J. W., Adamou, A. and Hassane, O. (2012). Water use and yield of millet under the zai system: Understanding the processes using simulation. In: J. Kihara, D. Fatondji, J.W. Jones, G. Hoogenboom, R. Tabo and A. Bationo (Eds), *Improving Soil Fertility Recommendations in Africa using the Decision Support System for Agrotechnology Transfer (DSSAT)*. Springer.

Giller, K. E. (2001). *Nitrogen Fixation in Tropical Cropping Systems*. CABI Publishing, New York.

Grabowski, P. P., Haggblade, S., Kabwe, S. and Tembo, G. (2014). Minimum tillage adoption among commercial smallholder cotton farmers in Zambia, 2002 to 2011. *Agricultural Systems*, 131: 34–44.

Haggblade, S. and Tembo, G. (2003). Conservation farming in Zambia. EPTD Discussion Paper No. 108. IFPRI, Washington DC.

Johansen, C., Haque, M. E., Bell, R. W., Thierfelder, C. and Esdaile, R. J. (2012). Conservation agriculture for small holder rainfed farming: Opportunities and constraints of new mechanized seeding systems. *Field Crops Research*, 132: 18–32.

Kaluzi, L., Thierfelder, C. and Hopkins, D. W. (2017). Smallholder farmer innovation and contexts in maize-based Conservation Agriculture Systems in Central Malawi. *Sustainable Agriculture Research*, 6(3): 85.

Lee, N. and Thierfelder, C. (2017). Weed control under conservation agriculture in dryland smallholder farming systems of southern Africa: A review. *Agronomy for Sustainable Development*, 37(5): 48.

Lewis, D., Bell, S. D., Fay, J., Bothi, K. L., Gatere, L., Kabila, M., Mukamba, M., Matokwani, E., Mushimbalume, M. and Moraru, C. I. (2011). Community Markets for Conservation (COMACO) links biodiversity conservation with sustainable improvements in livelihoods and food production. *Proceedings of the National Academy of Sciences*, 108(34): 13957–62.

Lobell, D. B., Burke, M. B., Tebaldi, C., Mastrandrea, M. D., Falcon, W. P. and Naylor, R. L. (2008). Prioritizing climate change adaptation needs for food security in 2030. *Science*, 319: 607–10.

Mashingaidze, N., Madakadze, C., Twomlow, S., Nyamangara, J. and Hove, L. (2012). Crop yield and weed growth under conservation agriculture in semi-arid Zimbabwe. *Soil and Tillage Research*, 124: 102–10.

Mhlanga, B., Cheesman, S., Maasdorp, B., Muoni, T., Mabasa, S., Mangosho, E. and Thierfelder, C. (2015a). Weed community responses to rotations with cover crops in maize-based conservation agriculture systems of Zimbabwe. *Crop Protection*, 69: 1–8.

Mhlanga, B., Cheesman, S., Maasdorp, B., Mupangwa, W. and Thierfelder, C. (2015b). Contribution of cover crops to the productivity of maize-based conservation agriculture systems in Zimbabwe. *Crop Science*, 55(4): 1791.

Mhlanga, B., Cheesman, S., Chauhan, B. S. and Thierfelder, C. (2016). Weed emergence as affected by maize (*Zea mays* L.)-cover crop rotations in contrasting arable soils of Zimbabwe under conservation agriculture. *Crop Protection*, 81: 47–56.

Muoni, T., Rusinamhodzi, L. and Thierfelder, C. (2013). Weed control in conservation agriculture systems of Zimbabwe: Identifying economical best strategies. *Crop Protection*, 53: 23–8.

Muoni, T., Rusinamhodzi, L., Rugare, J. T., Mabasa, S., Mangosho, E., Mupangwa, W. and Thierfelder, C. (2014). Effect of herbicide application on weed flora under conservation agriculture in Zimbabwe. *Crop Protection*, 66: 1–7.

Mupangwa, W. and Thierfelder, C. (2014). Intensification of conservation agriculture systems for increased livestock feed and maize production in Zimbabwe. *International Journal of Agricultural Sustainability*, 12(4): 425–39.

Ngwira, A. R., Thierfelder, C. and Lambert, D. M. (2013). Conservation agriculture systems for Malawian smallholder farmers: Long-term effects on crop productivity, profitability and soil quality. *Renewable Agriculture and Food Systems*, 28(4): 350–63.

Nyamapfene, K. (1991). *Soils of Zimbabwe*. Nehanda Publishers (Pvt) Ltd, Harare, Zimbabwe, pp. 75–9.

Oldrieve, B. (1989). Conservation tillage in action. In: M. Vowles (Ed.), *Conservation Tillage: A handbook for Commercial Farmers in Zimbabwe*. Commercial Grain Producers Association, Harare, Zimbabwe.

Oldrieve, B. (1993). *Conservation Farming for Communal, Small Scale, Resettlement and Co-operative Farmers of Zimbabwe: A Farm Management Handbook*. Prestige Business Services (Pvt) Ltd, Harare, Zimbabwe.

Powlson, D. S., Stirling, C. M., Thierfelder, C., White, R. P. and Jat, M. L. (2016). Does conservation agriculture deliver climate change mitigation through soil carbon sequestration in tropical agro-ecosystems? *Agriculture, Ecosystems and Environment*, 220: 164–74.

Sims, B., Thierfelder, C., Kienzle, J., Friedrich, T. and Kassam, A. (2012). Development of the conservation agriculture equipment industry in sub-Saharan Africa. *Applied Engineering in Agriculture*, 28(6): 813–23.

Smith, A., Snapp, S., Dimes, J., Gwenambira, C. and Chikowo, R. (2016). Doubled-up legume rotations improve soil fertility and maintain productivity under variable conditions in maize-based cropping systems in Malawi. *Agricultural Systems*, 145: 139–49.

Snapp, S. S., Rohrbach, D. D., Simtowe, F. and Freema, H. A. (2002). Sustainable soil management options for Malawi: Can smallholder grow more legumes? *Agriculture, Ecosystems and Environment*, 91: 159–74.

Snapp, S. S., Jones, R. B., Minja, E. M., Rusike, J. and Silim, S. N. (2003). Pigeon pea for Africa: A versatile vegetable – and more. *HortScience*, 38(6): 1–7.

Thierfelder, C. and Wall, P. C. (2009). Effects of conservation agriculture techniques on infiltration and soil water content in Zambia and Zimbabwe. *Soil and Tillage Research*, 105(2): 217–27.

Thierfelder, C. and Wall, P. C. (2010). Rotation in conservation agriculture systems of Zambia: Effects on soil quality and water relations. *Experimental Agriculture*, 46(3): 309–25.

Thierfelder, C., Cheesman, S. and Rusinamhodzi, L. (2012). A comparative analysis of conservation agriculture systems: Benefits and challenges of rotations and intercropping in Zimbabwe. *Field Crops Research*, 137: 237–50.

Thierfelder, C., Cheesman, S. and Rusinamhodzi, L. (2013). Benefits and challenges of crop rotations in maize-based conservation agriculture (CA) cropping systems of southern Africa. *International Journal of Agricultural Sustainability*, 11(2): 108–24.

Thierfelder, C., Matemba-Mutasa, R. and Rusinamhodzi, L. (2015a). Yield response of maize (*Zea mays* L.) to conservation agriculture cropping system in Southern Africa. *Soil and Tillage Research*, 146: 230–42.

Thierfelder, C., Rusinamhodzi, L., Ngwira, A. R., Mupangwa, W., Nyagumbo, I., Kassie, G. T. and Cairns, J. E. (2015b). Conservation agriculture in Southern Africa: Advances in knowledge. *Renewable Agriculture and Food Systems*, 30(04): 328–48.

Thierfelder, C., Matemba-Mutasa, R., Bunderson, W. T., Mutenje, M., Nyagumbo, I. and Mupangwa, W. (2016a). Evaluating manual conservation agriculture systems in southern Africa. *Agriculture, Ecosystems and Environment*, 222: 112–24.

Thierfelder, C., Rusinamhodzi, L., Setimela, P., Walker, F. and Eash, N. S. (2016b). Conservation agriculture and drought-tolerant germplasm: Reaping the benefits of climate-smart agriculture technologies in central Mozambique. *Renewable Agriculture and Food Systems*, 31(5): 414–28.

Thierfelder, C., Baudron, F., Setimela, P., Nyagumbo, I., Mupangwa, W., Mhlanga, B., Lee, N. and Gérard, B. (2018). Complementary practices supporting conservation agriculture in southern Africa: A review. *Agronomy for Sustainable Development*, 38(2): 16.

Valbuena, D., Erenstein, O., Homann-Kee Tui, S., Abdoulaye, T., Claessens, L., Duncan, A. J., Gérard, B., Rufino, M. C., Teufel, N., van Rooyen, A. and van Wijk, M. T. (2012). Conservation Agriculture in mixed crop-livestock systems: Scoping crop residue trade-offs in sub-Saharan Africa and South Asia. *Field Crops Research*, 132: 175–84.

Wall, P. C. (2007). Tailoring Conservation Agriculture to the needs of small farmers in developing countries: An analysis of issues. *Journal of Crop Improvement*, 19: 137–55.

Wall, P. C. (2009). Strategies to overcome the competition for crop residues in Southern Africa: Some light at the end of the tunnel. In: *Lead Papers. 4th World Congress on Conservation Agriculture*, 4-7 February 2009, New Delhi, India. Innovations for Improving Efficiency, Equity and Environment, pp. 65–70.

Wall, P. C., Thierfelder, C., Ngwira, A. R., Govaerts, B., Nyagumbo, I. and Baudron, F. (2014). Conservation Agriculture in Eastern and Southern Africa. In: R. A. Jat, K. L. Sahrawat and A. H. Kassam (Eds), *Conservation Agriculture: Global Prospects and Challenges*. CABI, Wallingford, UK.

Wheeler, T. and von Braun, J. (2013). Climate change impacts on global food security. *Science*, 341(6145): 508–13.

ZCATF (Zimbabwe Conservation Agriculture Task Force) (2009). *Farming for the Future: A Guide to Conservation Agriculture in Zimbabwe*. ZCATF, Harare, Zimbabwe.

Zingore, S., Manyame, C., Nyamugafata, P. and Giller, K. E. (2005). Long-term changes in organic matter of woodland soils cleared for arable cropping in Zimbabwe. *European Journal of Soil Science*, 56(6): 727–36.

References for the Brazil section

Albertengo, J., Belloso, C., Giraudo, M. B., Peiretti, R., Permingeat, H. and Wall, L. (2014). Conservation agriculture in Argentina. In: R. A. Jat, K. L. Saharawat and A. Kassam (Eds), *Conservation Agriculture: Global Prospects and Challenges*. CAB International, Wallingford, UK, pp. 352–74.

Aquastat. (2018). FAO. Available at: http://www.fao.org/nr/water/aquastat/data/query/results.html (accessed 15 June 2018).

Bolliger, A., Magid, J., Amado, T. J. C., Neto, F. S., Ribeiro, M., Calegari, A., Ralisch, R. and Neergaard, A. (2006). Taking stock of the Brazlian 'Zero-Till Revolution': A review of landmark research and farmers' practice. *Advances in Agronomy*, 91: 47–110.

Bunch, R. (2012). *Restoring the Soil: A Guide for Using Green Manure/Cover Crops to Improve the Food Security of Smallholder Farmers*. Canadian Foodgrains Bank.

Calegari, A. (2009). Alternative land uses and farm diversification strategies to strengthen CA. In: *Lead Papers – 4th World Congress on Conservation Agriculture*, 4-7 February 2009, New Delhi, India. Innovations for Improving Efficiency, Equity and Environment, New Delhi, India, pp. 117–21.

Calegari, A. (2010). No-till in Southern Brazil. In: B. Crabtree (Ed.), *Search for Sustainability with No-till Bill in Dry Land Agriculture*. Crabtree Agricultural Consulting, Australia, pp. 162–78.

Calegari, A.; Mondardo, A., Bulisani, E. A., Wildner, L. P., Costa, M. B. B., Alacântara, P. B., Miyasaki, S. and Amado, T. J. C. (1993). *Cover Crops in the South Brazil*. 2nd Edition. AS-PTA, Rio de Janeiro, RJ, 346p.

Calegari, A., Costa, A. and Ralisch, R. (2007). Sustainable agriculture with no-tillage including cover crops and crop rotation. *Annals of Arid Zone*, 46(1): 1–23.

Calegari A., de Araújo, A. G., Costa, A., Fuentes Lanillo, R., Casão Junior, R. and Rheinheimer dos Santos, D. (2014). Conservation agriculture in Brazil. In: R. A. Jat, K. L. Saharawat and A. Kassam (Eds), *Conservation Agriculture: Global Prospects and Challenges*. CAB International, Wallingford, UK, pp. 54–88.

Clark A. (2007). *Managing Cover Crops Profitably*. 3rd Edition. Sustainable Agriculture Network, Beltsville, MD.

CTIC. (2017). Report of the 2016-17 National Cover Crop Survey. Joint publication of the Conservation Technology Information Center, the North Central Region Sustainable

Agriculture Research and Education Program, and the American Seed Trade Association, West Lafayette, IN.

Derpsch, R., Sidiras, N. and Roth, C. H. (1986). Results of studies made from 1977 to 1984 to control erosion by cover crops and no-tillage techniques in Paraná, Brazil. *Soil and Tillage Research*, 8: 253–63. doi:10.1016/0167-1987(86)90338-7.

Dixon, J., Gulliver, A. and Gibbon, D. (2001). *Farming Systems and Poverty Farming Systems and Poverty: Improving Farmers' Livelihoods in a Changing World*. FAO and World Bank, Rome and Washington DC, 412pp. doi:10.1017/S0014479702211059.

Duiker S. and Thomason W. (2014). Conservation agriculture in the USA. In R. A. Jat, K. L. Saharawat and A. Kassam (Eds), *Conservation Agriculture: Global Prospects and Challenges*. CAB International, Wallingford, UK, pp. 26–53.

Ekboir, J. M. (2003). Research and technology policies in innovation systems: Zero tillage in Brazil. *Research Policy* 32(4): 573–86.

Florentin, M. A., Peñalva, M., Calegari, A. and Derpsch, R. (2011). Green manure-cover crops and crop rotation on small farms. Integrated Crop Management Vol 12-2010. Plant Production and Protection Division, Food and Agriculture Organization of the United Nations, Rome, 97p.

Friedrich, T., Kassam, A. and Corsi, S. (2014). Conservation agriculture in Europe. In: R. A. Jat, K. L. Saharawat and A. Kassam (Eds), *Conservation Agriculture: Global Prospects and Challenges*. CAB International, Wallingford, UK, pp. 127–79.

Landers J. N. (1994). *Fascículo de experiências de plantio direto no cerrado*. Associação de Plantio Direto no Cerrado – APDC, Goiânia, 261p.

Mariki, W. (2004). The impact of conservation tillage and cover crops on soil fertility and crop production in Karatu and Hanang Districts of northern Tanzania. TFSC/GTZ Technical Report 1999–2003, 38p.

Montgomery, D. R. (2018). To feed the world sustainably, repair the soil. *Scientific America*, 16 July 2018. Nature America, Inc. Available at: https://blogs.scientificamerican.com/observations/to-feed-the-world-sustainably-repair-the-soil/ (accessed 18 July 2018).

Nawaz, A., Farooq, M., Lal, R., Rehman, A., Hussain, T. and Nadeem, A. (2016). Influence of sesbania brown manuring and rice residue mulch on soil health, weeds and system productivity of conservation rice-wheat systems. *Land Degradation and Development*, 28(3): 1078–90. https://doi.org/10.1002/ldr.2578.

Rochecouste, J. and Crabtree, W. L. (2014). Conservation agriculture in Australian dryland cropping. In: R. A. Jat, K. L. Saharawat and A. Kassam (Eds), *Conservation Agriculture: Global Prospects and Challenges*. CAB International, Wallingford, UK, pp. 108–26.

Séguy, L., Bouzinac, S., Maronezzi, A. C., Belot, J. L. and Martin, J. (2001). A safrinha de algodão: opção de cultura arriscada ou alternativa lucrativa dos sistemas de plantio direto nos trópicos úmidos. Boletim técnico no 37 da COODETEC CP, Cascavel, PR, Brasil.

Wütke, E. B., Calegari, A., Wildner, L. P. and Amabile, R. F. (2008). Leguminosas alimentícias e adubos verdes. In A. C. S. Albuquerque and A. G. da Silva (Eds), *Agricultura Tropical: quatro décadas de inovações tecnológicas, institucionais e políticas, Vol. 1: Produção e produtividade agrícola*. Embrapa Informação Tecnológica, Brasília, DF, pp. 251–74.

Wütke, E. B., Calegari, A. and Wildner, L. P. (2014). Espécies de adubos verdes e plantas de cobertura e recomendações para uso. In: O. F. de Lima Filho, E. J. Ambrosano, F. Rossi and J. A. D. Carlos (Eds), *Adubação verde e plantas de cobertura no Brasil: fundamentos e prática*. Embrapa v. 1, Brasília, DF, pp. 59–168.

Chapter 7

Management of vegetable Conservation Agriculture systems

Jamil Abdalla Fayad, Agricultural Research and Rural Extension Company of Santa Catarina (EPAGRI), Brazil; Jucinei José Comin, Federal University of Santa Catarina State (UFSC), Brazil; Alvaro Mafra, Santa Catarina State University, Brazil; Claudinei Kurtz, Agricultural Research and Rural Extension Company of Santa Catarina (EPAGRI), Brazil; Samira Jamil Fayad, Santa Catarina State University, Brazil; Nuno Rodrigues Madeira, EMBRAPA, Brazil; Darlan Marcheze and Leandro do Prado Wildner, Agricultural Research and Rural Extension Company of Santa Catarina (EPAGRI), Brazil; Ademir Calegari, Agricultural Research Institute of Paraná State (IAPAR), Brazil; and Arcangelo Loss, Cledimar Rogério Lourenzi, Monique Souza, and Guilherme Wilbert Ferreira, Federal University of Santa Catarina State (UFSC), Brazil

http://dx.doi.org/10.19103/AS.2019.0048.08

1 Introduction

The conventional production system (CPS) for vegetables in Brazil uses large amounts of agrochemicals and synthetic fertilizers, and intensive soil tillage, leading to excessive soil erosion, nutritional imbalances, and environmental pollution (Fontanetto and Keller, 2001; Andrade, 2010). This production system has also increased production costs and contributed to farmers' indebtedness.

In the 1970s, two Brazilian farmer movements arose to address the problems of CPS. One was the 'no-tillage with straw' movement, using cover crop or crop residues with minimum soil disturbance, avoiding high soil losses, but limited by its singular focus on soil conservation. The other farmer movement was the alternative agriculture movement based on ecological principles. The 'no-tillage with straw' approach concentrated mainly on grain production and was still very dependent on chemical inputs such as fertilizers and agrochemicals. Despite efforts by many professionals, it had little application to vegetable production. On the other hand, the agroecological movement relied heavily on organic amendments, leading to soil nutrient imbalance among other problems.

It is in this context that the no-tillage system for vegetables (NTV) emerged as an alternative concept to integrate these two movements with a focus on soil and plant health. NTV is a production system based on the application of Conservation Agriculture (CA) principles (see Chapter 1)[1] that reduces soil, water, and nutrient losses, limits the use of pesticides and soluble fertilizers, decreases production costs while maintaining or even increasing productivity, reduces labor requirements, and is environmentally friendly. It aims to enhance the interconnection between individual smallholder production systems and integrates production areas, forests, and ecological corridors, supporting agroecosystems (Fayad et al., 2016; Fig. 1).

Exciting results from the first vegetable production experiments, conducted by researchers from the Agricultural Research and Rural Extension Service of Santa Catarina (EPAGRI) at the Experimental Station of Caçador (EECdor), have shown that it is possible to shift from conventional to minimum tillage and to NTV. EPAGRI, support by the Center and Rural Worker Consultancy (Cepagri), and Rural Workers Union of Caçador (Sitruc) began the program with the farmers of the Alto Vale do Rio do Peixe region, with assistance from Santa Catarina State University (UDESC/CAV) and the Federal University of Santa Catarina (UFSC/CCA).

Initially the research and extension focus was on tomato production, followed by other species such as sweet pepper, eggplant, some brassica species, cucurbitaceae, parsnips, leafy vegetables, and onion. In the last 20 years of research and extension-supporting subsistence family farming, a

1 Throughout this chapter, the local term NTV is used to mean CA.

Figure 1 Typical smallholder family farm managed with the principles of NTV (No-Tillage System for Vegetables). Crop areas, grassland for animals, and storage facilities are surrounded by ecological corridors, forests, and rivers. Shown are areas with cauliflower at planting, growing, and harvest stages; tomato at the harvesting stage; different cover crops; leafy vegetables; forage banks; pasture for dairy cattle and chayote (Source: Angelina, 2018).

number of scientific articles have been published, as well as guides on tomato, chayote (a type of squash), hybrid pumpkins (tetsukabuto), and brassica. The program has also resulted in a Practical Guide to NTV with recommendations for 21 vegetable species and a Practical Guide to Participatory Soil Quality Evaluation to help smallholders monitor and improve soil health.

In this chapter, we will discuss the principles and practices of NTV with detailed examples of the cultivation of tomato, onion, and parsley cassava cultivation. Tomato cultivation was selected because it is a challenging crop to grow, being highly susceptible to chemical, biological, and physical stresses, including pests and diseases. We also present research on a long-term experiment on NTV for onion combined with winter cover crops and its effects on both soil health and crop yield. Finally, we discuss NTV for parsley cassava, an indigenous Andean crop. With this choice of crops, we hope to show the initial steps that have been taken to enable the transition of all the vegetables commonly cultivated by subsistence family farmers in the state of Santa Catarina to NTV.

NTV activities start with individual and family farmer groups who receive support from the rural extension technical assistance (RETA) service and from university students. Farmers enter into a 'contract of employment' with RETA to administer study crop areas (SCA), which are areas of commercial crops to be managed under the principles of NTV. In this contract, a timeline of activities is drawn for a period of up to 1 year. This timeline covers activities such as courses on NTV and the implementation of SCAs. It also covers visits by technical staff, laborers, and other farmers, study trips, meetings to share the results, and, finally, the renewal or not of the contract for another year (Fig. 2). It is important to establish this mutual commitment, which can be altered and developed based on the discussions between the parties involved, and often goes far beyond just the implementation of the planned actions and activities. It is designed to mediate between the technical-scientific knowledge of 'experts' and the practical, local knowledge of farmers, ensure regular evaluation of the steps already taken and those to come, as well as build up trust between those involved.

It is essential that the RETA agent is also a researcher, undertaking experiments with both the SCA and the EPAGRI experimental stations. Technicians (RETA agents, professors, researchers, and students) need to work in both the experimental areas and SCA to ensure effective development, adaptation, and practical application technologies and practices to implement NTV. Resources at EPAGRI experimental stations, such as equipment, laboratories, field staff, and experimental area, as well as the structure of universities, should also be accessible to the RETA agent, who, being connected to the reality of food production systems of family agriculture, can meet the demands of research for the adoption of NTV.

Figure 2 RETA agents and a university student collecting plants to determine the daily rate of nutrient absorption (DRNA) and accumulation of biomass of broccoli (Source: Anitápolis, 2012).

The central principle of these NTV programs is the promotion of plant health, based on:

- The promotion of plant health through minimization of biotic stresses such as nutritional deficiencies, salinity, water, temperature, light, pH, oxygen diffusion rate, and so forth;
- The promotion of plant health (Fig. 3) in relation to managing growth characteristics, including root architecture, plant size, and fruit quantity, according to the nutritional needs of each crop; techniques include cucurbit pruning and the vertical trellis system in tomato cultivation, as well as the use of plant resistance inducers (PRI) and bio-based fungicides such as Bordeaux mixture among others to improve crop resistance to disease;
- Plant nutrition based on understanding of growth rates and absorption of nutrients by crops, including adapting nutrition to environmental conditions, soil nutritional reserves, and monitoring plants themselves for signs of nutritional deficiency;
- Crop rotation and cover crops (cultivated and spontaneous), evolving to rotation with animals as well as crops, managed according to Voisin rational grazing (PRV) principles (allowing intensive rotational grazing), to create a more self-sufficient crop-manure-crop rotation system;
- Trying to add more than 10 tons of phytomass (dry mass) per ha per year through crop rotation planning to promote soil health;
- Soil disturbance restricted to planting lines or sowing berms (mounds);

Figure 3 Cover crops protect soil and provide a suitable environment for cauliflower.

- Management of spontaneous cover crops (such as plants more adapted to local conditions) so that they can improve the system without harming the production of the economic crop, evolving to a *no-tillage planting green system* (aiming to eliminate the use of herbicides; Teasdale et al., 2007);
- Decreasing and, if possible, eliminating the use of highly soluble fertilizers and agrochemicals;
- Stratified soil sampling for chemical analysis and monitoring;
- Decreasing production and environmental costs, without reducing crop productivity.

These principles have been applied in the NTV cultivation of crops such as tomato.

2 No-tillage vegetable (NTV) cultivation of tomatoes: introduction

The tomato is a perennial plant originating in the Andes but is usually grown as an annual. Tomato varieties can be determinate or indeterminate. It is agreed that the lower leaves assimilate carbohydrates for the roots while the top leaves feed leaf and apical meristem growth. Continuous pruning is needed to keep only two stems to facilitate solar radiation absorption and air movement in the canopy, and for easy harvesting of tomato fruits. The plant has a root system consisting of a taproot, secondary roots, and adventitious roots. In soil with a good 'fluffy' structure (i.e. deep and well drained), the taproot can reach depths greater than 1.5 m, while two-thirds of the roots are typically concentrated in the top 30 cm of the soil. The supply of water and nutrients is an important factor in the root environment because it has a marked effect on the growth, development, and health of the plant, especially as tomato plants can carry approximately 60–90 fruits that drain huge amounts of assimilates, nutrients, and water (Fig. 4). Feeding all the fruits equally, avoiding competition between fruits located in the first and last bunches, is only possible with a deep and extensive root system in well-structured soil, and with plants having optimal access to light. The tomato plant has high demand for light, especially in the morning, while radiation also helps decrease the time when leaves are wet, thus helping reduce disease pressure. Light affects the amount of lycopene, which affects the intensity of the red color of the fruit. Temperature also influences plant growth, notably the difference of 6–7°C between day and night temperatures. Sunny skies and temperatures up to 30°C produce fruits of intense red color.

Figure 4 Experiment with tomato grown using NTV cultivation to determine its growth and absorption of nutrients (Caçador, Santa Catarina State, Brazil, 1999).

3 NTV cultivation of tomatoes: seedling cultivation

3.1 The production of seedlings

It is necessary to have cultivars adapted to NTV conditions. This requires using healthy seedlings, with good vigor, well-formed tissues, and an intact root system. The seeds used to produce the seedlings can be produced on the farm or can be purchased commercially. If farm-grown, it is important to choose mature, healthy fruit with good-quality characteristics from the mother plant, followed by crushing and fermenting the gelatinous seed coating for about 48–72 h to combat bacterial diseases. Seeds should then be washed, dried, and stored in a refrigerator.

It is common practice among farmers to sow in a tray of 128 cells and allocate one seed in each cell (Fig. 5). The tray should first be washed and immersed in 1% sodium hypochlorite solution for 30 min and rinsed again. A substrate must be used to fill the cells of the tray, which contains all the nutrients the seedling needs for 'full formation'. Generally, farmers use commercial substrate and apply nitrogen so that the seedlings grow quickly and assume a dark green color. It is, however, important to note that the appearance of the seedling does not necessarily imply a healthy plant.

There is currently insufficient research to determine the daily rate of absorption of nutrients by tomato seedlings, in order to provide nutrients according to their growth requirements. We adapted the experimental data from the work of Fayad et al. (2016), set out in Table 1, indicating the amounts

Figure 5 Seedlings ready for planting, with three definitive leaves.

Table 1 Nutrient content in freshly planted tomato seedlings with three final leaves

N	P	K	Ca	Mg	S	Fe	Mn	Zn	Cu	B
mg plant^{-1}						µg plant^{-1}				
270	33	254	187	37	51	2500	1250	238	108	545

of nutrients that should be available for seedlings. These nutrients can come from the substrate in a cell in the tray or, if necessary, be added to the substrate or sprayed on to the leaves.

The optimum temperature range for seed germination and seedling development is between 20°C and 25°C. Germination stops at a temperature lower than 9°C. In practice, a minimum of 13°C is used since, at that temperature, it takes more than 40 days to germinate 83% of the seeds. After germination, if the seedlings are subjected to a temperature close to 14°C, growth stops, leaves turn yellow, the stalk turns blue, leaves split, and the number of flowers increases while fruit is deformed. The occurrence of 'blind' plants occurring in the nursery or shortly after planting may be related to a sudden drop in temperature (≤5°C) and cold wind, making calcium mobility difficult and causing apical meristem death. In the cultivars of the group Salada or Caqui in particular, it is common to see fruit with open loci, especially in the first floral clusters, related to the exposure of the seedling in its embryonic stage of flower differentiation, at a temperature of 15°C/10°C during the day/night, for a period exceeding 1 week.

The tomato seedling developed in a tray forms a fasciculate root system. In the case of direct sowing, the plant maintains the taproot which can reach more than 1.5 m deep and, consequently, improves the capacity to avoid nutritional and water stress. One possibility of maintaining the taproot in transplanted seedlings would be to produce each seedling in a long tube to contain the root and then transplant it when it has only two leaves.

Acclimatization of the seedlings prepares them for establishment in the field. Typically, the shelter in which they have been grown is removed, and the seedlings are exposed to the environment. Sunlight and water deficits are carefully monitored to maintain the seedlings' photosynthetic potential, enabling them to establish rapidly and grow vigorously in the field.

Another form of seedling production is through a bud from a healthy plant, preferably in its juvenile phase. Scissors or a disinfected knife is used. The bud with two to three leaves is removed, leaving a stem of approximately 5 cm. Two-thirds of the stem are buried in the tray substrate and kept moist. It is essential to promote rapid rooting to avoid the establishment of pathogens at the cutting site of the shoot, and it is recommended that the tray is placed in a sunny place, allowing the leaves to wilt but without water being lost in the substrate. This treatment produces vigorous seedlings ready for planting in approximately 13 days, with optimal rooting and easily established in the field. Another advantage is the appearance of the first floral bunch close to the ground.

3.2 Mechanization shift for NTV

The area of Santa Catarina where NTV cultivation takes place is characterized by high variability in topography, size of cultivated area, crops produced, and amounts of crop residues on the soil surface. Equipment needs, therefore, to adapt to a wide variety of conditions. However, most farm equipment is manufactured for medium and large farms and uniform conditions. Most family farmers have stopped building and adapting equipment in their own or community workshops. However, some farmers and small-scale manufacturers have developed equipment for use in NTV cultivation (Figs. 6–8). This includes planting equipment which can be drawn by people, animals, or small-scale 'micro' tractors. The equipment should create adequate conditions for the development of the transplanted seedlings. To achieve this, the basic design of planting equipment includes a guide wheel, a cutting coulter, and a furrowing knife.

The guide wheel, which can be double or single, aims to regulate the working depth and puts pressure on the straw to facilitate the work of the cutting coulter. In double wheel equipment, the cutting coulter is located between and just behind press wheels and, in single wheel equipment, is usually mounted in front of the wheel. The cutting coulter needs to cut the straw. There are several types and sizes for different conditions, the most common being fluted which

Figure 6 Planting equipment (designed to be pulled by animals or adapted for tractor use) with depth guide wheel, cutting coulter, and furrowing knife, and with containers with fertilizer and seed.

Figure 7 No-till planter (designed to be pulled by animals), with 17" cutting coulter, furrowing knife with an 'ema' finger and fertilizer distribution, manufactured by Fitarelli.

Figure 8 Rolling-down fall/winter mix cover crops with a knife-roller for tomato planting (Caçador-SC, 2000).

is 17 inches in diameter. The furrowing knife should cut the soil at constant depth and crumble the compacted soil profile, facilitating fertilizer placement and seedling growth.

3.3 Suitable adjustments for planting area

In order to prepare for NTV cultivation, it is recommended that some initial tasks be performed. These include:

1. stratified soil sampling for chemical analysis;
2. identifying and eliminating any compacted layer in the soil profile;
3. leveling the terrain to allow planting; and
4. sowing a cover crop mix for later rolling, opening of furrows, and planting.

Stratified soil sampling should be done between 0 cm to 10 cm and 10 cm to 20 cm for chemical analysis with a view to making pH corrections, followed by application of phosphate and potassium fertilizer. It is also necessary to evaluate the soil structure for identification and elimination of possible compacted layers. It is recommended to open a trench of at least 30 × 30 × 30 cm and use a pocket-knife to evaluate soil compaction (Fig. 9). The evaluation should determine the presence of a layer that makes it difficult and/or prevents deep rooting of the crops and/or promotes the accumulation of water. Other

Figure 9 Soil profile for tomato cultivated in a conventional cultivation system, showing the presence of surface sealing, a compacted layer, as well as superficial and uneven root distribution.

indicators of soil quality such as soil structure (presence of visible aggregates is well formed and resistant to a slight pressure), color, and odor of the soil can be evaluated to get an idea of the organic matter content.

To break up a compacted layer, it is recommended to use a chisel plow or other equipment that adequately reaches below the layer to be loosened. Subsoilers normally work at depths of more than 30 cm and it is difficult to operate them at depths between 12–25 cm, where a compacted layer is often observed. After compaction has been addressed, it is time to start shaping the terrain meet the planting, cultivation, and harvesting requirements of the main crops and cover crops in the rotation system. This can either be in the form of ridges, beds, or 'on the flat' (leveling). Leveling is preferred to facilitate the handling of cover crops with a knife-roller, weed management with a sprayer, and the rotation of main crops with cover crops and livestock. Single species or mixtures of cover crops should be planted to produce the straw mat that will be rolled on the soil surface.

NTV cultivation, characterized by appropriate crop rotation, diversifies plant cultivation and/or spontaneous plant management, maintains and even improves soil fertility, organic matter, and soil structure (Calegari et al., 2013), contributing to a decrease in weeds, pests, and diseases (Ball et al., 2005), as well as increasing crop yield (Stanger and Lauer, 2008; Campbell et al., 2011). The improvement of soil structure is mainly due to the formation of biopores formed by growing and decomposing plant roots, and soil aggregate stabilization (Oades, 1993). Several studies have demonstrated that plant roots increase the stability of macroaggregates (Amézketa, 1999; Six et al., 2004). Benefits of NTV include soil resistance to erosion (Bolliger et al., 2006; Loss et al., 2017), increased water infiltration and retention (Balwinder-Singh et al., 2011; Gava et al., 2013), increased soil carbon and nutrient contents, higher cation exchange capacity (Santos et al., 2018), and increase of micro- and macrobiota (Gatiboni et al., 2009; Liu et al., 2016).

It is imperative in the crop rotation plan to avoid the cultivation of solanaceous plants twice in the same area for at least 3 years. Cover crops should be managed by a knife-roller near full bloom and/or at the milky grain stage. In the case of tomato cultivation, which is labor intensive and requires greater movement of people and machines, the rolling down of cover crop can be extended so that the dry matter (DM) becomes more lignified and, consequently, the residues cover the soil surface for longer.

3.4 Field preparation

The field where seedlings are to be planted should be prepared to minimize causing stress to plants while optimizing growth. The field can be prepared with a knife-roller drawn by animals, micro or standard-sized tractors to open the furrow after the lodging of the cover crop (Fig. 10). The furrow is opened with a 17-inch diameter cutting coulter and knife with the ema finger located at its base. This breaks the ground from below at the points of weakness in the soil

Figure 10 Field prepared with a Fitarelli no-till planter, after knife-rolling of the cover crop (Knaul Brothers, Ituporanga, 2007).

structure without causing sidewall compaction. At the same time, phosphate and organic fertilizers can be added, in the amounts shown in Table 2.

Soil with a good aggregate structure can be achieved after 3-5 years of NTV cultivation. In a study of onion NTV cultivation (including both single and multispecies cover crops), Loss et al. (2015) demonstrated an increase in soil aggregation indices after 5 years in comparison with a CPS (conventional tillage) system and even with an area of native forest (with an increase in the geometric mean diameter (GMD) in the layers at depths of 0–5 and 5–10 cm).

Examples of field planting by hand using a small hoe and seedling development can be seen in Figs. 11 and12. It is worth highlighting that the development of tomato seedlings in NTV cultivation is slower compared to the CPS, because of lower soil temperature, due to the presence of the thick straw layer (Fig. 13) and better moisture conditions. This allows the roots to absorb the nutrients needed to develop healthy plant tissues.

Table 2 Amount of chemical phosphate fertilizer and organic fertilizer for field application, in grams per meter of furrow

	Amount (grams/furrow linear meter^{-1})	
P_2O_5 and OM content	P_2O_5	Poultry manure
Low	35	400
Average	26	300
High	18	200

Figure 11 Seedlings planting in the nursery (Nilvo and Knaul, Três Barras community, Ituporanga-SC, 2005).

Figure 12 Tomato seedlings at the beginning of development in the field.

Figure 13 First stage of tomato growth in straw cover. As the soil has lower temperature than in CT, plant growth is slower but plants are healthy.

3.5 Criteria to determine plant spacing and staking

Experiments with tomato NTV cultivation at CEEdor from 1998 to 2001 have helped to identify the optimum number of flower clusters and spacing between plants. These involved a mix of oat + oilseed radish + vetch, using knife-roller, drip fertigation, staked tomatoes, and a predetermined number of stems. They confirmed that, for a plant with one stem, the ideal is to maintain up to five floral clusters per plant and spacing between plants of 0.35 m. In the case of a plant with two stems, up to nine flower clusters per plant were optimal, five on the main stem and four on the secondary one, and with 0.50 m spacing between plants. Row spacing should be 1.4–1.5 m (Fig. 14) to facilitate cultural operations such as mowing and spraying, as well as to improve aeration and solar radiation, thus reducing leaf wetness. This spacing also leads to the least number of hollow and cracked fruits.

In organic cropping systems where, instead of highly soluble fertilizers and agrochemicals, only organic fertilizers and Bordeaux mixture are used, it is recommended to manage the crop with fewer fruit clusters per plant, reducing the spacing between plants and maintaining or reducing the spacing of 1.5 m between rows. One option is pruning to two stems and three to four clusters or two stems with three clusters on the main and two clusters on the secondary stem. This results in vigorous plants and large fruit with a short time to harvest and good yield.

Figure 14 Plant thinned to two stems at 1.5 m row spacing and 0.5 m plant spacing between rows.

Once spacing and pruning methods are chosen, it is time to think about how to support plants by tying them up vertically, using bamboo sticks and wire. This task requires skill and time, avoiding lodging of the plants caused by wind and the weight of the tomato fruit. Equipment such as reinforced palisades at the head of the line, wire stretchers, maximum line length, and, especially, the lateral support struts for the line are fundamental. In the municipality of Ituporanga, in the Alto Vale do Itajaí, the Knaul brothers and the Hoffman family have developed a field technique for vertically tied tomatoes, with wire about 50 m long and with a platform at each end, connected by wires stretched by turnstiles installed at each upper end of the platform to maintain tension (Fig. 15). It is recommended to install bamboo stakes before or soon after planting the seedlings. If this operation is carried out later, damage to the roots of the plants may occur, which will favor entry of pathogens into damaged areas and the onset of disease.

4 NTV cultivation of tomatoes: improving soil fertility

4.1 Management of soil acidity, phosphorus, and potassium

Chemical analysis should be carried out to assess soil acidity as well as phosphorus and potassium content, with the aim of maintaining the pH between 5.5 and 6.0 and increasing levels of P_2O_5 and K_2O (Table 3). The incorporation of limestone, phosphorus, and potassium must be carried out before cover

Figure 15 Tomato plants tied to bamboo stakes.

Table 3 Amount of phosphorus and potassium incorporated in the soil for total correction

P and K content in soil	P_2O_5 (kg ha^{-1})	K_2O (kg ha^{-1})
Very low	120	120
Low	60	60
Average	30	30

Table adapted by manual of fertilization and liming for the states of RS and SC (CQFS-RS/SC, 2016).

crop and tomato planting. After that point, soil amendments should be applied on the soil surface only or incorporated between planting lines to avoid root damage.

If liming is needed, half should be applied before plowing and the other half before disc harrowing, at least 90 days before planting. However, some farmers have used chisel, harrow, or roto-tillers to incorporate limestone which leads to excessive pH levels in the 0–10 cm soil layer and low levels in the 10–20 cm layer. This causes the roots of cover crops, vegetable crops, and weeds to be concentrated in the top layer of the soil, leaving the plants more susceptible to water stresses, nutrient imbalance, higher acidity, and aluminum and/or manganese toxicity. These stresses reduce tomato productivity and quality and can contribute to the incidence of pests and diseases (Marschner, 2012).

Soil chemical analysis and visual analysis of plant health can sometimes identify very high levels of phosphorus and potassium in soils (CQFS-RS/SC, 2016). This is usually the result of excessive use of poultry manure and chemical fertilizers. Foliar fertilizers have also been used to correct micronutrient deficiencies, while frequent insecticide and fungicide applications are typically needed to address increased threats from pests and diseases.

4.2 Cover crops

The tomato harvest occurs in the winter period, and it is recommended to use mixtures of different winter cover crop species. Winter cover crops include black oat (*Avena strigosa* Schreb.), rye (*Secale cereale*), blue (*Lupinus angustifolius*) or white (*Lupinus albus*) lupin, common vetch (*Vicia sativa*), and radish (*Raphanus sativus*). Suggested mixtures are of three, four, or more species, with the composition of oat + vetch + radish, and so forth, in the amount of 60–80 + 20 + 3–5 kg ha^{-1}, respectively. In the summer period (Fig. 20), corn (*Zea mays*), pearl millet (*Pennisetum glaucum*), sunn hemp (*Crotalaria juncea*), sunflower (*Helianthus annuus*), and grass (*Brachiaria plantaginea*) are planted. An example is sunflower + millet + sunn hemp in the amounts of 15 + 8 + 3–4 kg ha^{-1}, respectively. Some other cover species and recommended mixes can be found in the literature on this topic (Calegari, 1989, 2010, 2014, 2016; Calegari et al., 1993, 2013; Florentin et al., 2010; Calegari and Costa, 2010a,b;

Balota et al., 2014; Wütke et al., 2008, 2014; Calegari and Donizeti Carlos, 2014; Bolliger et al., 2006). A useful and frequently used practice is to choose the highest value of a recommended amount and to add 50% to calculate the amount of green manure seeds needed (Monegat, 1991). This ensures fast soil cover, providing efficient control of weeds during cover crop growth, and high DM production to form a thick straw mat.

Roots perform several functions, among them plant support, water and nutrient absorption, as well as the capacity to synthesize, accumulate, and release exudates that promote plant growth. In the rhizosphere (the interface between roots and soil), roots alter soil structure, promoting aeration and biological activity, since they are the main source of inputs of organic and inorganic compounds. By exuding a wide variety of these compounds, the roots benefit the soil microbial community and stimulate symbiotic activity, including improving pest resistance, as well as improving soil chemical and physical properties, and inhibiting the growth of competing plant species such as weeds. The roots and fungal hyphae are important agents of soil aggregation, as they directly and indirectly influence the physical, chemical, and biological processes related to aggregate stability (Jones et al., 2004; Rillig and Mummey, 2006; Marschner, 2012). In order to take advantage of all these benefits, it is recommended to minimize soil tillage and to cultivate green manures and spontaneous plants alongside vegetable crops, including during the off-season. With a constant and permanent presence of plants on the soil, there will be a greater quantity and diversity of microorganisms colonizing the rhizosphere, which will promote the health and growth of plants.

Cover crops are normally managed by rolling them down with a 'knife-roller' to create a mat of straw on the soil, delaying the appearance of weeds that must then be managed with brush cutters. This practice promotes biodiversity in the cultivation area and avoids nutritional stress, inhibits disease-causing organisms, increases numbers of natural enemies of insect pests, and contributes to the improvement of soil structure. The long-term aim is to stabilize production and create uniform horizontal and vertical fertility in the soil profile. It is important to know which cover crops and their mixtures can help to break 'pathogen' cycles with a combination of the best C/N ratio, diversified root systems, amount of DM deposition, spontaneous plant smothering of competing plants, and attraction of natural enemies (Fig. 16).

5 NTV cultivation of tomatoes: fertilization

In order to promote healthy plant growth, it is important to adjust nutrition to match the daily rate of nutrient absorption (DRNA) by monitoring plants and the stock of nutrients in the soil. As an example, water and nutrient requirements in tomato will vary according to its stage of development. Crop growth can be

Figure 16 Mixing summer cover crops (pearl millet, sunflower, mucuna) cultivated by the Horst family in Ituporanga, 2006.

divided into three basic phases with a similar rate of biomass (G) and nutrient absorption (DRNA):

a) Initial Stage, which goes approximately from the planting of the seedling to the beginning of the fruiting (0–38 days after sowing (DAS));
b) Intermediate Stage, which runs from approximately the beginning of fruiting to close to the second fruit harvest (38–108 DAS);
c) Final Stage, which goes approximately from the second to the last harvest (108–130 DAS).

In the first stage, approximately 17% of the total nutrients absorbed during the tomato cycle are consumed, because the plant is small and is passing from the juvenile stage to the beginning of fruiting. At this stage, heavy fertilization with N, P, and K should be avoided. In the intermediate stage, the plant consumes around 70% of all nutrients absorbed in the cultural cycle, characterized by the adult or reproductive stage, with greater intensity of water and nutrient consumption. In the final stage, the plant consumes only 13% of the total nutrients absorbed, as it is at the filling stage of the last fruits to be harvested.

It is important to take account of climatic conditions, particularly temperature and light availability which affect photosynthetic and respiratory activity and thus growth. These variables can increase, decrease, and even eliminate weekly fertilization with a direct impact on the production and health of tomato plants. Days with good light availability will increase photosynthetic production and evapotranspiration, leading to greater nutrient and water requirements by plants. On the other hand, cloudy conditions and low temperatures mean that plants need fewer nutrients (and certainly less water). In the study area, we learnt to increase the supply of fertilizer via drip fertigation on days of particularly good weather when plant growth sometimes doubled. Failing to do this can lead to a loss of productivity and nutritional stress which can significantly damage plant health.

Another factor in making decisions about plant nutrition is the ability to monitor signals from the plants themselves which can indicate their health and nutritional requirements. These indicators include:

- plant architecture – a uniform shape in individual plants and along a row of plants suggests there are no areas of nutrient deficiency in the soil (Fig. 17);
- the difference in the size of the fruits between the bunches which also indicate if nutrition is matching growth (Figs. 18 and 19);
- the size, color, and color intensity of leaves (Fig. 20) – for example, yellowish coloration indicates insufficient nutrition;
- the difference in green color intensity between old and new leaves, which shows whether nutrition is keeping pace with plant growth (Figs. 20 and 21).

Figure 17 A small difference in the size of the fruits between clusters, meaning that the plant has received the correct supply of water and nutrients. The leaves of the 'skirt' are beginning to show a light green color, indicating equilibrium between fertilization and plant demand.

Figure 18 The first three fruit clusters show some uniformity in size, but with a significant difference in the fourth cluster. In this example, there has been a discrepancy between nutrient supply compared to DRNA and water requirements. This difference may also have been caused by below-optimal temperatures for fruit growth.

Experienced NTV practitioners, whether farmers or RETA agents, can assess the health and nutritional status of plants by looking at these indicators. In the case of roots, the main indicator is root system size, which should be proportional to the size of the above-ground part of the plant. In the case of an imbalance between the two parts of the plant, it is necessary to adjust the size of the vegetative part of the plant by apical pruning.

The absorption of nutrients by tomato plants can be affected by temperatures below 15°C and above 30°C, with the optimum absorption between 20°C and 25°C. Nutrient absorption by roots is also influenced by oxygen, because it requires energy to build the pH difference between the inner and outer part of the root cell membrane and trigger the nutrient uptake via 'carrier proteins'. Soil moisture is also important since it influences the rate of transpiration. There is therefore no point in fertilizing plants if the soil does not have an adequate structure which allows the plant to absorb nutrients and water in adequate quantities for its maintenance, growth, and development of fruit.

Figure 19 The small difference in the size of the fruits between bunches, forming a long pyramid with a narrow base, indicates that the fertilization is meeting DRNA (Anitápolis-SC, 2013).

Figure 20 The presence of approximately 20% of the total leaves of the plant with light green coloration (apical leaves and new shoots), which indicates that the fertilization is adjusted to the plant nutritional requirements.

Figure 21 The presence of a small gradient in fruit size between the plant bunches, the uniform and continuous horizontal foliar wall along the planting line, and the skirt of the plant with a darker green coloration in relation to the coloring of the new leaves (approximately 25% of the total with light green coloration) indicates optimal fertilization and plant health.

Inappropriate fertilization can have damaging effects. Insufficient fertilization reduces plant growth, health, and productivity. Excessive fertilization leads to exaggerated vegetative growth and encourages the incidence of disease and insect pests, increasing reliance on fungicides, acaricides, and insecticides (Marschner, 2012).

6 No-tillage vegetable (NTV) cultivation of onion: introduction

As with tomato, the NTV aims to deliver a number of benefits including the following:

- protecting and enhancing soil health;
- improving cycling of nutrients;
- improving onion yields and quality;
- elimination of use of herbicides and other pesticides through the use of cover crops;
- reducing production costs and the need for mechanization (e.g., by reducing the need for tillage or weeding).

In the case of onion, which has a relatively small stature, a small number of leaves, and a low leaf area index, the use of single cover crops or mixtures of cover crop species is fundamental, since it produces the high amounts of plant residues required for successful NTV cultivation.

Long-term onion NTV trials found that winter cover crops (*Avena strigosa, Raphanus sativus, and Secale cereale*) promoted increases of nutrient contents in topsoil and deeper in the soil profile over 8 years. The values of soil organic matter (SOM), P, K, Ca, and Mg increased as the system became fully established. pH values (6.0), Al (0.0 $cmol_c\ kg^{-1}$), ad CEC pH 7.0 (16 $cmol_c\ kg^{-1}$) remained stable (Souza, 2017). The release rates of C, N, P, K, Ca, and Mg are associated with the decomposition of cover crop DM. Plant residues of rye and black oat can remain for up to 90 days on the soil surface throughout the onion cycle, which increases protection and promotes release of nutrients (Bolliger et al., 2006; Oliveira et al., 2016a,b). Cover crops are efficient in cycling of nutrients by exploiting different layers of the soil, protecting the surface from the impact of rain (which can wash away nutrients and cause erosion), reducing temperature variations, maintaining humidity, thus contributing to maintenance and enhancement of soil fertility (Calegari, 1989; Melo et al., 2011). Cover crops help to recover and increase total organic carbon (TOC) and thus organic matter in soils. This promotes the microbial activity which improves N and humic substances such as humin in soil through mineralization of organic matter (Loss et al., 2015). Studies show more stable microbial communities with lower values of qCO_2 (metabolic quotient), which results in efficient microbial biomass use of the organic C, since less C is lost in respiration as CO_2 and more C is incorporated into microbial tissue (Vargas, 2012).

Cover crops also improve soil physical properties, including the formation and maintenance of soil aggregates, especially macroaggregates, improving measures such as the aggregate weighted average diameter (WMD) and GMD (Loss et al., 2015). Other physical parameters, such as total porosity, macro- and microporosity, and volumetric moisture, also increase in NTV cultivation compared to conventional systems (Loss et al., 2017). This is because of the constant presence and activity of roots and little soil disturbance. Soil disturbance is restricted to planting lines, since seedlings are produced in seedbeds and then transplanted into furrows prepared by seeding machines adapted for cutting the soil and cover crop residues.

Black oat (*Avena strigosa* Schreb.) and rye (*Secale cereale L.*) produce high amounts of DM, around 3.5–4.0 t ha^{-1}, with a C/N ratio greater than 26–28, which favors the persistence of crop residues on the soil surface and slower nutrient mineralization (Derpsch and Calegari, 1985; Calegari et al., 1993; Oliveira et al., 2016a,b; Souza, 2017). The rolling of the cover crops is done early to fit in with the planting date of the early onion variety used in the region, which does not allow cover crops to achieve their maximum DM production.

The mixture of these oat and rye with oilseed radish (*Raphanus sativus* L.) produced on average 4.5 t ha^{-1} generating a residue with C/N ratio close to 13, allowing rapid decomposition of the cover crop biomass and rapid nutrient release, improving nutrition of onion plants and bulb yield (Calegari et al., 1993; Oliveira et al., 2016a,b; Souza, 2017). The NTV system led to an increase in the total production of bulbs and classes of bulbs over the years. The system started in 2009 with a total production of 10 t ha^{-1}, mainly with box 2 bulbs (35–50 mm diameter) and, after 7 years, providing yields of 21–24 t ha^{-1}, with higher production of bulbs of box 3 (50–60 mm) and box 4 (70–90 mm; Souza, 2017).

NTV systems are also effective in dealing with weed species, such as white pickle, black pickle, viola, and wild peanut, which are a serious problem in onion production. This can be attributed to the lack of soil disturbance that favors these weed species under conventional cultivation. It is also the result of the large amounts of cover crop residue on the soil surface during the onion production period, which prevents weeds from becoming established, as well as the fact that some cover crops contain phenolic compounds with allelopathic properties which inhibit weed growth (Souza, 2017). In particular, weed pressure is reduced during the first 45 days after transplanting of onion seedlings, which corresponds to the onset of the critical onion development period.

In the Experimental Station at EPAGRI in Ituporanga (SC), Vilanova et al. (2014) tested different cover crop species on onion cultivation in a humic Haplumbrept soil. The treatments evaluated were: T1: weedy control; T2: oat (*Avena strigosa* Schreb); T3: rye (*Secale cereale* L.); T4: radish (*Raphanus sativus* L.); T5: radish + rye; and T6: radish + oat.

The results showed that radish + rye at 60 DAS, and radish + rye at 120 DAS, promoted highest cover crop biomass, while the lowest amount of weed DM was after radish + rye, followed by radish combined with early season onion. The highest yield of class 3 onion bulbs (50—70 mm), compared to the control, was also obtained using these two treatments (Table 4).

7 NTV cultivation of onions: seedling production

The production of seedlings is carried out from December to April in beds with cover crops that are rolled down. The beds for seedling production should be in a field that is easily accessible, preferably flat, north-facing and without shade, weed-free, and near a water source. It is recommended that seedling and crop production be carried out in the same field every 3 years to reduce potential disease problems (EPAGRI, 2013).

The beds should be planted in areas with a neutral soil pH, as well as sufficient phosphorus and potassium levels. Beds with a height of 10–15 cm and width of 1.0–1.2 m should follow the contours of the field so that they remain broadly level. It is recommended that the surface of the beds be even

Table 4 Onion production (bulbs and total yield) (t ha^{-1})

Treatments	Class 3	Class 2	Discard (*no commercial)	Total
T1	1.08c	9.33a	0.0b	10.41c
T2	1.40bc	9.47a	0.04b	10.91bc
T3	2.19abc	10.21a	0.07b	12.47abc
T4	3.64a	9.56a	0.33a	13.53a
T5	2.77ab	9.94a	0.17ab	12.88ab
T6	2.33abc	9.34a	0.15ab	11.82abc
CV	48.8	8.9	130.0	13.0

Source: Vilanova et al. (2014).

but with a slight slope to ensure sufficient drainage. The cover crop is broadcast or sown with a no-tillage planter (Fig. 22). When the seedlings are 20 cm high, an organic fertilizer can be applied to provide essential nutrients if needed.

Broadcast seeding is the method used by most farmers in Santa Catarina State. Beds can be prepared using small transverse grooves, 1–1.5 cm deep, open transversely or longitudinally on the bed, spaced 10 cm apart, in which the seeds will be evenly distributed. The recommended sowing rate is 2.5–3.0 g of seeds per m² of bed. About 2 kg of seed is needed to obtain seedlings for transplanting 1 ha of onion crop. It is important to note that smaller plant densities will facilitate better aeration to reduce disease pressure. Seeds are spread on the beds and covered with a maximum 2 cm of sawdust, incinerated rice husk, organic compost, or vermicompost (earthworm humus) to provide protection and promote growth.

Figure 22 Summer and winter cover crops planted on beds for seedlings production for transplanting. The system can be implemented with or without solarization. Sowing with solarization is carried out after removal of the plastic and, without solarization, after cover crop rolling, taking into account the sowing times recommended for each cultivar.

8 NTV cultivation of onions: photoperiod and nutrient requirements

The daily amount of available light (photoperiod) is a key factor in bulb formation by onion plants. Onion cultivars are classified into three groups based on the amount of light they need: short-, intermediate-, and long-day onions. Cultivars are also divided into early, medium, and late types, depending on the maturation cycle of the onion. Cultivars should be chosen which match local photoperiod conditions, bearing in mind factors such as latitude and time of the year, if production is not to be compromised (Alves et al., 2016).

Nutrient requirements vary according to the stage of development which can be divided into two basic stages:

- vegetative growth (from planting through to the beginning of bulbification, which accounts for approximately 13% of the plant's total nutrient requirement);
- bulbification (from the onset of bulbification to harvest), which accounts for the remaining 87%.

Key nutrient requirements in the whole plant, in descending order, are K, N, Ca, P, and Mg in amounts of 329, 288, 100, 53, and 30 mg plant^{1}, followed by the microelements Mn, B, Zn, and Cu in amounts of 472, 332, 314, and 158 μg plant^{1}, respectively. In the bulb the order is N, K, P, Ca, Mg, Mn, Zn, B and Cu in amounts of 232, 222, 49; 39 and 20 mg plant^{1} and 304; 293; 209 and 132 μg plant^{1}, respectively. In no-tillage systems with proper crop rotation, higher levels of organic matter, phosphorus, potassium, calcium, and magnesium are usually observed in the surface soil than in deeper layers (Sá, 1993; De Maria et al., 1999; Silveira and Stone, 2001). Other important changes in NTV systems are an increase in cation exchange capacity and the reduction of aluminum toxicity.

9 NTV cultivation of onions: planting

Spacing between planting lines will vary depending on the planting equipment (Fig. 23). In transplanted onions, the distance between rows is in the range of 33–40 cm with spacing of 7.5–10 cm between plants. Population densities will vary from 250 000 to 400 000 seedlings per ha, the most common being 300 000–330 000 plants per ha. In direct-seeded NTV, the spacings range from 25 to 30 cm between lines with 10–15 seeds per linear meter, achieving populations between 350 000 and 450 000 plants per ha.

In this system, onion seeds are planted directly in the production field.

Direct sowing in the field can be accomplished with machines and implements adapted to NTV cultivation, mainly by having a cutting coulter to

Figure 23 Transplanting of onion seedlings on black oat straw with tilled soil restricted to the planting line at Valdemar L. da Silva Farm in Alfredo Wagner (SC) (Photo: Edio Zunino Sgrott).

cut through straw. It is important to ensure a uniform sowing depth to ensure uniform emergence. This directly influences the final stand of the crop and, consequently, yield and bulb quality (Fig. 24). The line spacing used ranges from 25 cm to 35 cm, with an average of 30 cm, with a seeding density of 18 seeds per linear meter, aiming to establish 10–14 plants per linear meter, which will provide a population density of approximately 400 000 plants per ha

Figure 24 Onion direct seeding in NTV on millet straw with around 8 t ha^{-1} of dry mass.

(Fig. 25). The challenge is to improve the performance of direct onion seeding in the NTV system at high residue levels (10–12 t ha^{-1} dry mass). Other research priorities in the no-till system are the management of cover crops associated with sowing machines and weed management.

10 NTV cultivation of onions: fertilization

10.1 Transplant system

When onions are transplanted, research showed that the amount of N had a significant effect on bulb yield (Kurtz et al., 2013). Doses that provided the maximum economic return were found to be 131 kg ha^{-1} and 102 kg ha^{-1}, respectively. In the 2008/09 harvest, when there was a higher intensity of rainfall, postharvest losses were 25% when all the N was applied in one topdressing application, while losses were reduced to 15% when N topdressing was applied in three stages. Partitioning also influenced the architecture of plants with 5 cm taller plants in the three-split application than in the one-split application.

10.2 Direct seeding

In NTV with direct seeding, approximately 20% more N is needed because the growth cycle is 170 days instead of 120 days with transplanting. In addition, in the direct-seeded no-tillage system, larger plant populations are generally used than in the transplant system, thus increasing the N requirement (May et al., 2007).

Figure 25 Direct-sown onion in oat straw at harvest (170 days after sowing).

In the 2012/13 crop studied by Kurtz (2015), the average yield of the two cultivars used (Bola Precoce and Crioula) increased from 24.7 t ha^{-1} in the treatment without addition of N, to 29.0 t ha^{-1} at the dose of maximum technical efficiency (MET), that is, with the addition of 157 kg ha^{-1} of N, representing an increase of 17.3%. Yields increased most significantly in 2013/14 and 2014/15. The yield increased from 23.6 t ha^{-1} and 19.0 t ha^{-1} in the control without N, to 43.5 t ha^{-1} and 35.0 t ha^{-1} with N applications of 192 kg ha^{-1} and 200 kg ha^{-1}, respectively, representing an increase of 84% in the yield in both crops through addition of N. The required amounts of N for the maximum yield in this study were higher than those observed by May et al. (2007) in a study carried out in the state of São Paulo. This shows that the decision on the addition of N must consider, among other factors, the locality, cultivation conditions, and the cultivar. Recommendations for nitrogen addition based on SOM contents are shown in Table 5.

To produce yields greater than 30 t ha^{-1}, 4 kg of N ha^{-1} per additional ton of bulbs to be produced can be added (beyond the amounts shown in Table 5). In an established no-tillage system, where nitrogen cycling is better, N doses in Table 5 can be reduced by up to 25% due to better nitrogen cycling. In the transplant system, 15% of the N is applied at planting and the remaining 25%, 35%, and 25% at 35, 60, and 85 days after transplanting. In the direct sowing system, 20% higher doses (than those shown in Table 5) can be applied, with 20 kg ha^{-1} (~10%) applied at sowing and the rest of the dose divided equally over four applications at 45, 80, 110, and 135 DAS. These are minimum requirements and splitting fertilization over more stages is preferable. It is critical to ensure sufficient irrigation during fertilization to ensure sufficient foliar development for bulbification.

Research by Kurtz et al. (2016) found that phosphorus was the fourth most important nutrient for onion after N, K, and Ca with a concentration of 140.13 mg plant, particularly concentrated in the bulb. This requires application of 34.9 kg ha^{-1}or 80.3 kg ha^{-1} P_2O_5 among macronutrients. Onion crops respond particularly well to P addition, particularly as P is often present in low concentrations in soils (Ernani, 2016). The crop responds best to modest levels of P (4 mg dm^{-3}–extractor Mehlich-1) with no response to high levels of P in the soil (20 mg dm^{-3} and 32 mg dm^{-3}-Mehlich-1 extractor; Bissani et al., 2008). It is possible to reduce P fertilization once NTV systems are fully established since fewer nutrients are lost from the system.

Table 5 Nitrogen recommendation based on soil organic matter contents (CQFS-RS/SC, 2016)

Soil organic matter (%)	Nitrogen (kg de N ha^{-1})
≤2.5	120
2.6–5.0	100
>5.0	≤80

Potassium (mineral or organic) fertilizer can be applied prior to sowing or in the sowing/planting groove. It is not advisable to add doses of more than 60 kg ha^{-1} of potassium in the sowing or transplant line. For medium and high doses, it is best to apply 50% K at transplanting/sowing and the remainder with N at 60 and 85 days after transplanting or at 110 and 140 days after direct sowing. To achieve yields greater than 30 t ha^{-1}, 3 kg of K_2O per additional ton of bulbs is typically added, though it is important not to add excessive amounts.

10.3 Sulfur

Until recently, sulfur (S) deficiency was confused with the absence of other nutrients, mainly zinc (Kurtz, 2013). S deficiency, if not corrected immediately, can lead to a reduction of productivity of more than 50%. The lack of S is common in poorly managed sandy soils with low organic matter content, but also occurs in soils in NTV systems. Based on the research carried out by EPAGRI in partnership with UFPR on Cambisol soils (clay content <40%), it is recommended to add 60 kg ha^{-1} of S when S content in the soil is <15 mg dm^{-3}, 30 kg ha^{-1} of S when S contents are in the range of 15–20 mg dm^{-3}, while no S is needed at S content >20 mg dm^{-3}. For Nitisols (clay >50%), it is recommended to add 30 kg ha^{-1} of S when the S content is <10 mg dm^{-3} (Kurtz et al., 2018).

10.4 Micronutrients

Addition of zinc and manganese improves onion yields even in well-managed soils in NTV with medium and high levels of organic matter. In manganese-deficient areas, it is recommended to apply two to four foliar sprays with 0.5–1% manganese sulfate at intervals of 1–2 weeks. Soil application of 5–20 kg ha^{-1} Mn at planting is also possible but is less efficient due to nutrient immobilization.

It is recommended to incorporate 3–4 kg ha^{-1} Zn (possibly with NPK fertilizer) in the furrow prior to planting. Fertilization with Zn has a residual effect of at least 2 years. Another alternative to supply Zn is using manure and other organic fertilizers. Deficiencies of zinc during plant development can be alleviated with a 0.5% foliar zinc sulfate solution, applied two to four times in intervals of 1–2 weeks. Application of zinc sulfate can cause injury to the leaves when the environment is dry and hot, in which application should be late in the afternoon when conditions are cooler.

For intensively cultivated soils with low organic matter that have just been converted to NTV, boron (B) can be added to the soil at a rate of 1.5–2.0 kg ha^{-1} prior to planting. After 4–5 years of NTV cultivation, it may be possible to reduce B application. B can be applied by spraying or incorporation in the planting furrow together with NPK. Annual applications may be needed because its residual effect is small. Another alternative to supply B is the use of manure and other organic fertilizers. To overcome B deficiencies during plant development, two to four foliar applications of 0.25% boric acid or borax are recommended at intervals of 1–2 weeks.

11 NTV cultivation of onions: harvesting, curing, and storage

Since onion bulbs are highly perishable, they require optimal production techniques to ensure good-quality bulbs, timing harvest accurately, proper curing, care in handling and transportation, and efficient storage systems (EPAGRI, 2013). The onion harvest is one of the main factors influencing the quality of the product. The ideal point of harvest is when the bulb reaches physiological maturity, which is manifested by the tipping or popping of the plant due to the wilting of the pseudo stem (Fig. 26; EPAGRI, 2013).

The curing process consists of drying the external films and the pseudo stem (neck), making the bulb more resistant to damage and infections by microorganisms, improving the commercial quality of the onion, and increasing its storability. Curing serves to eliminate moisture in the outer surface of the bulbs and to dry the green parts, ensuring the cervix is closed, reducing the respiration rate and providing the characteristic color of each variety of onion. Well-cured bulbs can be stored for long periods of time and transported

Figure 26 Onion shoot indicating harvest point. In this field study, the leaves were cut to facilitate harvesting, transportation, and storage.

over long distances, while prematurely harvested or poorly cured onions are difficult to store or transport. Cure should be started in the field for a period of 3–14 days, depending on the weather. At the time of harvesting, the bulbs are arranged in rows with the leaves of some plants covering the others to protect them from direct sunlight, thus avoiding the development of green pigmentation and burns. Since the purpose of the cure is to dry the plants, it is important that the crop is free of weeds as shading makes it difficult to cure them. Light rain during the curing period does not normally affect the bulbs, but heavy rain means bulbs may not be cured properly or stored well. In rainy periods, the entire curing process can be carried out in sheds in well-ventilated sites. After curing, the bulbs should be collected and stored in wooden or plastic boxes, ventilation bags or bins (Fig. 27). In the store, the cure is complete in a few weeks, resulting in a dry, better-colored, and more easily preserved product. After curing, the bulbs can be immediately marketed or stored until sold (EPAGRI, 2013).

12 No-tillage vegetable (NTV) cultivation of parsley cassava (*Arracacia xanthorrhiza*, Bancroft)

Parsley cassava (*Arracacia xanthorrhiza*, Bancroft) originates from a region from north-central Peru to south of Colombia (Hermann, 1997). This humid tropical highland region is characterized by a very stable climate and the absence of distinct seasons, which allows continuous growth of the plant, with the opportunity for planting and harvesting at any time of the year. This has major

Figure 27 Collection of cured bulbs in a wooden box adapted for use in storage. Valdemar L. da Silva, Alfredo Wagner (SC).

implications for the physiology of the plant and the production system adopted in Brazil (Fig. 28). The tuberous roots are highly valued for their smooth texture and good taste. Parsley cassava has high levels of starch, good digestibility, and high levels of zinc, magnesium, calcium, potassium, phosphorus, iron, and vitamins A and B (Sediyama et al., 2005; Añes et al., 2002; Carmo and Leonel, 2012; Luengo, 2000). In countries such as Peru, the roots and leaves are a food source, while the stem and crown are also consumed in Puerto Rico and Brazil. The crop is also used as animal feed in countries such as Colombia and Brazil (Câmara, 1984)

Parsley cassava is traditionally cultivated at an altitude above 800 m in Southeast Brazil and above 650 m in South Brazil, with an annual average temperature varying between 15°C and 18°C. However, it is sometimes found at lower altitudes where the average annual temperature exceeds 20°C, although, at temperatures above 30°C, fungal and bacterial diseases become significant problems (Madeira et al., 2017). Parsley cassava has mild frost tolerance but does not survive severe and consecutive frosts, especially in the initial stage of growth when it is most vulnerable. Low temperatures and lack of water can lead to nutrient translocation (Fig. 29). These temperature constraints dictate planting.

Most planting in Southeast Brazil is from March to June to avoid losses during the hot and rainy seasons later in the year, and from June to August in South Brazil, avoiding frost damage in June and July, since the seedlings are still underground and thus protected. Planting using pre-rooted or pre-sprouted seedlings reduces the risk of losses and makes out-of-season production possible (Madeira and Carvalho, 2016).

Figure 28 Parsley cassava plants of the giant clone Angelina (A) and cv. Amaral Senador Amaral (B) harvested at 10 months of age, in the community of Rio Fortuna in Angelina-SC.

Figure 29 High relocation of mineral nutrients and photo assimilates is caused by low temperature and water deficit. Depending on the timing, intensity, and duration, this stress is beneficial to productivity.

13 NTV cultivation of parsley cassava: planting

The choice of management starts with the planting system, method of propagation, and spacing which will depend on the season, the characteristics of the area such as soil type and topography, variety, diversity of crops that the producer normally cultivates, and the equipment available. Seedling quality and soil management are fundamental for good crop performance (Fig. 30). Seedling quality contributes to an ideal stand and high parsley cassava root crop yield (Fig. 31). If grown on widely spaced ridges, it is possible to plant two lines of parsley cassava. In areas that have been in NTV cultivation for a long time, it is possible to plant seedlings on flat land without ridges. The spacing will depend in particular on the variety and the system. It is common in NTV systems to plant on ridges with cover crops of oat or rye followed by parsley cassava, and then maize and finally tobacco. The ridges are approximately 1.2 m apart, from center to center. In this case, it is possible to plant two rows of parsley cassava on the ridges, with row spacing of 50 cm and 30–40 cm between plants in the rows. The arrangement on the ridges should be in a zigzag pattern, in the case of the varieties of normal size such as 'Yellow of Senador Amaral', 'Rúbia', or 'Catarina' varieties. In the case of larger varieties such as Giant Angelina and Carijó, it is best to plant only one row on each ridge with spacing of 50 cm between plants in rows.

Figure 30 Planting on a large ridge with two planting lines spaced apart by 50 cm and 120 cm between the ridges.

Figure 31 Parsley cassava cultivated in an established NTV system on flat terrain without ridges.

14 NTV cultivation of parsley cassava: mechanization

There is a great need for better equipment for smallholder farmers using NTV systems. This equipment has to be able to work with different crops, topographies, and size of cultivated areas and large quantities of straw of over 10 tons of DM ha^{-1} $year^{-1}$. Some examples of equipment developed by farmers, extension personnel, and researchers specifically for NTV cultivation are shown in Figs. 32–35.

It is recommended to roll down and cut the cover crop, prepare a planting furrow, and plant the parsley cassava seedlings on the same day, aiming to keep the green cover crops for a longer period to feed the soil macrobiota, mainly insects that can feed on cultivated plants. It is also important to plant a green cover crop, because it is difficult to cut through partially wilted cover crop biomass. In the latter case, it is necessary to wait for the straw to dry, which takes approximately 20–30 days, when weeds will start to appear, requiring herbicide use. In the first situation, vegetable seedlings develop before weeds come up, so no herbicide is needed. Some producers, with advanced NTV systems,

Figure 32 Knife-rollers made out of a hollow metal tube filled with water (A and C), and a solid pipe of concrete (B). Type A and C cut the straw to quickly kill cover crops and accelerate straw degradation, and reduce the duration of ground cover. Knife-roller B only crimps the stems, prolonging the effect of the ground cover, keeping the flattened plants alive for a longer period.

Figure 33 Animal traction NTV equipment for hilly areas, both of which cut through cover crops and prepare a planting slot, and apply fertilizer in the furrow (second type). These provide high-performance equipment without the additional cost of machines for traction and fuel expenditure, in addition to minimizing soil compaction.

Figure 34 Micro tractor traction equipment developed to cut through cover crops and prepare a planting furrow. The first with only one line of work that performs fertilizer distribution and uses a 17″ rippled coulter and knife behind it; and the second with two working lines 80 cm between them, iron wheel and 32″ cutting coulter that replaces the tire, also with knife shanks. The equipment superficially exposes surface soil and when used in proper soil moisture condition does not cause sidewall compaction.

Figure 35 Details of two pieces of equipment. The first shows a bubbled coulter followed by a shank with 2 cm wide foot, which causes fracturing of the soil from the bottom up as an 'inverted ridge', with the advantage of avoiding exposure of the soil provided it is used at the correct working speed and soil moisture. The second shows a smooth cutting coulter and wheels that hold firm the straw to facilitate cutting.

perform the rolling of the material and planting of the seedlings without cutting the straw and without disturbing the soil in the planting line (Fig. 36).

15 NTV cultivation of parsley cassava: fertilization

Parsley cassava accumulates nutrients and DM throughout the crop cycle. The dynamics of nutrient absorption, particularly for nitrogen and potassium, suggest these nutrients should be supplied in installments. The accumulation of nutrients up to 90 days after planting (DAP) is less than 3% of the total, 24% of the total from 90 to 210 DAP, and 75% up to 360 DAP. This provides the basis to build fertilization programs according to the needs of the plant in its various stages of development. When starting a NTV cultivation, it is important to look for areas of nutrient/fertility deficiency in the soil, which can be corrected by adding organic fertilizers in the initial stage of development (Fig. 37).

Figure 36 Parsley cassava seedling planting without cutting the straw and without soil disturbance in planting line in Angelina-SC State, Brazil, 2018.

Figure 37 At this stage of cover crops, development is possible to identify fertility patches and correct them with organic fertilizers (Angelina - SC, 2013).

16 NTV cultivation of parsley cassava: harvesting, sorting, and storage

Harvesting can be entirely manual, which is more common in small and hilly areas of cultivation, or semi-mechanized in larger areas with a flatter terrain. Manual harvesting involves pulling out of the plants with the aid of a hoe or digger when the plants are more attached to the soil. In semi-mechanized harvesting, a moldboard plow, subsoiler, or blade (plate) is used below the roots to lift them, making them easier to harvest.

17 No-tillage vegetable (NTV) cultivation of other vegetables

To promote crop diversity in NTV, farmers developed a rototiller that cuts a slot and slashes the cover crop on the soil surface in order to plant or transplant vegetables (Breda Jr. and Factor, 2009). Mr Airton Benedito Feltran in São Jose do Rio Pardo, SP, started NTV cultivation with onion, and moved on to grow beet and other crops such as pumpkin, cabbage, and carrot, showing that it is possible to achieve superior vegetable yields with NTV systems compared with conventional production (Fig. 38). Padovan et al. (2007), in an experimental NTV organic system with different cover crops in Savannah, Brazil, found that using sunn hemp before cabbage promoted high cabbage yields (Table 6).

Girardello et al. (2017) grew lettuce using NTV cultivation with different cover crops: black oat (*Avena strigosa* Schreb), hairy vetch (*Vicia sativa* L.), white lupin (*Lupinus albus* L.), black oat + hairy vetch, black oat + white lupin and control (fallow) in Alpestre (RS), South Brazil (27°14′42″ S lat, 53°1′41″ W long, 471 masl, Koppen climate class (Cfa). They achieved highest lettuce yields

Figure 38 Cabbage in no till after transplanting and during growing season (São José do Rio Pardo, São Paulo State, Brazil).

Table 6 No-till cabbage in an organic system using different cover crops (Dourados Mato Grosso do Sul State, Brazil. 2006)

	Aerial part of the cabbage				
Treatments	Fresh biomass (kg ha^{-1})	Cabbage 'head' (kg per $plant^{-1}$)	Horizontal diameter (cm)	Vertical diameter (cm)	Cabbage yield (t ha^{-1})
Clean plot (no mulch)	1.25a	0.92a	13.70a	11.35a	28.58c
Sunn hemp (Crotalaria juncea)	2.04a	1.35a	14.08a	11.95a	44.95a
Jack bean (Canavalia ensiformis)	1.75a	1.04a	13.95a	11.58a	34.88abc
Dwarf pigeon pea (Cajanus cajan)	1.97a	1.22a	14.05a	11.78a	40.95ab
Pearl millet (Pennisetum glaucum)	1.50a	1.00a	14.00a	11.68a	33.96abc
Sunn hemp + Pearl millet	2.00a	1.30a	14.15a	11.98a	43.60ab
Fallow (spontaneous vegetation)	1.75a	1.09a	14.03a	11.40a	32.10bc

Source: Padovan et al. (2007). T (5%).

growing the lettuce in vetch + black oat (Girardello et al., 2017). Hirata and Hirata (2015) found that 10 t ha^{-1} DM of pearl millet (*Pennisetum glaucum*) over the soil provided adequate weed control for successful lettuce production in the summer season.

18 Conclusion and future trends

This chapter has shown how it is possible to make the transition from conventional vegetable cultivation based on significant use of agrochemicals (fertilizers and pesticides) to NTV systems based on ecological principles promoting soil health as the basis for successful crop production. This transition has been possible because of effective collaboration among farmers, between farmers and the research community, and between both these groups and RETA agents. The result has been the evolution of what has been called the green no-tillage system (Fig. 39).

To build on this foundation, there needs to be more research in such areas as substituting synthetic fertilizers for organic alternatives that fully meet daily plant growth requirements according to NTV principles. There needs, for

Figure 39 Characteristics of tomato cultivation in a green no-till system.

example, to be more research on the composition, durability, and efficacy of Bordeaux mixture. Another area of research is in understanding and developing fertilizers that perform different functions, whether promoting root and plant development (including stimulating the rhizosphere), or improving the sensory and nutritional properties of the crop. In the case of the cultivation of crops such as tomato, there also needs to be more research on how best to manage pests such as the small fruit borer in NTV systems through cultural and other means, and with less reliance on pesticides.

19 References

Alves, D. P., Wamser, G. H. and Oliveira, V. R. (2016), Escolha do cultivar. In: Menezes Junior, F. O. G. and Marcuzzo, L. L. (Orgs.), *Manual de Práticas Agrícolas: Guia para a sustentabilidade das lavouras de cebola do estado de Santa Catarina*. Florianópolis: Epagri, 143p.

Amézketa, E. (1999), Soil aggregate stability: a review. *Journal of Sustainable Agriculture*, 14(2-3), 83-151. doi:10.1300/J064v14n02_08.

Andrade, A. G. (2010), *Manejo e conservação do solo e da água no contexto das mudanças ambientais*. Rio De Janeiro: Embrapa Solos, pp. 25-40.

Añes, B., Espinoza, W. and Vásquez, J. (2002), Producción de Apio Andino em Respuesta al suministro de fertilizantes. *Revista Forestal Venezolana, Mérida*, 46(2), 39-45.

Ball, B. C., Bingham, I., Rees, R. M., Watson, C. A. and Litterick, A. (2005), The role of crop rotations in determining soil structure and crop growth conditions. *Canadian Journal of Soil Science*, 85, 557-77.

Balota, E. L., Calegari, A., Nakatani, A. S. and Coyne, M. S. (2014), Benefits of winter cover crops and no-tillage for microbial parameters in a Brazilian Oxisol: A long-term study. *Agriculture, Ecosystems and Environment*, 197, 31–40.

Balwinder-Singh, Eberbach, P. L., Humphreys, E. and Kukal, S. S. (2011), The effect of rice straw mulch on evapotranspiration, transpiration and soil evaporation of irrigated wheat in Punjab, India. *Agricultural Water Management*, 98, 1847–55.

Bissani, C. A., Gianello, C., Camargo, F. A. O. and Tedesco, M. J. (2008), *Fertilidade dos solos e manejo da adubação de culturas* (2ª. Ed.). Porto Alegre: Metrópole, 344p.

Bolliger, A., Magid, J., Amado, T. J. C., Skóra Neto, F., Santos Ribeiro, M. F., Calegari, A., Ralisch, R. and Neergaard, A. (2006), Taking stock of the Brazilian "zero till revolution": a review of landmark research and farmers' practice. *Advances in Agronomy*, 91, 47–110.

Breda Jr., J. M. and Factor, T. L. (2009), Oportunidades e dificuldades no plantio direto de hortaliças: O caso de São José do Rio Pardo. *Horticultura Brasileira*, 27(2 (Suplemento – CD Rural) agosto), 4033.

Calegari, A. (1989), Adubação verde. In: *Manual técnico do Subprograma de Manejo e Conservação de solo*. Curitiba, PR: Secretaria de Agricultura e Abastecimento, pp. 178–85.

Calegari, A. (2010), Alternativas de manejo de cobertura do solo em pomares. In: *"Agropecuária Catarinense – 9o SENAFRUT"*, 22 a 24 de julho. V. 23, no. 2, julho 2010. Florianópolis, SC, pp. 98–109.

Calegari, A. (2014), Perspectivas e estratégias para a sustentabilidade e o aumento da biodiversidade dos sistemas agrícolas com o uso de adubos verdes. In: Lima Filho, O. F. de, Ambrosano, E. J., Rossi, F. and Carlos, J. A. D. (Eds), *Adubação verde e plantas de cobertura no Brasil: fundamentos e prática*. Brasília, DF: Embrapa, v. 1, Cap. 1, pp. 21–36.

Calegari, A. (2016), *Plantas de cobertura. Manual Técnico* (3ª. Ed.). Uberaba, MG: Penergetic, 24p.

Calegari, A. and Costa, A. (2010a), Manutenção da cobertura melhora atributos do solo. In: Visão Agrícola. USP, ESALQ, Ano 6, Jul/Dez., Vol. 9. Piracicaba, SP, pp. 13–16.

Calegari, A. and Costa, A. (2010b), Sistemas conservacionistas de uso do solo. In: *"Manejo e conservação do solo e da água no contexto das mudanças ambientais". Organizado por Rachel Bardy Prado, Ana Paula Turetta e Aluísio Granato de Andrade*. Rio de Janeiro: Embrapa Solos, pp. 279–308.

Calegari, A. and Donizeti Carlos, J. A. (2014), Recomendações de plantio e informações gerais sobre o uso de espécies para adubação verde no Brasil. In: Lima Filho, O. F. de, Ambrosano, E. J., Rossi, F. and Carlos, J. A. D. (Eds), *Adubação verde e plantas de cobertura no Brasil: fundamentos e prática*. Brasília, DF: Embrapa, v. 2, Cap. 27, pp. 453–78.

Calegari, A., Mondardo, A., Bulisani, E. A., Wildner, L. Do P., Costa, M. B. B., Alcântara, P. B., Miyasaka, S. and Amado, T. J. C. (1993), *Adubação verde no Sul do Brasil* (2a. Ed.). Rio de Janeiro, RJ: AS-PTA, 346p.

Calegari, A., Tourdonnet, S., Tessier, D., Rheinheimer, D. S., Ralisch, R., Hargrove, W., Guimarães, M. F. and Tavares Filho, J. (2013), Influence of soil management and crop rotation on physical properties in a long-term experiment in Paraná, Brazil. *Communications in Soil Science and Plant Analysis*, 44, 2019–31. doi:10.1080/00103624.2013.790406.

Câmara, F. L. A. (1984). *Estudo de tecnologias objetivando precocidade de produção de batata-baroa (Arracacia xanthorrhiza Bancroft)*. Viçosa: UFV, 50p (Tese mestrado).

Campbell, C. A., Lafond, G. P., Vandenbygaart, A. J., Zentner, R. P., Lemke, R., May, W. E. and Holzapfel, C. B. (2011), Effect of crop rotation, fertilizer and tillage management on spring wheat grain yield and N and P content in a thin Black Chernozem: a long-term study. *Canadian Journal of Plant Science*, 91, 467–83.

Carmo, E. L. and Leonel, M. (2012), Composição físico-química e cor de clones de mandioquinha-salsa. *Revista Energia na Agricultura, Botucatu*, 27(1), pp. 62–81.

CQFS – Comissão de química e fertilidade do solo – RS/SC (2016), *Manual de calagem e adubação para os estados do Rio grande do Sul e de Santa Catarina* (11ª. Ed.). Santa Maria: Sociedade Brasília de Ciência do Solo – Núcleo Regional Sul, 376p.

De Maria, I. C., Nnabude, P. C. and Castro, O. M. de (1999), Long term tillage and crop rotation effects on soil chemical properties of a Rhodic Ferralsol in southern Brazil. *Soil and Tillage Research*, 51(1), 71–9.

Derpsch, R. and Calegari, A. (1985), *Guia de plantas para adubação verde de inverno*. Londrina: IAPAR, 96p (IAPAR, Documentos, 9).

EPAGRI (2013), *Sistema de produção para a cebola – Santa Catarina*. Florianópolis: Epagri, 106p (Sistemas de Produção, 46).

Ernani, P. R. (2016), *Química do solo e disponibilidade de nutrientes às plantas* (2ª. Ed.). Lages (SC), 256p.

Fayad, J. A., Comin, J. J. and Bertol, I. (2016), *Sistema de Plantio Direto de Hortaliças (SPDH): O cultivo de brássica: cove-flor, brócolis e repolho*. Florianópolis: Epagri, 86p (Epagri. Boletim Didático, 132).

Florentin, M. A., Peñalva, M., Calegari A. and Derpsch, R. (2010), Green manure/cover crops and crop rotation in conservation agriculture on small farmers. Integrated Crop Management (vol. 12). FAO, Rome, 97p, ISBN 978-92-5-106839-7; ISSN 1020-4555.

Fontanetto, H. and Keller, O. (2001), Efecto de diferentes labranzas sobre propiedades edáficas de un argiudol. In: Díaz Rossello, R. (Ed.), *Siembra directa em el Cono Sur*. Montevideo, Uruguay: PROCISUR, pp. 275–87.

Gatiboni, L. C., Coimbra, J. L. M., Wildner, L. P. and Denardin, R. B. N. (2009), Modificações na fauna edáfica durante a decomposição da palhada de centeio e aveia preta em sistema plantio direto. *Revista Biotemas, Florianópolis*, 22(2), 45–53.

Gava, R., de Freitas, P. S. L., de Faria, R. T., Rezende, R. and Frizzone, J. A. (2013), Soil water evaporation under densities of coverage with vegetable residue. *Engenharia Agrícola*, 33, pp. 89–98.

Girardello, R., Silva, D. M., Guerra, D., Lanzanova, M. E. and Souza, E. L. (2017), Produção de alface sob plantio direto em sistema de transição agroecológica [Lettuce production under plantation direct in agroecological transition system]. *Revista Verde de Agroecologia e Desenvolvimento Sustentável*, 12(2), 273–9. Pombal, PB: Grupo Verde de Agroecologia e Abelhas. Available at: http://www.gvaa.com.br/revista/index.php/RVADS. doi:10.18378/rvads.v12i2.4827.

Hermann, M. (1997), Arracacha (*Arracacia xanthorrhiza* Bancroft) Andean roots and tubers: Ahipa, arracacha, maca and yacon. Promoting the conservation and use of underutilized and neglected crops. No. 21. Gatersleben, Germany e Rome, Italy: EPAGRI, pp. 75–172.

Hirata, A. C. S. and Hirata, E. K. (2015), Manejo do milheto para plantio direto de alface americana no verão. *Pesquisa & Tecnologia*, 12(1).

Jones, D. L., Hodge, A. and Kuzyakov, Y. (2004), Plant and mycorrhizal regulation of rhizodeposition. *New Phytologist*, 163, 459–80.

Kurtz, C. (2013), Epagri identifica novo problema nutricional da cultura da cebola. *Jornal A Comarca - Ituporanga (SC)*, 10 a 16 de outubro, p. 11.

Kurtz, C. (2015), Acúmulo de nutrientes e métodos de diagnose nutricional de nitrogênio para a cultura da cebola. Curitiba, PR: Universidade Federal do Paraná - UFPR, 96p (Tese de Doutorado).

Kurtz, C., Ernani, P. R., Pauletti, V., Menezes Junior, F. O. G. and Vieira Neto, J. (2013), Produtividade e conservação de cebola afetadas pela adubação nitrogenada no sistema de plantio direto. *Horticultura Brasileira*, 31, 559–67.

Kurtz, C., Pauletti, V., Fayad, J. A. and Vieira Neto, J. (2016), Crescimento e absorção de nutriente pela cultivar de cebola Bola Precoce. *Hortivultura Brasileira*, 34, 279–88.

Kurtz, C., Menezes Junior, F. O. G. and Higashikawa, F. S. (2018), Fertilidade do solo, adubação e nutrição da cultura da cebola. Florianópolis: EPAGRI, 104p (EPAGRI, Boletim Técnico, 184).

Liu, T., Chen, X., Hu, F., Ran, W., Shen, Q., Li, H. and Whalen, J. K. (2016), Carbon-rich organic fertilizers to increase soil biodiversity: evidence from a meta-analysis of nematode communities. *Agriculture, Ecosystems and Environment*, 232, 199–207.

Loss, A., Basso, A., Oliveira, B. S., Koucher, L. P., Oliveira, R. A., Kurtz, C., Lovato, P. E., Curmi, P., Brunetto, G. and Comin, J. J. (2015), Carbono orgânico total e agregação do solo em sistema de plantio direto agroecológico e convencional de cebola. *Revista Brasileira de Ciência do Solo*, 39, 1212–24.

Loss, A., Schmitz, D., dos Santos Junior, E., Kurtz, C., Da Veiga, M. and Comin, J. J. (2017), Atributos físicos do solo sob sistemas de plantio direto e preparo convencional em cultivo de cebola. *Revista Colombiana de Ciencias Horticolas*, 11, 105–13.

Luengo, R. F. A. (2000), Tabela de composição nutricional das hortaliças. Embrapa Brasília: Ed. 2 al.]. ...[et Luengo 26) Documentos, Hortaliças, (Embrapa tabela Til.: 4p. 1415-2312 ISSN Série. III. Título. II. Tabela. T- nutricional Composição - Hortaliça 1. CDD 641.1 (19.ed.).

Madeira, N. R. and Carvalho, A. D. F. (2016) Produção de mudas de mandioquinha- salsa. In: Nascimento, W. M. and Pereira, R. B. (Eds), *Hortaliças de propagação vegetativa: tecnologia de multiplicação*. Brasília, DF: Embrapa Hortaliças, pp. 129–50.

Madeira, N. R., Carvalho, A. D. F., Silva, G. O., Pinheiro, J. B., Pereira, R. B., Michereff Filho, M., Feldberg, N. P., Moreira, S. O., Silveira, G. S. R. and Cássia, R. M. (2017), *Proposição de um sistema de produção de mudas de mandioquinha-salsa*. Brasília, DF: Embrapa Hortaliças, 24p (Circular Técnica 161).

Marschner, H. (2012), *Mineral nutrition of higher plants* (3ª. Ed.). London: Elsevier, 643p.

May, A., Cecílio Filho, A. B., Porto, D. R. Q., Vargas, P. F. and Barbosa, A. C. B. (2007), Produtividade de híbridos de cebola em função da população de plantas e de fertilização nitrogenada e potássica. *Horticultura Brasileira*, 25, 53–9.

Melo, A. V., Galvão, J. C. C., Braun, H., Santos, M. M., Coimbra, R. R., Silva, R. R. and Reis, W. F. (2011), Extração de nutrientes e produção de biomassa de aveia- preta cultivada em solo submetido a dezoito anos de adubação orgânica e mineral. *Semina: Ciências Agrárias*, 32(2), 411–20.

Monegat, C. (1991), *Plantas de cobertura do solo - características e manejo em pequenas propriedades*. Chapecó: Edição do Autor, 333p.

Oades, J. M. (1993), The role of biology in the formation, stabilization and degradation of soil structure. *Geoderma*, 56, 377–400.

Oliveira, R. A., Brunetto, G., Loss, A., Gatiboni, L. C., Kurtz, C., Muller, V. M., Lovato, P. E., Oliveira, B. S., Souza, M. and Comin, J. J. (2016a), Decomposição e liberação de nutrientes de plantas de cobertura e seus efeitos em atributos químicos do solo e produtividade da cebola. *Revista Brasileira Ciência do Solo*, 40, 1-17.

Oliveira, R. A., Brunetto, G., Loss, A., Gatiboni, L. C., Kurtz, C., Muller Junior, V., Lovato, P. E., Oliveira, B. S., Souza, M. and Comin, J. J. (2016b), Cover crops effects on soil chemical properties and onion yield. *Revista Brasileira de Ciência do Solo (Online)*, 40, 1-17.

Padovan, M. P., Cesar, M. N. Z. and Alovisi, A. M. T. (2007), Plantio direto de repolho sobre a palhada de adubos verdes num sistema sob manejo orgânico. Direct sowing of cabbage under straw of green manure in an organic management system. Resumos do V CBA - Manejo de Agroecossistemas Sustentáveis, Rev. Bras. de Agroecologia/ out. 2007. Vol. 2, No. 2.

Rillig, M. C. and Mummey, D. L. (2006), Mycorrhizas and soil structure. *New Phytologist*, 171(1), pp. 41-53.

Sá, J. C. M. (1993), Manejo da fertilidade do solo no sistema plantio doireto. In: EMBRAPA. Centro Nacional de Pesquisa de Trigo (Passo Fundo, RS). Plantio direto no Brasil. Passo Fundo: EMBRAPA- CNPT / Fundacep - FECOTRIGO / Fundação ABC, pp. 37-60.

Santos, L. H., Loss, A., Lourenzi, C. R., Souza, M., Gonzatto, R., Kurtz, C., Brunetto, G. and Comin, J. J. (2018), Total nitrogen and humic substances in aggregates of soils with onion crops under no-tillage and conventional tillage systems. *Revsta Colombiana De Ciencias Horticolas*, 12, 166-74.

Sediyama, M. A. N., et al. (2005), Cultura da mandioquinha-salsa ou batata-baroa. Boletim Técnico da Empresa de Pesquisa Agropecuária de Minas Gerais, Belo Horizonte, n. 77, pp. 1-28.

Silveira, P. M. da and Stone, L. F. (2001), Teores de nutrientes e de matéria orgânica afetados pela rotação de culturas e sistemas de preparo do solo. *Revista Brasileira de Ciência do Solo*, 25(2), 387-94.

Six, J., Bossuyt, H., Degryze, S. and Denef, K. (2004), A history of research on the link between (micro) aggregates, soil biota, and soil organic matter dynamics. *Soil and Tillage Research*, 79, 7-31.

Souza, M. (2017), Desempenho agronômico e estudo fitoquímico de plantas de cobertura em sistema de plantio direto agroecológico de cebola. Tese (Doutorado) - Universidade Federal de Santa Catarina, Centro de Ciências Agrárias, Programa de Pós-Graduação em Agroecossistemas, Florianópolis, 210p.

Stanger, T. F. and Lauer, J. G. (2008), Corn grain yield response to crop rotation and nitrogen over 35 years. *Agronomy Journal*, 100, 643-50.

Teasdale J. R., Brandsaeter, L. O., Calegari, A. and Skora Neto, F. (2007), Cover crops and weed management. In: Upadhyaya, M. K. and Blackshaw, R. E. (Eds), *Non-chemical Weed Management: Principles, Concepts and Technology*. Reading, UK, pp. 49-64.

Vargas, M. M. M. (2012), Atributos químicos e biológicos do solo e rendimento da cebola em sistema de plantio direto após cultivo com diferentes plantas de cobertura de inverno. Dissertação (Mestrado) - Universidade Federal de Santa Catarina, Centro de Ciências Agrárias. Programa de Pós - Graduação em Agroecossistemas, Florianópolis, 37p.

Vilanova, C. C., Comin, J. J., Kurtz, C., Müller Júnior, V., Uriarte, J. F., Ventura, B. S., Souza, M., Lovato, P. E. Leguizamon, E. and Brunetto, G. (2014), Interferência de plantas de cobertura sobre a incidência de plantas invasoras e a produção de cebola sob sistema de plantio direto- Interference of cover crops on the weed's incidence and onion yield under no-tillage. *Scientia Agraria*. ISSN 1519-1125.

Wütke, E. B., Calegari, A., Wildner, L. P. and Amabile, R. F. (2008). Leguminosas alimentícias e adubos verdes. In: Albuquerque, A. C. S. and da Silva, A. G. (Eds), *"Agricultura Tropical: quatro décadas de inovações tecnológicas, institucionais e políticas". Vol. 1 Produção e produtividade agrícola*. Brasília, DF: Embrapa Informação Tecnológica, pp. 251–74.

Wütke, E. B., Calegari, A. and Wildner, L. Do P. (2014). Espécies de adubos verdes e plantas de cobertura e recomendações para uso. In: Lima Filho, O. F. de, Ambrosanao, E. J., Rossi, F. and Carlos, J. A. D. (Eds), *Adubação verde e plantas de cobertura no Brasil: fundamentos e prática*. Brasília, DF: Embrapa (vol. 1), pp. 59–168.

Chapter 8

Managing perennial Conservation Agriculture systems: orchards, plantations and agroforestry

Amir Kassam, University of Reading, UK; Emilio J. González-Sánchez, University of Cordoba and Asociación Española de Agricultura de Conservación Suelos Vivos (AEACSV), Spain and European Conservation Agriculture Federation, Belgium; Sim Choon Cheak, Sime Darby Research Sdn. Bhd, Malaysia; Zaharah A. Rahman, formerly Putra University, Malaysia; Julio Roman-Vazquez and Francisco Marquez-Garcia, University of Cordoba, Spain; Rosa Carbonell-Bojollo, Centro IFAPA Alameda del Obispo, Spain; Oscar Veroz-Gonzalez, Asociación Española de Agricultura de Conservación Suelos Vivos (AEACSV), Spain; and Dennis P. Garrity, World Agroforestry Centre (ICRAF), Kenya

1 Introduction

The purpose of this chapter is to describe a selection of Conservation Agriculture (CA) systems that are based on perennial crops in association with other crops that could be perennial or annual. Such systems include orchards, plantations, agroforestry, including silvo-pastoral and pasture systems.

The concepts and principles of CA (Chapter 1, Volume 1) and their practical applications in CA systems have evolved and been tested over many years internationally. The initial impetus to move away from tillage agriculture came from the need to respond to the Dust Bowls and agricultural land degradation in the prairies in the mid-west of the United States during the 1930s as a result of the intensive tillage that followed the introduction of mouldboard ploughing which was followed by multi-year drought. The area affected had been opened

http://dx.doi.org/10.19103/AS.2019.0048.09

up for arable cropping, thus much of the interest in no-till system or CA was initially focussed on arable cropping systems. In the initial years, CA approach was adopted in North America and South America and over time, especially since 1990, CA systems for annual cropping have been adopted in all continents (Kassam et al., 2019).

However, over the last two decades, increasing attention has been directed to establishing CA-based systems with perennial crops particularly in orchard, plantation and agroforestry systems, including silvo-pasture systems in which woodland trees are combined with pasture and livestock. Permanent pasture and rangeland systems with livestock are the 'simplest' CA systems. These systems as well as silvo-pasture systems are not included in this chapter, as they are elaborated in Chapter 10, Volume 1.

The focus of this chapter is on management practices in CA-based perennial systems in different parts of the world and presents selected examples of perennial CA systems to illustrate their characteristics. These systems apply all the three principles of CA – no-till, soil cover and crop associations. These CA systems include perennial tree crops in orchards in Spain, with subtropical winter rainfall environment, such as olive orchards under sown with legumes such as hairy vetch or in association with sown or natural pastures or vegetation; perennial trees in plantation systems in Malaysia, with humid tropical environment, such as oil palm under sown with a legume such as mucuna, or rubber trees under sown with mucuna or with natural grasses; agroforestry systems in Africa, in semi-arid tropical environments, involving nitrogen-fixing perennial trees such as Faidherbia or Gliricidia in associations with cereals and legumes in Africa. There are other examples of perennial CA systems which are not covered in this chapter including vineyards and fruit tree orchards in Europe, citrus systems in association with under sown legume or grasses, or in association with annual crops or short-term perennials such as cassava in Argentina, agroforestry systems in Paraguay and shade coffee plantations in Central America.

2 CA-based olive orchard systems in Spain

This section presents results of many years of development of CA in orchards in Andalusia, Spain, comprising woody trees such as olives and vines. The Spanish Association of Conservation Agriculture Living Soils (in Spanish AEACSV) has widely published on the topic. The following sections present the essential ideas from the publications 'Cubiertas Vegetales en Olivar' (AEACSV, 2001; AEACSV-Junta de Andalucía, 2007), together with relevant results from other studies.

Woody orchard crop systems have undergone a significant development in the Mediterranean region (EUROSTAT, 2018), particularly the olive tree

(*Olea europaea*) system which has a widespread presence. The typology of the woody orchards is very diverse, although in many cases trees are rainfed, grown in relatively poor soils and with steep slopes. Such conditions make woody crops particularly sensitive to soil erosion and other degradation processes, with the possible loss of productivity and profitability which any kind of unchecked degradation entails (de Graaff and Eppink, 1999), resulting in a negative effect on the economy of farmers and their communities. This is of utmost importance, given the significant increase in production costs over the last decades (Gómez-Limón et al., 2012).

Soil erosion is a threat to sustainability in orchard production all over the world. Indeed, soil erosion is considered to be the greatest environmental problem for olive production in Spain (García-Ruiz, 2010). Consequently, soil conservation practices are essential as soil losses are high without them. Erosive rates higher than 400 Mg ha^{-1} in some extreme rainstorms have been measured in Spain (Vanwalleghem et al., 2010). This is due to the terrain characteristics of orchards or plantations, the frequent episodes of intense rain, the scarcity of soil mulch cover characteristic of tillage-based orchard systems (Pastor, 2008), together with the practice of intensive tillage itself, which is the soil management system most followed by olive farmers (Gómez et al., 2011; Taguas et al., 2011; Vanwalleghem et al., 2010).

To protect the soil from erosion, CA approach comprises the best soil management system, as it avoids tillage and promotes the maintenance of a permanent soil cover (FAO, 2018). Not only annual crops, but also woody perennial crops are considered to be suitable for CA, as reviewed below.

2.1 Groundcovers and their benefits in Conservation Agriculture

Groundcovers (GC) are an integral part of CA systems with woody perennial crops, whereby the soil surface between the rows of trees remains protected against erosion and water capture, infiltration and retention remain high. With this technique, at least 30% (minimum GC in CA systems) of the soil is protected either by sown cover crops, or natural vegetation or other covers, such as biomass from tree pruning or leaves from tree. For the establishment of sown cover crops and the spread of biomass covers, farmers should use methods that are compatible with CA principles (Gonzalez-Sanchez et al., 2015). GC are applicable to many woody crops, such as olives, vineyards, citrus, peach, apple and almond trees, among others.

Woody perennials keep the soil area under the canopy protected from the direct impact of raindrops and solar radiation throughout the year. However, the spaces between the rows of trees are exposed and prone to water erosion generated by the effect of these impacts. The majority of tree crops in Spain are rainfed, with water being the main limiting factor for production.

GC produce greater infiltration of rainwater, as well as reduce run-off and evaporation, and result in a more favourable distribution of water in the soil profile across the field. The overall soil moisture balance indicates that more water is usually available in a covered soil than in a tilled bare soil, provided the harvest operation is carried out at an appropriate time. In addition, the presence of dry plant biomass on the soil surface limits evaporation.

In the Andalusia region, landscapes with olive orchards have improved in terms of soil and water conservation due to CA (Gómez et al., 2011; Gómez-Limón et al., 2012). In many areas, immense gullies (Fig. 1) are the result of many years of soil and landscape degradation resulting from water erosion due to intensive tillage. Managing CA-based olive orchards in those areas during the past 20 years have changed the hydrology and the productivity of overall landscape for the better. Where the occurrence of gullies was frequent, more and more GC can now be found (Fig. 2).

The establishment of a vegetative GC results in a significant reduction in soil erosion that can reach up to 95%, and run-off, leading to a diminution of the loss of nutrients, resulting in an improvement of the quality of the surface water. Indeed, GC in orchards with woody perennial crops represent good agronomy as well as providing notable environmental benefits: soil structure is improved, organic matter is increased, atmospheric carbon is fixed into the soil, natural fertility is improved and soil moisture content and retention is increased. GC also lead to less pollution of the water and increase the biodiversity present in the soil and also above the ground. There are several studies that have demonstrated the effectiveness of GC in olive orchards to reduce run-off

Figure 1 Degraded soil in a typical olive orchard landscape in Cordoba, Spain.

Figure 2 Groundcover of grasses in an olive orchard in Cordoba, Spain.

and soil loss (Gómez et al., 2009a; Ordóñez et al., 2007; Francia et al., 2006), improve the hydrological balance for the crop (Alcántara et al., 2011; Durán-Zuazo et al., 2009), reduce the pollution of surface waters (Franklin et al., 2007; Ordóñez et al., 2007) and increase in the organic carbon in the soil (Márquez-García et al., 2013; Carbonell et al., 2010; Moreno et al., 2009), among others.

However, farmers are conservative by nature and any change in agronomic management is a challenge, and the implementation of the practice of establishing GC is no exception. The shift from tillage agriculture with bare soils to GC with no-tillage requires the farmer to make some important decisions, including the type of GC that best suits their farms and woody crops, within their investment possibilities. GC that farmers can use in their fields are described below.

2.2 Groundcover types and soil management

Some of the key characteristics of a GC are as follows. GC must have a low height development so that it does not hinder the transit through the farm. It must have a rapid growth capacity to be able to cover the soil surface as quickly as possible; its roots must be superficial, so that they do not explore the lower parts of the soil profile too much, and do not extract water that can be used by the tree; and should not host pests.

An adequate management of the GC is fundamental to optimize the consumption of water and to maintain the productivity of the olive orchard, especially in dry climates such as in the south of Spain where the annual regimen of precipitation is markedly seasonal, with rainy autumns and winters, and long dry summers. Therefore, the most important factors to be taken into account in the management of the GC will be determined by the balance of water in the soil. These are, first, the method of mowing to be used, and secondly, the time of mowing of the GC, in order to reduce or minimize the maximum consumption of water during the period of maximum water need for the crop, as well as to avoid competition for nutrients. It is therefore necessary to control the growth of the GC and eliminate or reduce these potential negative effects. There are different ways of managing the GC: mechanical, chemical and animal grazing.

To achieve an effective control of the GC by means of a mechanical mowing operation, it is necessary that the species used has on the one hand a low regrowth, to minimize transpiration after mowing and, on the other hand, a good capacity to produce more biomass so that it remains an effective cover for the soil and reduces evaporation, thus preserving the water content in the soil.

The capacity of regrowth of the species decreases as the harvest is later, when the phenological cycle of the species is more advanced; therefore, species with a short phenological cycle are the best adapted to this type of management. For example, in cruciferous species, white mustard showed the shorter phenological cycle and a low regrowth capacity after both early and late harvest, while the cultivated species Ethiopian radish and mustard reached significant rates of regrowth (Alcántara et al., 2004; Alcántara, 2005). Excessive regrowth reduces the water and nutrient content and affects the production of olives.

Chemical control is also an effective tool in the system of GC management for which low volume and non-residual herbicides with very low environmental impact are recommended. It is important to check the equipment with which the product is applied to ensure efficient and safe work. The date must allow the development of the cover to be sufficient to offer good soil protection and prolonged permanence of plant remains. This reference date, for the climatic conditions of Andalusia and 'for an average year', may be that of mid-March, which in the cover of grasses usually corresponds to the phenological state before the emergence of the flag leaf. This date may be varied according to the rainfall regime of each year. Glyphosate at the doses of 0.7–1 kg per ha have been very effective if the right timing of application is chosen.

Grazing is a widespread activity in mountainous areas of Andalusia, which in many cases is linked to the ecological cultivation of olive groves in disadvantaged areas. It is highly recommended, in steep areas, to let sheep graze at the beginning of spring, since they will control the cover and will not diminish the production of olives. The monitoring of soil moisture status is emphasized when animals enter the plot, since they can cause serious problems of compaction.

2.3 Classification of groundcovers

GC can be classified as follows:

- Spontaneous GC
    - spontaneous weeds
    - selected weeds
- Sown GC
    - grasses
    - legumes
    - cruciferous
- Inert GC
    - pruning residues
    - olive leaves.

2.3.1 Spontaneous groundcovers

They are the most widespread in Spain. These GC are composed of weeds present on the farm. The farmer can either let this flora grow without control, or manage it to select some more beneficial species. In this case, the most recommendable species are those widespread grasses in Andalusia that produce more biomass within a low competition with the olive grove.

On selected cover with grasses, farmers normally apply herbicides that select the plants to the beneficial species. This application of herbicides can be based on fluroxipir which tolerates grasses and eliminates broadleaf plants. It should be applied when the plants have developed to around 10–15 cm, and the recommended date to do so varies depending on the year, but it must be done in autumn or early winter. If the existing species are to be directed towards grasses, the use of selective grass herbicides must be made, and which will have to be repeated whenever hard-to-control weeds appear, such as malvae or conyza (AEACSV, 2001).

In terms of management, after the dry season and with the first rains, the plants present in the seed bank of the field in question are allowed to grow. This cover will protect the soil against the rains of autumn, winter and early spring. If the soil is degraded, it is advisable to fertilize this cover with 50 units of nitrogenous fertilizer, as otherwise few plants will grow. Around the end of March, it is necessary to impose a control on the GC growth to avoid a competition with the olive grove. This control can be either chemical, with herbicides or by controlled grazing. The choice of one method or another will depend on the plants present on the farm, the type of olive grove, whether livestock are available, and investment possibilities, among others (AEACSV, 2001).

If the spontaneous covers are controlled with chemical mowing, it is recommended to leave a band, which can alternate in the place (central or one of the sides of the cover) of approximately half a metre so that this small strip of the cover completes its growth cycle and seeding, and thus is able to continue covering the ground the following season.

In the catalogue of herbicide active ingredients, the most suitable for controlling the GC should be consulted, although the most widespread ones are based on glyphosate with some mixture depending on which weeds are present. In the case of mechanical grazing or mowing it can be difficult to maintain the band to produce seeds, so it would be necessary to monitor the presence of cover in later years, and if necessary, some species should be sown (AEACSV-Junta de Andalucía, 2007).

The spontaneous cover is recommended in soils where the topography of the terrain makes the planting of species difficult and in historically carved or dissected areas where often there are several species that permit the development of a dense protective cover. The advantage of such a cover is the saving in certain costs: seed and sowing operation and selective herbicide treatment. One of its disadvantages is that the plant species that make the cover are often very diverse (different growth habits, variable susceptibility to herbicides) (AEACSV-Junta de Andalucía, 2007).

The importance of mowing the cover should be mentioned, since leaving these herbs of spontaneous growth at will would result in lower olive yields. In hilly areas, GC control by grazing by sheep and goats is recommended, always allowing the animals to enter when the soil does not have excess moisture, otherwise the soil will be compacted. In olive groves of the countryside, with lower average slope and higher yields, the cover can be treated with herbicides or brush cutters. The main disadvantages that are found for the control of the cover here are:

- Chemical control. Higher doses of herbicides than those used for grass covers, with the consequent higher economic cost.
- Mechanical mowing. The vegetation can evolve towards perennial species, of easy regrowth; all of them of difficult control with brush cutters.
- Mowing with grazing. Grazing at the right time. Do not enter the field with moist soil that will lead to compaction.

2.3.2 Sown groundcovers

Grasses (barley, ryegrass, others): This type of cover has been well studied in Andalusia (Spain), showing successful experiences. The management of this GC is similar to the previous case described in Section 2.3.1. In fact, this case would be in practice very similar, if a selection of the spontaneous growth is made towards the grasses. This GC is composed of one or several species. The sowing is performed with a conventional seeder or what is more economical, broadcasting seeds with a fertilizer spreader. No certified seeds are required, as the aim is not to harvest the grains, so they can be obtained at an economical price (AEACSV, 2001).

Sown grasses are particularly recommended for olive orchards whose soils have a high degree of erosion or they have been previously managed in reduced or no-tillage but without any cover. In such cases, the presence of the weeds typical of the tillage conditions in spring, difficult to control, make it advisable to introduce the sown cover, which can compete with the species present in the field, thus decreasing the number of weeds in favour of the GC.

In the case of cultivated species, oats, barley and rye, the indicative dose is 100 kg per ha of vegetation cover (50 kg per ha of land). If we sow spontaneous species (ryegrass, barley, etc.), the dose recommended is about 15 kg per ha of cover, 7.5 kg per ha of land (AEACSV, 2001).

Legumes (vetch, clover, others): This type of GC has been less studied despite the fact that they have a high environmental value and interest to fix nitrogen and fertilize the crop naturally. In organic olive groves, this type of fixation can be used, being able to be controlled with mechanical means or with grazing (AEACSV-Junta de Andalucía, 2007). The difficulty they present is that their residues are not persistent enough to protect the soil all season long. This inconvenience makes them generally undesirable, since they will not be able to prevent high erosive processes during the rainy season, in the autumn-spring periods. However, this GC can be mixed with grasses (AEACSV, 2001).

Normally, the herbicide control of this GC is more difficult and expensive to carry out than in grasses. However, mechanical harvesting is carried out efficiently in legume covers, especially with species with low re-sprouting capacity such as vetch, and provided that the harvest is very late (after March and with the plant in bloom).

Cruciferae (white mustard, Sinapis alba): The use of GC of cruciferous species is a very suitable alternative, on the one hand due to the need to find species capable of entering rotation with the grass covers, since after several years of use of the same species as cover, there might be an inversion of flora. On the other hand, for the numerous advantages that these species present, such as:

- They are species known by farmers, and it is common to find many of them spontaneously in the olive groves.
- They tend to present rapid growth and abundant production of biomass, essential characteristics for an adequate control of erosion.
- Being species that mostly start their cycle in winter, they would coexist with the crop in a time of high rainfall and little evapotranspiration, and

therefore there would be low competition for water with the olive tree; in spring when the water balance begins to become negative, the cover would be eliminated.

- Many of them have a powerful root system that favours the infiltration of water, thereby increasing the storage of water in the soil and make them a very promising species for decompacting the soil down to deeper layers (Wolfe, 2000).
- Its enormous potential as species that control *Verticillium dahliae* (Davis et al., 1996; Shetty et al., 2000), currently the major phytopathological threat of olive groves, as well as other soil diseases (Smolinska and Horbowicz, 1999), weeds (Boydston and Hang, 1995; Al-Khatib et al., 1997) and nematodes (Mojtahedi et al., 1993) because of their content of glucosinolates, sulphur compounds derived from the secondary metabolism of plants with great herbicidal power, insecticide, nematocide and fungicide.

Determine the appropriate time to control the GC to avoid competition for water with the olive tree, which is complicated, especially with a mechanical mowing, as there are many factors to consider, so it is difficult to give common guidelines that serve for all places and climatic years. With mowing very early, there is a significant decrease in the water content in the soil during spring and summer, due to the regrowth in the covers, the rapid decomposition of the remaining biomass and the proliferation of weeds. On the other hand, late mowing results in little-or-no regrowth, in a high production of biomass and high persistence of the remaining biomass, in providing adequate coverage of the soil, in avoiding the appearance of weeds during the spring-summer cycle and in retaining the moisture content. However, it is not advisable to delay the control too much in order to avoid competition for water and nutrients with the tree. The global knowledge of all these factors is what will provide the technician or the farmer with an adequate criterion to define the harvest moment in each specific case (AEACSV-Junta de Andalucía, 2007).

In the management by mechanical mowing of *Sinapis alba* GC under no-till, when two cuts were made to the GC, the water content in the soil improved compared to a no-till system with bare soil (Alcántara, 2005). One cut was before flowering (around the month of February), leaving the cover at a height of 40–50 cm to limit transpiration, and the second a definitive one, mowing the GC at ground level and leaving the remaining biomass on the soil. This should be done depending on the meteorological conditions of the season but in no case before the month of April (AEACSV-Junta de Andalucía, 2007). In the case of chemical control, the guidelines will be similar to those in the previous cases described above – fundamentally with glyphosate-based herbicides, at low volume and with low environmental impact.

2.3.3 Inert covers (pruning residues and olive tree leaves)

Inert covers are composed of non-living elements, unlike those treated so far. They can be mainly pruning remains, or olive tree leaves (AEACSV-Junta de Andalucía, 2007).

They consist of the scattering of the crushed pruning remains or of some sub-product of the olive grove. It is convenient to keep in mind that they should not be incorporated into the soil with mechanical tillage, but left on the soil surface. By slowly decomposing, they provide a long-term and adequate protection of the soil. It is emphasized that the pruning remains increase the organic matter in the superficial layers of the soil, increase the water content in the soil and improve the structure of the soil profile in its first layers.

Vegetable remains from the cleaning of the olive in the mill (especially leaves) have also been used as cover material. This type of cover is very similar to crushed pruning remains. We must take care that the material does not contain insect pests or diseases. Farmers are sometimes reluctant to use this GC due to 'repilo', a disease produced by the *Cycloconium fungus oleaginum* (AEACSV, 2001).

In conclusion, the techniques of GC described above are part of the practical approach of CA in olive groves, and other woody perennial crops. GC is a sustainable and recommendable option for the Andalusian olive groves. The reported agronomic and environmental benefits they provide can favour them in the fulfilment of the new green requirements of the EU policies.

Undoubtedly, there is no ideal vegetation GC, nor will all be suitable for a specific purpose. The success of the system will depend on the choice of the species, and several factors must be taken into account. In the first place, it is necessary to determine the deficiencies that need to be corrected and to know the particularities of the crop in which the GC is going to be established. The chosen species must adapt to the crop cycle and not pose an obstacle to carrying out its management operations. Secondly, the preferences of the farmer towards those with which he is more familiar in terms of their growth cycles and management must be taken into consideration (Saavedra, 2003).

With these conditions, in most cases, grass GC is recommended for its biomass, good persistence of biomass remains in the soil, easy control with products of low toxicity and low doses and its knowledge by the farmer. According to recent studies, it would be interesting to involve rotating cruciferous species, especially white mustard, which can be controlled mechanically.

3 CA-based plantation systems in Malaysia

In Malaysia, plantations are being brought under CA management to comply with environmental quality legislation as well as for reasons of sustainability and the need to reduce costs of purchased production inputs. In the following sections, CA-based systems for oil palm and rubber plantations are described.

3.1 CA-based oil palm plantation system

CA practices adopted by most oil palm plantations in Malaysia typically begin from land preparation for field planting until the end of its 25 years of economic life cycle. Conventional clean clearing techniques of land preparation adopted till the early 1990s, which involved burning of vegetation, has been outlawed by legislation since 1974. Open burning, which can bring about quick return of vegetation into various gaseous carbon derivatives among others, can also lead to losses of valuable plant nutrients such as nitrogen, phosphorus, soil organic matter and so forth. The oil palm industry currently adopts an environmentally friendly zero-burning technique which also facilitates the recycling of valuable plant nutrients and organic matter into the soil.

In the zero-burning method, the standing or above-ground biomass containing a large amount of dry matter and nutrients (Table 1) are felled and sliced into <10 cm thick (Fig. 3) trunk chips. The chipped trunks are then spread in the inter-rows (Fig. 3) to decompose (Ooi, 2011). Decomposition generally takes place within 18 months, where about 70% of potassium content and bulk of other nutrients are released (Khalid et al., 1999c). Organic matter and nutrients released from the decomposing trunk chips significantly improves chemical properties of the top soils (Table 2). The soil nitrogen, phosphorus, potassium and magnesium content increased between 40% and more than 500% after the trunk chips decomposed. These large quantities of nutrients could supply more than adequate nutrients to meet the vegetative growth demand of young oil palms during the entire immature phase if they can be trapped and fully utilized (Sim et al., 2008). In order to benefit from the large amount of nutrients released, leguminous cover plants are planted soon after land clearing to minimize top soil erosion, mobilize nutrients released from the decaying palm biomass and recycle these nutrients to the soil and newly planted palms when the legumes dieback while at the same time reduce the loss of released nutrients through surface run-off and soil sediments.

Table 1 Dry matter and nutrients in palm biomass at replant

		Nutrients (kg ha^{-1})			
Palm biomass	Dry matter (t ha^{-1})	N	P	K	Mg
Above-ground	85	577	50	1255	141
Below-ground	16	65	8	129	15
Total	101	642	58	1384	156

Source: adapted from Khalid et al. (1999a,b).

Figure 3 Evenly spread oil palm trunk chippings after felling. Source: adapted from Ooi (2011). Inset: Close-up image of a trunk chip.

Table 2 Changes in soil chemical properties in top 15 cm soil after 12 months of decomposition

Treatment	Months after treatment	Total N (%)	Avail P (ppm)	meq/100 g soil Exch. K	meq/100 g soil Exch. Mg
Before chipping	0	0.15	0.15	0.15	0.26
Without trunk chip	12	0.20	90.80	0.50	0.32
With trunk chip	12	0.28	354.60	2.62	1.11
% difference		140	391	524	347

Source: adapted from Khalid et al. (1999c).

Among 80 legume species which were commonly used in CA practices worldwide (FAO, 2012), six were commonly planted in Malaysia. The characteristics of the legumes planted in Malaysia are described in Table 3 (Chiu et al., 2011). Upon achieving 100% ground coverage (approximately 8 months after planting, Fig. 4), the eroded top soils were reduced from 7.5 t ha^{-1} of soils to almost negligible (Arif et al., 2007). Run-off losses were also reduced, albeit by 30% in areas covered by legume compared with bare ground (Ng et al., 2008). Other major benefits of planting leguminous cover are the contribution of nitrogen fixation and addition of biomass turned soil organic matter into the soils. About 250 kg N ha^{-1} were fixed by legumes in addition to other nutrients immobilized (Ng et al., 2010). In addition, planting legume cover crops would result in better availability of phosphorus from the applied rock phosphate (Zaharah and Bah, 1997), improved soil structures, increased quantity, diversity and activity of soil microfauna as well as a form of weed control (Chiu et al., 2011).

Table 3 Characteristics of common leguminous cover plants planted in Malaysia

Species	Plant characteristics	Nitrogen (kg ha^{-1})			Seeding rate (kg ha^{-1})
		Accumulated	Fixed	%	
Pueraria javanica	Vigorous, thick and dominant without overhead shade, suppresses other undesirable plants	13–225	6–172	44–76	2.0–4.0
Calopogonium caeruleum	Slow initially, spreads vigorously later, shade tolerant	-	-	-	0.5–1.0
Calopogonium mucunoides	Growth rapid initially, but dies early, unpalatable to cattle	4–108	1–76	35–79	4.0–6.0
Centrosema pubescens	Fast establishing, very susceptible to pests	6–132	3–104	48–79	4.0–6.0
Mucuna cochinchinensis	Rapid coverage, but short-lived (6–9 months) and drought sensitive	22–305	6–224	36–86	4.0–8.0
Mucuna bracteate	Very vigorous, thick and dominant, suppresses other weeds, moderate shade and drought tolerant. Unpalatable to cattle when mature	180–522	126–365	33–88	0.1–0.2

Source: adapted from Chiu et al. (2011).

Figure 4 Good legume (*Mucuna bracteata*) cover plants in an immature oil palm field at 8 months after legume planting. Courtesy of Ng P.H.C.

As the oil palm trees mature, the good legume coverage achieved in early years will thin out gradually due to both operational circle weeding and shading by oil palm canopies (for certain species). Natural covers such as soft grasses and *Nephrolepis* sp. ferns are then encouraged to succeed the legumes. Although the soft grasses and *Nephrolepis* sp. could not replace legumes in fixing nitrogen, they were equally effective in minimizing soil erosion when compared to bare ground (Tables 4 and 5). Only ground vegetation which competes for nutrients or impedes estate operations such as those self-grown in the palm circles and harvester paths are eradicated by producers. Another soil conservation measure widely adopted in oil palm industry is soil mulching with palm residues such as pruned fronds, empty fruit bunches (EFB), dried decanter cake, pressed sludge cake and so forth onto the ground surface (Fig. 5). Stacking of pruned fronds as a form of conservation was also demonstrated to significantly reduce soil erosion, surface run-off and retain soil moisture in a mature oil palm field (Table 5) in the absence of legume covers (Fig. 7).

3.2 CA-based rubber plantation system

Influences of suitable GC to mitigate the long-term detrimental effects of planting of Hevea were well studied and adopted by most major plantations in Malaysia (Vrignon-Brenas et al., 2019). Maintaining GC plants in the inter-rows

Table 4 Effect of grasses and ferns on soil erosion under mature rubber

		Amount of soil eroded (t ha^{-1} yr^{-1})		
Soil series	Rainfall (mm)	Bare ground	Soft grasses	*Nephrolepis* sp.
Rengam	2920	103	44	Negligible
Serdang	3250	132	117	59

Source: adapted from Chui et al. (2011).

Table 5 Effect of pruned frond stacking on soil erosion and run-off losses

	Bare ground (B)	Frond stacked (F)	Reduction (B-F)
Rainfall (mm)		1426	
Soil loss (kg ha^{-1} yr^{-1})	14.9	1.1	13.8
% run-off	30.6	2.8	27.8
Rainfall (mm)		Na	
Soil loss (kg ha^{-1} yr^{-1})	30.8	4.9	25.9
% run-off	30.8	17.8	13

Source: adapted from Yacob et al. (2011).

Figure 5 Leaf litters formed underneath the legume cover plant. Source: adapted from Chiu et al. (2011).

(1 month) (12 months) (21 months)

Figure 6 Groundcover by legume (*Mucuna bracteata*) with time. From 1 month to 21 months after plantings. Courtesy of Ng P. H. C.

is deemed as an important measure to prevent deterioration and maintain as well as improve declining soil fertility status (Pushparajah et al., 1977).

Generally, there are three common groups of cover plants in the plantation: the leguminous creepers (e.g. *Pueraria, Calopogonium, Centrosema*), light or natural grasses (e.g. *Paspalum, Axonopus*) and ferns (e.g. *Nephrolepis*). Leguminous creeper is widely planted throughout the immature rubber fields (Figs. 6 and 8), to prevent soil erosion, enhance nitrogen fixing and recycling and increase the soil organic matter. Upon reaching maturity, heavy shading of closed canopy resulted in declining leguminous population and succession of natural grasses and ferns (Fig. 9).

Undoubtedly the nutrients immobilized and the enhanced nitrogen fixed by legume would eventually return to the soil as the litter decomposed.

Figure 7 An example of combination of pruned frond stacking and empty fruit bunching (EFB) mulching in oil palm field. Source: adapted from Yacob et al. (2011).

Leguminous creepers returned the highest amount of N and K, about 230–350 kg ha^{-1} N and 85–131 kg ha^{-1} K. Light grasses conversely returned the least amount, contributing about 24–65 kg ha^{-1} N and 31–86 kg ha^{-1} K (Table 6). In well-manured trees, the rubber trees produced approximately 3000 kg ha^{-1}

Figure 8 Good legume (*Mucuna bracteata*) cover plants in immature rubber fields. Courtesy of C. C. Sim.

Figure 9 Presence of 'natural grasses' and leaf litter in a young rubber fields. Courtesy of C.C. Sim.

Table 6 Comparison of nutrients returned to soil by various cover plants over the 5 years of immature period

	Nutrients (kg ha^{-1})				
Cover plant	N	P	K	Mg	Total Nutrients
Leguminous creepers	226-353	18-27	85-131	15-27	344-538
Grasses	24-65	8-16	31-86	9-15	72-182
Mikania cordata	74-119	9-14	63-99	9-24	155-256
'Naturals'	13-117	3-10	46-140	3-18	65-276

Leguminous creepers = Mixed *Pueraria phaseoloides, Centrosema pubescens* and *Calopogonium mucunoides*; Grasses = Mixed *Axonopus compressus* and *Paspalum conjugatum*.
Source: adapted from Pushparajah et al. (1977).

of dry leaf litter annually. The decomposing leaf litter recycled a considerable amount of nutrients of about 50 kg ha^{-1} N and 10 k g ha^{-1} K annually.

Soil carbon matter, apart from being the organic source of plant nutrients, also plays an important part in improving soil physical properties (Table 7). The types of covers directly influence the soil carbon status (Pushparajah and Chellapah, 1969). Soils under leguminous covers maintain the highest organic matter status compared to other GC. In addition, longer residual effect of organic matter was attained under leguminous covers (Fig. 10).

Table 7 Nutrients stored in rubber leaves and returned as litter

	Dry weight leaf litter	Nutrients in litter (kg ha^{-1} yr^{-1})				
Fertilization	(kg/ha/yr)	N	P	K	Mg	Ca
Nil fertilizer	2360	42	1.6	7.1	5.5	26
NPKMg fertilizer	2860	53	2.1	10.2	7.0	33

Source: adapted from Tan, 1975).

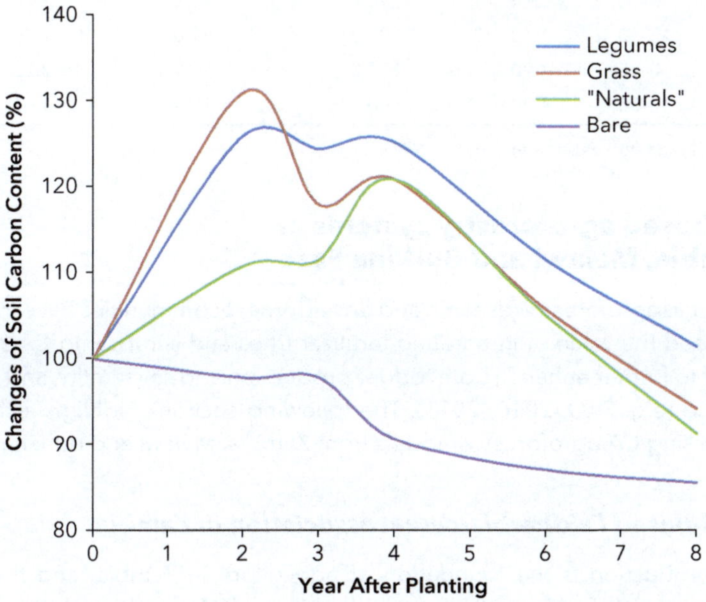

Figure 10 Percentages of changes in soil carbon content and residual effect upon 8 years after planting. Source: recomputed graph from Watson et al. (1963) after Pushparajah et al. (1977).

In the absence of a vegetative cover, surface soil is exposed to compaction and erosion, where the presence of cover plants alleviates such problems. The establishment of covers enhances the organic matter of the soils (Fig. 10), which indirectly contributed to the improvement of soil structure, internal drainage and moisture retention. Leguminous covers exerted much greater influence in improving soil structure than grasses, natural or *Mikania* (Pushparajah et al., 1977). Soils under leguminous and 'natural' covers had lower bulk density and higher pore space which resulted in better water infiltration and retention in the soil (Table 8).

Table 8 Residual effect of various types of groundcovers on soil physical properties

Cover	Soil depth (cm)	Bulk density (g cc^{-1})	Permeability (cm h^{-1})	Percentage aggregation (%)	Mean weight diameter (mm)
Grass	0–15	1.11	65.0	91.1	2.67
	15–30	1.16	77.5	82.6	1.04
Mikania cordata	0–15	1.21	52.5	88.3	3.00
	15–30	1.22	30.0	76.9	1.54
Legume	0–15	1.04	90.0	93.9	3.78
	15–30	1.12	87.5	89.0	2.26
'Naturals'	0–15	1.00	105.0	90.1	3.22
	15–30	1.07	60.0	75.4	1.56

Source: adapted from Pushparajah et al. (1977).

4 CA-based agroforestry systems in Zambia, Malawi and Burkina Faso

The CA and agroforestry research and development communities have mutually recognized the value of integrating fertilizer trees and shrubs into CA systems in order to further enhance both fodder production and soil fertility, and system resilience (e.g. FAO, 2010, 2011). The following sections highlight examples of promising CA-agroforestry models from Zambia, Malawi and Burkina Faso.

4.1 CA-based Faidherbia-cereal association in Zambia

Maize production is the foundation of agriculture in Zambia, and the basis for the country's food supply. But yields are low (1.1 t ha^{-1}), and three-fourth of farmers fail to produce enough maize to sell in the market. Since 1996, a coalition of stakeholders from the private sector, government, and the donor community has promoted a package of agronomic practices for smallholders in Zambia based on the principles of CA (Haggblade and Tembo, 2003). The effort is spearheaded by the Zambian Conservation Farming Unit (CFU) (www.conservationagriculture.org). To date, CA has been introduced over large areas of the country. In 2015/16, CA cropland area was estimated at 316 000 ha (Kassam et al., 2019).

One of the main goals attributed to CA is to stimulate biological activity and improve the soil structure. To do this, farmers are encouraged to keep the soil covered with organic matter throughout the year. This involves the retention of crop residues: Conventionally, farmers tend to burn them. And it involves the rotation of maize with other crops, particularly cotton and grain legumes, and/or the planting of cover crops such as sunn hemp *(Crotalaria juncea)*. The additional benefits of CA – including improved soil structure, gains from nitrogen-fixing crop rotations and reduced field preparation labour – occur gradually as soil health improves.

The value of conservation farming to Zambia agriculture is well recognized by the government, and receives strong policy and extension support by the Ministry of Agriculture.

As the Zambian CFU worked to develop solutions to make CA feasible for smallholders, they realized that more than two-thirds of their smallholder clientele could not afford inorganic fertilizers, and they had little-or-no access to manure or other nutrient sources. This fundamentally limited smallholder maize yields, and further depleted their soil fertility each year. To address the constraint, the CFU investigated the incorporation of *Faidherbia albida* trees into maize production systems.

Faidherbia is a nitrogen-fixing acacia species that is indigenous to Africa, and it is widespread throughout the continent. What makes it unique is its growth habit, known as 'reverse leaf phenology'. Faidherbia goes dormant and it sheds its foliage during the early rainy season, at the time when field crops are being established. Its leaves only regrow at the end of the wet season. This unusual phenology makes it highly compatible with food crops, since it does not compete with them significantly for light, nutrients or water during the growing season. On the contrary, annual crops in the vicinity of Faidherbia trees tend to exhibit improved performance and yield (Barnes and Fagg, 2003).

Numerous published reports have recorded increases in maize grain yield when grown in association with Faidherbia. These reports range from increases of 6% to more than 200% (Barnes and Fagg, 2003), depending on the age and density of trees, the agronomic practices used and the weather conditions. Faidherbia's effects tend to be most remarkable in conditions of low soil fertility. In Zambia, results of on-farm trials over 4 years found that maize yields in the vicinity of Faidherbia trees were double those of maize beyond the tree canopy (IRENA, 2019). The work also drew on observations in Malawi, where maize yields were increased up to 280% in the zone under the tree canopy compared with the zone outside the tree canopy (Saka et al., 1994).

The association between *Faidherbia albida* and increased crop yields is well documented. Barnes and Fagg (2003) noted in their comprehensive monograph on the species that 'there has been a huge amount published on the beneficial effect of *Faidherbia albida* on the soil once it is established'. Most of these studies have observed significant increases in yield beneath or near the trees. They observed that the tree is found over a wide range of soils and climates and with varied plant and animal associates, from desert to wet tropical climates. However, it does not tolerate competition from other plant species, and thus does not have invasive tendencies.

Establishing Faidherbia trees requires some patience on the part of the farmer and development-support institutions. It is one of the fastest-growing acacia species, but its initial growth is relatively slow as it develops a deep root system. It therefore takes a few years before the trees begin to provide substantial leaf biomass and fertility benefits. In a survey of 300 farmers with Faidherbia in their maize fields, one-third of the farmers indicated that the trees began to provide significant benefits to their crops in 1–3 years. Another 43% related that it took 4–6 years before they observed the benefits of planting Faidherbia (Phombeya, 1999). However, establishing Faidherbia does not preclude planting other nitrogen-fixing trees or shrubs in the same fields that have a more immediate impact on soil fertility and crop yields.

The Zambian CFU recommends that Faidherbia seedlings be planted in a grid pattern at 100 trees per ha. Fields with Faidherbia-maize systems managed with such a planting pattern (10 m × 10 m) can accommodate full mechanization. The result is a maize farming system under an agroforest of Faidherbia trees. The trees may live for 70–100 years, providing inter-generational benefits for a farm family, with a very modest initial investment. As the trees mature, and develop a spreading canopy, they are gradually thinned down to about 25–30 trees per ha.

There is increasing recognition of the opportunity to exploit the abilities of Faidherbia, and in recent years a more concerted effort has been made to improve and enhance this indigenous African agroforestry system in many parts of the continent (Garrity et al., 2010). Currently, the departments of agriculture in Zambia and Malawi are encouraging farmers to establish Faidherbia trees in their maize fields, the aim being to increase food production. The efforts are backed by national policy and supported by the Zambia National Farmers Union. The Zambian CFU estimates that the tree is now cultivated in CA systems over an area of 300 000 ha (http://www.new-ag.info/developments/devItem; http://www.agfax.net/radio/detail).

Further research is needed to quantify the time stream of benefits on nitrogen fixation and soil fertility of incorporating trees from when they are newly planted. Research is also needed into the genetic variation in the *Faidherbia albida* systems, so that strategies can be implemented to safeguard and use the species' genetic diversity, and superior germplasm can be made available to farmers.

There is evidence that the integration of fertilizer trees into smallholder maize production in Zambia, alone or in combination with CA systems, has resulted in greater productivity, food security and family income. These practices are, however, knowledge-intensive as opposed to being cash-intensive. Thus, sustained rural advisory services, through the public sector and private sector, are important to ensuring sustained uptake and expansion over the longer term.

4.2 CA-based intercropping maize with nitrogen-fixing trees in Malawi

The story of the development of practices to integrate soil regenerating trees into continuously cropped land began in earnest in the 1980s. The classic model of these systems was alley cropping. This model involved densely planted rows (or double rows) of N-fixing trees planted at intervals in crop fields. The annual crops were then cultivated in the alleyways (Kang, 1984).

The concept was inspired by the considerable use of such systems on the island of Flores in Indonesia, which continues today (van Noordwijk, 2010), and by observations of variants of the principle applied in limited areas elsewhere (e.g. in southeastern Nigeria). The trees are pruned frequently, and the biomass is spread in the cropped area and used as a surface mulch.

These systems often succeeded in raising and sustaining crop yields, particularly on infertile or degraded soils. The early results in experimental trials stimulated the development of a continent-wide network that tested the alley-cropping principle quite widely. This included various permutations of fast-growing N-fixing trees, and pruning regimes, and their effects on crop yields and systems sustainability.

Considerable work was also done on them in Asia and Latin America as well. However, the labour needed to maintain the tree hedgerows was a significant constraint. To date, the adoption of the practice has been limited, other than the locations mentioned above where it was practiced traditionally.

One important spinoff of the extensive work on hedgerow intercropping, however, was the development of the practice of deploying natural strips as an effective soil and water conservation system for croplands on slopes (Garrity, 2002). This simple and effective technology was developed from observations of indigenous practices. It has now been integrated into smallholder extension programmes in a number of Asian and African countries.

A second cycle of research, based on the alley-cropping concept, was initiated during the mid-1990s by ICRAF in Malawi. The alley-cropping model was modified to reduce the labour burden while still taking advantage of the biofertilizer and soil regenerating benefits of fast-growing fertilizer trees (Akinnifesi et al., 2010). The most successful system that the team developed was to plant the fertilizer trees at a spacing of 1 m × 2 m throughout the maize field, and pruning the plants two to three times per year, leaving a short-coppiced stump about 25–30 cm tall.

These systems demonstrated the ability to produce substantially higher maize yields, improve soil quality, and provide high returns to both land and labour (Gudeta et al., 2009). The most popular system in southern Malawi, where land holdings are very small (<0.5 ha), is intercropping maize with nitrogen-fixing tree species of *Gliricidia sepium*, *Sesbania sesban*, *Tephrosia candida* or pigeon peas. A long-term experiment spanning more than a decade, involving the continuous cultivation of maize with Gliricidia at Makoka Research Station, Malawi, yielded more than 5 t ha^{-1} in good years, and an average of 3.7 t ha^{-1} overall, in the absence of mineral fertilizers. This compared favourably with an average of 0.5–1.0 t ha^{-1} in control plots without Gliricidia or mineral fertilizer (Akinnifesi et al., 2007; Makumba et al., 2006). Also, Tephrosia (*T. vogelii* and *T. candida*) and pigeon peas are often relay-intercropped with maize (Snapp et al., 1998; Akinnifesi et al., 2008).

The success with these coppiced fertilizer shrub systems led to the development of the National Agroforestry Food Security Programme in Malawi, which was launched in 2006. That programme scaled up the practice with *Gliricidia sepium* as the main shrub species. It also promoted the practice of relay-planting fast-growing woody perennials in the maize, which would be incorporated at the end of the following dry season, having produced a store of N-rich biofertilizer.

4.3 CA-based agroforestry system in Burkina Faso

In Burkina Faso, the 1968–73 Sahelian drought caused an acute human and environmental crisis. The densely populated Yatenga Province on the northern Central Plateau of Burkina Faso was particularly affected. Failure of the agricultural systems prompted massive labour migration and caused social disruption (Monimart, 1989). Between 1975 and 1985, some villages lost up to 25% of their population as they migrated south to areas of higher rainfall. In the early 1980s, groundwater levels in the Central Plateau dropped an estimated 50–100 cm/year (Reij, 1983). Many wells and boreholes went dry just after the end of the rainy season and had to be deepened.

Average sorghum and millet yields decreased to slightly below 300 kg/ha (Matlon and Spencer, 1984; Matlon, 1990). As a result, the majority of farm households had annual food deficits of 50% or more (Broekhuyse, 1983). Most land was cultivated continuously on shallow lateritic, crusting soils with very low natural fertility, creating extensive areas of barren, unusable land. As the landscape became denuded and exposed to severe water erosion, the land and the people became increasingly vulnerable to drought.

During the 1980s, farmers began experimenting with traditional techniques to reclaim severely degraded farmland that water could not penetrate. They developed variations of the practice digging a grid of planting pits (*zaï*) across their fields. The depth and diameter of the pits was increased, and organic matter added to the basins to regenerate an environment where plants could grow. The pits concentrated nutrients in the plant root zone, and retained water for extended periods of time, allowing crops to better survive dry spells. The pits are prepared during the dry season, enabling early planting for increased yields. The use of new and improved planting pits spread rapidly among farmers with the support of NGOs and the extension service. Another practice that became popular with farmers was the construction of contour stone bunds that reduce surface run-off from the fields. By 2001, well over 100 000 ha of strongly degraded land had been rehabilitated by projects and by farmers on the northern part of the Central Plateau alone (Reij and Thiombiano, 2003).

Taking into account what has been achieved in this region since then, and on other parts of the Central Plateau, it is now estimated that the total area rehabilitated over the past three decades is somewhere between 200 000 and 300 000 ha (Botoni and Reij, 2009; Reij et al., 2009). A recent study shows that in villages with a long history of soil and water conservation, 72–94% of the cultivated land has been rehabilitated with one or more conservation techniques. In villages with a shorter history, this figure ranges from 9% to 43% (Reij et al., 2009).

Zaï planting pits alone usually have a greater impact on yields than stone bunds alone, but the greatest returns occur from using both together. The effect of the practices was synergistic with added manure. With these increases, farm households that suffered from food deficits of 6 months or more during the early 1980s have been able to reduce their deficit periods from 6 months to 2–3 months, or to zero in some cases (Reij et al., 2009).

The *zai* planting pits have been used to intensify cereal production as well as to produce trees, or to combine cereal and tree production in agroforestry systems. By stimulating tree production in combination with planting pits, biomass production is dramatically increased for soil amelioration as well as livestock fodder.

On the Central Plateau, rehabilitated fields now average 126 trees per ha, compared with 103 trees per ha on control plots. The trees on rehabilitated land are larger and represent a wider range of species. These modified traditional agroforestry, water and soil management practices have transformed barren agricultural landscapes into complex agricultural systems with more vegetation and more varied vegetation (Reij et al., 2009).

Some common indigenous tree species, such as *Combretum glutinosum* and *Piliostigma reticulatum*, can be managed as coppiced stumps at a density of at least several 100 per hectare, under a management system much like the coppiced trees of Gliricidia that are integrated into maize systems in Malawi. If this practice is also combined with full-canopy Faidherbia trees, several tonnes of additional biomass can be generated annually per ha to accelerate soil fertility replenishment, provide additional livestock fodder and increase yields. These developments have also brought changes in how rural people earn their livelihoods. After the harvest, men once commonly migrated to urban areas for employment, but this pattern is changing as more men remain in the villages because they can now earn sufficient income from agriculture.

5 Conclusion

This chapter provides a glimpse into a small variety of CA-based perennial systems. Such systems exist in all continents, but their spread and rate of expansion is not being recorded systematically. CA-based perennial systems offer a range of benefits to the producers and society that cannot be harnessed to the same extent with conventional tillage-based perennial systems. In fact, conventionally managed perennial monocrop systems, that undergo regular mechanical soil disturbance and do not maintain permanent mulch cover, suffer from severe degradation and erosion over time, losing soil and landscape health, productivity, output and profit, and delivery of ecosystem services to producers and society contributing also to environmental pollution outside the production areas.

CA systems, including perennial systems, aim at mimicking natural forest landscapes. Thus, perennial CA systems provide a reliable approach to rehabilitating soil health and land productivity as well as re-establish or strengthen system sustainability and resilience. All the examples of CA perennial systems illustrated in this chapter provide evidence in support of this regenerative ability of perennial CA systems, whether orchard, plantation or agroforestry systems.

Similar to what happens in annual CA systems, perennial CA systems are better adapted to climate change due to improved water and nutrient retention and supply, and they contribute to climate change mitigation though greater carbon sequestration in the soil and much reduced use of fossil fuel. These aspects contribute greatly to reducing purchased input requirements and to stable and greater profit. In fact, it is becoming increasingly apparent that many perennial CA systems can be managed with very low level of external inputs.

What is important to note is that CA perennial systems are being managed in the low-income regions in the tropics as well as industrialized regions such as Europe, on large or small area, and by mechanized and non-mechanized farmers. They offer soil health, productivity and ecosystem benefits to smallholders as well as to larger-scale operators and are contributing to raising household incomes and sustainability and to making land-use climate-smart.

In light of the above, it is expected that perennial CA systems will continue to expand worldwide in the coming decades.

6 Where to look for further information

The literature on perennial CA systems is not well organized nor is there a comprehensive database to indicate the land area covered by such systems. As noted at the beginning of this chapter, perennial CA systems globally are receiving increased attention for reasons of productivity, economic, environmental and social benefits similar to those that apply to annual cropping systems. There is a large economic and environmental potential for sustainable land use, improved livelihoods and enhanced food security with perennial CA systems including their better adaptability to climate change and hilly or marginal environments. Consequently, much greater multi-disciplinary research and development efforts need to be made to accelerate their adoption and optimum management in all continents and agroecologies.

A general source for Conservation Agriculture is the Food and Agriculture Organization of the UN webpage which carries information on CA systems (http://www.fao.org/conservation-agriculture/en/). Research systems in South and Central America do promote mixed CA production systems involving annual and perennial crops including pastures. The Brazilian Agricultural Research Corporation (EMBRAPA) has held several international conferences on integrated crop-livestock-forest systems with a focus on CA approach (see e.g.: https://www.embrapa.br/en/tema-integracao-lavoura-pecuaria-floresta-ilpf/nota-tecnica).

Similarly, national research systems in the Mediterranean regions of Europe, North Africa and West Asia are actively researching on perennial CA systems that include fruit and nut tree orchards and vineyards. Regional CA organizations such as the European CA Federation (ECAF) website (www.ecaf.org) and African Conservation Tillage (ACT) Network (www.act-africa.org) offer some technical publications on the subject. World Agroforestry Centre (ICRAF) has been active in research and development work in this area in Africa, Asia and Latin America and has several publications on the topic of Conservation Agriculture with trees (CAWT) including a Guide on CAWT (see: www.icraf.org). Similarly, the French Centre for Cooperation on International Research on Agriculture Development (CIRAD) is a source of literature on the subject (see: www.cirad.org). However, perennial CA systems in general are not the main topic of papers or books. They are sometimes presented as subtopics in globally oriented publications as in CIRAD (2012) and Kassam et al. (2013, 2016).

7 References

AEACSV. 2001. *Asociación Española Agricultura de Conservación Suelos Vivos (AEAC/SV). Agricultura de Conservación en el olivar: Cubiertas vegetales.* AEAC/SV, Córdoba, 36pp.

AEACSV-Junta de Andalucía. 2007. Cubiertas vegetales en olivar. In Spanish. Available at: https://www.juntadeandalucia.es/export/drupaljda/1337161293CUBIERTAS_VEGETALES_9-07_mod_por_JMD.pdf.

Akinnifesi, F. K., Makumba, W., Sileshi, G., Ajayi, O. C. and Mweta, D. 2007. Synergistic effect of inorganic N and P fertilizers and organic inputs from *Gliricidia sepium* on productivity of intercropped maize in Southern Malawi. *Plant and Soil* 294, 203–17.

Akinnifesi, F. K., Chirwa, P. W., Ajayi, O. C., Sileshi, G., Matakala, P., Kwesiga, F. R., Harawa, H. and Makumba, W. 2008. Contributions of agroforestry research to livelihood of smallholder farmers in Southern Africa: 1. Taking stock of the adaptation, adoption and impact of fertilizer tree options. *Agricultural Journal* 3, 58–75.

Akinnifesi, F. K., Ajayi, O. C., Sileshi, G., Chirwa, P. W. and Chianu, J. 2010. Fertilizer tree systems for sustainable food security in the maize-based production systems of East and Southern Africa Region: a review. *Journal of Sustainable Development* 30, 615–29. doi:10.1051/agron/2009058.

Alcántara, C. 2005. Selección y manejo de especies crucíferas para su uso como cubiertas vegetales en olivar. Tesis Doctoral. Departamento de Ciencias y Recursos Agrícolas y Forestales, Universidad de Córdoba.

Alcántara, C., Sánchez, S. and Saavedra, M. 2004. Siega mecánica y capacidad de rebrote de cubiertas de crucíferas en olivar. *Phytoma* 155, 14–7.

Alcántara, C., Pujadas, A. and Saavedra, M. 2011. Management of cruciferous cover crops by mowing for soil and water conservation in southern Spain. *Agricultural Water Management* 98, 1071–80.

Al-Khatib, K., Libbey, C. and Boydston, R. 1997. Weed suppression with *Brassica* green manure crops in green pea. *Weed Science* 45, 439–45.

Arif, S., Shahrakbah, Y. and Kee, K. K. 2007. Impact of leguminous covers and palm chips on soil nutrient losses in oil palm replants. In: *Proceedings of International Conference on Oil Palm and Environment (ICOPE)*, Bali, Indonesia, Preprint, 16pp.

Barnes, R. D. and Fagg, C. W. 2003. Faidherbia albida. Monograph and annotated bibliography. Tropical Forestry Papers No. 41. Oxford Forestry Institute, Oxford, UK.

Botoni, E. and Reij, C. 2009. *La transformation silencieuse de l'environnement et des systèmes de production au Sahel: L'impacts des investissements publics et privés dans la gestion des ressources naturelles*. Comité Permanent Inter-Etats de Lutte Contre la Secheresse dans le Sahel (CILSS) and Vrije University Amsterdam, Amsterdam, the Netherlands.

Boydston, R. A. and Hang, A. 1995. Rapeseed (*Brassica napus*) green manure crop suppresses weeds in potato (*Solanum tuberosum*). *Weed Technology* 9, 669–75.

Broekhuyse, J. T. 1983. *Transformatie van Mossi land*. Koninklijk Instituut voor de Tropen, Amsterdam, the Netherlands.

Carbonell, R., Ordóñez, R. and Rodríguez, A. 2010. Influence of olive mill waste application on the role of soil as a carbon source sink. *Climatic Change* 102, 625–40.

Chiu, S. B., Gatot, A. R. and Saiful, R. A. 2011. Agronomic principles and practices of groundcover management of oil palm. In: *Seminar on Agronomic Principles and Practices of Oil Palm Cultivation*. Agricultural Crop Trust, Sibu, pp. 199–231.

CIRAD. 2012. *Proceedings of the 3rd International Conference on Conservation Agriculture in Southeast Asia: Conservation Agriculture and Sustainable Upland Livelihoods -- Innovations for, with and by Farmers to Adaptable to Local and Global Changes* - Hanoi, Vietnam, 10–12 December 2012. CIRAD, France.

Davis, J. R., Huisman, O. C., Westermann, D. T., Hafez, S. L., Everson, D. O., Sorensen, L. H. and Schneider, A. T. 1996. Effects of green manures on Verticillium wilt of potato. *Phytopathology* 86(5), 444–53.

De Graaff, J. and Eppink, L. A. A. J. 1999. Olive oil production and soil conservation in southern Spain, in relation to EU subsidy policies. *Land Use Policy* 16, 259–67.

Durán-Zuazo, V. H., Rodríguez-Pleguezuelo, C. R., Arroyo-Panadero, L., Martínez-Raya, A., Francia-Martínez, J. R. and Cárceles-Rodríguez, B. 2009. Soil conservation measures in rainfed olive orchards in South-Eastern Spain: impacts of plant strips on soil water dynamics. *Pedosphere* 19, 453–64.

EUROSTAT. 2018. Agricultural production – crops. Available at: http://ec.europa.eu/eurostat/statistics-explained/index.php/Agricultural_production_-_crops.

FAO. 2010. An international consultation on integrated crop-livestock systems for development. *Integrated Crop Management* (vol. 13). FAO, Rome, 79p.

FAO. 2011. *Save and Grow: A Policymaker's Guide to Sustainable Intensification of Smallholder Crop Production*. FAO, Rome, 102pp.

FAO. 2012. Some commonly used cover crop species. Available at: http://www.fao.org/ag/ca/2b.html (accessed on 15 June 2013).

FAO. 2018. Conservation Agriculture website. Available at: http://www.fao.org/conservation-agriculture/en/.

Francia, J. R., Durán, V. H. and Martínez, A. 2006. Environmental impact from mountainous olive orchards under different soil-management systems (SE Spain). *Science of the Total Environment* 358, 46–60.

Franklin, D., Truman, C., Potter, T., Bosch, D., Strickland, T. and Bednarz, C. 2007. Nitrogen and phosphorus runoff losses from variable and constant intensity rainfall simulations on loamy sand under conventional and strip tillage systems. *Journal of Environmental Quality* 36(3), 846–54. doi:10.2134/jeq2005.0359.

Garcia-Ruiz, J. M. 2010. The effects of land uses on soil erosion in Spain: a review. *Catena* 81(1), 1–11. doi:10.1016/j.catena.2010.01.001.

Garrity, D. P. 2002. Increasing the scope for food crop production on sloping lands in Asia: contour farming with natural vegetative strips in the Philippines. In: Uphoff, N. (Ed.), *Agroecological Innovations: Increasing Food Production with Participatory Development*. Earthscan, London, pp. 221–32. Chapter 11.

Garrity, D. P., Akinnifesi, O., Ajayi, O., Sileshi, G. W., Mowo, J. G., Kalinganire, A., Larwanou, M. and Bayala, J. 2010. Evergreen Agriculture: a robust approach to sustainable food security in Africa. *Food Security* 2(3), 197–214.

Gómez, J. A., Llewellyn, C., Bash, G., Sutton, B., Dyson, J. S. and Jones, C. A. 2011. The effects of cover crops and conventional tillage on soil and runoff loss in vineyards and olive groves in several Mediterranean countries. *Soil Use and Management* 27, 502–14.

Gómez-Limón, J. A., Picazo-Tadeo, A. J. and Reig-Martínez, E. 2012. Eco-efficiency assessment of olive farms in Andalusia. *Land Use Policy* 29, 395–406.

Gonzalez-Sanchez, E. J., Veroz-González, O., Blanco-Roldan, G. L., Marquez-Garcia, F. and Carbonell-Bojollo, R. 2015. A renewed view of conservation agriculture and its evolution over the last decade in Spain. *Soil and Tillage Research* 146, 204–12.

Gudeta, S., Akinnifesi, F., Ajayi, O. and Place, F. 2009. Evidence for impact of green fertilizers on maize production in sub-Saharan Africa: a meta-analysis. ICRAF Working Paper. World Agroforestry Centre, Nairobi, Kenya.

Haggblade, S. and Tembo, G. 2003. Early evidence on conservation farming in Zambia. EPTD Discussion Paper 108. International Food Policy Research Institute, Washington DC.

IRENA. 2019. *Sustainable Harvest: Bioenergy Potential from Agroforestry and Nitrogen-Fixing Wood Crops in Africa*. International Renewable Energy Agency, Abu Dhabi.

Kang, B. T., Wilson, G. F. and Lawson, T. L. 1984. *Alley Cropping: A Stable Alternative to Shifting Cultivation*. International Institute of Tropical Agriculture, Ibadan, Nigeria.

Kassam, A., Basch, G., Friedrich, T., Shaxson, F., Goddard, T., Amado, T., Crabtree, B., Hongwen, L., Mello, I., Pisante, M. and Mkomwa, S. 2013. Sustainable soil management is more than what and how crops are grown. In: Lal, R. and Stewart, R. A. (Eds), *Principles of Soil Management in Agro-Ecosystems. Advances in Soil Science*. CRC Press.

Kassam, A., Mkomwa, S. and Friedrich, T. (Eds). 2016. *Conservation Agriculture for Africa: Building Resilient Farming Systems in a Changing Climate*. CABI Publishing, Wallingford, UK.

Kassam, A., Friedrich, T. and Derpsch, R. 2019. Global spread of Conservation Agriculture. *International Journal of Environmental Studies* 76(1), 29–51. doi:10.1080/00207233.2018.1494927.

Khalid, H., Zin, Z. Z. and Anderson, J. M. 1999a. Quantification of oil palm biomass and nutrient value in a mature plantation. I. Above-ground biomass. *Journal of Oil Palm Research* 11(1), 23–32.

Khalid, H., Zin, Z. Z. and Anderson, J. M. 1999b. Quantification of oil palm biomass and nutrient value in a mature plantation. I. Below-ground biomass. *Journal of Oil Palm Research* 11(2), 63–71.

Khalid, H., Zin, Z. Z. and Anderson, J. M. 1999c. Effects of oil palm residues management at replanting on soil nutrient dynamics and oil palm growth. In: Arrifin, D., Chan, K. C. and Sharifah, S. R. (Eds), *Proceedings of the 1999 PORIM Integrability Palm Oil Congress*. Palm Oil Research Institute of Malaysia, Kuala Lumpur, pp. 235–46.

Makumba, W., Janssen, B., Oenema, O., Akinnifesi, F. K., Mweta, D. and Kwesiga, F. 2006. The long-term effects of a Gliricidia-maize intercropping system in southern Malawi, on Gliricidia and maize yields, and soil properties. *Agriculture, Ecosystems and Environment* 116, 85–92.

Matlon, P. J. 1990. Improving productivity in sorghum and pearl millet in semi-arid Africa. *Food Research Institute Studies* 22(1), 1–43.

Matlon, P. J. and Spencer, D. S. 1984. Increasing food production in sub-Saharan Africa: environmental problems and inadequate technical solutions. *American Journal of Agricultural Economics* 66(5), 672–6.

Márquez-García, F., González-Sánchez, E. J., Castro-García, S. and Ordóñez-Fernández, R. 2013. Improvement of soil carbon sink by cover crops in olive orchards under semiarid conditions. Influence of the type of soil and weed. *Spanish Journal of Agricultural Research* 11, 335–46.

Mojtahedi, H., Santo, G. S., Wilson, J. H. and Hang, A. N. 1993. Managing meloidogyne chitwoodi on potato with rapeseed as green manure. *Plat Disease* 77, 42–6.

Monimart, M. 1989. *Femmes du Sahel: La désertification au quotidien*. Editions Karthala/Organisation for Economic Cooperation and Development Club du Sahel, Paris.

Moreno, M., Garcia, S., Cañizares, R., Castro, J. and Benítez, E. 2009. Rainfed olive farming in south-eastern Spain: long-term effect of soil management on biological indicators of soil quality. *Agriculture, Ecosystems and Environment* 131, 333–9.

Ng, P. H. C., Goh, K. J., Gan, H. H., Yacob, S., Kee, K. K. and Teo, C. B. 2008. Introduction of shade-tolerant cover crop, *Mucuna bracteata*, enhances the sustainability of oil palm (*Elaeis guineensis*). In: *Proceedings of 5th International Crop Science Congress 2008*, 13–18 April 2008, Jeju South Korea, Preprint.

Ng, P. H. C., Goh, K. J., Gan, H. H., Yacob, S. and Zaharah, A. R. 2010. The use of phosphate rocks for growing *Mucuna bracteata* in oil palm legume systems to enhance sustainability. In: *Proceedings of MSSS 2010 International Conference*, 12–16 April 2010, Kuantan Malaysia. Preprint.

Ooi, L. H. 2011. Agronomic principles and practices of replanting oil palm. In: *Seminar on Agronomic Principles and Practices of Oil Palm Cultivation*. Agricultural Crop Trust, Sibu, pp. 171-97.

Ordóñez, R., Rodríguez-Lizana, A., Espejo, A. J., González, P. and Saavedra, M. 2007. Soil and available phosphorus losses in ecological olive groves. *European Journal of Agronomy* 27, 144-53.

Pastor, M. 2008. Sistemas de manejo de suelo. In: Barranco, D., Fernández-Escobar, R. and Rallo, L. (Eds), *El cultivo del olivo*. Mundi Prensa, Madrid, Spain, pp. 19-35.

Phombeya, H. S. K. 1999. Nutrient sourcing and recycling by *Faidherbia albida* trees in Malawi. PhD Dissertation. Wye College, University of London, 219pp.

Pushparajah, E. and Chellapah, K. 1969. Manuring of rubber in relation to covers. *Journal of Rubber Research Institute of Malaya* 21(2), 126.

Pushparajah, E., Tan, K. H. and Soong, N. K. 1977. Influence of covers and fertilizers and management on soil. In: *Soils Under Hevea in Peninsular Malaysia and Their Management*. Rubber Research Institute of Malaysia, Kuala Lumpur, pp. 75 - 93.

Reij, C. 1983. *L'évolution de la lutte anti-érosive en Haute Volta: Vers une plus grande participation de la population*. Institute for Environmental Studies, Vrije University, Amsterdam, the Netherlands.

Reij, C. and Thiombiano, T. 2003. *Développement rural et environnement au Burkina Faso: La réhabilitation de la capacité productive des terroirs sur la partie nord du Plateau Central entre 1980 et 2001*. Ambassade des Pays-Bas, German Agency for Technical Cooperation- ATECORE, and U.S. Agency for International Development, Ouagadougou, Burkina Faso.

Reij, C., Tappan, G. and Smale, M. 2009. Agroenvironmental transformation in the Sahel: another kind of "Green Revolution". IFPRI Discussion Paper 00914. International Food Policy Research Institute, Washington DC.

Saavedra, M. 2003. El manejo de la cubierta vegetal en el control de la erosión en Olivar. En. In: Bienes, R. and Marqués, M. J. (Eds), *Perspectivas de la Degradación del Suelo. I Simposio Nacional sobre el Control de la Erosión y Degradación del Suelo, Madrid*, pp. 43-54.

Saka, A. R., Bunderson, W. T., Itimu, O. A., Phombeya, H. S. K. and Mbekeani, Y. 1994. The effects of *Acacia albida* on soils and maize grain yields under smallholder farm conditions in Malawi. *Forest Ecology and Management* 64, 217-30.

Shetty, K. G., Subbarao, K. V., Huisman, O. C. and Hubbard, J. C. 2000. Mechanism of broccoli-mediated Verticillium wilt reduction in cauliflower. *Phytopathology* 90(3), 305-10. doi:10.1094/PHYTO.2000.90.3.305.

Sim, C. C., Teo, C. B. and Tey, S. H. 2008. Replanting for sustainable high yield. *The Planters* 84(992), 739-51.

Smolinska, U. and Horbowich, M. 1999. Fungicidal activity of volatiles from selected cruciferous plants against resting propagules of soil-borne fungal pathogens. *Journal of Phytopathology* 147, 119-24.

Snapp, S. S., Mafongoya, P. L. and Waddington, S. 1998. Organic matter technologies for integrated nutrient management in smallholder cropping systems of southern Africa. *Agriculture, Ecosystems and Environment* 71, 185-200.

Taguas, E. V., Moral, C., Ayuso, J. L., Pérez, R. and Gómez, J. A. 2011. Modelling the spatial distribution of water erosion within a Spanish olive orchard microcathment using SEED model. *Geomorphology* 133, 47-56.

Tan, K. H. 1975. Hevea leaf litter fall and chemical changes during litter decomposition. In: *3rd ASEAN Soil Conference*, Kuala Lumpur.

Van Noordwijk, M. 2010. Personal communication between Dr. Dennis Garrity and Dr. Meine van Noordwijk, Chief Scientist, World Agroforestry Centre, Bogor, Indonesia.

Vanwalleghem, T., Laguna, A., Giráldez, J. V. and Jiménez-Hornero, F. J. 2010. Applying a simple methodology to assess historical soil erosion in olive orchards. *Geomorphology* 114, 294–302.

Vrignon-Brenas, S., Gay, F., Ricard, S., Snoeck, D., Perron, T., Mareschal, L., Laclau, J.-P., Gohet, É. and Malagoli, P. 2019. Nutrient management of immature rubber plantations. A review. *Agronomy for Sustainable Development* 39(1), 11. doi:10.1007/s13593-019-0554-6.

Watson, G. A., Wong, P. W. and Narayanan, R. 1963. Effects of cover plants on soil nutrient status and on growth of Hevea. II. Influence of applications of rock phosphate, basic slag and magnesium limestone on the nutrient of leguminous cover plants. *Journal of Rubber Research Institute of Malaya* 18, 28.

Wolfe, D. 2000. Summer covers relieve compaction. In: Clark, A. (Coord.), *Managing Cover Crops Profitably* (2nd edn.). Sustainable Agriculture Network, Beltsville, p. 84.

Yacob, S., Arif, S. and Kee, K. K. 2011. Soil and water conservation under oil palm cultivation. In: *Seminar on Agronomic Principles and Practices of Oil Palm Cultivation*. Agricultural Crop Trust, Sibu, pp. 483–509.

Zaharah, A. R. and Bah, A. R. 1997. Effect of green manures on P solubilization and uptake from phosphate rocks. *Nutrient Cycling in Agroecosystems* 48, 247–55.

Chapter 9

Integration of crop-livestock in Conservation Agriculture systems

John N. Landers, Independent Consultant, Brazil; Pedro Luiz de Freitas, Embrapa Solos, Brazil; Luiz Carlos Balbino, Embrapa Cerrados, Brazil; Júlio César Salton, Embrapa Agropecuária Oeste, Brazil; and Robélio Leandro Marchão, Embrapa Cerrados, Brazil

1 Introduction

Alternation of crops and livestock has been good farming practice for at least 500 years in modern Europe (Crittall, 1959). An integrated crop-livestock system was demonstrated by Jethro Tull in England (1751), who showed 'How to feed turnips to sheep and other animals; direct grazing with moveable fences to control the rate at which the turnips are grazed'. In Italy, during the Roman empire, fallows were enriched with grasses and legumes to form a ley, inserted in arable farming to re-generate soil fertility for the next arable phase (Kron, 2000; White, 1970). Sir George Stapledon in his book *Ley Farming* (Stapledon, 1941) described twentieth-century ley farming systems alternating a short temporary pasture phase with a crop phase, as a contribution to the Second World War effort, converting low-yielding pastures into crop land.

http://dx.doi.org/10.19103/AS.2019.0048.10

This chapter will examine the state-of-the-art of different Integrated Crop-Livestock systems in various eco-regions worldwide under Conservation Agriculture (CA). It is not an exhaustive literature review, but selects research work and farmer experiences which are illustrative of the principles involved in.

Integrated Crop-Livestock management systems under Conservation Agriculture (ICL-CA) with zero tillage (ZT) have evolved rapidly over recent years, led principally by innovative farmers and inquisitive scientists examining their solutions plus their unsolved problems and proposing more innovations. This combination triggered the deciphering of the scientific principles underlying the lessons of ICL-CA practice today, in its present sustainable modes. Soussana et al. (2014) indicated a pre-occupation with a lack of agro-ecological principles in CA; however, the authors note an increasing historical tendency to involve ecological questions in ICL-CA studies, but no inclusion of ZT in FAO's ten principles of agroecology; perhaps a too-strict definition of agro-ecology is responsible for this apparent dichotomy.

This chapter draws heavily on Brazilian experience, not only because of author affiliations, but due to the fact that Brazil is a current world leader in ICL-CA and the same system with forestry (ICL-CA (F)) adoption and research, currently with 11.5 million ha under these systems (83% under ICL-CA) and a government-sponsored Low Carbon Agriculture programme (Cordeiro et al., 2015) supporting research and extension towards achieving Brazil's international goals to reduce GHG emissions. The process of pasture degradation will not be covered here, since it is dealt with in detail by Macedo et al. (2013) and Boddey et al. (2004). However, the process is considerably reduced or eliminated by ICL-CA.

Various recent systems of ICL-CA are outlined, tracing the common principles involved and the comparative merits of the various systems and their applicability to different environments. The scientific basis for the synergies and interactions in the rotation of crops and livestock is examined separately, as are the impacts on soil health and our environment. The latter include increases in soil organic matter (SOM) and biological activity/diversity, mitigation of the demand for clearing native vegetation through land use intensification (LUI), increased aquifer recharge and reduced flooding, and reductions in the emission of greenhouse gases (GHGs). Also, opportunities for synergies with ICL-CA can be achieved with innovative technologies, such as biological pest and disease controls (BC) and use of natural rock meal (both substituting agricultural chemicals, controlled traffic farming (CTF) to avoid deep soil compaction and precision agriculture, to rationalize the use of resources). ICL-CA also provides better cover and winter feed for wildlife and, most important, almost eliminates erosion losses on farms. Other beneficial effects off-farm occur in lower silting of water bodies and lower water treatment costs (Bragagnolo et al., 1997).

1.1 Background

Faulkner (1943) in his book *Plowman's Folly* argued that no scientific proof had ever been produced to justify ploughing. At the time, he was preaching shallow minimum tillage. But from the outset of ZT in Kentucky, USA (Phillips and Young, 1973), it was popularized by the name 'No-Till' (NT) and was introduced with the non-systemic 'burn-down' chemical Paraquat, to renovate pastures on hilly land without causing erosion. This partnership between practice and theory has proved very effective in developing CA, ICL-CA and (ICL-CA(F)) worldwide.

The agronomic importance of ICL-CA in the United States, Brazil and other countries was not immediately accepted by the majority of stakeholders. There was considerable resistance to this fundamental upheaval of the sacrosanct paradigm of ploughing. University lecturers and professors had to re-write their lecture notes, extensionists had to change their teachings and the machinery lobby actively campaigned against CA, fearing loss of sales for their conventional tillage equipment.

After the arrival of the game-changing systemic desiccant Glyphosate, CA, ZT or NT developed vertiginously. Led by the more inquisitive individuals, American, Brazilian, Australian and other New World scientists forged ahead to unveil the mysteries of natural law which now underpin CA worldwide. This permits us to foresee fully sustainable ICL-CA and ICL-CA(F) systems as the best potential solution to meeting increasing world food demand and conserving the planet's natural resources (Landers, 2007; Landers and Freitas, 2001). Without CA, especially ZT/NT it cannot be overstressed that there is a gradual attrition of SOM, soil carbon and soil itself occurring under conventional tillage (CT) systems, through oxidation of SOM (Lal et al., 2018) and erosion (Reicosky, 2018) that will eventually compromise their sustainability. Here, we deal only with CA systems (see discussion of origins and definition below). This innovation was incorporated into ICL, making pasture renovation easy and without risk of erosion, while with CT this is less feasible. In our terminology, CA is here added to ICL to emphasize that it is a necessary adjunct to sustainability namely: ICL-CA.

André Voisin (1957) gave the following definition 'Ley farming is the regular introduction into a crop rotation of a herbage crop of varying duration, which should be grazed as much as possible'. The reason for the last statement is that cutting and carrying forage eliminates the recycling effect of nutrients being returned as dung and urine to the herbage crop. This stimulates various kinds of biological activity. Voisin noted that earthworm activity in an old pasture brought to the surface 57 ton $ha^{-1} yr^{1}$ of worm casts, six times more than a normal application of animal manure. Since ICL-CA was not known at that time, it is reasonable to assume that most of the 'feeblenesses' in ley farming described by Voisin (1953) were caused by ploughing (full inversion tillage). His use of the term 'herbage crop' encompasses both grasses and other forage species, generally legumes, forbs and mixtures of same. At that time, pure legume leys of Sainfoin (*Onobrychis viciifolia*) or Lucerne (*Medicago satina*) were also common in Europe and Australia. Sainfoin leys are used for green chop, hay and grazing (Carbonero et al., 2011) and this also applies to lucerne. In the United States, Sulc and Franzluebbers (2014) note that phases in ley farming (sod-based farming) typically involved 2–10 years of perennial forages, rotated with 1–8 years of cropping, but this has been supplanted by intensive continuous cropping.

Following the First World Congress on Conservation Agriculture (Madrid, 2001), these concepts were enshrined within the term 'Conservation Agriculture' (Vaneph and Benites, 2001). The three principles originally embodied in CA are essentially the same as those of CA as shown in Box 1:

Box 1 Definition of Conservation Agriculture

- **Continuous no, or minimum, soil disturbance** (almost universally zero tillage prior to crop establishment);
- **Permanent soil cover** (crop residue and/or live crop covering the surface);
- **Diversified cropping with annual and perennial crops including legumes** (sequences/associations in pluri-annual rotations, that is not repeating the same crop on the same land in the same or following year).

This definition of CA also covers systems such as the Zai hole planting of maize in Zambia.

Today, we are privileged to participate in the biggest revolution ever seen in world agriculture (Borlaug, pers. comm., 1993), bigger even than the introduction of the tractor, the evolution in plant genetics and agricultural chemicals. Working with nature, instead of against her, has led to a fundamental understanding of the factors which maintain and improve soil health (Reicosky, 2018). The principles of CA are universal, but the solutions are local.

2 Basic principles of integrated crop-livestock systems

These ICL-CA systems combine crop and livestock activities in different ways. The most important system is ley farming, but other, non-ley farming systems exploit stubble and fallow grazing or silage, hay, root and other fodder plus cover crops or green chop on, or from, arable areas. The efficient use of nutrients, agrochemicals and energy, together with the reduction of GHG emissions are key factors to enhance eco-efficiency of the system (Pacheco et al., 2015).

Area, or spatial, integration occurs when graziers utilize communal areas, or stubble grazing on the farm, with cattle brought from outside, and when livestock feeds and supplements are exported to a livestock enterprise outside the farm. It also covers the application of manure from confined animals outside the farm and other off-farm wastes (e.g. sewage sludge) applied to crops on-farm. As this covers a plethora of situations, area integration will not be treated in depth in this chapter, which will deal principally with on-farm systems. Box 2 lists the synergies identified in ICL-CA systems:

Box 2 Synergies identified in ICL-CA on-farm systems

- a crop or pasture phase leaves residual nutrients for pasture establishment or crop nutrition;
- livestock manure and urine enhance nutrient cycling and soil biological activity;
- the pasture phase renovates soil structure through improved crumb structure and mesofauna activity, which increases rainfall infiltration rates (under good grazing management);
- ICL-CA systems increase soil and above-ground biodiversity, empowering biological controls of pests, weeds and diseases;
- improved oxygen supply to roots and sequestering of carbon, provided the system accumulates SOM;
- the crop phase supplies supplemental grazing in the crop area (of stubble and/or cover/pasture crops), at a time when permanent pasture carrying capacity is low;
- when rainfed pasture carrying capacity is low, supplying grazed or cut forage from cropped areas effectively increases the farm herd size through enhanced off-season carrying capacity (Vasconcelos and Landers, 1993);
- livestock manure and urine enhance nutrient cycling and soil biological activity, this is especially true in promoting the increase in soil microbial biomass (Souza et al., 2010);
- the role of non-crop plant roots should be stressed, in improving soil structure (via root exudates, mainly of grasses, that cement soil crumbs, and macropores left from desiccated crop and pasture roots) soil porosity, soil resilience and water-holding/ infiltration capacity; this latter is also enhanced by macrofauna burrows, principally earthworms and Coleoptera larvae (Marchão, 2007);
- ICL-CA systems increase soil biodiversity of all species, especially microfauna communities important for the empowerment of biological controls and solubilization of nutrients;

- labour peaks in different activities are evened out over the year;
- the mix of enterprises reduces financial risks (due to low prices or low yields);
- a whole-system fertilization approach is applied, spreading the cost of fertilization over all the enterprises;
- use of agricultural chemicals is reduced and environmental impacts minimized (Landers and Freitas, 2001);
- maintaining a mixed farming system instead of a specialized cropping or livestock farming system provides a greater range of possible future land use pathways (Moore, 2014);
- LUI reduces the demand for horizontal expansion on to native vegetation (Landers and Freitas, 2001).

Franzluebbers et al. (2014) stated that ICL-CA can 'reclaim ecological interactions that make agro-ecosystems more efficient at cycling nutrients, preserving natural resources, improving soil quality and enhancing biodiversity'. On the other hand, with integration of crop and livestock activities, demands on management, labour skills and investment are increased (Sulc and Franzluebbers, 2014).

3 Basic principles of ley farming

'Ley farming' under ICL-CA is defined as: 'Rotation of a Zero Tillage (ZT) crop phase with a short or long pasture ley, also established under CA'. Typically, these phases are for 1–4 years and the crop phase of similar length. Given adequate rainfall, or irrigation, up to three different crops a year may be possible, including a temporary catch crop ley sown into, or after, the main or second arable crop, for grazing, or as a cover crop until the next annual crop is planted.

With the advent of short-season soybeans in Brazil, a second crop of maize began to supplant fallow or millet and sorghum as cover crops. Later, the undersowing or intersowing of pasture grasses in an arable crop allowed cattle to be integrated into the system and also the increase of a cover crop to substitute for unproductive fallow over winter. This is depicted in Fig. 1 for the high rainfall region of northern Mato Grosso (ca. 1800 mm $ha^{-1} yr^{-1}$). In South Africa, Grain (2018) recommend 'After maize or soybeans, or late in the growing season (as a winter cover crop or relay crop) cereal rye, annual ryegrass, wheat, oats or triticale as grasses can be planted to accumulate soil organic matter, recycle nutrients and reduce soil compaction. Brassicas (oilseed or radish or turnips) is used to recycle nutrients, reduce compaction and promote weed and disease suppression. Vetch can be planted as a nitrogen source'. See Chapter 5 for more on cover crops.

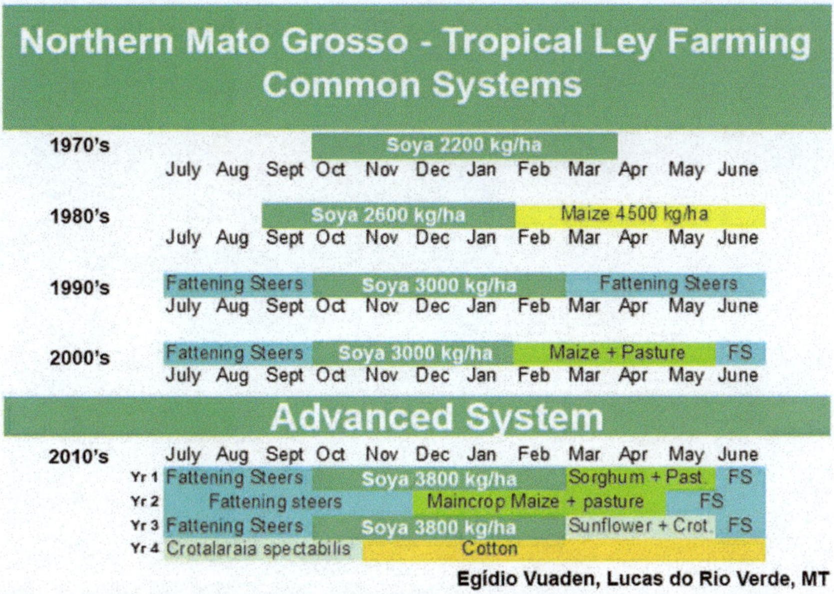

Figure 1 Evolution of CA rotations in northern Mato Grosso. Source: J. N. Landers.

3.1 Sowing of annual crops into degraded pasture land

Before planting the first crop into a degraded pasture, soil cultivation may be necessary to build up the soil profile by incorporating lime and/or (rock) phosphate or to remove a compacted layer, thereafter planting is with CA and all soil amendments are surface-applied (Freitas, 2005). The Embrapa São Mateus system (Salton et al., 2013), for degraded (tropical) pastures, carries out all soil amendments and physical operations to improve soil structure and then sows a temporary grass ley (5–9 months) followed by soybeans (see Fig. 2). Immediate planting of soybeans is not recommended, as adequate lead time is required for the amendments to be effective.

3.2 Establishing pasture or cover crops in or after annual crops

The original Santa Fé system (Kluthcouski and Aidar, 2003) has been widely adopted in tropical Brazil and was developed on non-degraded soils, undersowing maize with a pasture (*Urochloa* spp.). At maturity, the concept of fertilizing the system should be followed and all nutrients, except possibly N, should be applied for the system (Bernardi et al., 2003). To permit this, before planting the soil has to have enough P to supply the current crop. A summary of the planting/sowing methods used in various ICL-CA systems is as follows:

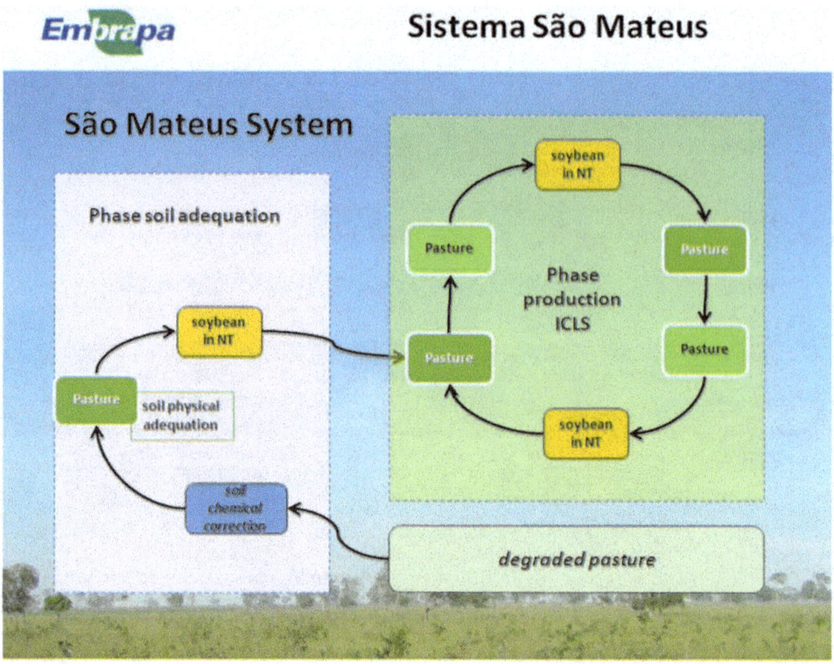

Figure 2 The São Mateus system for pasture renovation and cropping soybeans. Source: Salton et al. (2013).

- Undersowing
  - Pasture or cover crop seed is sown in the row of the annual crop, superficially for small seeds (ca.1–2 cm) or a slightly greater depth for larger seeds (ca. 3–4 cm) – a third seed hopper is required on the drill for the legume/cover crop and the small seeds can be mixed with 50–100 kg ha^{-1} of single superphosphate, to assist metering;
  - If seeds are mixed with the fertilizer at planting, use the mix immediately and avoid using KCl, to eliminate depression of germination.
- Intersowing
  - The seed is sown between the crop rows, either with an extra seed hopper on the crop planter/drill, a separate pass of a drill or planter or a manual jab planter for small farmers, for example *Crotalaria* spp. (N + nematode control) or other legume spp. in maize or other crop.
- Oversowing
  - Cover crops and grasses oversown into soya, especially millet, sorghum and pasture spp.: seed is distributed over the crop just prior to leaf fall, either by a pendular fertilizer spreader (better distribution than a centrifugal type) or by spray plane (in this case the direction should be

transverse to the rows for the seed to lodge in the shade of the row), also seed should have the capacity to germinate easily on the surface, for example oats, sorghum, millet and Townsville stylo (*Stylosanthes humilis*) – the leaf fall covers the seed and helps with moisture retention. This system gains planting time but is more risky than planting into soil;
    - Forage peanut (*Arachis pintoi*) *oversown* into Bermuda grass (Sanchez et al., 2018);
    - Manual broadcasting is applicable for small farmers.
- Post-harvest sowing
    - Implies less rainfall and less stored soil moisture when compared to the two previous systems but ensures good seed/soil contact;
    - May be with drill or planter or hand-jab manual planter for a second grain crop (adequate moisture) or cover crops, single or multiple species (the latter preferred);
    - Pasture ley establishment with sown grass species or grass/legume mix;
    - Cover crops or pasture seed broadcast on to stubble (high risk);
    - Self-seeding of legume spp. in leys, for example *Medicago* spp. in Australia;
    - Perennial forage legume under- or intersown in annual crop to establish a permanent legume sward cover for continuous annual CA cropping plus dry-season high-protein grazing (Vasconcelos and Landers, 1993).

Technologies which establish a following crop before harvest of a main or second crop are termed relay cropping.

3.3 CA sowing of crops or forage legumes into living pasture

An example is planting pigeon pea (*Cajanus cajan*) into *Urochloa* spp. (Balbino and Di Stefano, 1999). Comparing, over a 3-month dry season, two early ICL-CA systems in Central Brazil with extensive grazing on degraded pasture and native cerrado vegetation, for a year 1 rice/pearl millet: year 2 soybean/pearl millet/pasture sequence showed a 19.9 kg ha^{-1} (417%) incremental year 2 gain in liveweight over the control 4.77 kg ha^{-1}, while a mixed *P.maximu* cv.Tanzania + pigeon pea pasture, in year 2 gave an incremental gain of 30.20 kg ha^{-1} (723%). This, and other, noteworthy results contributed to a huge increase in beef yield per hectare and widespread adoption of ICL-CA in the region.

Competition from the pasture plants can be checked with pre-plant in-row desiccation or post-emergent sub-lethal doses of selective herbicide. Some legumes, when established long enough to have good root reserves, for example perennial soybean (*Neonotonia wightii*) can re-sprout after a 2 lit ha^{-1} application of commercial Glyphosate; however, two 1.5 lit ha^{-1} of commercial Paraquat with a 10-day interval did not kill 1-year-old pigeon pea, but checked it sufficiently to not compete with intersown maize. Vasconcelos and Landers (1993) demonstrated planting of maize into an established legume sward of Siratro (*Macroptilium atropurpureum*) with equivalent yield to the farmer's CT maize (6000 kg ha^{-1}) but vigorous perennial soybean in the same system suppressed the maize yield.

3.4 De-coupling of crop and livestock enterprises

Moore (2014) showed a recent trend in southern Australia towards increased cropping areas on mixed farms and this is also the case in Europe. In the United States, Sulc and Franzluebbers (2014) observed that 'specialization in agricultural systems has resulted in decoupling of crop and livestock production, disrupting within-farm nutrient cycling, leading to large nutrient imbalances; excessive nutrient accumulation occurs on large confined animal feeding operations, while grain farms rely heavily on purchased fertilizers'. Powell et al. (2004) show a similar trend in lower rainfall areas of sub-Saharan Africa (SSA) and note that the economic incentive for a mixed farm enterprise to provide its own inputs diminishes as specialization becomes more profitable. This analysis is extensive to other countries with specialized, separate crop and confined livestock enterprises and emphasizes the sustainability of ICL-CA systems.

4 Integrated crop-livestock management (ICL-CA) systems: ley farming systems

4.1 Tropical systems: Brazilian tropical ley farming systems

Examples include:

- Soya-maize succession undersown with grass – 3–5 months grazed ley of *U. ruziziensis*;
- Soya/*U. brizantha* oversown or sown post-harvest for dry-season pasture – up to 5 months pasture without the grass seeding;
- Soya/maize succession (last year) – Forage sorghum (early bite) + *U. brizantha* 3–5-year ley;
- Late-sown soya – oversown forage sorghum, pearl or finger millet (*P. americanum or E. corocana*) /grazing 5 months max;
- Soy late-sown – oversown with Ruzi grass (*U. ruziziensis*) , hayed, followed by pasture of re-growth by weaners;

- Soy/rainfed rice/mixed cover crops (1 year): soy-maize/cover crops mix (2 years), very advanced system in high RF areas with stubble grazing (little uptake due to high management requirement);
- Santa Fé system (irrigated): soybeans for seed, maize for silage, undersown with *U. brizantha* for green chop;
- Santa Fé system (dryland): pasture grasses under- or intersown in (usually second crop) maize, to establish a pasture ley or cover crop (short or long duration). Most common grasses *U. ruziziensis, U. brizantha* and *P. Maximum* varieties (the latter on fertile soils).

The latter system is the most prominent dryland system in tropical Brazil and is being superseded on degraded pastures by the new São Mateus system. Merola (2002), the farmer who developed the system with Embrapa, enunciated the advantages of ICL-CA in Box 3:

Box 3 Advantages of ICL-CA listed by the tropical pioneer Ricardo Merola

- Reduced weed infestation;
- Soil de-compaction (after grass);
- Control of soil fungal pathogens and nematodes;
- Greater erosion control;
- Diversification of crops in the rotation;
- Reduction of consumptive use of water in irrigated crops;
- Higher bean and other crop yields;
- Lower use of fungicides and herbicides.

Source: Merola (2002).

U. ruziziensis or U. brizantha sown after the longest daylength do not receive the trigger for flowering and stay green, giving vegetative growth for several months after the end of the rains, exploring deep-stored soil water; winter temporary stocking rates of over 3 AU[1] ha^{-1} have been reported in West Bahia on corn stubble with *B. brizantha* for a period of 90 days (Döwich, pers. comm., 2004) and above 4 AU ha^{-1} on P. maximum cultivars in Matto Grosso do Sul (van der Vinne, pers. Comm., 2009).

Increases in crop yields following a pasture phase are a widespread phenomenon (Fernández et al., 2015; Kluthcouski et al., 2015; Garcia-Prechac, 1992; Garcia-Prechac et al., 2004; Stapledon, 1941). Both Kluthcouski and

1 Animal unit of 450 kg liveweight.

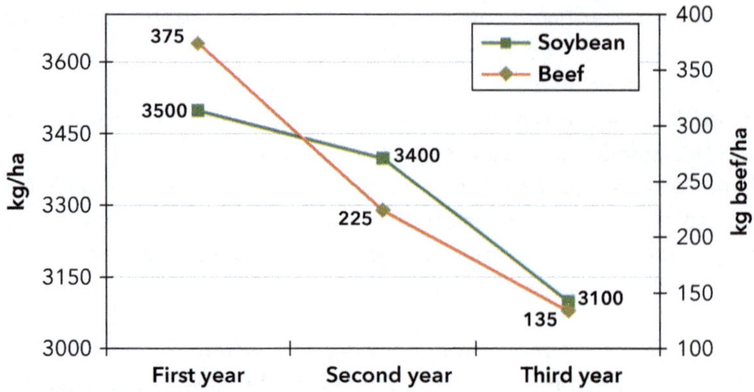

Figure 3 Progressive decrease of soybean yields after 3 years of pasture and in beef production after 3 years of soybeans. Source: Fundação. Mato Grosso do Sul.

Stapeldon also observed higher animal productivity in ley farming systems; Fernández et al. (2015) indicated that this was accompanied by high SOM and good grazing management. Figure 3 illustrates this well.

Since stocking rates fall off rapidly, the option for a 1-year ley and high-quality grasses, for example. *P. maximum* cultivars (on high-fertility-status soils), ensures the highest possible winter stocking rate and the smallest proportion of farm area in winter pasture. Under this system, fat grass-fed steers are being sold at 24 months old (van der Vinne, pers. comm., 2004). Oversowing pasture seed in soybeans has had varied success due to irregular rainfall, while intersowing and undersowing have been achieved without yield depression, at the cost of one or two applications of an under-dose of graminicide, to suppress competition without killing the grass (Cobucci and Portela, 2003). Detailed work in the maize/*Urochloa* spp. association by the same authors has shown minimal depression of maize yield when recommended planting instructions are followed. In well-managed ICL-CA(F) systems in Brazil, Marchão and Vilela (2018) showed 'positive impacts of ICL-CA(F) systems include: a) increases by 15 to 20% in soil organic matter in comparison to organic matter levels of native Cerrado; b) increases of 90% in phosphorous use efficiency, in the long term, in comparison to that verified in soybean-corn rotation; c) yield gains in soybeans of 10 to 15% when succeeding fertilized and high productivity pastures; d) average increases in animal productivity in the fattening stage by about 4 times (600 kg of live weight/hectare/ year in relation to the breeding-fattening stage of traditional livestock with 120-150 kg of live weight/hectare/year); e) animal productivity average increases in the breeding stage by about three times (300 kg of weaned calves/ hectare/ year in comparison to traditional livestock with 85-110 kg of weaned calves/ hectare/year)'.

4.2 Tropical systems: other countries

Examples include:

- Coarse grains/legume ley – Australia;
- Extensive woodland summer grazing/crop residues in dry season – Australia;
- Interplanting of maize with velvet bean (*Stizolobium aterrimum*), cowpea (*Vigna unguiculata*) and lablab (*Dolichos lablab*) – Zimbabwe.

The benefits of a pasture ley will generally increase with the length of the ley. In the humid tropics of Brazil, short-season soybeans permitted second-crop maize inter- or undersown with modern pasture grasses such as Palisade grass and Ruzi and new *P. maximum* cultivars. The previous drought-resistant second-crop options after soybeans, of forage sorghum (*S. bicolor*) and pearl or finger millets (*Pennisetum americanum*, *Eleusine corocana*) are now reserved for the very end of the rains, either as a cover crop or pasture. The system used in northern Australia is more extensive, combining summer native woodland pasture and crop residues in the dry season. Attempts at using Townsville stylo (*Stylosanthes humilis*) as a legume ley in rotation with sorghum in this arid region have not met with widespread success (McCown, 1996). Interplanting of maize with legumes for dry-season grazing of enriched maize stover has shown some success in Zimbabwe.

In Acre (Amazon biome), Valentim et al. (2001, 2003) associated forage peanut(*Arachis pintoi)* cv Belmonte with Bermuda grass cv. Coastcross (*Cynodon dactylon*) and cv. Massai (*P. Maximum*). In the Brazilian tropical Cerrado, the mix of two *Stylosanthes* spp., *S. capitata* and *S. macrocephala*, was launched as a single cultivar 'Campo Grande', but management difficulties, high seed price had inhibited uptake. However, this cultivar is being used successfully as a late-season cover crop (Pradella, pers. comm., 2019) in West Bahia, Brazil (1100–1300 mm yr^{-1}). McCown (1996) drilled grass into an established legume in semi-arid tropical Australia, but concluded that a legume ley was not efficient in that climate.

Vasconcelos and Landers (1993) demonstrated legume sward establishment undersown in maize, with only 2–4 kg ha^{-1} of seed, allowing direct drilling of grass seed at the start of the next rains, when the legume tap root would allow it to compete with newly sown grass. The system was discontinued for lack of a herbicide to control these legumes in soybeans. It is now possible to control excessive competition in glyphosate-resistant crops with sub-doses of glyphosate, which broadens the choice of legume to more vigorous species, able to compete better with grass, for example Perennial Soybean cv. IRI 1394 (Tinaroo) and Centro (*Centrosema pubescens* IRI 1282). This new technology can establish grass/legume associations cheaply and accelerate their adoption, provided the forages are locally adapted.

Results of intercropping research in Zimbabwe (Mapiye et al., 2007) with velvet bean (*Mucuna pruriens*), cowpea (*Vigna unguiculata*) and lablab (*Lablab purpureus*) in maize, Napier (*Pennisetum purpureum*), Bana grass (*Paspalum notatum*), millet or sorghum (short dry-season ley) were successful, except where the legume was too vigorous for the companion crop. Intercropping with legumes increased herbage yield by 40–60%, although the cereal grain yields were depressed by up to 20%. In Brazil, intersowing of Signal grass (Urochloa decumbens) and cowpea after soybeans showed an 8.43% gain in beef production (3 kg beef ha^{-1}) over a pure stand of Signal grass for 90 days and 10.3% on the following soybean yield (+420 kg ha^{-1}), where these gains were attributed to the cowpea fixing N in the soil and improving protein content of the pasture.

4.3 Subtropical systems: Brazilian subtropical systems

Examples include:

- Soya/second-crop maize/undersown with pasture, winter-grazed (lower latitudes);
- Soya/annual ryegrass (*L. italicum*), winter-grazed (higher latitudes);
- Soya/black oats (*Avena strygosa*)/ *L. italicum,* winter-grazed;
- Irrigated rice/winter stubble + native pasture or *L. italicum,* winter-grazed (higher latitudes).

4.4 Subtropical systems: other countries

Examples include:

- Alternating wheat/legume ley systems with sheep in the wheat/sheep belt (Australia);
- 2 years pasture/peanut/cotton/2years pasture (Florida, USA);
- Fish pond with livestock manure (Asia, Africa);
- Manure slurry from on-farm dairy/feedlot or off-farm sewage sludge/ compost distributed onto pasture or arable area (many countries);
- Manure biodigestor residue for crop fertilization (India, Brazil and others);
- Rice/stubble grazing (Asia) is covered in an earlier chapter.

In southern Brazil, the soya/wheat succession leaves little room for grazing after wheat harvest. Cold winters also limit the choice of forage. Here, successful

ICL-CA often consists of a rotation or succession of soybean, corn, black beans (*Phaseolus vulgaris*) or rice under CA throughout the summer, combined with winter annual grazing of mixed or individual black oat and ryegrass crops and native pasture (Mendes et al., 2018); variable yields of winter wheat are exchanged for winter pasture.

Many innovative ley farming systems are found in Australia's wheat-sheep belt (Nichols et al., 2012). The northern belt is in interior subtropical eastern Australia, with annual rainfall between about 580 and 750 mm. It transitions through all year to summer rainfall. In Table 1, GRDC (2010) recommends grass and legume species for leys in rotation with wheat in subtropical Queensland.

Pasture species must also be matched to appropriate climate and soil conditions to maximize performance. Farmers in the southeastern United States are seeing potential in a sod-based (ley) rotation of 2 years with Bahiagrass (*Paspalum notatum*) or Bermuda grass, (*Cynodon dactylon*), followed by 1 year of peanut (*Arachis hypogaea*) and then 1 year of cotton (*Gossypium hirsutum L.*). This overcomes stagnant cash-crop yields, disease and nematode problems and the risk of cash-crop damage from late-season hurricanes (Franzluebbers, 2007).

4.5 Temperate systems

Typical temperate systems include:

- 2–6-year perennial ryegrass/clover leys (*Lolium perenne, Trifolium pratense*) undersown in cereals preceding the ley (Europe);
- 1-year leys of annual ryegrass (*L. italicum*) intercalated in the preceding cropping system (Europe);
- Arable crops and mixed grass/legume leys in 2–5-year phases (New Zealand);
- Mixed grass/legume or forb leys (Europe);
- Maize, 2–3 years/lucerne or lucerne-grass mix 2–4 years (USA);
- Windrowing hay and crop residues for grazing under snow (USA);
- Legume and grass spp. leys rotated with wheat in the wheat/sheep belt (Australia);
- Sewage sludge and manure slurry applied on pastures (various countries);
- Fish ponds/integration with animal wastes (Asia);
- A 1-year ley in a cereal rotation with recycling of pig manure (Sweden).

In Australia, lucerne, red and white clovers (*Medicago sativa, Trifolium pratense*) are still important, but the 1930s introduction of annual medics, for example *M. littoralis* and other *Medicago* spp.), were widely adopted as legume leys

Table 1 Benefits and non-benefits and suitable species for different pasture or forage rotations in cropping systems in southern Queensland

Pasture/forage rotation type	Length	Benefits	Non-benefits	Suitable species
Non-legume forage crops	<1 year	Easy establishment Rapid production of feed Enable rapid shift between crop and livestock production	Non nitrogen fixing Little effect on soil OM Little disease break or weed control benefits	*Oats* *Forage sorghum* *Grazing millets*
Annual legumes	<1 year	Boost soil nitrogen available for sub-sequent crops, some disease and weed management benefits Enable rapid shifts between crop and livestock production	Provide limited ground cover, OM Can be weeds in subsequent crops (esp. hard-seed medics)	*Snail Medicago scutellata or barrel medic M. truncatula* *Lablab* *Cowpea* *Common or purple vetch, Vicia sativa & V. benghalensis*
Perennial legume phases	1–4 years	Boost soil nitrogen available for subsequent crops Good weed control and disease break benefits	Limited improvements in soil OM Lower cropping intensity Provide limited ground cover	Lucerne Butterfly pea[a] *Clitoria ternatea* Burgundy bean[a] *Macroptilium bracteatum* Sulla[a] or French Honeysuckle, *Sulla coronaria*
Short-term grass/legume pastures	1–4 years	Reduce weed populations Provide good ground cover Increase in soil OM and crumb structure greater than for pure legume or annual forages	Lower cropping intensity N fixed less and not as rapidly available as from pure legumes Difficulties controlling persistent pastures in subsequent crops	Shorter-lived grasses (e.g. Silk sorghum, Rhodes grass *Chloris gayana*, Purple pigeon grass) Less persistent legumes (e.g. Lucerne, burgundy *bean*[a], snail medic)
Longer-term grass / legume pastures	>4 years	Good ground cover and weed control benefits Greatest increase in soil OM and total soil N	Less flexibility in shifting between crop and livestock production Difficulties controlling persistent pastures in subsequent crops	Persistent perennial grasses (e.g. bambatsi *Panicum coloratum*, biss bluegrass, digit grass *Digitaria eriantha*) Persistent legumes (e.g. burr *M. polymorpha* and barrel MEDCs, *Desmanthus* sp, Caatinga Stylo, *Stylosanthes seabrana*).

[a] indicates newer species where understanding of their role and characteristics in cropping systems is still developing.
Source: adapted from GRDC (2010).

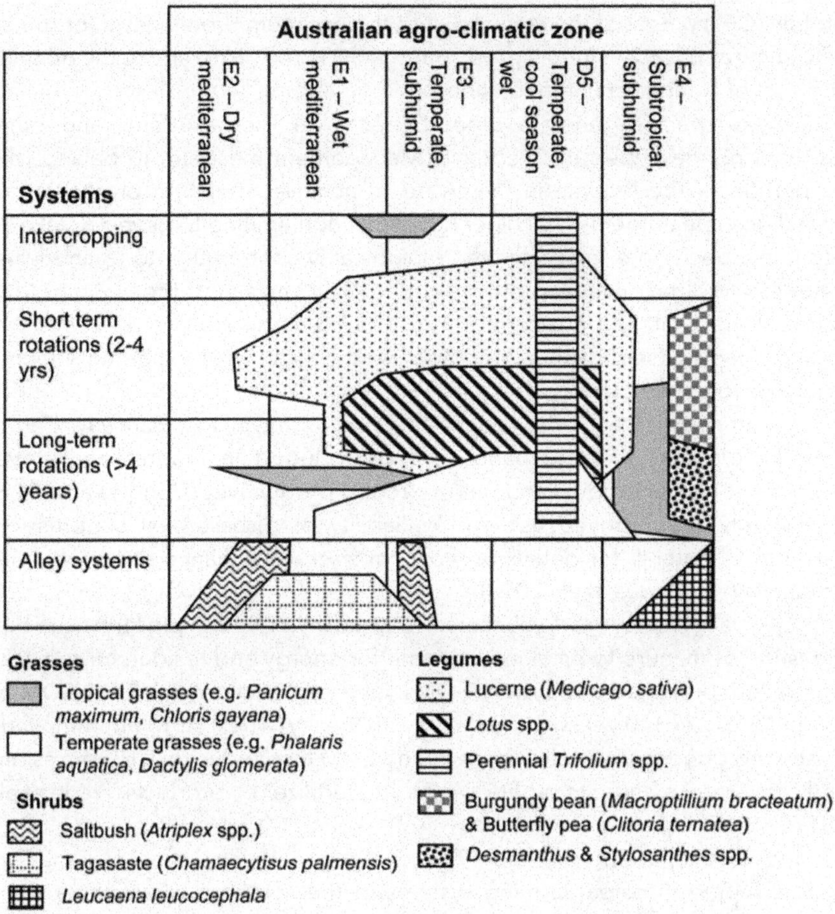

Figure 4 Australian CA systems by ecological zone. Source: CSIRO.

alternated with wheat (Nichols et al., 2012). In Fig. 4, Bell et al. (2014) show the adaptation of principal pasture species to Australian climate zones. This farmer information was developed over 50 years or more of, mainly, CSIRO, research and illustrates the benefits of matching plant genetics to climate and soils.

Dairy production systems in Europe are mainly based on ley-arable rotations (Cong et al., 2017). In these systems, lucerne, red and white clovers are still important.

4.6 Temperate systems: mixed grass/legume or forb leys

These are common in both subtropical (including Mediterranean) and temperate climates. They have two principal functions: (i) to raise the crude protein (CP) content of the animal diet to the optimum protein level for rumen efficiency (about 6% – Hutton, pers. comm., 1975) and (ii) to reduce the quantity and cost of N fertilizer in the rotation.

Mixed species generate other benefits via root exudates and other mechanisms described in Chapter 5. Management must aim to balance the proportion of the legume in the sward to achieve maximum production of total digestible nutrients (TDN) per hectare at non-limiting CP levels or take an economic decision to maximize the biological N contribution to optimize net returns. N fertilization favours the grass, K fertilization favours the legume (John Landers' grandfather, pers. comm., 1952) as do moderate sward heights that reduce shading (see Section 7.3). Mob or strip grazing of exuberant legume growth favours the grass.

In Denmark, three forbs (chicory, *Cichorium intybus L.;* caraway, *Carum carvi L.;* plantain, *Plantago lanceolata L.*) included in ryegrass-red clover mixtures enhanced above- and below-ground productivity (Cong et al., 2017), presumably by improved nutrient availability or disease/pest suppression. See also Chapter 5 for details on how cover crops enhance soil health and productivity (Mendes et al., 2018).

A 1-year grass ley in a 4-year cereal rotation on a Swedish pig farm used bio-digestion of manure to produce methane for energy and residue for fertilizer. This system projected steady state SOM increases from the original 2% SOM, to 3% and 5% over 140 years, without and with the ley (wheat straw not removed), illustrating good sustainability by reverting SOM loss, reducing GHG emissions and converting manure to profitable gas and fertilizer by-products (Prade et al., 2017).

In Australia, during the mid-1930s, self-regenerating annual legume leys were adopted in cereal farming systems of the southern wheat-sheep belt. This is predominantly a Mediterranean climate, with about 400–700 mm of winter rain. Self-seeding annual medics (*Medicago* spp. and others) are used in rotation with wheat or pasture. In New Zealand, arable crops (cereals, brassicas and pulses) are grown for 2–5 years in rotation with 2–5 years of grazed grass-legume ley.

In the United States, the ley (sod-based) crop rotation that has not lost favour is corn (*Zea mays*) with lucerne and forb or grass mixtures on dairy farms, in which forage is mechanically harvested for 3–4 years and then rotated with corn. Forage is fed to dairy cows and nutrients are cycled back to the land via application of manure. This is an example of spatially separated crop-livestock integration on the same farm (Sulc and Franzluebbers, 2014). With direct sod-seeding of winter grains into dormant pasture, Franzluebbers (2007) noted a depressing effect on spring forage yield if grain harvest was delayed and that supplemental irrigation assisted winter grain establishment. However, grain yields were low at approximately 1.5 Mg ha^{-1}. This situation requires the farmer to make a decision on this trade-off with inadequate information on future prices or rainfall.

In Sri Lanka, Jayasundra et al. (1997), at 8°N on coastal sand with 1500 mm annual rainfall, showed a 27% increase in forage N of *Setaria sphacelata* cv. Splendida, due to soil transfer from associated *L. leucocephala or G. sepium.* Louarn et al. (2015) showed transfers of 19.7 and 49 kg N ha^{-1} from lucerne and white clover to the grass, respectively. Thilakarathna et al. (2016), also in Sri Lanka, stated: 'In grass/legume mixes, the selection of plants with compatible traits offers the best path forward to improving below ground nitrogen transfer from legumes to companion crops'. This is a new science and merits further definition of the most efficient grass/legume combinations.

4.7 Smallholder farmer ICL-CA systems

This is an area of little progress worldwide. Systems for pigs or fowls and rice are treated in Chapter 7. There is a plethora of literature on smallholder ICL systems, but a dearth in systems applying CA. Examples include the following.

- *Kandy Garden system, Sri Lanka and India*: In high rainfall forested areas, smallholders on plots of <2.0 ha plant a mix of trees, bananas, local fruits, spices, annual and root crops and hedgerows of fodder shrubs for small livestock, or green manuring in a traditional and proven sustainable system. Average size of holdings in Kandy, Sri Lanka is 0.4 ha and this may be augmented by rice paddies in the bottomlands. Annual rainfall is 2000–2500 mm, at altitudes between 400 and 1050 metres with canopy cover of 45–98%, result in very low erosion losses of 0.05 ton $ha^{-1}yr^{-1}$ (native forest <0.02), on deep soils. This compares with losses of 18–70 ton $ha^{-1}yr^{-1}$ on commercial crops such as tea and tobacco without conservation contours, on slopes of 10–40%. Total species number ranges from 37 to 143, woody species number from 11 to 39 and planting is done by hand with minimal soil disturbance (Pushpakumara et al., 2012). Fowls and small livestock are fed on discarded or cut vegetable matter and household scraps.
- *Faidherbia and Teff system (highlands of Ethiopia)*: In this region of Ethiopia, the native *Faidherbia albida* tree is maintained by farmers to give shade to coffee or in plantations of Teff and other cereals plus livestock (Laike, 1992), where traditional planting uses the digging stick, de facto CA; trees are pollarded every 3–4 years to give firewood. This species tolerates the waterlogged vertisols of the river plains and produces most leaves in the dry season (700–1200 mm annual rainfall and about 2000 m elevation).

- *Soybeans/maize/dairy system in Brazil*: Small farmers in Brazil are concentrated in the mainly subtropical South. A survey of 50 municipalities in West Paraná state showed less than 0.5% adoption of ICL-CA (F) and some planting of eucalypts on contour banks.

The first two ICL-CA systems are archetypes, practiced for hundreds of years and are truly sustainable. The Kandy Garden is a special ICL-CA case, as planting follows CA principles and the system includes fowls, small animals and fish culture. Besides erosion control, ecosystem services are listed as air quality maintenance, climate regulation, water regulation, water purification, regulation of human diseases, reduction of flood peaks, biological and mechanized pest controls and crop pollination. The second system should be of high potential for introducing animal traction CA, if the constraint of competition for crop residues from livestock can be attenuated by providing alternative dry-season feed. The Brazilian small farmer has a very tight system on old fertile soils and is fully CA, copying technology from larger farmers, mainly as an erosion control practice on sloping lands. Typically, he augments his income with outside work. This system and similar CA ones are common also in neighbouring Paraguay.

5 Systems involving forestry (ICL-CA(F))

5.1 Basic principles

The integrated crop-livestock-forest system (ICL-CA(F)) with ZT combines trees, field crops and animals on a synergic basis. It is predicted on scientific principles, being ecologically desirable, operationally feasible and socially acceptable by farmers (Macedo, 2010; Balbino et al., 2011; Pacheco et al., 2013). This section covers agri-silvi-pastoral systems (a mix of annual cropping, grazed pastures and tree planting) and most systems can also be broadly classified as alley cropping. The interactions between trees, animals, pastures and crops will be analysed in Section 7.5.

Alley cropping was defined by USDA (1997) as 'the planting of trees or shrubs in two or more sets of single or multiple rows with agronomic, horticultural, or forage crops cultivated in the alleys between the rows of woody plants'. Alley cropping as such, without animals or without CA, will not be covered here; integration with trees for timber will be dealt with under agri-silvi-pastoral systems. Alley cropping species are normally limited to shrubs or small trees harvested for forage or green mulching. Common fodder trees in alley cropping are: *Gliricidia sepium*, Leucaena (*Leucaena leucocephala*) and hybrids, Algarroba (*Prosopis juliflora*) and *Faidherbia albida* (Africa).

Bell et al. (2014) noted that competition between the alley crop for water and nutrients and an alley grain crop needs to be managed as a trade-off according to external factors affecting profit (price, rainfall). In semi-arid regions, when alley species are tapping groundwater, there is little competition with annual crops. In areas of high water table, the alley species can be important in lowering water tables through their evapotranspiration. Legume trees vary greatly in their nitrogen-fixing capacity.

5.2 Typical systems in Brazil

In Brazil, grain crops commonly occupy the first 1–4 years of establishment (wider tree rows allow more years of cropping), with an early soybean variety followed by maize, sorghum or millet, with pasture establishment in, or after, the last annual crop and grazing thereafter (Kluthcouski et al., 2015). CA is usual in the second crop onwards, after initial soil correction.

Pacheco et al. (2015), at 17°S, planted an ICL-CA(F) experiment using first year soybeans followed by second-crop maize undersown with *U. brizantha* cv. Marandu with a *Eucalyptus urophylla* clone planted coincidently in four rows, 22 metres apart and 2 m between trees (645 trees ha^{-1}) Starting in year 2, cross-bred beef cattle showed 270 kg carcass weight gain $ha^{-1} yr^{-1}$ to slaughter age of 2–2.5 years, while prior farm average was 105 kg/$ha^{-1} yr^{-1}$ and slaughter at 3–4 years. Average raw timber increase for the eucalypt was 40 $m^3 ha^{-1} yr^{-1}$ over 6 years, using only 36% of the area as compared to 100% with 1667 trees ha^{-1} in commercial pure stands. For more detailed information on Brazilian ICL-CA (F) systems, see Balbino et al. (2011).

5.3 Species planted in Brazil

Eucalyptus grandis, E. urophylla, E. saligna and other *Eucalyptus* spp. are predominant in the wet-dry tropics and Amazon biome, due to fast growth rates, less shade and availability of selected clone seeds. *E. citriodora* is slower-growing and used for timber. Care must be exercised in the climate/soil characteristics as the clones are very sensitive to local condition. Other tree species used in this region for ICL-CA(F) are acacia (*Acacia mangium*), *paricá* (*Schizolobium amazonicum*), teak (*Tectona grandis*), African mahogany (*Khaya ivorensis*) and Caribbean Pine (*Pinus caribaea*). Harvest interval for most eucalypts is 6–7 years (biofuel, fence or electricity posts) and for the slower-growing *E. citriodora* and teak (*Tectona grandis*) for timber, about 10–12 years. There is also research with Brazilian mahogany (*Swietenia macrophylla*), neem tree (*Azadirachta indica*), *mulateiro* (*Calicophylum spruceanum*), *amarelão* (*Aspidosperma vargassi*), kapok (*Ceiba pentandra*), *taxi-branco* (*Sclerolobium paniculatum*), balsa tree (*Ochroma pyramidale*), or quickstick (*G. sepium*) and Australian cedar (*Toona ciliata*). In subtropical Brazil, *E. benthamii*, *E. dunnii* and *Pinus elliottii* predominate, also used are canafistula (*Peltophorum dubium*) and silky oak (*Grevillea robusta*). Drumond (2012) recommends for semi-arid conditions (500–700 mm yr^{-1} Leucena, Gliricídia, Algarroba, Sabiá (*Mimosa caesalpiniaefolia*) and some *Eucalyptus* spp.).

5.4 Species planted in other countries

In the Southeast United States, Florida especially, Loblolly pine (*Pinus taeda* L.), Longleaf pine (*Pinus palustris* Mill.), Slash pine (*P. elliottii*) are suitable for silvi-pastoral systems. Haying between young trees is recommended until the trees are old enough to better withstand pressure from livestock presence and grazing (Nowak et al., 2002). For a review of legume species worldwide used in combination with pastures, see Dubeux et al. (2017).

5.5 Tree spacing and planting arrangements

General principles include: (i) trees for sawn timber require wider spacing in single rows to achieve adequate trunk diameter in a shorter time to harvest, of 10–12 years in the tropics, recommended population 150–300 trees ha^{-1}, (ii) multiple rows are not recommended for timber production as phototropism distorts tree architecture; (iii) closer in-row spacings can permit a small income from an early thinning to 50% of final population and sold for dryer fuel, fence posts, house poles or scaffolding; (iv) wider spacing between single or multiple rows favours pasture and crop productivity; (v) spacing between trees for pasture/crops varies from 10–30 m and (vi) wider spacing favours productivity of crops and pasture by reducing the shade and nutrient competition factors.

5.6 Discussion

There is always a delicate balance in ICL-CA(F) systems:

1. lack of flexibility versus long-term secure financial gains;
2. denser plantings for thinning sales for fuel or fence posts and so forth versus greater individual tree growth at lower populations;
3. early grazing versus risk of tree damage;
4. loss of income on early annual crops due to shading and competition from tree roots needs to be attributed as a cost to the timber enterprise;
5. cost of terminal de-stumping needs to be factored in to returns on timber sales;
6. shade benefits, especially to milk cows, need to be attributed to the timber enterprise.

6 Non-ley integrated farming systems

6.1 Cut and carry systems

- Dairy operation using maize silage + root crops and kale from crop area (UK);
- Dairy operation grazing summer grass leys and winter supplements of beet pulp or silage + enriched cereal straw (Denmark);
- Maize silage from irrigated area for feedlot fattening (Merola, 2002, Brazil);
- Rotationally grazed root crops with sheep, in an arable rotation (mostly UK and New Zealand);
- Cut and carry green chop, silage, grain, hay, root crops or straw (USA, Europe, New Zealand, Australia and others);
- Fish ponds in China use animal wastes and on-farm green chop.

These systems involve high labour inputs and are opted for where winter or dry-season stocking rates are low.

6.2 Haymaking, silage or straw for feed

Examples include:

- Excess summer biomass in pasture hayed or made into silage or haylage (crop partly dried) for winter or dry-season feed (all climates);
- Cover crop (gramineous, leguminous or other) hayed for dairy or feedlot;
- Wheat or rice straw supplemented with urea for confined dairy cows (Indian subcontinent).

Integration of crop and livestock enterprises does not always include a ley. Rotational grazing of root, or other non-pasture forage crops in an arable system is quite common for winter feed in Europe and New Zealand, especially with sheep. Unlike other fodder roots, unrestricted grazing of sugar beet tops causes bloat and may be lethal.

Cut and carry systems may be on-farm, from a separate arable area for feedlot, dairy or other livestock (Europe and developed countries) or of the area integration type (see Section 7). Soil cover is difficult to maintain with root crops.

Haymaking may be highly mechanized (baled hay), now rarely manual. Many small Indian subcontinent farmers have used CA for wheat, now beginning for rice, but have no room for pasture, keeping dairy cows in confinement, when (preferably) wheat straw is used for feed, with urea supplement (Roy and Rangnekar, 2006) and dried dung is sold for fuel. In Nepal, Paudel et al. (2016) showed an 8.6% increase in milk yield for chopped rice straw + 4% urea solution over the control with only un-chopped rice straw.

7 Area integration systems

These systems involve use of off-farm resources. Cattle and sheep operations have historically involved specialization of breeding operations on the poorest land, with young animals migrating once or twice to rearing and fattening areas of successive levels of fertility. Due to their complexity, area integration of ICL-CA systems will not be dealt with in detail. Some examples are as follows:

- Arable area stubble-grazed by itinerant herdsmen (Africa);
- Soya/maize succession, grass undersown in maize – dry-season pasture by rental or joint venture animals (Brazil);
- Use of manure from animals in confinement, biodigestor residue or sewage sludge to return organic fertilizer to the crop or pasture land (many countries);
- Share cropper arable crop area stubble-grazed by landowner on share cropper's land (declining, NE Brazil);
- Irrigated lucerne dried and ground as feed component for off-farm livestock enterprises (California, UK and others);
- Purpose-planted irrigated Sudan grass or other for hay to export inter-farm or internationally (USA);
- Use of urea-enriched wheat and rice straw for animal feed rations or direct feeding to dairy cattle (Indian sub-continent).

8 Interactions in integrated crop-livestock systems

The complex interactions in ICL-CA systems are common to most systems and will be generically analysed here. Grazing animals have important influences on the efficiency of soil functions in ICL-CA. The essential difference between CA and CT is that the latter has very different interactions in the system, due to residue incorporation in the soil and generally lower biological activity.

8.1 Soil properties

8.1.1 Soil fertility and nutrient cycling

Apart from the fertility in the soil (natural or applied), plants fix CO_2 and N_2 from the air, to furnish cellulose, lignin and proteins to the animals in the first link of the food chain. Plants also provide an ecosystem service by sequestering CO_2

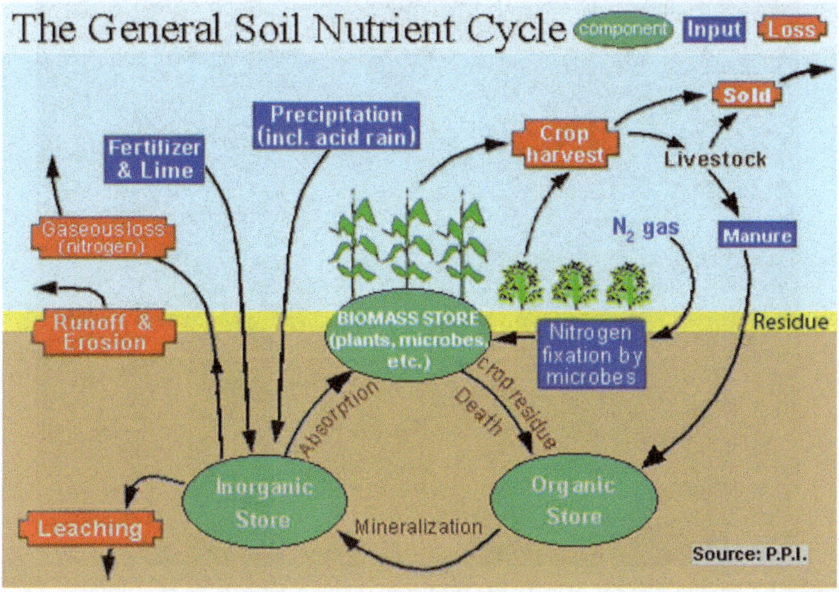

Figure 5 Complex cycling of nutrients under ICL-CA.

from the atmosphere (see Section 8.5). Soil organic carbon (SOC) represents 58% of SOM, which is a good indicator of soil fertility within a given soil type. Figure 5 shows the complex cycling of nutrients under ICL-CA.

8.1.2 The nitrogen cycle

The principal cause of tropical pasture degradation is N deficiency (Macedo, 2009; Haynes and Williams, 1993) and this is also important in temperate regions. The latter authors observe that ruminant animals are essential to fertility restoration in a ley as some 70% of grazed nutrients are returned via excreta in assimilable forms.

Legumes as a source of symbiotic N fixation in ICL-CA may be tree, shrub or herbaceous species. Predominantly, tree species are included in agri-silvi-pastoral or silvi-pastoral systems, shrubs and small trees are found in alley cropping and herbaceous species in mixed grass/legume or pure legume or forb leys in ICL-CA (agri-pastoral systems). There are two principal sources of N in leys: fertilizer N and biological (symbiotic) N fixation (BNF), principally by *Rhizobium* spp. or *Bradyrhizobium* spp. and less by *Azospirillum* spp. (Döbereiner, 1990); however, N is also released by degradation of SOM and pasture residues, built up during the ley. Non-symbiotic fixation is a minor contributor, via free-living *Azotobacter* spp. in soil, while atmospheric nitrogen

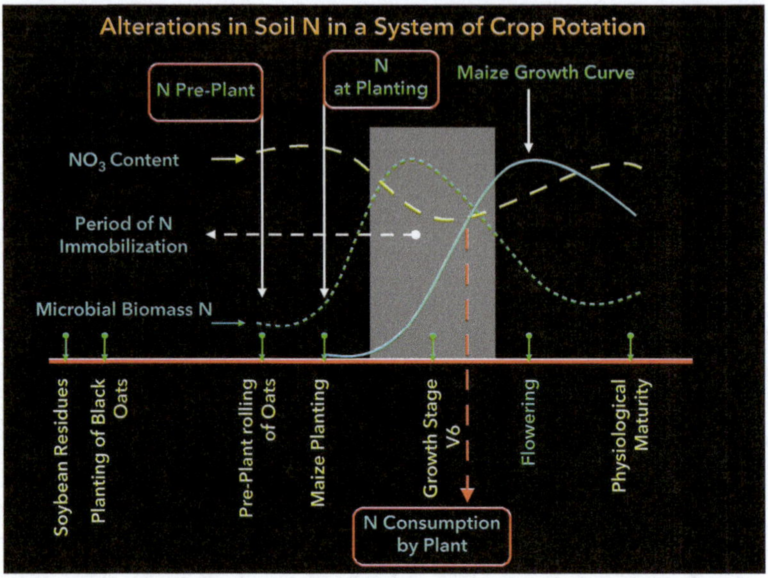

Figure 6 Alterations in soil N in a system of crop rotation. Source: Sá et al. (1999).

via lightning is minimal (Russell and Bisinger, 2015). Soil N additions are affected by grazing management (Haynes and Williams, 1993), with low intensity, approximately 90% of N returns to the soil as manure and urine. However, apart from varying with grazing intensity and herbage quality, Oenema (2005) estimate that only 40–50% of urine N remains available to plants. In Australia, Traill et al. (2018) showed that 'after retention, rather than removal, of butterfly pea (*Clitoria ternatea*), centro (*Centrosema pubescens*) and lablab (*Dolichos lab-lab*) biomass, maize accumulated an additional 80–132 kg N ha^{-1}. After fodder removal, only centro provided N benefits to the maize crop (equivalent to 33 kg urea-N ha^{-1})'. The farmer must gauge the trade-off between retaining biomass or using it for animal feed.

In the early years (1-3 years) of adoption of CA in any form, extra N is required to compensate that taken up by soil microbes as they decompose SOM to build their body proteins. This can affect early crop growth, but the N is liberated to the soil later as the microbes die back, as illustrated in Fig. 6.

8.1.3 Dynamics of phosphorus, other nutrients and micronutrients

Phosphates can be fixed by free aluminium in acid soils; this is generally ameliorated by liming, and gypsum, which also furnish Ca and S (Scheid Lopes, 1996). Thus, nutrient efficiency of fertilizer P can be very low, especially in soils with low biological activity. However, soil-fixed P can be made available to crops via root exudates of (some) plants and soil microbial action (see Section 8.3). Neumann et al. (2014) found that 'observed differences (in lettuce growth) may be the result of plant interactions with the soil-specific microbiomes' and the diversity of cover crop mixes can be expected to maximize these positive effects. For a more detailed discussion of this topic, see Schilling et al. (1998).

With ICL-CA in Brazil's tropical savannahs (16°S), Eberhardt et al. (2017) recommends broadcasting P on pastures as opposed to upfront remedial (incorporated) application and showed that this transfers effectively to a following CA soybean crop after pasture desiccation. Recently, in Brazil, there has been a revival of the use of natural rock meal from different minerals (e.g. mica-schist, basalt, sedimentary phosphates and others) which are alleged to 're-mineralize' the originally very acid and P-fixing Cerrado soils (Martin, 1996). This has been widely corroborated by farmers, e.g. Flavio Faedo of Rio Verde in Goiás State in Brazil, a CA pioneer who has dispensed with chemical fertilizer for 5 years and also applied soil microbe inoculation, suffering no noticeable drop in soybean or second-crop maize yield, with reduced costs (Faedo, pers. comm., 2018). This practice draws on the build-up of fixed P over 30 years of CA: albeit this example confounds rock meal and soil microbial effects, but the net effect for the hugely positive. The scientific explanation of this phenomenon has not been elucidated.

Micronutrients are generally more available under the high biological activity in CA soils, except when remedial lime is applied superficially in more than minimal doses, which then creates a high pH in the surface humus, when certain micronutrients, especially manganese, become insoluble and their deficiency (easily removed by foliar application) suppresses yields. Differences in K, Ca, Mg and S nutrition are small between ICL-CA and CA cropping without the pasture phase, except that leaching may not be exacerbated by improved internal drainage, if there is a significant increase in SOM and hence cation exchange capacity (CEC) to act as a buffer.

8.1.4 Manure and sewage sludge applied as fertilizer

In countries where confined livestock produce manure in considerable quantities that are returned to the land, there is a need to avoid nitrate contamination of water tables and to safeguard public health from pathogens, such as *Escherichia coli*. Regulations limit the amount per hectare per year that may be applied to pastures or crops and the pre-treatment applied. Sewage sludge is normally de-watered and treated for pathogens and odour and may be composted or also mixed with compostable urban waste for this purpose.

Use of biodigestors using animal wastes can furnish fertilizer for crop production and gas for cooking or electricity generation. This is common on pig and chicken farms in Brazil, where the meat packers supply the biodigestors and receive carbon credits to pay off the investment; after this the equipment is the farmer's. Integration of fish pond production with manure from ducks, geese, chickens, sheep, cattle, pigs, goats or human waste increased fish production by 2–3.9 times in China (Chen et al., 1995) and the animals would be fed from grain or forage grown on the farm or from outside.

8.1.5 Soil physical properties

There is widespread agreement that a pasture phase improves soil physical characteristics, such as bulk density, penetration resistance, infiltration rate, aggregate stability and permeability to air and water (Marchão, 2007; Donagemma et al., 2016); the former stressed the role of grasses, particularly Palisade, Ruzi and Signal grasses (*Urochloa* brianthia, *U. ruzizienssis* and *U. decumbens*) for improving soil physical parameters in the tropics, due to the effects of grass roots and incremental biomass production, possibly also to increased microbiological activity. For example, root exudates from pearl millet form tubes of aggregated soil around the roots (Landers, personal observation, 1996). The effects of pasture appear to be independent of soil texture. This can be seen in studies such as Macedo et al. (2001) in a long-term experiment in Mato Grosso, Brazil, on clayey textured soil (Dark red latosol, Latossolo vermelho-escuro or Ferralsol) and Vilela et al. (2001) on a sandy textured soil (Neossolo Quartzarênico – Arenosol) who observed a similar tendency of improvement in soil physical properties. Franzluebbers et al. (2001) showed that soil bulk density was inversely related to SOC, and Sá et al. (2018) in a detailed study of six agricultural systems demonstrated that SOC stock and biological activity were correlated inversely with soil degradation and correlated positively with system biodiversity and soil resilience. These are therefore useful indicators of soil health.

Increased SOM promotes improvements in crumb structure (Castro Filho et al., 1991), while Australian McGarry (2000) indicated that cultivating soil above its plastic limit is responsible for degradation of physical properties. In Uruguay, soil physical conditions were degraded under continuous annual cropping, with significant recovery after 2–4 years of a well-managed pasture (Garcia-Prechac et al., 2004). Aggregate stability, as an indicator of soil quality, is directly related to susceptibility to erosion and compaction; it depends on the content of organic matter and biological activity. In a very light-textured tropical Brazilian Cerrado soil (Neossolo quartzarênico, Entisol or Quartzipsamment), Vilela et al. (2001) reported that, after 4 years of Palisade grass var. Marandu, pasture aggregates of >2 mm equivalent diameter showed similar values to those under natural vegetation.

Figure 7 illustrates up to a fourfold increase in water conductivity (rainfall infiltration rate) after 5 years of a pasture phase following 30 years of CT cropping. This is indicative of improvements in other physical parameters in the

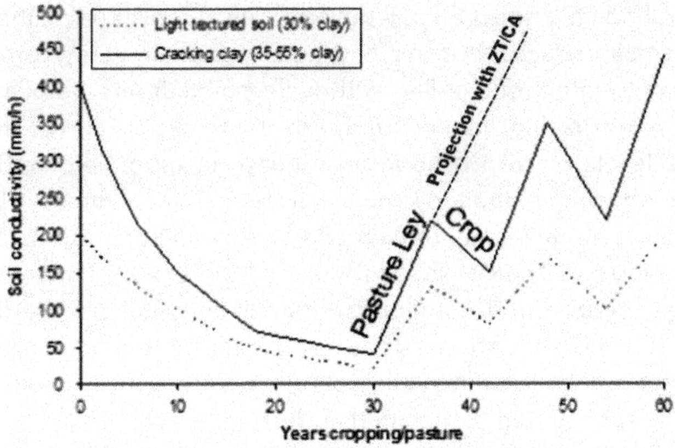

Figure 7 The reduction in soil infiltration rate after 30 years of CT cropping. Source: Silbrn et al. (2007).

long ley, and CA in the crop phase would considerably reduce the deterioration in soil structure (GRDC, 2010). Mendes et al. (2018) observed: 'However, despite signs of soil compaction in ICL-CA fields, many studies show that soil hydraulic conductivity is consistently higher, suggesting the presence of a more functional pore system for soil water and air fluxes than in tilled fields'.

There is a dearth of information in CA ley farming on physical parameters > 30 cm, which is especially serious with ever-increasing machinery axle loads; subsoil compaction is more correlated with these than with wheel pressure on soil (Chamen, 2006; Prade et al., 2017). Stoner et al. (1991) showed significant differences in physical parameters in the 30–40 cm layer in a Cerrado latosol after only 13 years of CT, when axle weights were less than half of those today. Faust et al. (2018) in the Northern Great plains, USA, noted that shallow compaction occurred under grazing livestock, which increased run-off. In New South Wales, Australia, Hunt et al. (2016) found that sheep grazing on crop residues did not reduce crop yields in a no-till, CTF system in an equi-seasonal rainfall environment, provided that soil cover was greater than 70% and summer weeds were controlled by herbicide. Farmers' fear that this would damage soil and reduce soil structure, crop yield and profitability were unfounded. Wheat yield decreases were due to grazing of green wheat and the plant response to de-foliation and not to small alterations in soil properties. In subtropical Brazil, Morães et al. found minimal loss of soil physical condition, provided moderate stocking rates were used.

Allan et al. (2016) concluded for wheat-sheep systems in subtropical Australia: 'The results show that light grazing of sheep on crop residues over summer had a negligible effect on crop growth and yield as well as weeds and soil water. The effect of grazing on soil mineral N was inconsistent. Grazing residues also had little effect on soil bulk density at most sites, although some sites had a small increase, but this had no effect on the crop, probably due to the alleviating effect of seeding with knife-points in the no-tillage system. However, a high grazing intensity through winter and summer significantly reduced the levels of crop residue and following crop yield on heavy soil, probably by reducing infiltration'. The lower the residual biomass, the lesser is the soil's ability to retain water (mulch effect). Animal hoof compaction would appear restricted to 10 cm for sheep (Bell and Moore, 2012; Hunt et al., 2016). When soils are grazed under dry or low rainfall conditions in the dry season, cattle compaction is not serious, does not exceed 15 cm and is removed by (thin) tines (shanks, knife points) to apply fertilizer in the planter or drill, with no noticeable effect on crop growth (Landers, personal observation in Brazil; Cerrado 1995). Strip grazing (with movable electric fences) provides long rest periods and helps to maintain plant growth and water infiltration in pastures (Russell and Bisinger, 2015).

The predominance of fine and very fine sand in superficial layers of a Neossolo Quartzarênico (Arenosol) favours the formation of surface sealing (Donagemma et al., 2016; Scopel et al., 2012). High fine sand content and the heterogeneity of sand particles are also responsible for the increase in cone penetration resistance (Abrahão et al., 1998). The advent of CA cropping upgraded the agricultural aptitude class of light soils in Brazil (>20% sand to 75 cm depth) from inapt or with restrictions for cropping to apt or with moderate restrictions for cropping (Donagemma et al., 2016).

8.2 Soil biological activity

Microorganisms are the most numerous living things in the soil and are divided between microflora (fungi, bacteria, algae and viruses) and microfauna (protozoa and micro arthropods; Araújo and Hungria, 1994). However, earthworms generally represent the largest proportion of soil fauna biomass. Advantages of enhanced biological activity in soils under ICL-CA include:

- A win-win situation with multiple benefits that redound in improved yields;
- Greater nutrient efficiencies via livestock dung/urine recycling, CA stimuli to enzyme solubilization and earthworm and dung beetle digestion;
- Suppression of nematodes by crop rotation with non-host and antagonistic plants;
- Suppression of some soil diseases by antagonistic fungi, for example *Trichoderma* spp.;
- More favourable conditions for multiplication of soil biota;
- Increased soil micro-aggregation improves internal drainage, aeration and rainfall infiltration to depth;

- Surface infiltration of rainwater is improved by earthworm and dung beetle burrows and aided by longer periods on the surface when held by residues;
- Root development is also enhanced by these burrows;
- Reduced moisture stress through greater RAM (higher rainfall capture and increased water-holding capacity);
- Improved legume inoculants and innovative Azospirillum, Mycorrhizal and bocage inoculants.
- Lower environmental impacts from reduced chemical fertilizer use, faster degradation of agro-chemicals and lower GHG emissions.

8.2.1 Soil fauna

Earthworms are divided into epigeic species (topsoil and surface dwellers which make worm casts) and endogeic species (which make deep vertical burrows). Earthworms have multiple, mostly favourable, impacts. Blouin et al. (2013) summarized these as follows: (i) earthworms digest surface and SOM and recycle N (11–30% of crop uptake), (ii) macro-aggregate content may be positively or negatively affected, according to species, (iii) earthworm burrows assist internal drainage, soil porosity, rainfall infiltration and aeration, (iv) earthworm burrows encourage root penetration. Also, Brown et al. (1999) showed an average gain in biomass from earthworm activity in 14 crop species to be 35.8 ± 8.9%, significant at $p < 0.08$; highest increases were observed in soils with sandy textures, poor in organic matter and with a moderately acid pH. An example of how earthworm burrows aid root penetration is seen in Fig. 8.

In Norway, Riley et al. (2008) measured '...overall increases in earthworm density (84%), earthworm biomass (80%) and the density of earthworm channels (132%) in the topsoil between 1994 and 2004. Most of these increases were found in systems with 1 to 3 years of ley in the 4-year rotation' and organic matter decline under CT. Blanchart et al. (1999) indicate that 'the joint presence of both "compacting" and "decompacting" epigeic species (large and small respectively) and organic residues in tropical agroecosystems appears to be necessary to sustain physical soil fertility'. In kaolinitic soils, endogeic earthworms are important contributors to soil structure, but not so in (cracking) vertisols.

Dung beetles are very effective in recycling animal faeces in pasture. There are thousands of species worldwide, but cattle dung species were absent in Australia and New Zealand; from the 1930s onwards, many species were imported for use in different eco-regions, also to reduce fly populations (Bornemissza, 1976). They and their larvae feed on dung and they make tunnels underneath dung pats (paracoprid spp.) or roll balls of dung to a burrow some distance away (roller spp.). Their digestion of dung returns nutrients in more

Figure 8 Roots following an endogeic earthworm burrow. Source: S. Ball, Nuffield Australia.

assimilable forms to the pasture (Bornemissza, 1976). Faria and Hernandes (2017) in Brazil studied these two groups of dung beetles: the paracoprids increased SOM, but the rollers did not, with the latter species preferring sandy soil.

8.2.2 Synergies in soil microbiology

Higher activity of soil enzymes, micro- and mesofauna occurs under ICL-CA, enhanced by grazing animals (Silva et al., 2015). This synergy results in higher nutrient availability and higher efficiency of nutrient use. Also, soil diseases are controlled by antagonistic fungi, such as Trichoderma spp. and other organisms; the former is now available as a commercial seed or soil treatment. The net effect of microbial activities on crop yields is complex, since activities of some soil enzymes and their effect on nutrient availability are affected significantly by liming (pH), tillage systems (Ekenler and Tabatabai, 2003) and other external factors.

The β-glucosidase (BG) enzyme is related to the carbon cycle, acting in the cleavage of cellobiose into glucose molecules (Almeida et al., 2015); phosphatase (PHA) is responsible for converting fixed phosphate in soil to plant-available organic forms (Margalef et al., 2017): aryl-sulfatase acts in the degradation of composite molecules containing sulphur; cellulase degrades cellulose (lignin) in crop residues; and urease catalyses the hydrolysis of urea to ammonia and carbon dioxide, having a negative effect on the efficiency of this fertilizer (Mendes et al., 2018). CA crop rotations with permanent soil cover in Cambodia significantly enhanced BG and arylsulfatase activities in the surface soil layer, as a result of very high biomass-C input and the absence of soil disturbance (Hok et al., 2018). Highly productive soils present high levels of microbial biomass and respiration and also high activity levels of cellulase, BG, arylsulfatase and acid phosphatase (Lopes et al., 2012).

8.2.3 Degradation of pesticides

Activation of soil biological processes under ICL-CA has been shown to accelerate the decomposition of insecticides (see Table 2). In a ley of Palisade grass cv. Marandu in Mato Grosso do Sul, Brazil (22°S), Portilho et al. (2015) clearly demonstrated large positive increments for BG over the control (CT), of 356% and 252% in two ICL-CA systems and reductions in the half-life of insecticides Bifenthrin and Permethrin of 44 to 14/25 and 47 to 9/22 years, respectively, or reductions of 81% and 43% for the first and 80% and 55% for the second ICL-CA system, as compared to CT. This is due to the enhanced adsorption by higher SOM levels and, possibly, to elevated soil enzyme activity, under a CA system.

8.2.4 Inoculants

Inoculation of legumes with *Rhizobium* and *Bradyrhizobium* spp. has been practiced for many years, is standard practice in most legume (pulse) crops

Table 2 Biodegradation of insecticides with Signal grass cv. Marandu under ICL-CA

Management system	Glucosidase	Acid phosphatase	Bifentrina	Permetrina
	μg PNF g soil^{-1} h^{-1}		TD 50 = soil half-life	
ICL-CA-1	356	1206	14	9
ICL-CA-2	282	649	25	21
CA	188	612	25	22
CT	99	291	44	47

Source: adapted from Portilho et al. (2015).

and the crop range is being extended. Recent developments have made inoculation of gramineous species with *Azospirillum brasiliense* and other species commercially possible (Hungria, 2011). Van Geel et al. (2016), reviewing 435 experiments, found an average increase in crop biomass of 49% from mycorrhizal inoculation, but also indicated that a single proven strain was superior to a cocktail. The modern use of 'Bocage', a cocktail of soil used by organic farmers is also increasingly being practiced in Brazil during the crop phase of ICL-CA. Soil conditions under CA are more favourable to the multiplication of these microbes (no soil inversion, more amenable surface temperatures, better aeration and internal drainage).

8.3 Crop rotations

8.3.1 Benefits of crop rotation in ley farming

Some typical Brazilian tropical rotations are found in Figs. 1 and 9. The various roles of the pasture phase are to (i) diversify income from crops by grazing cattle, (ii) to restore soil condition, (iii) improve bio-availability of nutrients, (iv) improve biodiversity and (v) provide phytosanitary controls. Pluri-annual rotations with high biodiversity of crop species reduce insect pests, weeds and plant diseases. Rotations can also improve following crop yields through several synergies, including disease (Kheyrodin, 1994), weed (GRDC, 2010) or pest (Brust and King, 1994) suppression by a preceding crop, or through silicon nutrition (Fauteux et al., 2005). The art of a successful rotation is that plants which favour diseases, pests or weeds should be followed by those which are not susceptible, always maintaining profit.

8.3.2 Control of diseases, pests and weeds

To illustrate rotational control of disease, Franzluebbers (2007) noted that 'peanut seed yield was often greater following 2 yr of Bahiagrass (Paspalum notatum) than following annual crops and was much greater than in continuous peanut, as a result of a reduction in stem rot (Sclerotinium rolfsii)'. This raises the question of how long a ley is needed to control specific pests, weeds and diseases: the answer will be area-, site-, sward-, and species-specific.

One function of grass-legume leys or legume banks in a rotation is to raise the protein level of the ruminant diet in order to efficiently digest the cellulose-lignin of (low protein) feed. This has largely been sidelined by urea supplements in grazing or feeding urea-treated crop residues (Mesfin and Ktaw, 2010; Tengyun, 2000), especially in the Indian subcontinent. A grass-legume or pure legume ley in a rotation also has the function of furnishing residual biologically fixed N to a following crop phase, showing the potential to supply 100–200 kg N ha^{-1} in a 0–200 cm profile, with lucerne showing the highest levels, albeit variable with rainfall (Peoples et al., 1998); more N and other nutrients will be liberated slowly if SOM degrades during the crop phase. This phase can leave residual fertilizer nutrients for pasture renovation and also residual N, if a legume is the last crop before the ley.

Cattle of all species utilize forage and/or crop residues and enhance nutrient cycling within the system (Franzluebbers, 2007). According to Bonaudo et al. (2014), the animal's role is to recycle and increase the use efficiency of resources. Bortolini (2005, pers. comm.), a cover crop specialist in Mato Grosso in Brazil, stressed the need for high diversity, and pluri-annual rotations, warns of declining returns with soybeans repeated every year, especially recommending ICL-CA rotations as being the most diverse and hence most effective in promoting biological controls. However, the most common cropping patterns for ICL-CA in Brazil are successions with soybeans repeated year after year in summer (see Figs. 1 and 9). ICL-CA represents less than a third of soybean area in Brazil, where the crop phase is followed by a 2- to 4-year ley. On old cropland with a high-nutrient status, very high feed quality guinea grass cultivars can be used, but on lower-nutrient-status soils, Ruzi grass (Urochloa ruziziensis) or Palisade grass *(Urochloa brizantha)* are used; in lower rainfall areas due to greater rusticity, buffel grass (*Cenchrus ciliaris*) is grown. Rotations will function most efficiently when grazing livestock are integrated into the farming system; Ruzi grass is being favoured, especially for short winter leys, due to a considerably lower rate of Glyphosate for desiccation (see Section 14 on nematode suppression by *Urochloa* spp).

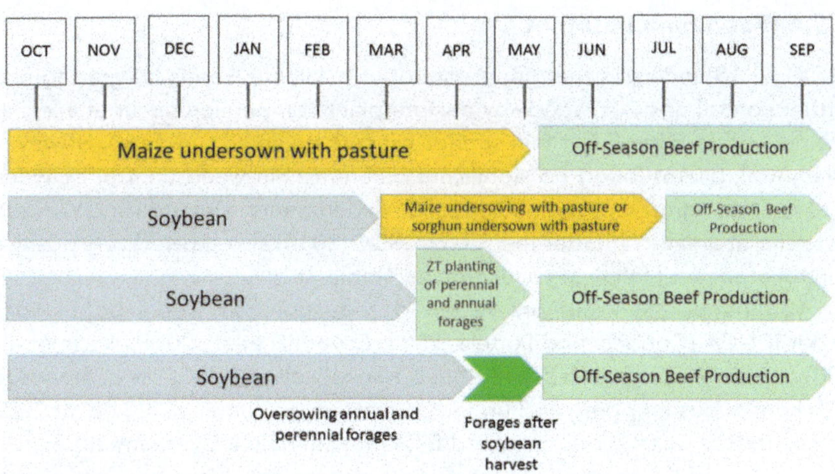

Figure 9 Alternative tropical crop successions in ICL-CA for dry-season grazing or soil cover. Source: Vilela (2005).

8.3.3 Control of nematodes

Nematode control by rotation with crops or pasture leys, which either suppress or kill nematodes, is complex. This requires careful diagnosis of the nematode species and their populations present in order to prescribe the optimum rotation to minimize crop damage. The susceptibility to, and capacity for, control of nematodes varies considerably between crop, cover crop species and nematode species, affecting yield of the crop at different levels (Asmus, 2003).

The efficiency of cover crops for nematode control depends on the crop rotation practiced: Abreu (2015) demonstrated a variation of 30.85–87.60 in percentage reduction of the population of the nematode *Pratylenchus brachyurus in* 17 cover crop options in a soybean monocrop succession. Rotation with maize or *C. ochroleuca* reduced the population of *Rotylenchulus reniformis* when compared to monocropping of soybeans, but there is no advantage of other cover crops over a fallow of Ruzi grass in the control of this nematode. Salton et al. (2014) registered 3424 individuals per 300 cm^3 of the latter species under CT (soybean/fallow), but a maximum of 24 for ICL-CA and CA systems including gramineae. The use of gramineae and Crotalaria spp. to control nematodes is now general practice in Brazil. A reduction of 99% in the population of *Pratylenchus brachyurus* after 3 years of Palisade grass as a winter cover crop in a soybean/maize succession was accompanied by a maize yield increase of 21.6% when compared with the most common cover crop, millet (Abreu and Borges, 2015). Similar results were demonstrated in maize (Abreu and Borges, 2015). The greater the diversity of crops in the rotation, the higher the control efficacy for nematodes and other pests and diseases (possibly extensive to weeds).

8.3.4 Weed management

GRDC (2010) indicates that good control of palatable weeds by grazing and further control of other weeds by pasture plant competition occur in a ley, as exemplified in Table 3 but thin strands exacerbate weed problems. However, Allan et al. (2016) found no significant differences in weed control between grazed and ungrazed plots in 30 out of 31 experiments in Victoria and Western Australia, at greater latitudes than 30°S. GRDC (2010) also showed very positive control of wild oat following a lucerne ley (Table 3).

Various Brazilian authors reported reductions in weed populations under ICL-CA (Cobucci and Portela, 2003; Severino et al., 2006; Ikeda et al., 2007a,b). The latter studied weed seed bank levels in ICL-CA in comparison with continuous cropping and under CA rotation. Salton et al. (2014), at 22°S, observed that weed cover occupied 87% more area in a CT treatment than in the average of CA and ICL-CA treatments and grazed areas had less weeds than pure cropland.

Table 3 Difference in wild oat populations between a continuous wheat cropping system and in crops after a lucerne ley

		Wild oat (plants/ha)
Continuous wheat		11600
After a lucerne pasture	First crop	85
	Second crop	350
	Third crop	1700
	Fourth crop	3100

Source: adapted from GRDC (2010).

8.3.5 Interactions with crop residues

The rate of decomposition of three arable crop residues in tropical Brazil is shown in Table 4.

Crop residues can have an important allelopathic effect on emergence of crop plants and weeds of related families and also in disease suppression, either by smothering the pathogen and impeding its sporulation or, by not providing a favourable substrate for multiplication. Ferraz et al. (2011) demonstrated that antagonistic soil fungi and complete mulch cover suppressed the development of white mould sclerodia (*Sclerotinia sclerotiorum*) in soybeans.

Residues in Africa are often used as livestock feed, fuel and construction material and provide income through their sale (McIntire et al., 1992). Residues that remain in fields provide mulch and erosion protection but are commonly

Table 4 Decomposition and loss of soil cover with time in maize, rice and soybean stover

Crop residue	No. of days after the first rain	Dry-matter weight of residue (kg/ha)	Soil cover (%)
Maize	30	7.500	82
	60	4.300	54
	90	2.800	30
	120	1.400	22
Rice	30	6.200	85
	60	3.100	46
	90	2.200	38
	120	1.700	26
Soybean	30	1.700	35
	60	540	16
	90	240	7
	120	-	-

Source: adapted from Seguy et al. (1992).

grazed in developing countries, or burnt to facilitate soil preparation and disease control (Powell et al., 2004). This exacerbates the emissions of GHGs. A trend from extensive pastoralism towards ICLS, due to increased food demand from a rising population, will likely be followed by the reverse trend towards more specialized separate crop and livestock enterprises, when higher inputs, such as tractors and fertilizers, are introduced. At this point, the economic incentive for a mixed farm enterprise to provide its own inputs diminishes and specialization becomes more profitable (Powell et al., 2004; Duttargavni, 2013). These processes are typical of historical agricultural development as land becomes less productive and scarcer, common in developing countries. Franzluebbers (2007) also described a similar historical process in the United States.

In SSA, it is still possible to improve biomass production with CA (Tarawali et al., 2013). However, where there is livestock integration in dry and semi-arid areas based on pastoralists coming into cropland from the outside, after crop harvest, there has to be a mutually agreed community-level management of crop residues so that there is a win-win situation operating instead of a 'tragedy of commons'. In lower rainfall areas of SSA, Powell et al. (2004) noted that animal manure was essential to slow the decline of soil fertility under extensive cropping systems and was even recognized by contractual agreements between pastoralists and crop farmers. Manure-bound nutrients are more plant available than crop residue-bound nutrients, while in the early stages of degradation, fibrous residues sequester N for microbial biomass generation, later liberated (Somda et al., 1995; Powell et al., 1999).

Referring to Africa, McIntire et al. (1992) stated: 'Farmers use various methods to feed crop residues to their livestock. In increasing order of labour requirements, these include (i) livestock having open access to whole residues on harvested fields; (ii) harvest and removal of stalks with subsequent open access to stubble on harvested fields; (iii) harvest and removal of stalks with subsequent restricted access to stubble on harvested fields; (iv) transport and storage of residues for feed or sale and (v) harvest of crop thinnings from fields for selective feeding before the harvest of the main crop residue. As more intensive modes of agricultural production are adopted in densely populated areas, such as in India and China, all crop residues may be harvested for livestock feed and/or for sale'. This is not yet the case in India, where most rice straw is still burnt (Landers, 2017).

8.3.6 Effects of rotations on farm yields

Weston et al. (2002) in subtropical Queensland (ave. rainfall 681 mm yr^{-1}) over 10 years showed a 24% gain in wheat yield after an annual medic ley, but only

5% after lucerne and a steady decline in continuous wheat yield over 50 years. These authors also explain that accumulation of SOM and nutrients in a ley will be very much reduced when rainfall limits growth and emphasizes the trade-off in pasture management between grazing and nutrient accumulation in the ley for following crops as a management decision, guided by relative profit expectancy. In Matto Grosso do Sul in Brazil, beef production and soybean yields fell significantly for a number of years after switching from one to another, as shown in Fig. 10.

In Table 5, Marchão and Vilela (2018) showed an 18% increase in soybean yield planted in a 17-year-old ley of Signal grass (Urochloa decumbens) under ZT/CA, with a 45% reduction in fertilizer use over this period, illustrating the benefits of alternation of crops and pastures. In this experiment, the advantage of ZT/CA over CT was minimal. The yield effect on originally infertile soils is temporary and in Fig. 10 can be seen as the reduction in soybean and beef yields over 3 years following the alternation of crop and pasture, with no fertilizer added to the pasture. Incremental beef yield after soybeans in the longer term is due to increased pasture growth from residual fertility from the

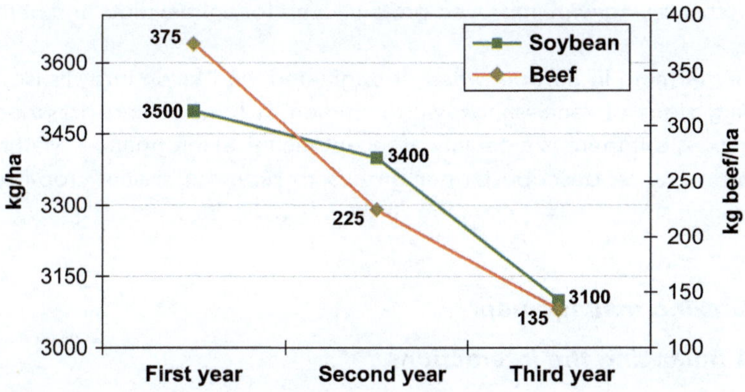

Figure 10 Decrease of soybean yields after 3 years of pasture and in beef production after 3 years of soybeans. Source: FundaÇâo MS.

Table 5 Soybean yields (kg/há) under continuous soybeans (CS) and a crop-pasture (ley) system (CP) under zero and conventional tillage, on a clay-textured red latosol, in Planaltina-DF, Brazil

	2007/2008b		
Cropping sqequence	Conventional tillage	Zero tillage	Average
Soya - sorghum- soya (CS)	3.079	3.044	3.061 b
Brachiaria brizanthab (CP) Soya	3.540	3.603	3.571 a
Difference (%)	14.97	18.36	16.66

Source: Adapted from Marchão & Vilela (2018).

crop phase and N fixation residual from the soya (first year). The increased soybean yield is more complex, deriving from reduced pest, disease and weed levels, improved soil structure and greater nutrient availability, due to recycling through the animals and liberation of available nutrients from the decomposition of the desiccated pasture (short term) and from SOM (longer term). This will not be so true in inherently fertile soils and where pastures are fertilized. On old cropped soils, as in Europe, the ley's principal function is to restore fertility for a crop phase.

In tropical Brazil, a heavily fertilized annual crop phase on low fertility soils generates residual fertility in the system, used to renovate pasture at low cost.

In subtropical Queensland, as shown by Armstrong et al. (2003), on a shallow vertisol, the effects of pulses in a CA rotation was short-lived. They opined: '... if soil fertility levels are to be maintained or improved, perennial pasture rotations will need to be used'. These included grass-legume leys and perennial legume leys with Siratro cv. Aztec (*Macroptilium atropurpurem*), pure grass leys and annual legume *Vigna trilobata*: only the perennial leys increased soil N and SOC'. In subtropical Brazil, Castro Júnior (1998) proved the viability of pasture following crop in ICL-CA; benefits were reduced N application in crops, cost saving with improved protein levels for animal diets and better soil cover.

Ley farming in Europe, where it originated, had fallen into disuse, but is showing signs of renaissance, with commercial firms selling grass/legume/forb mixes, but there is a paucity of recent technical information. Martin et al. (2016) in France describe farmer groups to promote spatial crop-livestock integration.

8.4 Grazing management

8.4.1 Balancing the interactions

Grazing management in ICL-CA will have mostly the same parameters as in other systems, but modified by the carryover effects on nutrients from the crop phase, generating higher stocking rates in the ley. Among the more common systems of managing grazing are rotational grazing, fixed stocking, grazing height or forage offer control, strip grazing and mob grazing. Their comparison is complex and will not be evaluated here.

In Fig. 11, the balance between compartments in an ICL-CA system depends critically on grazing intensity (Carvalho et al., 2010; Soussana and Lemaire, 2014). Moderate grazing intensities favour the coupling between C and N by cycling N via urine, stimulating leaf area renewal and favouring rhizo-deposition by grazing, often related to increases in soil C stocks (Balesdent

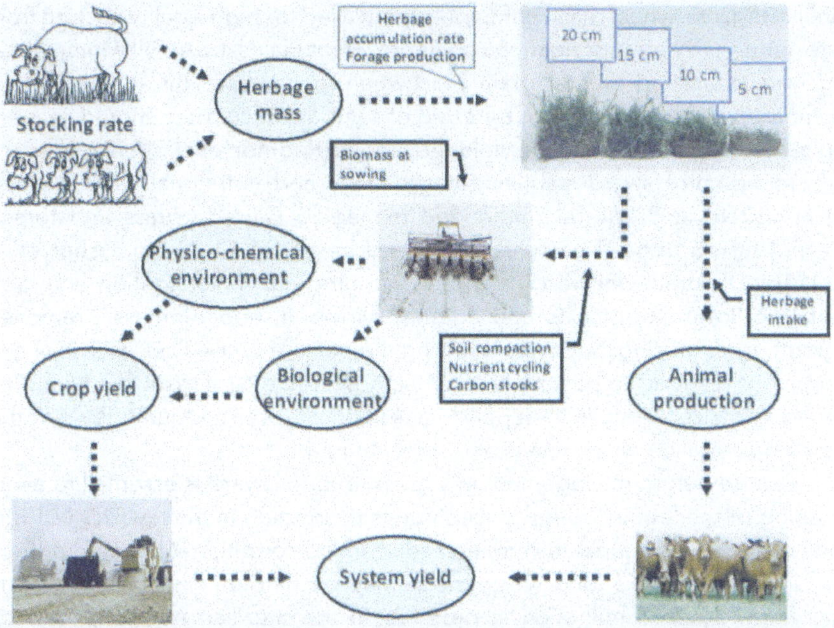

Figure 11 Conceptual diagram of how grazing intensity affects the ICL-CA system (Source: adapted from Carvalho et al., 2010).

and Balabane, 1996). Evers and Nelson (2000) in southeastern United States identified a tendency to lower sward heights entering a summer crop phase, due to higher grazing intensities, that reduced seed production in Italian ryegrass (*L. multiflorum*) and reduced the possibility of re-seeding. This is probably extensive to other annual grass spp. The conceptual presentation in Fig. 11 explains the interrelationships between management factors in an ICL-CA system.

Examples include the following. Allan et al. (2016) concluded for wheat-sheep systems in subtropical Australia: 'The results show that light grazing of sheep on crop residues over summer had a negligible effect on crop growth and yield as well as on weeds and soil water. The effect of grazing on soil mineral N was inconsistent. Grazing residues also had little effect on soil bulk density at most sites, although some sites had a small increase, but this had no effect on the crop, probably due to the alleviating effect of seeding with knife-points in the no-tillage system. However, a high grazing intensity through winter and summer, as in a pasture phase, significantly reduced the levels of crop residue and following crop yield on heavy soil, probably by reducing (rainfall) infiltration'. The lower the residual biomass, the lower the soil's ability to retain water (mulch and infiltration effects). Franzluebbers (2007) indicated that cereal grains (e.g. rye, wheat) have commonly been used for high-quality pasture from late autumn to early spring in the southern Great Plains, USA. Tall winter wheat cultivars often had greater grain yield when grazed until the jointing stage, compared with no grazing, but grazing of semi-dwarf cultivars should cease at an earlier stage to avoid grain yield reductions (Redmon et al., 1995).

To optimize sward stability, forage intake and nutritional value, Gomes et al. (2018) in Brazil, recommended managing Palisade grass and forage peanut mixed tropical pastures at a sward height of 24–30 cm. Fisher et al. (2010) in Australia showed that additional nitrogen mineralization and crop uptake following tropical forage legumes is lower if shoot biomass is removed by cutting or grazing. Soybean following the pasture phase fixes its own N and might not respond to pasture N residual; however, a high response would be found in maize, which is N dependent, especially after a legume ley, but this could be curtailed at low levels of residual biomass.

Pasture height management in a grass-legume sward is essential to avoid shading of the legume and reduction of its proportion in the ley. This will vary with every grass/legume mixture and soil/climate condition. Also, strip or mob grazing that provide long rest periods may be helpful in maintaining plant growth and water infiltration in pastures. In the Brazilian Amazon, Andrade et al. (2006) determined the ideal pre-grazing sward height for Massai grass (*P. maximum* × *P. infestum*) and forage peanut (*Arachis pintoi cv Ac 01*) was below 65–70 cm, to avoid excessive shade for the legume.

8.4.2 Microbial biomass and grazing management

Soil microbial biomass carbon (MBC) concentrations in pasture were favoured by the introduction of (large) animals to the system (Silva et al., 2015), and CA produced more MBC and less relative carbon loss than CT. Increases in MBC are equivalent to carbon sequestration (Balota et al., 1998). Excessive grazing intensity (10-cm residual pasture height) during the pasture cycle may cause a decrease in soil MBC and have a negative effect on the microbial biomass during the succeeding crop. Of all the enzymes analysed, only arylsuphatase and cellulase activities were altered by ICL-CA management, with differences between the moderate grazing intensity (grazing interval 20 days) and no grazing treatments (Silva et al., 2015).

8.4.3 Legume bloat susceptibility

Majak et al. (1995) in Canada, note that cattle are more susceptible to bloat when grazing young succulent legume leys or mixes and that lucerne (*Medicago sativa*) and clovers (*Trifolium* spp.) are most likely to cause bloat;

delaying grazing until dew is off the ley also helps to prevent bloat. Sainfoin (*Onobrychis viciifolia*), Birdsfoot Trefoil (*Lotus corniculatus*) and Cicer milk vetch (*Astragalus cicer*) are generally not dangerous in this respect. In Australia, lucerne and clovers are also considered problematic for bloat. Grazing of sugar beet tops is highly susceptible to cause bloat if not restricted (Landers, personal experience, 1960).

8.5 Inclusion of trees in the system

8.5.1 Pros and cons

The farmer must, first of all, accept the long-term investment represented by the tree component of the ICL-CA (F) farming system and determine what his objectives are in adopting the system.

8.5.2 Shading effects on crops and pasture

Schreiner (1987) registered average decrease of dry-matter yields of the order of 5%, 41% and 78% relative to the control, when four grasses were subjected to 25%, 50% and 80% shading. From the second year onwards, *Hemarthria altíssima* outperformed two Signal grass cultivars and Pensacola grass.

Carvalho et al. (2017) compared ICL-CA(F) with ICL-CA (without trees); from April 2013 to April 2014, stocking rates were 1.44 and 2.5 AU ha^{-1}. The difference was due to shading (and competition for nutrients) from the double rows of *E. urophylla* at 22 m apart. This also affected the sorghum yield in the rotation (1416.11 kg ha^{-1} versus 3927.5 kg ha^{-1}), a 64% reduction in grain yield, as compared to a reduction of 42% in stocking rate. The business case for ICL-CA(F) would have to show that the return from the timber species more than compensated for these losses. The conundrum is that timber species of higher value (e.g. teak, mahogany or *E. citriodora*) take much more time to harvest, extending the payback period and reducing management flexibility. Also, for a return to arable cropping, the cost of stump removal after the final cut is considerable.

Eriksen and Whitney (1981) in Hawaii, studying six tropical forages conclude that 'In general, shading reduced dry matter yields when N fertilizer was applied, but increased yields when N was deficient. Decreasing light intensity decreased the dry matter content of the forage, but increased the N content, the levels of most minerals, plant height, and leafiness'. Percentage DM was reduced by approximately 1% for every 10% reduction in light intensity'. It was also observed that reduced carbohydrate reserves in the roots and stubble under shade extended the recovery period of grasses after de-foliation, which '...would require careful management to avoid excessive depletion of root reserves, either by lenient grazing to maintain high leaf areas, or by allowing an extended recovery period'. In Brazil (18°S), hand-planted forage peanut between teak at 3.5 × 3.0 m tree spacing tolerated shade at full canopy and spread considerably (pers. comm., A.V. Campos, ICL-CA(F) practitioner for 20 years).

Lin et al. (1999) studying cool- and warm-season temperate forages found: 'Cool-season grasses showed more shade tolerance when grown during the summer–fall than when grown during the spring–early summer. Seven of the selected cool-season grasses grown during the summer – fall did not display significant reductions in mean dry weight (MDW) under 50% shade as compared to full sun. Smooth bromegrass (*Bromus inermis*) grown under 50% shade showed a significantly increased MDW production compared to growth in full sun. With the exception of Justus orchardgrass (*Dactylis glomerata L.*) and smooth bromegrass, growth of cool-season grasses was inhibited at 80% shade. Among the legumes harvested during the fall, the dry weights of both Desmodium species tested and Hog Peanut (*Amphicarpaea bracteata L.*) increased significantly under 50% and 80% shade. In addition, 'Cody' alfalfa (M. sativa), White Clover, Slender Lespedeza (L. virginica) and 'Kobe' Lespedeza (Lespedeza sp.) showed no significant reductions in MDW under 50% shade'. Shade tolerance in forages is quite variable, requiring careful species selection for this character and adaptation to climate.

8.5.3 Shading effects on animals

Shade can be natural from trees or artificial. Higgins et al. (2011) at the University of Florida, indicated that dairy cows (Bos taurus) provided with shade produced 10–19% more milk than cows without shade. Citing Virginia Tech researchers, who found that high levels of heat stress, experienced when temperatures exceed 90°F, can reduce milk production by 20–30% (10–25 lb of milk per day). Cows that were cooled peaked at 9 lb more milk per day as compared to cows that had not been cooled. Research at the University of Florida found that cattle provided with shade had conception rates of 44.4% as compared to conception rates of 25.3% for cattle without shade. At the University of Missouri, researchers found that shade increased the overall pregnancy rate of cows by nearly 40%. Cows with shade had an overall pregnancy rate of 87.5%, while the pregnancy rate was only 50% for cows without shade. Research has shown that bulls with access to shade have increased semen counts. The higher the humidity, the greater the stress at any given temperature. Brahman breeds (*Bos indicus*), predominantly for beef, are much more tolerant of high temperatures.

8.6 Environmental impacts of ICL-CA

8.6.1 Carbon sequestration and net GHG emissions

Carbon is intimately linked to soil fertility, represents the major element in SOM, which, in turn, is an important source of available nutrients and binding material for soil aggregates. Mendes et al. (2018) stated: 'The high complexity of ICL-CA systems, including different soil types, baseline C stocks, grazing intensities, crop species and time of adoption, all contribute to the variable results associated with soil C response under this system'. Coskun et al. (2016), in a wide review, showed that nitrogen availability is essential for increase in SOM and carbon sequestration.

The sustainable practices of ICL-CA and ICL-CA(F) offer multiple technical, economic, environmental and social benefits (Macedo, 2010). Bayer et al. (2011), in a high C biomass and N fertilizer input rotation in subtropical Brazil, noted a decrease in oxidation of CH_4, due to the encouragement of methanogenic bacteria, which resulted in increased CH_4 emissions. They suggested an additive effect of NH_4-N (derived from fertilizer) and dissolved organic carbon (DOC) in suppressing CH_4 oxidation.

Primavesi (2008) in tropical Brazil (22°S) showed a reduction in kg CH_4 emitted per kilogram of beef from 0.708 to 0.447 (36.8%) that occurred by reducing slaughter age from 39 to 26 months, through ICL-CA pasture renovation (see Fig. 12). Also, in tropical Brazil, Esteves et al. (2010) at the same location showed that CH_4 emissions per kilogram of liveweight were inversely correlated with increased stocking rate, which liberated half of the area of previously degraded pastures for cropping, with a deforestation mitigation index (DMI) of 2:1 (see next Section).

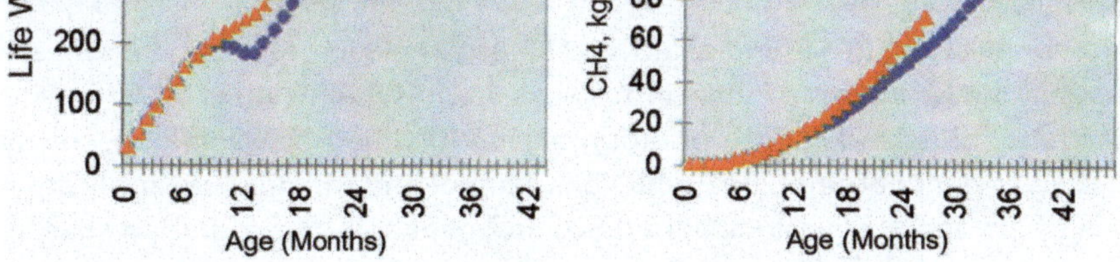

Figure 12 Relative weight gains and reductions in kg CH_4 emitted per kilo of beef for 26 and 39 month-old steers. Source: Primavesi (2008).

Piva et al. (2014), with sheep in subtropical Brazil, showed system NO_2 emissions were progressively higher, at 75 and 150 kg ha−[1] of fertilizer N than the control with zero N, with a similar trend between grazed and ungrazed winter pasture (black oats + annual ryegrass). Also in this region, comparing systems planting pure sorghum under ICL-CA with sorghum in the alleys of ICL-CA(F), de Carvalho et al. (2017) measured system fluxes per unit of grain where the former showed one third of the level in ICL-CA(F) (76 mg N_2O m^{-2} kg^{-1} as opposed to 224 mg N_2O m^{-2} kg^{-1} grain, significant at $p < 0.05$) due to 64% higher grain yield, but overall system fluxes were similar during the sorghum cycle. Application of fertilizer N raised the N_2O flux up to 2 months after application, especially with rainfall. A lower annual N_2O flux in ICL-CA(F) was attributed to phenolic compounds in *E. urophylla* leaf litter inhibiting microbial activity; that would not be true for deciduous species.

When we express GHG emissions per unit of yield, we demonstrate to the world that efficient agricultural practices reduce global warming, because emissions per unit of yield are smaller than per hectare figures and are related directly to consumer concerns. As a general principle, the gains in productivity of all agricultural products can show a decrease in GHG emissions per unit.

Segnini et al. (2007), comparing 27 years of a Signal grass (*U. decumbens*) pasture with mature Brazilian native (cerradão, light forest) vegetation, measured an incremental sequestration of carbon between 1.7 and 3.5 Mg C ha^{-1} yr^{-1} for un-fertilized and full liming plus fertilization treatments. These effects should be repeatable in a ley in similar soils and climate, possibly to a lesser degree.

The potential time to C-sequestration equilibrium in arable cropping worldwide was reported by Jarecki and Lal (2003) as 25–50 years. Gonçalves et al. (2019) indicated that soil C levels had reached equilibrium after 30 years of CA in a wheat/soybean ×2/black oats /maize system, where the average annual biomass input was 14.5 Mg ha^{-1} yr^{-1}, highly dependent on the black oat/maize contribution of 20.5 Mg ha^{-1} yr^{-1} This biomass requirement increases as latitude nears the equator, due to faster oxidation of SOM (Sá, pers. comm., 2019).

Scopel et al. (2012) indicated that C accumulation rates under CA are generally lower in France than in tropical Brazil. When climate in Europe is optimal for grassland canopy photosynthesis (10°C and 1200 mm yr^{-1}), critical stocking rate for net zero GHG emissions without and with manure application equals 0.85 and 1.2 livestock units (LSU) ha^{-1} over a 200 days grazing season. This illustrates the conflict between pasture intensification in ICL-CA and GHG emissions, while emphasizing the value of returning organic N in manure from confined animals to pastures (Soussana et al., 2014).

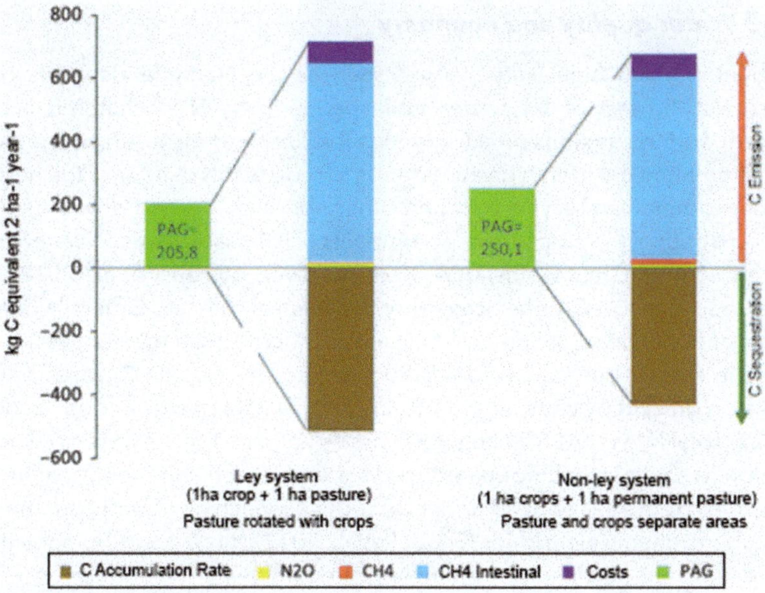

Figure 13 Global warming potential (GWP) for beef and grain production with ICL-CA and without and their components: soil C accumulation rate for 0–30 cm, nitrous oxide (N_2O) and enteric methane (CH_4) emissions and costs.

Brazilian studies indicate compensation for bovine methane emissions under ICL-CA(F) systems, with a positive GHG emission balance, resulting primarily from very high total tree biomass and C accumulation in the trees (Tsukamoto Filho et al., 2003; Cerri et al., 2007; Segnini et al., 2007; Macedo, 2009)Also, Tsukamoto Filho et al. (2003) showed that carbon sequestration under an ICL-CA(F) system with eucalyptus sequestered between 5.98 tC $ha^{-1}yr^{-1}$ in year 1 and 84.38 tC $ha^{-1}yr^{-1}$ in year 10–11. This more than compensates the enteric emissions of CH_4 of the cattle and exceeds those sequestered by a pure stand of eucalyptus. Figure 13 is self-explanatory and shows the breakdown and balance between GWP and sequestering mechanisms.

8.6.2 Land use intensification (LUI)

Mitigation of the demand for clearing native vegetation occurs through LUI, that is vertical expansion, by recuperation of degraded pastures, incentivating ley farming and generally increasing productivity through technical improvements. LUI also increases agricultural biodiversity and aquifer recharge, while reducing the emission of GHGs per unit of product (see Fig. 13). LUI generates a DMI of between 0.8 and 2.5 per ha in the ICL-CA system, depending on the rotation (Landers et al., 2005).

8.6.3 Water quality and economy

In countries with high water tables, such as the Netherlands, parts of the America's Midwest or UK, water table pollution by N leachates is a major concern, with restrictions on nitrogenous fertilizer and manure/sewage sludge use. Alternating crops with pastures tends to reduce this danger as the pasture's deeper rooting can recycle N, K and other nutrients leached down below the rooting depth of the crop phase. The impacts of ICL-CA systems on water quality in the Northern Great Plains, USA, were reviewed by Faust et al. (2018). They concluded that ICL-CA practices, when compared with CT, generally reduced losses of suspended solids and nutrients; CA crop management reduced N losses in surface run-off by 65% but increased the soluble P losses by 57%. However, grazing by cattle under ICL-CA increased total suspended solids loss by 22%, total N loss by 85% and total P loss by 25%; no figures were given for the rotation as a whole. In Europe, reduced surface run-off and binding of nutrient molecules by increased SOM under CA reduced leaching of both nutrients and agricultural chemical residues (Scopel et al., 2012). It is evident from the above discussion that leaching is a complex phenomenon, depending on several interrelated local factors and each case must be investigated separately.

The surface residue retention under CA creates a mulch effect which reduces evapotranspiration, retards overland flow and hence improves rainfall infiltration. Stone et al. (2006) showed that the variation of evapotranspiration in irrigated common beans (*P. vulgaris* var. Perola) in eight cover crop residues ranged from 259.8 to 343.7 mm, which was inversely proportional to soil cover and residue mass. The best cover crops were Palisade grass and Guinea grass cv. Mombaça (*Panicum maximum*) giving between 100% and 74/60% cover, respectively, at planting and harvest, the poorest were legumes, *S. guyanensis* cv. Mineirão with 22% and pigeon pea, 11% (at harvest). Irrigation water economy in common beans was measured at 20% by do Nascimento et al. (2001). The mulch also increases rainfall infiltration, thus increasing the total water available in the profile, by reductions in run-off and evapotranspiration (Et). Also, it prevents raindrop splash taking fungal spores to plant leaves, reducing yield depression from disease.

As organic matter and pore size distribution improve under a ley, water-holding capacity of the soil rises. To emphasize why ICL systems under CT are inferior, Marchão (2007) showed a 20.1% increase in readily available moisture (RAM) with the adoption of ICL-CA in a rotation of 4 years *P. maximum*/4 years soybeans. Higher cover will ensure better yields in dry years, which is an important risk reduction factor in ICL-CA adoption.

8.6.4 Erosion control

Soil erosion losses are a major source of surface water pollution worldwide. The Queensland government, Australia (2017), reported the percentage of soil loss from treatments of 87%, 69% and 6% soil cover as 0.03, 0.3 and 22 ton $ha^{-1} yr^{-1}$ and percentage of rainfall run-off as 3%, 26% and 70%, respectively, showing the crucial importance of maintaining soil cover in all forms. These data emphasize the impacts of ICL-CA systems on aquifer recharge and reduction of silting of water bodies.

Renovation of pastures with CA on sloping lands is practiced, in Brazil, desiccating low productivity pastures with Glyphosate and using CA to sow a higher-performing grass, producing a uniform, fertilized sward (Cardoso, rancher, pers. comm., 2018). Erosion with conventional preparation would be serious on such undulating land. Ribaski et al. (2005) also observed positive erosion control effects in silvi-pastoral systems in the subtropical pampa gaúcha, with *Eucalyptus grandis* and *Pinus elliottii*.

9 Adoption of ICL-CA and ICL-CA(F) systems

FAO (2001) registered a generalized survey of worldwide constraints to CA, of which the most important are summarized below and cover three eco-regions:

1 The wet-dry tropical savannah environments in the tropics;
2 Moist semi-arid savannah environments in the tropics and summer rainfall subtropics, and moist Mediterranean-type environments; and
3 Humid forest margins (including very moist savannahs).

Principal constraints to adoption identified:

- Value chain for crop production not well organized in the tropics, applies especially to the Amazon and other equatorial forests, where upland rice, an ideal rotation crop for soybeans, is less planted due to the unreliability of small traders and indifference of the larger ones;
- Technology not accessible or not affordable to smallholders and lack of credit for investment (less of a constraint if organized in cooperatives or farmer associations);
- Shifting cultivation needs to be stabilized (by outside inputs and rotation within a fixed property).

Adoption of a more complex (ICL-CA) system will be limited by farmer management capacity (Sulc and Franzluebbers, 2014). Gil et al. (2015) indicated that 89% of adopters in Mato Grosso state did not include tree planting and that credit was not a constraint to adoption, which was more dependent on knowledge availability and cultural factors (risk aversion). The system design depended on production strategy. Valentim et al. (2003), in Acre state, Amazônia, noted that slow establishment of forage peanut (*Arachis pintoi*) was a limiting factor to adoption and pasture performance.

With agri-silvi-pastoral systems, besides technical considerations, such as soil type, knowledge of the technology and qualified assistance, Vilela et al. (2001) and Kichel and Miranda (2002) indicated several socio-economic factors affecting adoption, namely availability of capital, access to input and output markets and the possibility of rental (or partnership) agreements between cattle owners and arable farmers, promoting pasture regeneration after a crop phase. The long time to payout of the tree component (6–8 years for most *Eucalyptus* spp., more for other spp.) is an important consideration, also reducing operational flexibility. Reasons given for adoption by cattle farmers in Brazil were (i) adjustment to market environmental demands and (ii) pasture renovation; for grain farmers (i) increased yields and (ii) reduced financial risk (Embrapa, 2016).

In spite of the huge potential of ICL-CA for smallholders, the paradigm change and even small investments in machinery required (at minimum a hand-jab plater costing about US$50–65) pose huge challenges to resource-poor farmers. In addition, in semi-arid areas, there is serious competition for crop residues to be used as livestock feed in detriment to their maintenance as soil cover. In subtropical Brazil, there is almost universal adoption of CA by smallholders, but rainfall patterns are favourable to double cropping and winter pasture, while many farm sizes are over 10 ha, which implies some capacity for investment in specialized planters, as opposed to <2–5 ha in Asia.

10 Policy considerations

Franzluebbers (2007) noted that 'to change agricultural production from its current state of specialization to one of integrated crop–livestock production will require major changes in: (i) farming attitudes and how farmers might utilize relevant scientific information, (ii) agribusiness developments and marketing, and (iii) government support to balance agricultural production with natural resource conservation'. Financial incentives to farmers for ICL-CA systems in general require a trade-off for environmental services which is a hard sell to the voting public.

On policy to reduce GHG emissions from European grasslands, Soussana et al. (2015) indicated: 'Priority should be given to the restoration of degraded pastures which offers win-win possibilities by combining increases in plant productivity, in soil carbon stocks and in animal production. Such a scheme has already been successfully introduced in Portugal (see www.terraprima.pt/pt/ accessed 20.3.19) using phosphorus fertilization and species-rich grass-legume mixtures'. Bell and Moore (2012) observed that a single parameter to measure the level of integration of farm enterprises is not viable and cited Sumberg (2003), who also observed the omission of resource allocation and management focus constraints in work published on the bio-physical aspects of ICL-CA and recommended analysis on the basis of four dimensions: space, time, ownership and management. A common method is required to guide worldwide policy incentives for ICL-CA, especially to combat global warming.

11 References

Abrahão, W. A. P., Costa, L. M., Mello, J. W. V. and Neves, J. C. L. (1998). Distribuição de frequência de tamanho da fração areia e compacidade relativa de solos desenvolvidos de sedimentos do grupo geológico Barreiras. *Revista Brasileira de Ciência do Solo* 22, 1–9. doi:10.1590/S0100-06831998000100001.

Abrecht, K. L. and Bristow, D. G. (1996). Coping with soil and climatic hazards during crop establishment in the semi-arid tropics. *Australian Journal of Experimental Agriculture* 36(8), 971–83.

Abreu, A. B. L. (2015). Comportamento do nematoide das lesões radiculares nas culturas de soja e milho em sucessão a diferentes coberturas de solo na entressafra. Master's thesis, Universidade Federal de Mato Grosso do Sul, Câmpus de Chapadão do Sul, Programa de Pós-Graduação em Agronomia.

Abreu, A. B. L. and Borges, E. P. (2015). Redução de Pratylenchus brachyurus na cultura de soja em sucessão a diferentes coberturas de solo na entressafra. Comunicado: Fundação Chapadão, MS, Brazil.

Allan, C. J., Jones, B., Falkiner, S., Nicholson, C., Hyde, S., Mauchline, S., Ferrier, D., Ward, P. and Siddique, K. H. M. (2016). Light grazing of crop residues by sheep in a Mediterranean-type environment has little impact on following no-tillage crops. *European Journal of Agronomy* 77, 70–80.

Almeida, R. F. de, Naves, E. R. and da Mota, R. P. (2015). Soil quality: Enzymatic activity of soil β-glucosidase. *Global Journal of Agricultural Research and Reviews* 3(2), 146–50.

Anderson, R. L. (2007). A changing perspective with weed management in semi-arid cropping systems. *Annals of Arid Zone* 46, 1–15.

Andrade, C. M. S. de, Garcia, R., Valentim, J. F. and Pereira, O. G. (2006). Grazing management strategies for massaigrass-forage peanut pastures. 3. Definition of sward targets and carrying capacity. *Revista Brasileira de Zootecnia* 35(2). doi:10.1590/S1516-35982006000200004.

Araújo, R. S. and Hungria, M. (Eds) (1994). Microrganismos de Importância Agrícola, Introdução. Embrapa CNPAF Documentos 44, 236p.

Araújo, V. M., Basso, R. F., Basso, C., Mendina, S. S. S., Oliveira, D., Nunes, P. A. A., Kunrath, T. R. and Carvalho, P. C. F. (2018). A biomassa residual do pasto não determina o rendimento da soja em um sistema integrado de produção agropecuária. In: *55ª Reunião anual da Sociedade Brasileira de Zootecnica e 28° Congresso Brasileiro de Zootecnica*, 27 a 30 de agosto de 2018, Goiânia, GO.

Assis, P. C. R., Stone, L. F., Silveira, A. L. R., Oliveira, J. M., Wruck, F. J. and Madari, B. E. (2017). Biological soil properties in integrated crop-livestock-forest systems. *Revista Brasileira de Ciência do Solo* 41.

Balbino, L. C. and Di Stefano, J. G. (1999). Projet Procitropicos: intégration de l'agriculture et de l'élevage par les systèmes de semis direct. In: Rasolo, F. and Raunet, M. (Eds), *Gestion agrobiologique des sols et des systèmes de culture*, Actes de l'atelier international du 23 au 28 mars 1998, Antsirabe, Madagascar. Cirad, Montpellier, pp. 409–17.

Balbino, L. C., Barcellos, A. O. and Stone, L. F. (Eds) (2011). *Marco Referencial: integração lavoura-pecuária floresta*. Embrapa, Brasília, DF, 130p.

Balbino, L. C., Cordeiro, L. A. M., Oliveira, P., Kluthcouski, J., Galerani, P. R. and Vilela, L. (2012). Agricultura sustentável por meio da Integração Lavoura-Pecuária-Floresta (ILPF). Informações Agronômicas IPNI, no. 138, pp. 1–18.

Balesdent, J. and Balabane, M. (1996). Major contribution of roots to soil carbon storage inferred from maize cultivated soils. *Soil Biology and Biochemistry* 28(9), 1261.

Ball, S. (2011). Investigation of Zero-Till Farming Systems in high rainfall cropping zones and its impact on soil biology and nutrient cycling. Nuffield Australia Project. No. 1006, 44p.

Balota, E., Colozzi Filho, A., Andrade, D. and Hungria, M. (1998). Biomassa microbiana e sua atividade em solos sob diferentes sistemas de preparo e sucesso de culturas. *Revista Brasileira de Ciência do Solo* 22, 641-9. doi:10.1590/S0100-06831998000400009.

Balota, E. L., Nogueira, M. A., Mendes, I. C., Hungria, M., Fagotti, D. S. L., Melo, G. M. P., Souza, R. C. and de Melo, W. J. (2013). Enzimas e seu Papel na Qualidade do Solo. *Tópicos em Ciência do Solo* 8, 221–78.

Bayer, C., Amado, T. J. C., Tornquist, C. G., Cerri, C. E. P., Dieckow, J., Zanatta, J. A. and Nicoloso, R. S. (2011). Estabilização do carbono no solo e mitigação das emissões de gases de efeito estufa na agricultura conservacionista. In: Klauberg Filho, O., Mafra, A. L. and Gatiboni, L. C. (Eds), *Tópicos em ciência do solo* (vol. 7), pp. 55–118.

Bell, L. W. and Moore, A. D. (2012). Integrated crop–livestock systems in Australian agriculture: Trends, drivers and implications. *Agricultural Systems* 111, 1–12.

Bell, L. W., Moore, A. D. and Kirkegaard, J. A. (2014). Evolution in crop-livestock integration systems that improve farm productivity and environmental performance in Australia. *European Journal of Agronomy* 57, 10–20.

Bernardi, A. C. de C., Machado, P. L. O. de, Freitas, P. L. de, Coelho, M. R., Leandro, W. M., Júnior, J. P. de O., Oliveira, R. P. de, Santos, H. G. dos, Madari, B. E. and Carvalho, M. da C. S. (2003). *Correção do solo e adubação no sistema de plantio direto nos cerrados*. Documentos, n. 46. Embrapa Solos, Rio de Janeiro, 22p.

Blanchart, E., Albrecht, A., Alegre, J., Duboisset, A., Pashanasi, B., Lavelle, P. and Brussaard, L. (1999). Effects of earthworms on soil structure and physical properties. In: Lavelle, P., Brussaard, L. and Hendrix, P. (Eds), *Earthworm Management in Tropical Agroecosystems*. CAB International, Wallingford, UK, pp. 139–62.

Blank, D. (2007). *A Fresh Look at Life Below the Surface*. Fort Meyers Lauderdale Echo Inc., USA.

Blouin, M, Hodson, M. E., Delgado, E. A., Baker, G., Brussaard, L., Butt, K. R., Dai, J., Dendooven, L., Peres, G., Tondoh, J. E., Cluzeau, D. and Brun, J.-J. (2013). A review of earthworm impact on soil function and ecosystem services. *European Journal of Soil Science* 64(2).

Bluemling, B. and Hu, C. S. (2011). Vertical and system integration instead of integrated water management? *Measures for mitigating NPSP in rural China. Groundwater Conference 2011*. South African Water Commission, Rietfontein, Pretoria, South Africa, pp. 269–77.

Boddey, R. M. and Dobereiner, J. (1995). Nitrogen fixation associated with grasses and cereals: Recent progress and perspectives for the future. *Fertilizer Research* 42(1-3), 241–50.

Boddey, R. M., Macedo, R., Tarré, R. M., Ferreira, E., de Oliveira, O. C., Rezende, C. de P., Cantarutti, R. B., Pereira, J. M., Alves, B. J. R. and Urquiaga, S. (2004). Nitrogen cycling in Brachiaria pastures: the key to understanding the process of pasture decline. *Agriculture, Ecosystems and Environment* 103, 389–403.

Boeni, M., Bayer, C., Dieckow, J., Conceição, P. C., Dick, D. P., Knicker, H., Salton, J. C. and Macedo, M. C. M. (2014). Organic matter composition in density fractions of Cerrado Ferralsols as revealed by CPMAS 13C NMR: influence of pastureland, cropland and integrated crop-livestock. *Agriculture, Ecosystems and Environment* 190, 80–6. doi:10.1016/j.agee.2013.09.024.

Bolliger, A. (2009). Taking stock of the Brazilian "Zero-Till Revelution". World Association of Soil and Water Conservation.

Bonaudo, T., Bendalun, A. B., Sebatier, R., Ryschawy, J., Magda, D. and Tichit, M. (2014). Agroecologic principles for the re-design of integrated crop-livestock systems. *European Journal of Agronomy* 57, 43–51.

Borghi, E. (2017). Física do Solo, Estoque de Carbono e Qualidade da Matéria Orgânica e Sistemas de Produção Envolvendo Culturas Gramíferas e Capim Braquiária Plantio Direto, utilizando as Condições de Solo de Cerrado como Referência. In: *Relatório Parcial (Projeto Agrisus 2038/17): Física do Solo e Matéria Orgânica em Sistemas de Produção Agrícola de Cerrado*. Embrapa Milho e Sorgo, Sete Lagoas/MG.

Borghi, E., Costa, N. V., Crusciol, C. A. C. and Mateus, G. P. (2008). Influence of the spatial distribution of maize and *Brachiaria brizantha* intercropping on the weed population under no-tillage. *Planta Daninha* 26(3).

Bornemissza, G. E. (1976). The Australian dung beetle project 1965–1975. Australian Meat Research Committee review no. 30, 32p.

Bortolini, C. G. (2005). Rotação de culturas no sistema plantio direto. In: *Encontro De Plantio Direto No Cerrado 8., Tangará da Serra 2005. Anais... Tangará da Serra, Gráfica e Editora Sanches*, 2005, pp. 115–18.

Bragagnolo, N., Pan, W. and Thomas, L. C. (1997). *Solo: uma experiência em manejo e conservação*. Ed. do autor, Curitiba, Paraná, Brazil, 102p.

Broch, D. L., Pitol, C. and Borges, E. P. (1997). Integração agricultura-pecuária: plantio direto da soja sobre pastagem na integração agropecuária. Fundação MS. Informativo Técnico, 01/97. Fundação MS, Maracaju, 24p.

Brown, G. G, Pashanasi, B., Villenave, C., Patron, J. C., Senapati, B. K., Giri, S., Barois, I., Lavelle, P., Blancharf, E., Blakemore, R. J., Spain, A. V. and Boyer, J. (1999). Effects of earthworms on plant production in the tropics. In: Lavelle P., Brussaard L. and Hendrix P. (Eds), *Earthworm Management in Tropical Agroecosystems*. CABI, Wallingford, UK, pp. 87–147.

Brust, G. E. and King, L. R. (1994). Effects of crop rotation and reduced chemical inputs on pests and predators in maize agroecosystems. *Agriculture, Ecosystems and Environment*, 48, 77–89.

California State Board of Forestry (2018). *A Handbook for Eucalyptus Planters*. Franklin Classics Trade Press, San Francisco CA, 46p.

Capowiez, Y., Cadoux, S., Bouchand, P., Roger-Estrade, J., Richard, G. and Boizard, H. (2009). Experimental evidence for the role of earthworms in compacted soil regeneration based on field observations and results from a semi-field experiment. *Soil Biology and Biochemistry* 41(4), 711–17.

Carbonero, C. H., Mueller-Harvey, M., Brown, T. A. and Smith, L. (2011). Sainfoin (*Onobrychis viciifolia*): A beneficial forage legume. *Plant Genetic Resources* 9(1), 70–80. doi:doi:10.1017/S1479262110000328.

Carvalho, M. M., Xavier, D. F. and Alvim, M. J. (2001). Caracte rísticas de algumas espécies arbóreas adequadas para associação com pastagens. Embrapa Gado de Leite. Circular Técnica, 64. Embrapa Gado de Leite, Juiz de Fora, 24p.

Carvalho, M. M., Alvim, M. J., Xavier, D. F., et al. (2002). Estabelecimento de sistemas silvipastoris: ênfase em áreas montanhosas e solos de baixa fertilidade. Circular Técnica 68. Embrapa Gado de Leite, Juiz de Fora.

Carvalho, J. L. N., Avanzi, J. C., Silva, M. L. N., de Mello, C. R. and Cerri, C. E. P. (2010). Potential of soil carbon sequestration in different biomes of Brazil. *Revista Brasileira de Ciência do Solo* 34(2).

Carvalho, D. D. C., Geraldine,A. M., Lobo Junior, M. and Marques, S. C. (2015). Biological control of white mold by *Trichoderma harzianum* in common bean under field conditions. *Pesquisa Agropecuária Brasileira* 50(12), 1220–4.

Carvalho, A. M. De, Oliveira, W. R. D. De, Ramos, M. L. G., Coser, T. R., Oliveira, A. D. de, Pulrolnik, K., Souza, K. W. De, Vilela, L. and Marchao, R. L. (2017). Soil N_2O fluxes in integrated production systems, continuous pasture and Cerrado. *Nutrient Cycling in Agroecosystems* 108(1), 69–83.

Castro Filho C., Muzilli, O. and Podanoschi, A. L. (1991). Estabilidade dos agregados e sua relação com o teor de carbono orgânico num latossolo roxo distrófico, em função de sistemas de plantio, rotações de culturas e métodos de preparo das amostras. *Revista Brasileira de Ciência do Solo* 22, 527–38.

Cerri, C. E. P., Easter, M., Paustian, K., Killian, K., Coleman, K., Bernoux, M., Falloon, P., Powlson, D. S., Batjes, N., Milne, E. and Cerri, C. C. (2007). Simulating SOC changes in 11 land use changes from Brazilian Amazon with Roth C and Century models. *Agriculture, Ecosystems and Environment* 122, 46–57.

Chamen, W. C. T. (2006). Controlled traffic farming on a field scale in the UK. Paper to be presented at the *International Soil Tillage Research Conference*, August/September 2006, Kiel, Germany.

Chen, H., Hu, B. and Charles, A. T. (1995). Chinese integrated fish farming: a comparative bioeconomic analysis. *Aquaculture Research* 26, 81–94.

Choocharoen, C., Neef, A., Preechapanya, P. and Hoffmann, V. (2014). Agrosilvopastoral systems in Northern Thailand and Northern Laos: Minority peoples' knowledge versus government policy. *Land* 3, 414–36. doi:10.3390/land3020414.

Cobucci, T. and Portela, C. M. O. (2003). Manejo de herbicidas no Sistema Santa Fé e na braquiária como fonte de cobertura morta. In: Kluthcouski, J., Stone, L. F. and Aidar, H. (Eds), *Integração lavoura-pecuária*. Embrapa Arroz e Feijão, Santo Antônio de Goiás, pp. 443–58.

Cong, W., Jinying, J. J., Rasmussen, J., Søegard, K. and Eriksen, J. (2017). Forbs enhance productivity of unfertilised grass-clover leys and support low-carbon bioenergy. *Scientific Reports* 7. doi:10.1038/s41598-017-01632-4.

Corbeels, M., Siqueitra Neto, M., Marchão, R. L. and Madari, B. E. (2016). Evidence of limited carbon sequestration in soils under no-tillage systems in the Cerrado of Brazil. *Scientific Reports.*

Cordeiro, L. A. M. C., Vilela, L., Marchão, R. L. Kluthcouski, J. and Martha Júnior, G. B. (2015). Integração Lavoura-Pecuária e Integração Lavoura-Pecuária-Floresta: Estratégias Para Intensificação Sustentável do Uso do Solo. *Cadernos de Ciencia e Tecnologia* 32, 15–53.

Coskun, D., Britto, D. T. and Kronzucker, H. J. (2016). Nutrient constraints on terrestrial carbon fixation: The role of nitrogen. *Journal of Plant Physiology* 203, 95–109. doi:10.1016/j.jplph.2016.05.016.

Costa, J. L. Da S. and Rava, C. A. (2003). Influência da braquiária no manejo de doenças do feijoeiro com origem no solo. In: Kluthcouski, J., Stone, L. F. and Aidar, H. (Eds), *Integração lavoura-pecuária*. Embrapa Arroz e Feijão, Santo Antônio de Goiás, pp. 523–33.

Crittall, E. (Ed.) (1959). *A History of the County of Wiltshire: Volume 4, British History*. Victoria County History, London. Available at: http://www.british-history.ac.uk/vch/wilts/vol4 (accessed on 22 February 2019).

Crusciol, C. A. C., Mateus, G. P., Nascente, a. S., Martins, P. O., Borghi, E. and Pariz, C. M. (2012). An innovative crop-forage intercrop system: Early cycle soybean cultivars and palisadegrass. *Agronomy Journal* 104, 1085–95.

Derpsch, R., Friedrich, T., Landers, J. N., Rainbow, R., Reicosky, D. C., Sa, J. C. M., Sturny, W. G., Wall, P., Ward, R. C. and Weiss, K. (2011). About the necessity of adequately defining no-tillage - a discussion paper. In: *Proceedings of the 5th World Congress of Conservation Agriculture*, 26–29 September 2011, Brisbane, Australia.

Dervash, M. A., Bhat, R. A., Mushtaq, N. and Singh, V. (2018). Dynamics and importance of soil mesofauna. Available at: https://www.researchgate.net/publication/325119590.

Donagemma, G. K., Freitas, P. L. De, Balieiro, F. De C., Fontana, A., Spera, S. T., Lumbreras, J. F., Viana, J. H. M., Araújo Filho, J. C. De, Santos, F. C. Dos, Albuquerque, M. R. De, Macedo, M. C. M., Teixeira, P. C., Amaral, A. J., Bortolon, E. and Bortolon, L. (2016). Caracterização, potencial agrícola e perspectivas de manejo de solos leves no Brasil. *Pesquisa Agropecuária Brasileira, Brasília* 51(9), 1003–20. Available at: https://seer.sct.embrapa.br/index.php/pab/article/view/22572/13571.

Doolette, J. B. (1980). The Australian ley farming system in North Africa and the Middle East. In: *Proceedings of International Congress in Dryland Farming*, Adelaide, South Australia. Department of Agriculture, Adelaide, Australia.

Dubeux, J. C. B, Muir, J. P., Apolinário, V. X. de O., Nair, P. K. R,, Lira, M. de A. and Sollenberger, L. E. (2017). Tree legumes: an underexploited resource in warm-climate silvopastures. *Revista Brasileira de Zootecnia* 46(8), 689–703.

Dupraz, C. and Liagre, F. (2008). *Agroforesterie des arbres et des cultures*. Editions France Agricole, Paris, France, 473p.

Duttargavni, S. (2013). Integrated crop livestock systems: A key to sustainable intensificatification. Available at: https://pt.slideshare.net/.../integrated-crop-livestock-systemsa-key-....

Eberhardt, D. N., Becquer, T., Marchão, R. L. Vendrame, P. R. S., Vilela, L., Corazza, E. J. and Guimarães, M. F. (2017). Phosphorus bioavailability in soybean grown after pasture under different fertility regimes. *Semina: Ciências Agrárias, Londrina* 38(2), 71–580. Available at: https://www.researchgate.net/publication/323582224_Phosphorus_bioavailability_in_soybean_grown_after_pasture_under_different_fertility_regimes (accessed on 09 May 2019).

Ehsanul, H. M. D. (2016). Suitable integrated crop-livestock production system in Char area of Bangladesh. *Journal of Fisheries and Livestock Production* 4, 156. doi:10.4172/2332-2608.1000156.

Ekenler, M. and Tabatabai, M. A. (2003). Responses of phosphatases and arylsulfatase in soils to liming and tillage systems. *Journal of Plant Nutrition and Soil Science* 166, 281–90.

EMBRAPA (2018). Espaço Tématico: Integração Lavoura-Pecuária-Floresta. Available at: embrapa.br (accessed on 04 February 2018).

Eriksen, F. I. and Whitney, A. S. (1981). Effects of light intensity on growth of some tropical forage species. I. Interaction of light intensity and nitrogen fertilization on six forage grasses. *Agronomy Journal* 73, 427–33.

Eriksen, J., Søegaard, K., Askegaard, M., Hansen, E. M. and Rasmussen, J. (2010). Forage legume impact on soil fertility and N balance. NJF Report, No. 3. Department of Agroecology and Environment, Faculty of Agricultural Sciences, Aarhus University, Danmark, pp. 61–5.

Equipe Beef Point (2001). As vantagens da consorciação de capins com leguminosas, Espaço Aberto, 26 Janeiro 2001. Beef Point, São Paulo, Brazil.

Esteves, S. N., Bernardi, A. C. C., Vinholis, M. M. and Primavesi, O. (2010). Estimativas da emissão de metano por bovinos criados em sistema de integração lavoura-pecuária em São Carlos, SP. Circular Técnica. 65. Embrapa Pecuária Sudeste, São Carlos, SP, 7p.

Evers, G. and Nelson, L. R. (2000). Grazing termination date influence on annual rye- grass seed production and reseeding in the South-Eastern USA. *Crop Science* 40, 1724–8.

FAO (2001). Chapter 2: Characterization of mixed farms. In: *Mixed Crop-Livestock Farming - A Review of Traditional Technologies Based on Literature and Field Experience*. FAO Animal Production and Health Papers 152. FAO, Rome, Italy.

Faria, P. M. de and Hernandes, M. I. M. (2017). Dung beetles associated with agroecosystems of Southern Brazil: Relationship with soil properties. *Revista Brasileira de Ciência do Solo* 41. doi:10.1590/18069657rbcs20160248.

Faulkner, E. H. (1943). *Plowman's Folly*. Univ of Oklahoma Pr (Trd), 174p.

Faust, D. R., Kumar, S., Archer, D. W., Hendrickson, J. H., Kronberg, S. L. and Liebig, M. A. (2018). Integrated crop-livestock systems and water quality in the northern great plains: Review of current practices and future research needs. *Journal of Environmental Quality* 47, 1–15.

Fauteux, F., Remus-Borel, W., Menzies, J. G. and Bélanger, R. B. (2005). Silicon and plant disease resistance against pathogenic fungi. *FEMS Microbiology Letters* 249, 1–6.

Fernández, P. L. Alvareza, C. R. and Taboada, M. A. (2015). Topsoil compaction and recovery in integrated no-tilled crop–livestock systems of Argentina. *Soil and Tillage Research* 153, 86–94.

Ferraz, L. de C. L., Nasser, L. C. B. and Café-Filho, A. C. (2011). Viabilidade de escleródios de Sclerotinia sclerotiorum e incidência de fungos antagonistas em solo de Cerrado. *Summa Phytopathologica* 37(4).

Ferrazi, L. C. L., Nasser, L. C. B. and Café-Filho, A. C. (2011). Viability of sclerotia of *Sclerotinia sclerotiorum* and incidence of antagonistic fungi in Cerrado soils. *African Journal of Food Science and Technology* 2(1), 1–9. Available at: http://www.interesjournals.org/AJFST.

Ferreira, A., Serra, A. and Laura, V. (2016). Influence of spatial arrangements on silvicultural characteristics of three Eucalyptus clones at integrated crop-livestock-forest system. *African Journal of Agricultural Research* 11(19), 1734–42.

Fischer, G., Tubiello, F. N., van Velthuizen, H. and Wiberg, D. A. (2006). Climate change impacts on irrigation water requirements: Effects of mitigation, 1990-2080. *Technological Forecasting and Social Change* 74, 1083–107. doi:10.1016/j.techfore.2006.05.021.

Fisher, J., Tozer, P. and Abrecht, D. (2010). *Review of Livestock Impacts on No-Till Systems*. Grains Research and Development Corporation, Canberra, Australia.

Fontana, A., Teixeira, W. G., Balieiro, F. C., Moura, T. P. A., Menezes, A. R. and Santana, C. I. (2016). Característica e atributos de latossolos sob diferentes usos em Luis Eduardo Magalhaes, Bahia. *Pesquisa Agropecuária Brasileira* 51, 1457–65.

Fornara, D. A. and Tilman, D. (2008). Plant functional composition influences rates of soil carbon and nitrogen accumulation. *Journal of Ecology* 9, 314–22.

Franzluebbers, A. J. (2007). Integrated crop-livestock systems in the southeastern USA. *Agronomy Journal* 99, 361–72.

Franzluebbers, A. J. (2010). Achieving soil organic carbon sequestration with conservation agricultural systems in the southeastern United States. *Soil Science Society of America Journal* 74, 347–57.

Franzluebbers, A. J. and Stuedemann, J. A. (2008). Soil physical responses to cattle grazing cover crops under conventional and no tillage in the Southern Piedmont USA. *Soil and Tillage Research* 100, 141–53.

Franzluebbers, A. J., Stuedemann, J. A. and Wilkinson, S. R. (2001). Bermudagrass management in the Southern Piedmont USA: I. Soil and surface residue carbon and sulfur. *Soil Science Society of America Journal* 65, 834–41.

Franzluebbers, A. J., Paine, L. K., Winsten, J. R. and Krome, M. (2012). Well-managed grazing systems: A forgotten hero of conservation. *Journal of Soil and Water Conservation* 67(4), 100A–4A.

Franzluebbers, A. J., Lemaire, G., Carvalho, P. C. F., Sulc, R. M. and Dedieu, B. (2014). Toward agricultural sustainability through integrated crop-livestock systems. II. Production responses Introduction. *European Journal of Agronomy* 57, 1–3. doi:10.1016/j.eja.2014.05.004.

Freitas, P. L. de (2005). Sistema Plantio Direto: Conceitos, Adoção e Fatores Limitantes. Comunicado Técnico, 31. Embrapa Solos, Rio de Janeiro, RJ, 9p. ISSN 1517-5685. Available at: www.infoteca.cnptia.embrapa.br/infoteca/bitstream/doc/855711/1/comtec312005plantiodireto.pdf.

Gabb, S., Bell, L. W., Dalgliesh, N., Wixon, A., Guppy, C. and Ramony, L-M. (2015). Additional nitrogen mineralisation and crop uptake following tropical forage legumes is lower if shoot biomass is removed. *Proceedings of the 17th Australian Agronomy Conference 2015: "Building Productive, Diverse and Sustainable Landscapes"*, 21–24 September 2015, Hobart, Tasmania, Australian Society of Agronomy Inc.

Garcıa-Préchac, F. (1992). Propiedades f'ısicas y erosión en rotaciones de cultivos y pasturas (Soil physical properties and erosion in crop-pastures rotations). *Revista INIA-Uruguay Inv. Agr.* 1, Tomo I, 127–40.

Garcia-Prechac, F., Ernst, O., Siri-Prieto, G. and Terra, J. A. (2004). Integrating no-till into crop-pasture rotations in Uruguay. *Soil and Tillage Research* 77, 1–13.

Giddens, J., Adams, W. E. and Dawson, R. N. (1971). Nitrogen accumulation in fescuegrass sod. *Agronomy Journal* 63. doi:10.2134/agronj1971.00021962006300030031x.

Gil, J., Siebold, M. and Berger, T. (2015). Adoption and development of integrated crop-livestock-forestry systems in Mato Grosso, Brazil. *Agriculture, Ecosystems and Environment* 199, 394–406.

Gomes, F. K., Oliveira, M. D. B. L., Homem, B. G. C., Boddey, R. M., Bernardes, T. F., Gionbelli, M. P., Lara, M. A. S. and Casagrande, D. R. (2018). Effects of grazing management in brachiaria grass-forage peanut pastures on canopy structure and forage intake. *Journal of Animal Science* 96, 3837–49.

Gonçalves, D. R. P., Sá, J. C. M., Mishra, U., Furlan, F., Fornari, A., Ferreira, L. A., Inagaki, T. M., Romaniw, J., de Oliveira Ferreira, A. and Briedis, C. (2019). Conservation agriculture based on diversified and high-performance production system leads to soil carbon sequestration in subtropical environments. *Journal of Cleaner Production* 219, 136–47.

Gould, N. S., Peake, D. C. I. and Dalgliesh, N. P. (1996). No-tillage planters for heavy-textured alfisols in the semi-arid tropics of Australia. *Australian Journal of Experimental Agriculture* 36(8).

Grain, S. A. (2018). Integrated crop and pasture-based livestock production systems Part 26: Integration of pasture leys with grain crop systems. Pretoria, South Africa, 16p.

Grains Research and Development Corporation, Australia (2010). *GRDC Update Papers. Pastures – Ley fit in cropping systems*. Toowoomba, QLD, Australia.

Guo, L. B. and Gifford, R. M. (2002). Soil carbon stocks and land use change: a meta analysis. *Global Change Biology* 8, 345–60.

Haynes, R. J. and Williams, P. H. (1993). Nutrient cycling and soil fertility in the grazed pasture ecosystems. *Advances in Agronomy* 49, 119–99.

Higgins, S. F., Agouridis, C. T. and Wightman, S. J. (2011). Options for Grazing Cattle. Cooperative Extension Service, University Of Kentucky College Of Agriculture, Lexington, KY. Available at: http://www2.ca.uky.edu/agcomm/pubs/aen/aen99/aen99.pdf (accessed on 05 July 2019).

Hilimire, K. (2011). Integrated crop/livestock agriculture in the United States: A review. *Journal of Sustainable Agriculture* 35, 376–93. doi:10.1080/10440046.2011.562042.

Hok, L., Moraes Sá, J. C. de, Reyes, M., Boulakia, S., Tivet, F., Leng, V., Kong, R., Clever, B., Cruz, H. D. da, Ferreira, L. A., Massao, I., Inagaki, T., Potma, G. D. R. and Bressan, P. T. (2018). Enzymes and C pools as indicators of C build up in short-term conservation agriculture in a savanna ecosystem in Cambodia. *Soil and Tillage Research* 177, 125–33. doi:10.1016/j.still.2017.11.015.

Humphreys, L. R. (1994). *Tropical Forages: Their Hole in Sustainable Agriculture*. Longman Scientific &Technical, Harlow, Essex, UK, 414p.

Hungria, M. (2011). Inoculação com Azospirillum brasilense: inovação em rendimento a baixo custo. Documents 325. Embrapa Soja, Londrina, PR, Brazil.

Hunt, J. R., Swan, A. D., Fettell, N. A., BreustP. D., Menz, I. D., Peoples, M. B. and Kirkegaard, J. A. (2016). Sheep grazing on crop residues do not reduce crop yields in no-till, controlled traffic farming systems in an equi-seasonal rainfall environment. *Field Crops Research* 196, 22–32.

Ikeda, F. S., Mitja, D. and Vilela, L. (2007a). Banco de sementes em solo sob cerrado "sensu stricto" depois de queimada. In: *Congresso Nacional De Botânica*, 58., 2007, São Paulo. A botânica no Brasil: pesquisa, ensino e políticas públicas ambientais: resumos. Sociedade Botânica do Brasil, São Paulo.

Ikeda, F. S., Mitjia, D., Vilela, L. and Carmona, R. (2007b). Banco de sementes no solo em sistemas de cultivo lavoura-pastagem. *Pesquisa Agropecuária Brasileira* 42, 1545–51.

Jarecki, M. and Lal, R. (2003). Crop management for soil carbon sequestration reviews. *Plant Sciences* 22(5), 471.

Jayasundra, H. R. S., Dennett, M. D. and Sangakkara, V. R. (1997). Biological nitrogen fixation in *Gliricidium sepium* and *Leucaena leucocephala*. *Tropical Grasslanda* 31,529-39.

Johnston, A. E. (2018). The importance of long-term experiments in agriculture: their management to ensure continued crop production and soil fertility; the Rothamsted experience. *European Journal of Soil Science* 69(1), 1-216.

Jones, R. K., Probert, M. E., Dalgliesh, N. P. and McCown, R. L. (1996). Nitrogen inputs from a pasture legume in rotations with cereals in the semi-arid tropics of northern Australia: experimentation and modelling on a clay loam soil. *Australian Journal of Experimental Agriculture* 36(8), 985–94.

Kassam, A., Friedrich, T., Derpsch, R. and Kienzle, J. (2015). Overview of the worldwide spread of conservation agriculture. *The Journal of Field Actions* 8. FAO, Rome, Italy.

Kendall, A and Rodríguez, A. (2002). Chapter X: Las qochas andinas: una solución para mitigar el riesgo agropecuario y doméstico en la sierra del Perú. In: *Antología sobre pequeño riego, vol. III Sistemas de riego no convencionales editor: queira*, J.P. Colegio de Postgraduados, 372p.

Kheyrodin, H. (1994). Crop rotations for managing soil-borne plant diseases. *Agriculture, Ecosystems and Environment* 48(1), 77–89.

Kichel, A. N. and Miranda, C. H. B. (2002). Sistemas de integração pecuária e lavoura como formas de otimização do processo produtivo. Embrapa Gado de Corte. Comunicado Técnico, 74. Embrapa Gado de Corte, Campo Grande, 5p.

Kirkegaard, J. A., Hunt, J. R., Lilley, J. M., Sprague, S. J., Bell, L. W. and Dove, H. (2015). Improving the water-use efficiency of Australian mixed farming systems. In: *World Congress on Integrated Crop-Livestock-Forest Systems 2015*, Brasília, Brazil.

Kluthcouski, J. and Aidar, H. (2003). Implantação, condução e resultados obtidos com o sistema Santa-fé. In: Kluthcouski, J., Aidar, H. and Stone, L. F. (Eds), *Integração lavoura-pecuária*. Embrapa Arroz e Feijão, Santo Antonio de Goiás, pp. 407–41.

Kluthcouski, J., Cordeiro, L. A. M. and Marchão, R. L. (2015). Potencial para Adoção da Estratégia de Integração Lavoura–Pecuária-Floresta para o Uso Sustentável de Solos Arenosos. In: Cordeiro, L. A. M., Vilela, L, Kluthcouski, J. and Marchão, R. L. (Eds), *Integração Lavoura-Pecuária-Floresta: o produtor pergunta, a Embrapa responde*. Coleção 500 Perguntas, 500 Respostas. Embrapa, Brasília, pp. 319–31.

Kron, G. (2000). Roman ley-farming. *JRA* 13, 277–87.

Laike, A. (1992). Faidherbia albida in the traditional farming systems of central Ethiopia. In: Vandenbeldt, R. J. (Ed.), *Faidherbia Albida in the West African Semi-arid Tropics: Proceedings of a Workshop*, 22–26 April 1991, Niamey, Niger. International Crops Research Institute for the Semi-Arid Tropics, Patancheru, A. P., India and International Centre for Research in Agroforestry, Nairobi, Kenya, pp. 39–41.

Lal, R., Smith, P., Jungkunst, H. F., Mitsch, W., Lehmann, J., Nair, P. K. R., McBratney, A. B., Sá, J. C. M., Schneider, J., Zinn, Y. Z., Skorupa, A. L. A., Zhang, H. L., Minasny, B., Srinivasrao, C. And Ravindranath, N. H. (2018). The carbon sequestration potential of terrestrial ecosystems. *Journal of Soil and Water Conservation* 73(6), 145A–52A.

Landers, J. N. (2007). Tropical crop–livestock systems in conservation agriculture: The Brazilian experience. Integrated Crop Management (vol. 5). FAO, Rome, Italy, 88p.

Landers, J. N. and Freitas, P. L. de (2001). Preservação da Vegetação Nativa nos Trópicos Brasileiros por Incentivos Econômicos aos Sistemas de Integração Lavoura x Pecuária com Plantio Direto. In: *Simpósio sobre Economia e Ecologia*, Belém, Brazil. Associação Brasileira Economia-Ecologia, Piracicaba-SP, Brazil.

Leite, L. F. C., Doraiswamy, P. and Madari, B. (2008). Seqüestro de Carbono em Solos sob Plantio Direto e Integração Lavoura-Pecuária: Uma estimativa do simulador computacional CQESTR. *Anais da XVII RBMCSA*. Rio de Janeiro.

Liebig, M. A., Tanaka, D. L., Kronberg, S. L., Scholljegerdes, E. J. and Karn, J. F. (2012). Integrated crops and livestock in central North Dakota, USA: agroecosystem management to buffer soil change. *Renewable Agriculture and Food Systems* 27(2), 115–24.

Lin, C. H., McGraw, R. L. and George, M. F. (1999). Shade effects on forage crops with potential in temperate agroforestry practices. *Agroforestry Systems* 44, 109–19.

Lobato, J. F. P., Freitas, A. K., Devincenzi, T., Cardoso, L. L., Tarouco, J. U., Vieira, R. M., Dillenburg, D. R. and Castro, I. (1998). Brazilian beef produced on pastures: Sustainable and healthy. *Meat Science* 98(2014), 336–45.

Lopes, A. S. (1996). The use of phosphate rocks to build up soil phosphorus and increase food production in acid soils: the Brazilian experience. In: Johnston, A. E. and Syers, J. K. (Eds), *Nutrient Management for Sustainable Crop Production in Asia*. Proceedings of a International Conference, Bali, Indonésia.

Lopes, A. A. C., de Sousa, D. M. G., Chaer, G. M., dos Reis Jr, F. B., Goedert, W. J. and Mendes, I. C. (2012). Interpretation of microbial soil indicators as a function of crop yield and organic carbon. *Soil Science Society of America Journal* 77, 461–72.

Louarn, G., Pereira-Lopès, E., Fustec, J., Mary, B., Voisin, A. S., Carvalho, P. C. F. and Gastal, F. (2015). The amounts and dynamics of nitrogen transfer to grasses differ in alfalfa and white clover-based grass-legume mixtures as a result of rooting strategies and rhizodeposit quality. *Plant Soil* 389, 289–305.

Macedo, M. C. M. (2010). Crop and livestock integration: the state of the art and the near future. *Revista Brasileira de Zootecnia* 38(spe), 133–46. doi:10.1590/S1516-35982009001300015.

Macedo, M. C. M. (2009). Integração lavoura e pecuária: O estado da arte e inovações tecnológicas. *Revista Brasileira de Zootecnia* 38, 133–46. doi:10.1590/S1516-35982009001300015.

Macedo, M. C. M., Bono, J. A., Zimmer, A., Costa, F. P., Kanno, T., Miranda, C. H. B. and Kichel, A. (2001). Preliminary results of agropastoral systems in the cerrados of mato grosso do sul - Brazil. In: *Workshop on Agropastoral System, South America 2001, Japan. Proceedings*. Working Report, 19. JIRCAS, Japan, pp. 35–42.

Macedo, M. C. M., Zimmer, A., Kichel, A. N., Almeida, R. G. de and Araújo, A. R. de (2013). Degradação de pastagens, alternativas de recuperação e renovação, e formas de mitigação. Available at: http://www.alice.cnptia.embrapa.br/alice/handle/doc/976514.

Magalhães, R. T. (1997). Evolução das propriedades físicas e químicas de solos submetidos ao manejo pelo sistema barreirão. (Dissertação Mestrado) Universidade Federal de Goiás, Goiânia, 86f.

Majak, W., Hall, J. W. and McCaughey, W. P. (1995). Pasture Management Strategies for Reducing the Risk of legume bloat in Cattle. *Proceedings of Symposium "Toxic Legumes"*, ASAS 86th Annual Meeting, Minneapolis, MN, pp. 493–8.

Mapiye, C., Mwale, M., Mupangwa, J. F., Mugabe, P. H., Poshiwa, X. and Chikumba, N. (2007). Utilization of ley legumes as livestock feed in Zimbabwe. *Tropical Grasslands* 41, 84–91.

Marchão, R. L. (2007). Integração lavoura-pecuária num latossolo do cerrado: impacto na física, matéria orgânica e macrofauna. Tese (Doutorado) - Universidade Federal de Goiás, Escola de Agronomia e Engenharia de Alimentos, 153p.

Marchão, R. L. and Vilela, L. (2018). Integrated crop-livestock systems as an alternative for sustainable intensification on land use in Brazil. In: *Brazilian Soil Science Society - SBCS Bulletim*, May/August 2018, pp. 58–9.

Marchão, R. L., Balbino, L. C., da Silva, E. M., dos Santos Jr, J. D. G., de Sá, M. A. C., Vilela, L. and Becquer, T. (2007). Qualidade física de um Latossolo Vermelho sob sistemas de integração lavoura-pecuária no Cerrado. *Pesquisa Agropecuária Brasileira* 42(6), 873–82. doi:10.1590/S0100-204X2007000600015.

Margalef, O., Sardans, J., Fernández-Martínez, M., Molowny-Horas, R., Janssens, I. A., Ciais, P., Goll, D., Richter, A., Obersteiner, M., Asensio, D. and Peñuelas, J. (2017). Global patterns of phosphatase activity in natural soils. *Scientific Reports* 7, 1337. doi:10.1038/s41598-017-01418-8.

Martin, C. C. (1996). Weed control in tropical ley farming systems: a review. *Australian Journal of Experimental Agriculture* 36(8) 1013–23.

Martin, G., Moraine, M. Ryschawy, J., Magne, M.-A. and Asai, M. (2016). Crop–livestock integration beyond the farm level: a review. *Agronomy for Sustainable Development* 36(3), 53.

McCown, R. L. (1996). Being realistic about no-tillage, legume ley farming for the Australian semi-arid tropics. *Australian Journal of Experimental Agriculture* 36, 1069–80.

McGarry, D. (2000). Optimising soil structure condition for cropping without tillage. Soil and Tillage Conference Paper. ISTRO, July 2000.

McIntire, J., Bourzat, D. and Prabhu, P. (1992). Crop – livestock interaction in sub-Saharan Africa. World Bank regional and sectoral studies. The World Bank, Washington DC. Available at: http://documents.worldbank.org/curated/en/505681468768678913/Crop-livestock-interaction-in-sub-Saharan-Africa.

Mendes, I. C., Tormena, C. A., Cherubin, M. R. and Karlen, D. L. (2018). Soil health assessment and maintenance in Central and South-Central Brazil. In: Reicosky, D. (Ed.), *Managing Soil Health for Sustainable Agriculture: Monitoring and Management* (1st edn., vol. 2). Burleigh Dodds Science Publishing, Cambridge, UK, pp. 379–417. doi:10.19103/AS.2017.0033.35.

Mendonça, V. Z., Mello, L. M. M., Andreotti. M., Pariz, C. M., Yano, E. H. and Pereira, F. C. B. L. (2015). Liberação de nutrientes da palhada de forrageiras consorciadas com milho e sucessão com soja. *Revista Brasileira de Ciência do Solo* 39, 183–93.

Merola, R. (2002). A experiência do Cerrado com inverno seco em sistema de irrigação com Sistema Santa Fé. Available at: www.brasil.ipni.net (accessed on 29 October 2018).

Mesfin, R. and Ktaw, G. (2010). Effect of feeding urea treated wheat straw based diet on biological performances and economic benefits of lactating Boran-Friesian crossbred dairy cows. *Livestock Research for Rural Development* 22(12).

Moore, A. (2014). Integrated farming systems for a sustainable future – what is the role of mixed farming in sustainable agriculture? CSIRO Agriculture Flagship, Canberra, 26 August 2014. Available at: https://grdc.com.au/resources-and-publications/grdc-update-papers/tab-content/grdc-update-papers/2014/08/integrated-farming-systems-for-a-sustainable-future-mixed-farming.

Moraes, A., Carvalho, P. C. de F., Lustosa, S. B. C., Lang, C. R. and Deiss, L. (2014a). Research on integrated crop-livestock systems in Brazil. *Revista Ciência Agronômica* 45(5).

Moraes, A., Carvalho, P. C. F., Anghinoni, I., Lustosa, S. B. C., Costa, S. E. V. G. and Kunrath, T. R. (2014b). Integrated crop-livestock in Brazilian subtropics. *European Journal of Agronomy* 57, 4–9.

Nascimento, J. L. do, Stone, L. F. and de Oliveira, L. F. C. (2001). Demanda Total de Água do Feijoeiro nos Sistemas de Plantio Convencional E Direto. *Pesquisa Agropecuária Tropical* 31(2), 159–61.

Neumann, G., Bott, S., Ohler, M. A., Mock, H.-P., Lippmann, R., Grosch, R. and Smalla, K. (2014). Root exudation and root development of lettuce (*Lactuca sativa* L. cv. Tizian) as affected by different soils. *Frontiers in Microbiology* 5, 2. doi:10.3389/fmicb.2014.00002 (accessed on 01 August 2019).

Nichols, P. G. H., Revell, C. K., Humphries, A. W., Howie, J. H., Hall, E. J., Sandral, G. A., Ghamkhar, K. and Harris, C. A. (2012). Temperate pasture legumes in Australia–their history, current use, and future prospects. *Crop and Pasture Science* 63(9), 691–725. doi:10.1071/CP12194.

Nicoli, C. M. L., Pacheco, A. R., Reis, C. F. and Venturoli, F. (2017). Income Diversification through a crop-livestock-forest integration system in the midwest Brazilian Region. *Journal of Agricultural Science and Technology B* 7, 374–85. doi: 10.17265/2161-6264/2017.06.002.

Nicoloso, R. S., Lovato, T., Amado, T. J. C., Bayer, C. and Lanzanova, M. E. (2008). Balanço do carbono orgânico no solo sob integração lavoura-pecuária no sul do Brasil. *Revista Brasileira de Ciência do Solo* 32(6), 2425–33. doi:10.1590/S0100-06832008000600020.

Nowak, N., Blount, A. and Workman, S. (2002). *Integrated Timber, Forage, and Livestock Production, Benefits of Silvopasture*. Circular 1430. University of Florida Institute of Food and Agricultural Sciences, Gainesville. Reviewed January 2017. Available at: http://edis.ifas.ufl.edu.

O'Donnell, J. J., Pala, M. C., Crossman, S. M. A., Kowalski, J. A. and Bulbulla, A. (1994). Growth and biomass production from four hedgerows species. *Agronomy Abstracts*, p.72.

Oenema, O. (2005). Nitrogen in global animal production and management options for improving nitrogen use efficiency. *Science in China Series C LifeSciences* 48(Suppl. 2), 871–87.

Pacheco, L. P., Pires, F. R., Monteiro, F. P., Procópio, S. O., Assis, R. L., Cargnelutti Filho, A., Carmo, M. L. and Petter, F. A. (2009). Soybean crop oversowing used as a technique to suppress weed emergence. *Planta Daninha* 27(3).

Pacheco, A. R., Chaves, R. Q. and Nicoli, C. M. L. (2013). Integration of crops, livestock, and forestry: a system of production for the Brazilian Cerrados. In: Hershey, C. H. and Neate, P. (Eds), *Eco-efficiency: From Vision to Reality*. CIAT, Cali.

Pacheco, A. R., Nicoli, C. M. L., Reis, C. F., Monteiro, L. A., Wander, A. E. (2015). Adoption of crop-livestock-forest systems: What comes next? *Proceedings of the World Congress on Crop, Livestock and Forestry Systems*. Embrapa, Brasilia, DF, Brazil, p. 84.

Parr, J. F., Papendick, R. I., Youngberg, I. G. and Meyer, R. E. (1990). Chapter 4. Sustainable agriculture in the United States. In: *Sustainable Agricultural Systems*. Soil and Water Conservation Society, Ankeny, IA.

Paudel, K. P., Subedi, Y. R. and Tamang, S. (2016). Rebuilding farm-forestry interface: An opportunity to address food security and environmental sustainability in Nepal. *Universal Journal of Agricultural Research* 4(6), 254–9. doi:10.13189/ujar.2016.040605.

Peoples, M. B., Gault, R. R., Scammell, G. J., Dear, B. S., Virgona, J., Sandral, G. A., Paul, J., Wolfe, E. C. and Angus, J. F. (1998). Effect of pasture management on the contribution of fixed N to the N economy of ley-farming systems. *Australian Journal of Agricultural Research* 49, 459–74.

Pettit, R. E. (2019). Organic matter, humus, humate, humic acid, fulvic acid and humin: their importance in soil fertility and plant health. Available at: https://humates.com/pdf/organicmatterpettit.pdf.

Phillips, S. H. and Young Jr, H. M. (1973). *No-tillage Agriculture*. Reiman Associates, Milwaukee, WI.

Piva, J. T., Dieckow, J., Bayer, C., Zanatta, J. A., Moraes, A., Tomazi, M., Pauletti, V., Barth, G. and Piccolo, M. C. (2014). Soil gaseous N_2O and CH_4 emissions and carbon pool due to integrated crop-livestock in a subtropical Ferralsol. *Agriculture, Ecosystems and Environment*, 190, 87–93.

Portela, C. M. O. (2003). Manejo de herbicidas no Sistema Santa Fé e na braquiária como fonte de cobertura morta. In: Kluthcouski, J., Stone, L. F. and Aidar, H. (Eds), *Integração lavoura-pecuária*. Embrapa Arroz e Feijão, Santo Antônio de Goiás, pp. 443–58.

Portilho, I. I. R., Scorza Júnior, R. P., Salton, J. C., Mendes, I. C. and Mercante, F. M. (2015). Persistence of insecticides and microbiological attributes in a soil under different management systems. *Ciência Rural* 45(1), 22–8.

Powell, J. M., Ikpe, F. N. and Somda, Z. C. (1999). Crop yield and fate of nitrogen and phosphorus after application of plant material or feces to soil. *Nutrient cycle in Agroecosystem* 54, 215–26.

Powell, J. M., Pearson, R. A. and Hiernaux, P. H. (2004). Crop-Livestock Interaction in Sub-Saharan Africa. *Agronomy Journal* 96, 469–83.

Prade, T., Kätterer, T. and Björnsson, L. (2017). Including a one-year grass ley increases soil organic carbon and decreases greenhouse gas emissions from cereal-dominated rotations – a Swedish farm case study. *Biosystems Engineering* 164, 200–12.

Primavesi, O. (2008). ILPD reduz metano bovino. (ICLZT reduces methane emissions in cattle). Direno no Cerrado n 50, Janeiro/Março 2008. Associação de Plantio Direto no Cerrado, Brasília, DF, p. 10.

Puckridge, D. W and French, R. J. (1983). The annual legume pasture in cereal—Ley farming systems of southern Australia: A review. *Agriculture, Ecosystems and Environment* 9(3), 229-67. Available at: https://doi.org/10.1016/0167-8809(83)90100-7.

Pushpakumara, D. K. N. G., Marambe, B., Silva, G. L. L. P., Weerahewa, J. and Punyawardena, B. V. R. (2012). A review of research on home gardens in Sri Lanka: the status, importance and future perspective. *Tropical Agriculturist* 160, 55–125. Available at: https://www.cabdirect.org/cabdirect/abstract/20133171150/.

Queensland Government (2017). Available at: https://www.qld.gov.au/environment/coasts-waterways/plans/resources.

Rao, A. V., Singh, K. C. and Gupta, J. P. (1997). Ley farming—an alternate farming system for sustainability in the Indian arid zone. *Arid Soil Research and Rehabilitation* 11(2), 201–10.

Redmon, L. A., McCollum, F. T., Horn, G. W., Cravey, M. D., Gunter, S. A., Beck, P. A., Mieres, J. M. and San Julian, R. (1995). Forage intake by beef steers grazing winter wheat with varied herbage allowances. *Journal of Range Management* 48, 198.

Reicosky, D. (2011). *Managing Soil Health for Sustainable Agriculture Volume 1: Fundamentals*. Burleigh Dodds Science Publishing, Cambridge, UK.

Reicosky, D. (Ed.) (2018). Managing soil health for sustainable agriculture – volume 1. *Revista Brasileira de Ciência do Solo* 41, Viçosa 2017.

Ribaski, J., Dedececk, R. A., Mattei, V. L., Flores, C. A., Vargas, A. F. C. and Ribaski, S. A. G. (2005). Sistemas silvipastoris: estratégias para o desenvolvimento rural sustentável para a metade sul do Estado do Rio Grande do Sul. Comunicado Técnico 150. Embrapa Florestas, Colombo/PR/Brazil, 8p. Available at: http://ainfo.cnptia.embrapa.br/digital/bitstream/item/37775/1/com-tec150.pdf.

Riley, H., Pommeresche, R., Korsaeth, A., Eltun, R. and Hansen, S. (2008). Soil structure, organic matter and earthworm activity in a comparison of cropping systems with contrasting tillage, rotations, fertilizer levels and manure use. *Agriculture, Ecosystems and Environment* 124(3):275–84. April 2008.

Roy, S. and Rangnekar, D. V. (2006). Farmer adoption of urea treatment of cereal straws for feeding of dairy animals: a success in Mithila milkshed, India. *Livestock Research for Rural Development* 18(8).

Russell, J. R. and Bisinger, J. J. (2015). Grazing system effects on soil compaction in Southern Iowa pastures. Iowa State University, Animal Industry Report: AS 661/ASL R2987. Available at: https://lib.dr.iastate.edu/cgi/viewcontent.cgi?referer=https://www.google.com/&httpsredir=1&article=2096&context=ans_air.

Sá, J. C. M. (1999). Manejo da fertilidade do solo no sistema plantio direto. In: Siqueira, J. O., Moreira, F. M. S., Lopes, A. S., Guilherme, L. R. G., Faquin, V., Furtini Neto, A. E. and Carvalho, J. G. (Eds), *Soil Fertility, Soil Biology, and Plant Nutrition Interrelationships*. Soc. Bras. Ci. Solo. UFLA, Lavras, MG, Brazil, pp. 267–320.

Sá, J. C. M., Gonçalves, D. R. P., Ferreira, L. A., Mishra, U., Inagaki, T. M., Furlan, F. J., Moro, R., Floriani, N., Briedis, C. and de Oliveira Ferreira, A. (2018). Carbon fractions and biological activity based indices can be used to study the impact of land management and ecological successions. *Ecological Indicators* 84, 96–105.

Sá, C. P. de, de Andrade, C. M. S. and Valentim, J. F. (2010). Análise Econômica para a Pecuária de Corte em Pastagens Melhoradas no Acre. Circular Técnico no. 51. Embrapa Acre.

Salton, J. C. and Lamas, F. M. (2011). Integração lavoura pecuária e o cultivo do algodoeiro nos cerrados. In: Freire, E. C. (Ed.), *Algodão no cerrado do Brasil* (2. ed.). Mundial, Aparecida de Goiânia, GO, pp. 473–94.

Salton, J. C., Mielniczuk, J. Bayer, C., Boeni, M., Conceição, P. C., Fabrício, A. C., Macedo, M. C. M. and Broch, D. L. (2008). Agregação e estabilidade de agregados do solo em sistemas agropecuários em Mato Grosso do Sul. *Revista Brasileira de Ciência do Solo* 32, 11–21.

Salton, J. C., Kichel, A. N., Arantes, M., Kruker, J. M., Zimmer, A. H., Mercante, F. M. and Almeida, R. G. de (2013). Sistema São Mateus – Sistema de integração lavoura-pecuária para a região do Bolsão Sul-Mato-Grossense. Comunicado Técnico 186. Embrapa Agropecuária Oeste, Dourados, MS, Brazil. Available at: http://ainfo.cnptia.embrapa.br/digital/bitstream/item/84887/1/COT2013186.pdf.

Salton, J. C. Mercante, F. M., Tomazi, M. Zanatta, J. A., Concenço, G., Silva, W. M. and Retore, M. (2014). Integrated crop-livestock system in tropical Brazil: Toward a sustainable production system. *Agriculture, Ecosystems and Environment* 190, 70–9.

Sanchez, J. M. D., Vendramini, J. M. B., Sollenberger, L. E., Silveira, M. L., Dubeux Jr, J. C. B., Moriel, P., Kuwahara, F. A., Cecatoe, U., Yarborough, J. K., Soares Filho, C. V. and de Oliveira, F. C. L. (2018). Forage characteristics of bermudagrass pastures overseeded with pintoi peanut and grazed at different stubble heights. *Crop Science* 58(4). doi:10.2135/cropsci2018.01.0007.

Santos, H. P., Ignaczak, J. C. and Wobeto, C. (1996). Produtividade Cultural de sistemas de culturas em rotação com trigo, sob Plantio Direto. *Pesquisa Agropecuária* 31(4), 277–82.

Schilling, G., Gransee, A., Deuhel, A., Ležoviž, G., Ruppel, S. (1998). Phosphorus availability, root exudates, and microbial activity in the rhizosphere. *Journal of Plant Nutrition and Soil Science* 161(4), 465–78. doi:10.1002/jpln.1998.3581610413.

Schreiner, H. G. (1987). Tolerância De Quatro Gramíneas Forrageiras A Diferentes Graus De Sombreamento. *Boletim de Pesquisa Florestal, Colombo* 15, 61–72.

Schuster, M. Z., Moraes, A, Pelissari, A., Carvalho, P. C. F., Harrison, K., Gazziero, D. L. P., Bonatto, R. A., Campos, B. M., Szymczak, L. S., Portugal, T. B., Lustosa, S. B. C., Silva, D. F. F., Deiss, L. and Gonzales, N. M. (2015). Effects of grazing management and crop rotation strategies on Conyza bonariensis on Integrated Crop-Livestock System after 11 years continuous glyphosate use. In: *World Congress on Integrated Crop-Livestock-Forest Systems*, Brasília, Brazil.

Scopel, I., Suertegaray, D. M. A., Souza, M. S., Peixinho, D. M. and Ferreira D. M. (2012). Neossolos Quartzarênicos órticos das áreas de areais do sudoeste do Rio Grande do Sul: características físicas e morfológicas. In: Suertegaray, D. M. A., Silva, L. A. P. Da and Guasselli, L. A. (Org.), *Arenização: natureza socializada*. Compasso Lugar-Cultura Imprensa Livre, Porto Alegre, pp. 489–528.

Scopel, E., Triomphe, B., Affholder, F., Silva, F. A. M., Corbeels, M., Xavier, J. H. V., Lahmar, R. Recous, S., Bernoux, M., Blanchart, E., Mendes, I. C. and Tourdonnet, S. (2013). Conservation agriculture cropping systems in temperate and tropical conditions, performances and impacts. A review. *Agronomy for Sustainable Development* 33(1), 113–30.

Segnini A., Martin Neto, L., Primavesi, O., Tadeu, W., da Silva, W. T. L. and Simõesl, M. I. (2007). Seqüestro de Carbono em Solos com Gramíneas. Circular Técnico No. 41.

Seguy, L., Bouzinac. S., Matsubara, M., Trevisan, E. and Galvagni, E. (1992). Gestão dos Solos e das Culturas nas Fronteiras agrícolas dos Cerrados Úmidos do Centro-Oeste. Convênio RPA/CIRADF-CA Fazenda Progresso, Lucas de Rio Verde-MT, CIRAD-CA, Montpellier, France.

Severino, F. J., Carvalho, S. J. P. and Christoffoleti, P. J. (2006). Interferências mútuas entre a cultura do milho, espécies forrageiras e plantas daninhas em um sistema de consórcio. III – Implicações sobre as plantas daninhas. *Planta Daninha* 24, 53–60.

Silburn, D. M., Freebairn, D. M. and Rattray, D. J. (2007). Tillage and the environment in sub-tropical Australia–Tradeoffs and challenges. *Soil and Tillage Research* 97, 306–17.

Silva, A. R., Sales A., Veloso, C. A. C. and Carvalho, E. J. M. (2015). Growth the African mahogany (*Khaya ivorensis*) in Oxisol yellow in the integrated Crop-Livestock-Forest system in eastern Amazon. In: *World Congress on Integrated Crop-Livestock-Forest Systems*, Brasília, Brazil.

Somda, Z. C., Powell, J. M., Fernàndez-Rivera, S. and Reed, J. D. (1995). Feed factors affecting nutrient excretion by ruminants and the fate of nutrients when applied to soil. In: Powell, J. M., Fernandez-Rivera, S., Williams, T. O. and Renard, C. (Eds), *Proceedings of an International Conference of International Livestock Centre for Africa (ILCA) on Livestock and Sustainable Nutrient Cycling in Mixed Farming Systems of Sub-Saharan Africa*, Addis Ababa, Ethiopia, 22–26 November 1993, pp. 227–43.

Soussana, J.-F. and Lemaire, G. (2014). Coupling carbon and nitrogen cycles for environmentally sustainable intensification of grasslands and crop livestock systems. *Agriculture, Ecosystems and Environment* 190, 9–17.

Soussana, J.-F., Klumpp, K. and Ehrhardt, F. (2014). The role of grassland in mitigating climate change. In: *Grassland Science in Europe, volume 19 - EGF at 50: the Future of European Grasslands*.

Soussana, J.-F., Tichit, M. and Lecomte, P. (2015). Chapter 12: Agroecology: Integration with livestock. In: *Agroecology for Food Security and Nutrition*. FAO, Rome.

Souza, E. D., Costa, S. E. V. G. A., Anghinoni, I., Lima, C. V. S., Carvalho, P. C. F. and Martins, A. P. (2010). Biomassa microbiana do solo em sistema de integração lavoura-pecuária em plantio direto, submetido a intensidades de pastejo. *Revista Brasileira de Ciência do Solo* 34(1), 79–88. doi:10.1590/S0100-06832010000100008.

Souza, C. A., Reis Junior, F. B., Mendes, I. C, Lemainski, J. and Silva, J. E. (2019). Lodo de esgoto em atributos biológicos do solo e na nodulação e produção de soja. *Pesquisa Agropecuária Brasileira* 44(10). doi:10.1590/S0100-204X2009001000016.

Sparavigna, A. C. (2011). Qochas on Andean highlands. Cornell University, 25 April 2011. Available at: https://arxiv.org/pdf/1104.4748.pdf (accessed on 22 February 2019).

Stapledon, R. G. (1941). *Ley Farming*. Penguin, Harmondsworth, UK, 160p.

Stone, L. F. and Moreira, J. A. A. (2000). Efeitos do sistema de preparo do solo no uso da água e na produtividade do feijoeiro. *Pesquisa Agropecuária Brasileira* 35, 835–41.

Stone, L. F., Silveira, P. M. da, Moreira, J. A. A. and Braz, A. J. B. P. (2006). Evapotranspiration of irrigated common bean under no-tillage on different cover crop mulches. *Pesquisa Agropecuária Brasileira* 41(4).

Stoner, E., Freitas Jr, E., Macedo, J., Mendes, R. C. A., Cardoso, I. M., Mabile, R. F., Bryant, R. B. and Lathwell, D. J. (1991). Physical constraints to root growth in savanna oxisols. TropSoils Bulletin, 91-01. NCSU, Raleigh, 28p.

Sulc, R. M. and Franzluebbers, A. J. (2014), Exploring integrated crop-livestock systems in different ecoregions of the United States. Publications from USDA-ARS/UNL Faculty. 1476. Available at: http://digitalcommons.unl.edu/usdaarsfacpub/1476 (accessed on 09 Novemeber 2018).

Sumberg, J. (2003). Towards a dis-aggregated view of crop-livestock integration in Western Africa. *Land Use Policy* 20, 253–64.

Tarawali, S., Duncan, A., Thorne, P., Valbuena,D., Descheemaeker, K. and Homann-KeeTui, S. (2013). Integrated crop livestock systems: A key to sustainable intensification in Africa. *22nd International Grassland Congress Sydney*, Australia, 15–19 September.

Tengyun, G. (2000). Review: Treatment and utilization of crop straw and stover in China. *Livestock Research for Rural Development* 12(1). Available at: http://www.fao.org/livestock/AGAP/FRG/lrrd/lrrd12/1/gao121.htm.

Thilakarathna, M. S., McElroy, M. S., Chapagain, T., Papadopoulos, Y. A. and Raizada, M. N. (2016). Belowground nitrogen transfer from legumes to non-legumes under managed herbaceous cropping systems. A review. *Agronomy for Sustainable Development* 36, 58. doi:10.1007/s13593-016-0396-4.

Tomaz, G. A. (2017). Barreiras a adoção da estratégia de integração lavoura pecuária floresta por agricultores e pecuaristas do estado de Goiás. Dissertação (Mestrado em Agronegócio) - Universidade Federal de Goiás, Goiânia, 91f.

Tomazi, M., Salton, J. C., Retore, M., Bayer, C. and Santos, D. S. dos (2015). Mitigation of greenhouse gases with integrated crop. Available at: https://www.embrapa.br/en

/busca-de-publicacoes/-/publicacao/1022442/mitigation-of-greenhouse-gases-with-integrated-crop.

Traill, S., Bell, L. W., Dalgliesh, N. P., Wilson, A., Ramony, L. M. and Guppy, C. (2018). Tropical forage legumes provide large nitrogen benefits to maize except when fodder is removed. *Crop and Pasture Science* 69, 183-93.

Tsukamoto Filho, A. A., Silva, M. L., Couto, L. and Muller, M. D. (2003). Análise econômica de um plantio de teca submetido a desbastes. *Revista Árvore* 27(4), 487-94.

Tull, J. (1751). *Horse-hoeing Husbandry: Or, An Essay on the Principles of Vegetation and Tillage* (3rd edn.). A. Millar, London.

Valentim, J. F., Carneiro, J. Da C., Sales, M. F. L. Valentim, J. F., Carneiro, J. Da C., Sales, M. F. L. (2001). Amendoim forrageiro cv. Belmonte: leguminosa para a diversificação das pastagens e conservação do solo no Acre. Circular técnica, 43. Embrapa Acre, Rio Branco, AC.

Valentim, J. F., de Andrade, C. M. S., de Mendonça, H. A. and Sales, M. F. L. (2003). Velocidade de estabelecimento de acessos de amendoim forrageiro na Amazônia Ocidental. *Revista Brasileira de Zootecnia* 32(6 Suppl. 1). doi:10.1590/S1516-35982003000700005.

Van Geel, M. De Beenhouwer, M., Lievens, B., Honnay, O. (2016). Crop specific and single species mycorrhizal inoculation is the best approach to improve crop growth in controlled environments. *Agronomy for Sustainable Development* 36(2). doi:10.1007/s13593-016-0373-y.

Vaneph, S. and Benites, J., (2001). *Report of the First World Congress on Conservation Agriculture: A World-wide Challenge*. European Conservation Agriculture Federation, Brussels, Belgium.

Vasconcelos, H. P. and Landers, J. N. (1993). Agricultura Sustentável no Cerrado. In: Arantes, N. E. and de Souza, D. M. (Eds), *Cultura da Soja nos Cerrados*. KP Potafos, Piracicaba, SP, Brazil, pp. 523-42.

Vertés, F., Hatch, D., Velthof, G., Taube, F., Laurent, F., Loiseau, P. and Recous, S. (2007). Short-term and cumulative effects of grassland cultivation on nitrogen and carbon cycling in ley-arable rotations. In: *Proceedings of the 14th Symposium of the European Grassland Federation, Permanent and Temporary Grassland: Plant, Environment and Economy*, pp.227-46.

Vilela, L., Barcellos, A. de O. and Sousa, D. M. G. de (2001). *Benefícios da Integração entre Lavoura e Pecuária*. Embrapa Cerrados, Planaltina, DF, Brazil, 21p.

Vilela, L., Martha Junior, G. B., Marchão, R. L., Guimarães Junior, R., Barioni, L. G. and Barcellos, A. de O. (2008). Integração lavoura-pecuária. In: Faleiro, F. G. and Farias Neto, A. L. De (Eds), *Savanas: desafios e estratégias para o equilíbrio entre sociedade, agronegócio e recursos naturais*. Embrapa Cerrados, Planaltina, pp. 933-62.

Vilela, L., Martha Junior, G. B., Macedo, M. C. M., Marchãol, R. L., Guimarães Júnior, R., Pulrolnik, K. and Maciel, G. A. (2011). Sistemas de integração lavoura-pecuária na região do Cerrado. *Pesquisa Agropecuária Brasileira* 46(10). doi:10.1590/S0100-204X2011001000003.

Voisin, A. (1953). "Grandeurs et faiblesses du Ley-Farming" ("Strength and weakness of Ley farming"). Bulletin Technique d'Information des Ingénieurs des Services Agricoles, no. 82.

Voisin, A. (1957). *Productivité de l'herbe*. Flammarion, Paris, 467p.

Weston, E. J., Dalal, R. C., Strong, W. M., Lehane, K. J., Cooper, J. E., King, A. J. and Holmes, C. J. (2002). Sustaining productivity of a Vertisol at Warra, Queensland, with fertilisers, no-tillage or legumes. 6. Production and nitrogen benefits from annual medic in rotation with wheat. *Australian Journal of Experimental Agriculture* 42, 961-9. doi:10.1071/EA01083.

White, K. D. (1970). *Roman Farming*. Cornell University Press, 536p.

Wohlfarth, G. W. and Schroeder, G. L. (1979). Use of manure in fish farming. A review. *Agricultural Wastes* 1(4), 279-99.

Chapter 10

The status of mechanization in Conservation Agriculture systems

Augusto Guilherme de Araújo, Agricultural Research Institute of Paraná State (IAPAR), Brazil; Brian Sims, Independent Consultant, UK; Jack Desbiolles, University of South Australia, Australia; Denizart Bolonhezi, Agronomic Institute of Campinas (IAC), Brazil; Enamul Haque, Murdoch University, Australia; He Jin, China Agricultural University, China; Jamil Abdalla Fayad, Independent Consultant, Brazil; Josef Kienzle, Food and Agriculture Organization of the United Nations (FAO), Italy; Leandro do Prado Wildner, Agricultural Research and Rural Extension Corporation of Santa Catarina State (EPAGRI), Brazil; Li Hongwen, China Agricultural University, China; Marcelo Zanella, Agricultural Research and Rural Extension Corporation of Santa Catarina State (EPAGRI), Brazil; Oussama El Gharras, National Agricultural Research Institute (INRA), Morocco; Richard Bell, Murdoch University, Australia; Roberto Peiretti, Global Farmer Network (GFN), Argentina; Saidi Mkomwa, African Conservation Tillage Network (ACT), Kenya; Sjoerd Willem Duiker, Pennsylvania State University, USA; and Theodor Friedrich, Food and Agriculture Organization of the United Nations (FAO), Italy

1 Introduction

In the last two decades, the expansion of Conservation Agriculture (CA) throughout the world has been accompanied by continual development of new machine designs and their adaptation to regional differences in soil, climate and crop production system.

http://dx.doi.org/10.19103/AS.2019.0048.11

The availability of machines with suitable technical efficiency and field performance, at a price that farmers and other users can afford, has become a fundamental requirement for the greater adoption and continuing success of CA systems at all scales of production.

However, there is a shortage of technical information on machinery used either in regions where the CA is consolidated, as in South America, or is still under development as is the case, for example, of Africa.

This chapter aims to contribute to knowledge systematization around machinery and technologies for CA systems, and to expand access to information on the main technical specifications of the machines used in CA systems across the Americas, Asia, Africa and Australia.

To achieve this aim, 17 experts with wide experience of the subject and knowledge of the theory and practice of selecting and using CA machinery were invited to share their experiences.

The chapter is organized into four sections. The first discusses the needs of CA in terms of mechanization inputs; then the machines for both small- and large-scale operations on the four continents are described, and the final section presents a vision of how the mechanization of CA is likely to evolve in the near future.

One of the challenges of organizing a comprehensive text such as this is to standardize the terminology used by authors from different countries. An effort has been made to maintain the authors' original terms, but with minor changes for clarity where necessary.

As is found in other texts on the subject (e.g. Baker et al., 2007; Casão Junior et al., 2012), here the terms 'Conservation Agriculture' and 'no-tillage (or no-till)' are used as synonyms to refer to 'any method of crop establishment whereby seed is placed into previously untilled soil in contrast to conventional sowing' (Ashworth et al., 2010). Although it is important to note that CA involves more than just no-till, as permanent soil cover and enhanced crop biodiversity (through crop rotations and associations) are equally important pillars.

Other terms such as direct seeding, direct sowing and direct drilling are also used by the authors to refer to CA crop establishment practices. Generally, no-till seeders refer to machines designed to directly sow a continuous flow of small seeds (such as cereals), while no-till planters refer to precision planters designed to precisely sow seeds at specified intervals along the row. Stubble refers to the standing remains of a crop after harvesting and constitutes part of what are termed crop residues, which include other parts (e.g. chaff) of the harvested crop as well. Vertical discs, used primarily to cut through surface residue, can also be called coulters. Animal traction and animal draught refer to the use of draught animals to pull implements such as seeders or sprayers under the umbrella term of draught animal power.

2 Mechanization needs for Conservation Agriculture systems

The most significant differences in cropping practices between conventional tillage-based agriculture and CA are in land preparation and seeding (Li, 2005; Baker et al., 2007), with follow-on effects on weed management and energy requirements.

2.1 Land preparation

The use of tillage as a standard periodic operation is completely eliminated in a fully functioning CA system and mechanical soil disturbance remains only for very specific tasks (e.g. maintaining irrigation furrows between permanent raised beds).

In the early years of CA adoption on severely degraded and hard-setting soils, it might be necessary to rip the planting lines (with a narrow chisel tine) in order to open the soil until crop roots and soil life have established a more amenable soil structure. With the exception of conditions of light surface soil compaction, ripping is best performed with tractor (rather than animal) power because of the relatively high-draught requirement. Once the soil structure has improved, ripping can be discontinued. At this point, the harvest operation under CA becomes the most important part of land preparation for seeding the next crop. It is advisable to adopt harvest techniques that leave the residues in place to facilitate a quick turnaround between crops. The management of crop residues, that is their distribution and handling (standing, chopped or un-chopped) during harvest, has direct influence on the ease, potential problems and quality of the subsequent planting operation and on the weed population (Baker et al., 2007). A specific CA tool for cover crop, weed and residue management is the knife-roller, crop crimper or vegetation crusher.

2.2 Seeding and planting: technical features

The most significant change of techniques from tillage-based farming to CA is seeding and planting, which in CA systems often require specific equipment to bring the seed into soil contact in an undisturbed soil with residue cover (Baker et al., 2007).

One manually operated implement is the jab planter, in effect a mechanized planting stick. Animal traction and tractor-powered no-till seeders and planters use chisels or discs to open a narrow seed slot with minimum soil disturbance. In the case of chisel openers, vertical cutting discs for the residue cover might be required to reduce blockages. No-till direct seeding equipment is more robust and heavier than conventional seeders. Chisel or tine furrow openers are cheap and robust and can be used on lighter weight equipment but create more soil disturbance than discs. Disc openers, particularly single- and double-disc types, produce less soil disturbance and have a better capacity to manage un-chopped crop residues but require higher equipment weight for soil penetration and come at a higher cost. However, discs inclined to the vertical (tilt angle) improve digging ability and can reduce the required weight to make this technology eligible for animal traction or single-axle (or two-wheel - 2WT) tractors (Murray et al., 2006; Ashworth et al., 2010).

Different options for low-cost no-till planters and seeders for 2WTs are now available. These developments have spread recently to several African countries where local adaptation is taking place (Li, 2005). For extreme, difficult to manage residues (such as rice), no-till planters with a built-in straw chopper cleaning the planting rows ahead of the furrow openers have been developed (Sims et al., 2012).

2.3 Weed and cover crop management: technical features

Soil tillage has traditionally been seen as a tool for weed management. Weed management under CA is done with as little soil engagement as possible, that is no deep-digging hoes or cultivators. The purpose of weed management is to avoid competition with the crop, not to clean the field of all traces of weeds which can be manually pulled, slashed with a cutlass or scraped at the soil surface, rolled down by a roller crimper or simply be suppressed by mulch, the crop or a cover crop.

For starting CA in conditions of high weed infestation and limited availability of soil cover, the use of herbicides to control weeds is a useful option for reducing labour requirements. For small-scale farmers, simple weed wipers are an option as they carry a lower risk of drift and crop damage than spraying the herbicide (Sims et al., 2012). The most popular equipment for herbicide application is still the lever-operated knapsack sprayer. Under CA conditions, multi-nozzle spray booms for wider area coverage can be used, or a spray shield can be fitted to allow inter-row weed management in row crops (AAB, 2006). Additionally, boom sprayers for herbicide application in CA with widths from 2 to 40 m are available for manual or animal traction, for 2 or 4WTs.

2.4 Power sources

Without the energy-intensive tillage operation, the required peak farm power is roughly halved. Small-scale farmers using only manual labour or animal traction may no longer need to rely on tractor services. In tractor-powered systems, the energy required per ha is reduced by 50% through eliminating tillage and tractor size can be reduced by 40% (Kassam et al., 2017). In Africa, this has led to a serious reconsideration of the 2WT as a suitable power source for smallholder farmers.

3 Mechanization of small-scale Conservation Agriculture systems

3.1 Latin America and the Caribbean

Small-scale farming systems in Latin America and the Caribbean (LAC) are, in general, managed by farmers and their families, without any full-time employees, and comprise almost 17 million units and about 60 million people, for whom agriculture is the main economic activity (Schneider, 2016). There is a huge diversity of small-scale farming systems between regions. Cultivated land area and quality, climate, crops, access to credit and markets, along with other factors, determine the yield potential and economic importance of the sector. Schneider (2016) points out their fundamental role in food production and the high proportion (>79% of the total) of family units, and also the contrast in average cultivated area between sub-regions (Table 1).

Agricultural practices in LAC are an important cause of soil degradation (Gardi et al., 2014). Unsuitable soil tillage methods, often on steep hillsides vulnerable to high intensity rains, burning of residues and overgrazing, are factors leading to soil loss and degradation, the effects of which are exacerbated by climate change (Montiel and Ibrahim, 2016).

Concerns over the need to develop CA machinery for small-scale farmers were raised in southern Brazil in the 1980s as a way to rationalize manual labour and reduce soil degradation (Casão Junior et al., 2012). As a result, a wide variety of machinery was designed and developed for this sector (Figs. 1–4). Included are different models of hand-jab planters for no-till (Araújo, 2016),

Table 1 Main characteristics of small-scale family farming units in Latin America and the Caribbean[a]

Sub-region	Number of family units	Family units as a proportion of total number of establishments (%)	Average cultivated area (ha)
Caribbean	1 507 757	88	1.3
Central America/Mexico	5 883 205	79	3.1
Andean countries	4 051 342	80	3.1
Southern cone	5 154 533	84	47.0
Total	16 596 837	81	13.6

[a] Source: extracted from Schneider (2016).

Figure 1 Manual tools for CA seeding in Latin America (Guatemala): the hand-jab planter (the red implement in the foreground and to the right of the main picture) and the traditional 'chuzo' (a wooden stick with a metal tip). Photos: A. Araújo.

Figure 2 Typical no-till seeder for animal-drawn used in Brazil. Photos: A. Araújo.

various designs of animal-drawn CA seeders for different field conditions, knife-rollers to deal with cover crops, lime spreaders and animal-drawn and manual sprayers of different tank capacities (Bolliger et al., 2006; Casão Junior et al., 2012; Ribeiro et al., 2007; Calegari et al., 2014).

According to Sims and Kienzle (2009), a number of private manufacturers appeared on the Brazilian market in the 1990s, mainly in the southern region, where the supply chain was more developed in terms of research, finance and technical support.

At least eight different designs of animal-drawn CA seeders were commercially available for small farmers in this period. Most of them were equipped with a flat vertical disc for residue cutting, seed and fertilizer metering devices, tine opener for placing fertilizer deeper than the seed, in the same slot, double-disc opener for seeds and V-shaped wheels to cover and compact soil slot. The field performance of these CA seeders depends on its careful adjustment, mainly regarding the weight transfer to the cutting disc to enhance its soil penetration and residue cutting performance (Casão Junior et al., 2012).

Figure 3 Brazilian animal-drawn knife-rollers with wooden (left) and steel (right) drums. Photos: A. Araújo.

Figure 4 Brazilian animal-drawn and manual sprayers with tank capacities of 250 L (left) and 20 L (right). Photos: R. Casão.

Average work rates for animal-drawn CA seeders and hand-jab planters are, respectively, 12 and 22 h ha^{-1} for sowing beans (*Phaseolus vulgaris*) at a row spacing of 0.5 m (Araújo, 1990). However, these rates may show high variability according to local conditions (especially of soil type, moisture content, vegetative cover and surface slope).

In the second half of the 1990s, credit supply for small-scale machinery increased due to the growth of the agricultural sector. This stimulated the purchase of tractors (up to 75 hp) by smallholder farmers and reduced the demand for animal-drawn machinery in southern Brazil. At the same time, manufacturers started to produce CA machinery for the smallholder sector using technology similar to that employed in the design and development of more sophisticated machinery for large-scale mechanization.

Brazil has a diverse set of small- and medium-size agricultural machinery manufacturers which had supported the evolution of small-scale CA with technological innovations. By the end of the 1990s, about ten Brazilian industries fully served the national and LAC markets with CA machinery for animal-drawn and small tractors (up to 75 hp) principally comprising seeders, sprayers and knife-rollers.

There is, however, a huge potential for the use of smaller (and cheaper) machinery in the north-eastern and northern regions of Brazil, as well as throughout the LAC region, which already has several successful smallholder mechanized CA experiences.

Despite the success of CA machinery for manual use and for draught animal power, there is still potential for design innovations to reduce costs and maintenance requirements and to increase the rate of adoption by smallholder farmers. For instance, further, detailed research should be done on new seed and fertilizer metering devices and soil openers to simplify power transmission, minimize soil disturbance and reduce seeder size and weight.

The quality of no-till seeding is impaired in animal draught systems when soil vegetative cover is above 4 tonne ha^{-1} due mainly to incomplete straw cutting and hairpinning. It is important to investigate new soil engaging tools for different conditions and cover crops. Also, there is a need to conduct further studies on knife-roller design and operation to establish the timing window for cover crop management with a view to eliminating, or at least reducing, herbicide usage for weed control.

Public policies to encourage the acquisition of CA machinery by small farmer groups such as PRONAF in Brazil (Sims and Kienzle, 2009) should be promoted in the LAC region in order to reduce hourly costs of machines with low annual usage such as the knife-roller, sprayers and others. This will make their use more accessible to the smallholder farming sector.

3.1.1 No-till system for vegetables

In the state of Santa Catarina, southern Brazil, no-till for small-scale vegetable farmers is a remarkable success, where machinery has been developed in a joint effort between farmers, research institutions and extension services. Such machines include knife-rollers and soil openers (a shallow furrow or seed slot) with devices for cover crop cutting and crop fertilization prior to manual seedling transplanting of tomato, pepper, eggplant, watermelon, pumpkin, cauliflower, cabbage, onion and broccoli.

Knife-rollers for 2 or 4WTs are available, but models with curved knifes are most highly recommended because they have proven to give better results with regard to reducing the decomposition rate of cover crops (Fig. 5).

No-till seeders for animal traction and 2 or 4WTs are used with devices such as flat, wavy or fluted discs for cutting surface vegetation and tines or double

Figure 5 Knife-rollers used in no-till system for vegetables with staggered straight (left) and curved (right) knives. Photos: M. Zanella.

discs for seed slot opening and fertilizer delivery (Fig. 6) for later seedling transplanting. Rotary blades on 2WTs are also used for strip tillage (Fig. 7) but cause high soil disturbance and smearing, resulting in a lack of porosity in the seed slot.

Field experience has shown that the knife-rolling, cutting, soil opening and transplanting operations should be done on the same day. This is in order to keep the cover crops green for as long as possible to feed soil biota, mainly insects, and to provide better conditions for seedlings to compete with weeds.

Figure 6 Seed slot opening alternatives for animal-drawn, two- and four-wheel tractors, for no-till vegetable production. Photos: M. Zanella.

Figure 7 Two-wheel tractor with rotary blades for cutting cover crop (left) and simultaneously opening a furrow (right) for transplanting of vegetables. Photos: M. Zanella.

3.2 Sub-Saharan Africa (SSA)

Throughout sub-Saharan Africa (SSA), smallholder farmers are the major producers of food; in many countries their contribution can be as high as 80% of national food production. This is achieved against a background of small farm sizes - of around 2 ha - and often difficult and erratic access to good-quality inputs - especially seeds and fertilizer, but also crop-care agrochemicals and mechanization. This situation is frequently superimposed on one of imperfect rural infrastructure, especially rural roads which can become difficult to negotiate or impassable in the wet season; and rural-urban migration. Migration to urban centres is an ongoing phenomenon as young, fit people (predominantly males) are drawn to towns and cities in search of improved economic fortune and an escape from the unrewarding hard physical work and drudgery of farming mainly powered by human muscles. This results in the situation that those remaining to manage the family smallholding are the elderly, children and wives. In this scenario, farm power availability becomes a major production constraint and drudgery increases, further driving the exodus to the perceived material benefits of town life.

One answer to the question of how to enhance farm power sources has been to access tractor-powered ploughing services to replace the alternative of hand-hoe primary soil cultivation. The use of ploughs has a serious deleterious impact on both soil and water - the two most important natural resources from the point of view of crop production - and this has led to the realization that to produce sufficient food for SSA's burgeoning population, sustainable mechanization options are a vital input for smallholder farming systems.

The concept of sustainable agricultural mechanization covers the entire production cycle from input supply, crop and animal production, to transport, processing and marketing (FAO, 2018). Sustainable mechanization protects natural resources and mitigates the impacts of climate change, while alleviating drudgery, increasing farm power supply and improving the productivity and timeliness of agricultural and related operations. Essentially, this means that CA practices are required for sustainability and so we will now turn to the mechanization options appropriate for SSA smallholders.

3.2.1 Options for CA mechanization

All three power sources (human, animal and motorized) are employed in smallholder CA systems and there are appropriate implements for each of them.

Direct planting is achieved by hand, using jab planters (or even a pointed stick, perhaps with a metal tip to reduce wear). As with all direct planting, the requirement is to penetrate the surface organic cover (such as crop residues or managed cover crops) and place seed and fertilizer in the soil at the desired depth and lateral position. Animal- and tractor-powered direct planters use several methods to cut through surface residues and place seed and fertilizer. In low surface residue conditions, an inclined narrow tine can create a slot into which seed and fertilizer can be placed – usually from ground-wheel driven metering systems designed to place seed above the fertilizer. A vertical disc can be located in front of the narrow tine to cut through surface residue if necessary (Fig. 8).

Under conditions of heavier residue cover, penetration to the soil without residue accumulation on the planter tine shanks usually requires disc openers. The most frequently used configuration is double-disc openers. These are of slightly different diameters, offset (one slightly behind the other) and inclined to the direction of travel to form a V-shape. Seed (and fertilizer usually in a second set of inclined offset discs) is metered between the discs and the seed slot is

Figure 8 A two-wheel tractor-mounted direct planter showing (from left to right) the forward mounted vertical disc for cutting through residue, the inclined narrow tine to open the seed and fertilizer slot and the drive/compaction wheel. Photo: B. Sims.

closed by a trailing press wheel. Other available options include single discs inclined to the direction of travel, cross-slot (inverted T) openers and star-wheel openers. Generally, these require less vertical weight on the planting lines than that required to ensure disc penetration. Offset discs inclined to the vertical are also employed for seed and fertilizer placement with minimum soil disturbance.

To keep costs low, one popular practice in SSA is to open rip lines with a narrow-tined chisel plough, pulled by draught animals or tractors (Fig. 9). Seed and fertilizer can then be manually placed in the rip line or simple seed and fertilizer boxes can be mounted behind the chisel tine.

Weed management in CA systems cannot employ conventional cultivation machinery. The goal should be to minimize competition with the crop (for nutrients, water and light) rather than 100% elimination of weeds. Management options include mechanical, chemical and biological possibilities - with the last category always preferred where feasible.

Mechanical methods include manual control by means of hand-pulling (rogueing) in situations of light infestation, surface scraping with sharp hand-hoes or slashing with a machete. Cover crops can be managed at their optimum stage of susceptibility (often just before flowering) by rolling and crimping them with knife-rollers.

For chemical weed and cover crop management, there is a range of conventional boom sprayers available for manual, draught animal and tractor power sources. For manual power, the knapsack sprayer is the most frequently encountered option, but herbicide wick weed wipers, such as the Zamwipe, are also a possibility, as are controlled droplet application (cda) applicators. To make the knapsack sprayer easier, safer and more accurate to use, it can be

Figure 9 Opening rip lines for manual seed and fertilizer placement under heavy surface residue conditions. Photo: C. Thierfelder.

mounted on a wheeled chassis where the pump is driven by a drive wheel via an eccentric. A spray boom can be added (usually with four nozzles) to give a uniform spray bandwidth of 2 m, applied well behind the operator (Fig. 10).

Biological weed management is best achieved by increasing the competition between crops and weeds. Increased competition can be achieved in a variety of ways:

- By increasing the seed rate and reducing the inter-plant distance;
- Through the application of multi-cropping systems, which include intercrops, cover crops, relay cropping and agroforestry;
- Crop rotation is, of course, a fundamentally sound farming practice for many reasons, including breaking weed establishment cycles;
- Mulching with crop and cover crop residues will suppress weed seed germination and weed development.

Development of CA equipment for smallholder farmers is taking off in SSA (Sims et al., 2012) – albeit from a modest starting point. Imported CA equipment (especially from Brazil) has been inspirational for some local manufacturers who have perceived the potential market in the region. East African manufacturers such as Nandra Engineering Works Ltd and Intermec Engineering Ltd, Tanzania; Ndume Ltd, Kenya; Agricultural Engineering and Appropriate Technology Research Institute (AEATRI), Uganda, have produced draught animal-powered rippers and no-till (direct) seeders, manual jab planters and sprayers. The industry continues to grow and now encompasses equipment for two-wheeled tractors, which perhaps will fill an important intermediate power niche in SSA in the future.

Figure 10 A knapsack sprayer mounted on a wheeled chassis with the pump operated from an eccentric off a drive wheel. The four-nozzle boom has a spray width of 2 m and is well behind the operator. Photo: B. Sims.

In southern Africa, direct planters have been developed in Zimbabwe (e.g. Golden Valley Agricultural Research Trust - GART, Zambia; Piket Implements and Dragon Engineering, South Africa; Hästt, Zimbabwe) for manual and draught animal power. South Africa has huge potential for producing CA equipment for smallholders, but this has yet to be fully developed. Progress has been made with the development of planters and sprayers; the Haraka rolling jab planter (manufactured by Eden Equip; Fig. 11) is an example of well-directed efforts aimed at the most farm power-constrained smallholders.

Three-point hitch mounted rippers are available commercially (e.g. Rovic Engineering (Pty) Ltd, South Africa; Ndume Ltd, Kenya; Fig. 12) and robust

Figure 11 The Haraka hand-pushed rolling jab planter manufactured by Eden Equip in South Africa. Photo: Growing Nations/A. Basson

Figure 12 A two-row chisel plough used as a ripper and subsoiler (Rovic). Photo: B. Sims.

no-till planters are available from John Deere (although manufactured in Mexico) (Fig. 13).

3.2.2 Delivering CA mechanization to smallholders

The adoption of CA in SSA, especially by the smallholder sector, has not been as dramatic as in other regions of the world (Kassam et al., 2019). The reasons for modest adoption rates are many, but include:

- the problem of weeding, often with handheld tools, in direct seeding situations where ploughing is not an option for soil inversion and weed burial;
- a lack of locally available, affordable, CA machinery. Unlike in Brazil where government policies actively promoted the development of the smallholder CA machinery industry (Casão Junior et al., 2012) – in Africa there has been no such initiative.

Smallholder CA adoption has been most noticeable in eastern and southern African countries and continentally there are 1.5 million ha of CA, representing just over 1% of the arable area. This level of adoption is expected to accelerate

Figure 13 Two-row tractor-mounted direct planter (John Deere 1025, manufactured in Mexico). Photo: B. Sims.

in the coming years as more experience is gained, farmer support organizations are strengthened and government policies supporting development of the agriculture and manufacturing sectors become both better financed and more focussed.

We have seen the need for smallholders to produce more food for growing urban populations while sustaining natural resources. Improved access to farm power and mechanization is a vital part of equipping farmers to transition to sustainable production increases.

Resource-poor farmers will often find it difficult to invest in expensive farm power and CA mechanization options. Mechanization technologies are usually 'lumpy' in the sense that they require heavy upfront investment and usually quite lengthy payback periods. Specialized mechanization service providers are one way out of this conundrum (FAO and CIMMYT, 2018). By ensuring – through appropriate training campaigns where necessary – that service-providing entrepreneurs are knowledgeable and well equipped to supply sustainable mechanization inputs at a fair price and with due regard for natural resource conservation, smallholders can transition to CA without incurring the heavy costs of investment in the required equipment.

3.3 North Africa

The predominant farming system in North Africa (which includes six countries, Algeria, Egypt, Libya, Mauritania, Morocco and Tunisia) is based on the integration of crop and livestock production. The region is one of the most vulnerable to the appalling effects of climate change, with, among other phenomena, increases in temperatures and drought severity and frequency (Agoumi, 2003).

Algeria, Morocco and Tunisia share similar cropping systems and farming environment within their rainfed arable lands. More than 85% of these lands rely on rains and 80% of farms are less than 10 ha; with wheat and barley as the main crops. The cropping season is from October–November until May–July and a very few spring crops are grown from the end of January–February to May–June (on less than 10% of arable land). Continuous cereals cropping is dominant due to government policies that promote wheat production through subsidies of inputs (mainly certified seeds) and guaranteed markets. Rotation crops (legumes and oil seed) are not so developed due to weak support systems. Actually, certified seeds for legumes and oil crops are not as accessible as those for cereals. Also, the mechanization of these crops is not as well developed.

In North Africa, mechanization on more than 70% of farms is carried out by small-scale farmers who are also service providers; they own tractors and implements, and their income relies on providing mechanization services. CA mechanization relies upon two main technologies, that is seed drills or planters and agrochemical sprayers.

Traditional mechanization in North Africa was based on human and animal power until the introduction of tractors and steel ploughs by the colonial farmers in the early twentieth century. The traditional wooden plough with a steel point was adapted to each environment (soil type, slope and tillage operation) and source of power (horses, mules, donkeys, dromedaries and oxen). This traditional mechanization allowed North African agriculture to develop and prosper for hundreds of years, but it has now completely disappeared due to labour scarcity and the low comparative efficiency of the technology.

The introduction of tractors and tillage implements (mainly ploughs and disc harrows) combined with the harsh climatic conditions (drought and sunshine for over 200 days a year) has rapidly depleted the quality of soil structure, fertility and health. As a result, attention has turned to CA with research based initially on vertical tillage but soon focussing on no-till since the early 1980s. Direct seed drills adapted to local needs (especially with regard to size and cost-effectiveness) were launched in the early 1990s in Morocco (El Gharras et al., 2017).

In Algeria, Morocco and Tunisia, CA development projects started in the 2000s with the import of small (2–4 m width) no-till seeders with disc openers in order to promote CA for rainfed field crops.

The high cost of these machines and their advanced technology were the main factors that limited their adoption in the region. Indeed, in addition to the price, disc-opener seeders are not suitable for the first transition years from conventional to no-tillage mainly on rocky soils and when residues are not retained on the soil surface. Moreover, they require relatively powerful tractors (above 80 hp) because they are heavy (more than 1 tonne per metre width) and need too much maintenance (greasing and spare parts). On the other hand, seed drills with tine openers give better performances (Bahri, 1992) being better adapted to local soil conditions and the most common tractors in use within the region (55–75 hp).

An experimental no-till seed drill was developed by the National Agricultural Research Institute (INRA) in Morocco based on soil scientists' recommendations that it should cut through residues and be able to follow surface contours. This prototype was then re-designed in collaboration with a private company that manufactured about 50 units (Fig. 14) over 10 years. The machine's performance on experimental stations as well as in farmers' fields was excellent and the sale price was about US$8000 for the 3 m wide machine. Five commercial generations were produced, over the 10 years, by way of incorporating design improvements.

An animal draught version was developed and tested with farmers in the mountain regions under forest trees where tractors do not have access (Fig. 15).

Figure 14 First commercial no-till seed drill developed in Morocco. Photo: O. El Gharras.

Figure 15 Animal-drawn no-till seed drill at work in Morocco. Photo: O. El Gharras.

The main objective was to control soil erosion resulting from the tillage tools introduced in the colonial period.

Despite the improvement efforts, the tractor-powered seed drill did not find the expected success for two main reasons. The first, and most important, is that officials at the ministry level and extension services were still promoting 'conventional' tillage practices and, second, is that farmers found the maintenance requirements too burdensome (e.g. the machine has 26 grease points requiring daily attention). New specifications for a low-cost and simpler seed drill were drawn up in collaboration with the manufacturing industry and based on farmers' requirements, within an Australian-funded project with the International Center for Agricultural Research in Dry Areas (ICARDA) (CANA, 2011–15). The CANA project covered the three North African countries (Algeria,

Morocco and Tunisia) and aimed to promote CA mechanization by means of introducing imported tine machines and developing designs with the support of local manufacturers.

The imported machines introduced by the project were:

- two models of a low-cost seed drill from Syria (Figs. 16 and 17) manufactured by private workshops. The designs were based on robust chisel-tined openers but the metering units and transmissions were poorly designed with low material quality and the machinery did not last long;

Figure 16 Al Ashbal seed drill with 14 seeding elements. Photo: O. El Gharras.

Figure 17 Errachid AlBab seed drill with 13 seeding elements. Photo: O. El Gharras.

- the John Shearer seed drill from Australia (Fig. 18) performed excellently, but the adjustment for seed metering was too difficult for farmers and spare parts were not available on national markets;
- the European Gil (Sembradoras Gil, Spain) which is very common in Morocco and Tunisia (Fig. 19) for conventional seeders. More than 50% of conventional seed drills in Morocco are Gil, but no-till seeders were not introduced until the 2012–13 cropping season.

Figure 18 John Shearer seed drill at field tests. Photo: O. El Gharras.

Figure 19 Gil no-till seed drill sold in Morocco and Tunisia. Photo: O. El Gharras.

In Algeria, a private company (CMA) is manufacturing Sola seeders under an agreement with the Spanish company (Maquinaria Agrícola Solà). With support from Algeria's Technical Institute for field Crops (ITGC), CMA has developed a prototype of a simple seed drill (Fig. 20) that is still under evaluation. This model is not yet available commercially, but has high potential.

A low-cost, chisel-opener, seed drill was also designed based on a tool frame manufactured by a private company in partnership with INRA, in Morocco (Fig. 21). Its performance is outstanding and the farmer can add options such

Figure 20 BOUDOUR tine seed drill developed in Algeria. Photo: N. Louahdi.

Figure 21 Low-cost no-till seed drill at work on a rocky field. Photo: O. El Gharras.

as press wheels and coulters. The commercial price of this seed drill is about US$7500 for a 3 m wide machine.

In Tunisia, the National Institute for Field Crops (INGC) has developed, in collaboration with a private manufacturing workshop, a no-till seed drill prototype (Fig. 22) which was evaluated on an experimental station and presented in 2016 at the Tunisian International Agriculture Fair (SIAMAP). There is no commercial prototype and there has been no follow-up on this project since then.

In parallel with the design and evaluation of CA machinery, there is a need to establish public policies that encourage CA adoption and consider the economic and social conditions of each country or region. Service providers, for example, look at no-till as a threat to their businesses since all tillage practices before sowing are eliminated. Indirectly, they express this concern while they witness the benefits and advantages of no-till CA farming.

3.4 East Asia

Small-scale farms in East Asia comprise more than half of the world's smallholders (with farming areas of less than 2 ha) (Rigg et al., 2016; Otsuka et al., 2016). In this region, CA is being adopted rapidly in multi-cropping systems (e.g. wheat-maize, wheat-rice and rice-rice-rape).

Machinery for small-scale CA systems mainly includes straw choppers and spreaders, no-till seeders and subsoilers.

The straw chopper powered by tractor or harvester is widely used for crop residue management. This machine uses blades to hit, cut, rub and tear residues, and then spreads the chopped residues over the field (Wenying et al.,

Figure 22 Tunisian no-till seed drill prototype in the SIAMAP in Tunis. Photo: M. Jadlaqui.

2009). In high residue areas, these small-size straw choppers normally need to improve their working performance; for example, by producing residues of equal chopped length and evenly spread, by reducing power consumption through the optimization of blade design and performance and improved straw-spreading devices (Zhiqiang et al., 2017). Some models are shown in Table 2.

Table 2 Typical small-size straw choppers used in East Asia

Model/ manufacturer	Type	Design	Technical features
1JH-180[a]/ HENAN HAOFENG	Straw chopper		The sharp and wear-resistant blades can be of the bent, straight or hammer claw types. Working width is 1.8 m and work rate is 0.45-0.90 ha h^{-1}. Required power is 51.45-66.15 kW.
1JH-150[b]/ SHANDONG DAHUABAOLAI	Straw chopper		Hammer and knife blades are interchangeable on shredding mechanism. The speed of the chopping shaft is 2000-2490 rpm and working width is 1.5 m. Chopped stubble height must be less than 75 mm and required power is above 36.75 kW.
TDC 1400–0S[c]/ MATSUYAMA PLOW MEFG. CO., LTD	Straw chopper		An electric hydraulic cylinder eliminates the need for external hydraulic equipment and requires no additional investment. It can be steplessly adjustable within the range of +70° ~ −55° above and below the right offset up to 0.2 m. Working width is 1.40 m and speed is 14~23 km h^{-1}. The rotation speed of the chopping shaft is 1994 rpm. Power requirement is between 30 and 45 kW.
4YZ-3 (PRO106Y)[d]/ HEBEI SINO-Agr	Stalk chopper mounted underneath maize harvester		Chopper height can be adjusted to compensate for uneven ground conditions and blades with self-grinding function are adopted. It is powered by the maize combine harvester and requires 81 kW. Working width is 2.4 m work rate is 0.3-0.5 ha h^{-1}.

[a] http://www.hfjtworld.com/.
[b] https://www.dhbl.net/.
[c] https://www.niplo.co.jp/en/.
[d] http://www.kubota.com.cn/kams.

Small-scale no-till seeders used in East Asia can be categorized as manual, animal traction or tractor-powered. Typically, manual and animal-drawn seeders are widely used in hilly areas. For tractor-powered seeders, effective residue anti-blocking capability is a key aspect for good sowing performance with two main approaches followed to solve the problem: power-driven and improved straw-flow (Sharma, 2014). In the first approach, residues on or beside the seeding rows are chopped, cut or thrown by power-driven anti-blocking devices and, in the latter, seed slot openers are arranged along multiple cross-beams to increase clearance spacing. These models of seed slot openers can be categorized as having passive anti-blocking devices which operate by the movement of the seeder and are not power-driven. Currently, many machines may still experience problems such as blockage under heavy residue conditions, resulting in uneven seeding depth, and more work is required to upgrade technologies to achieve more effective anti-blocking and more precise seeding. Some models are shown in Table 3.

Small-scale subsoilers can be chisel-tined with straight or curved shanks for routine soil loosening, or the bent-leg tine deep subsoiler type for severely degraded soils. The tip of the chisel-ripper type penetrates the soil to lift and loosen any hard pan usually to a depth up to 0.3 m. Deep subsoilers should only be employed in CA systems to start the rehabilitation process of severely compacted soils below 0.3 m depth. In this case, the tines cut and lift the entire soil profile (Gan et al., 2014). Small-size subsoilers usually have no residue handling and soil consolidation devices, consequently modifications are required to improve anti-blocking characteristics and create a levelled field surface after subsoiling (Evans et al., 1996). Some models are shown in Table 4.

3.5 South and Southeast Asia

In various forms, CA is now being practised over 180 million ha globally (Kassam et al., 2019) but mostly in large mechanized farms and rainfed and supplementary irrigation areas. Although most of the CA area is on large farms, more small farmers practise CA globally.

However, the wide range of 4WT options for mechanized CA are not available to those regions where 2WTs are the predominant power source for mechanized tillage and crop establishment. Rapid adoption and scaling out of the use of the 2WT for primary tillage has occurred in many Asian countries and increasing interest has been created in Africa (Haque et al., 2017a). As a result, there are now large numbers of 2WTs operating in Asia (Fig. 23), Africa

Table 3 Typical small-scale no-till seeders in East Asia

Model/ manufacturer	Power source	Crops	Design	Technical features
Li seeder[a]/ YUNFAN (He et al., 2014)	Human	Maize, soybean		A manual seeder for no-till on small farms suitable for a wide range of conditions, including wet soils. The handle contains the seed and a shoulder bag carries the fertilizer. Through a chopping action it can deliver one or more seeds simultaneously, while fertilizer can be applied separately. Work rate of 0.2–0.3 ha^{-1} day^{-1} $person^{-1}$.
YT-2B-1 (Zhu et al., 2005)	Animal	Beans, maize		This animal traction, hand-controlled planter was specifically designed for beans and maize. The operator walks behind the machine and controls the operation with handles. It is suitable for seed and fertilizer, with working width of 0.6 m and working depth of 30–50 mm.
ARC GONGLI[b] (Esdaile, 2011)	2WT	Rice, wheat		Variable tine layouts are available for different soil and residue conditions. Integral fertilizer box allows for accurate fertilizer placement in seed rows. Integral press wheels firm the soil to improve soil-seed contact and plant emergence.
Powered disc (He et al., 2014)	4WT	Maize		The anti-blocking device of this no-till seeder is a scalloped powered disc to cut roots and residues. They are followed by shark tooth row clearance devices. The following narrow tine opens the slot for seed and fertilizer.
2BYCF-3[c]/ YUCHENG LEYUAN (He et al., 2014)	4WT	Maize		Residues in front of the tines are moved to the side by the free rollers (in yellow) mounted in the seeding line. In operation, the residue in front of the tines is separated laterally by the roller. The seeding tine and precision metering wheel are shielded to prevent the residue from falling back into the seed furrows.
2BMQF-6/12[d]/ LUOYANG XINLE	4WT	Wheat, maize soybean, millet		Anti-blocking device rotary serrated blades cut through stalks and roots so that the seed slot openers can pass through easily. Working width is 1.94 m, required power is 59–66 kW and work rate is 0.4–0.67 ha h^{-1}.

[a] http://www.qybzg.com/en.

[b] https://sites.google.com/site/twowheeltractorgroup/arc-gongli-brochures.

[c] http://www.agro-china.net/Corn-Planter-with-Fertilizer-pd6059924.html.

[d] http://julinmachine.diytrade.com/sdp/2388061/4/pd/19132597-2822357/No-till_seed_drill_for_conservation_tillage.html.

Table 4 Small-size subsoilers in East Asia

Model/ manufacturer	Type	Design	Technical features
SV2-BD[a]/ KAWABE (Tanaka et al., 2007)	Vibrating subsoiler	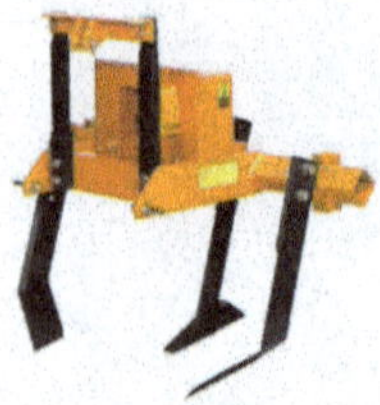	One central straight shank and two lateral curved shanks. Maximum operating width and depth are 1.6 and 0.5 m, respectively. Ground surface is lifted 0.03–0.05 m by subsoiler and required power is 18.4–31.6 kW.
1SQ-340[b]/ CHUNXIANG	V-blade typed subsoiler		Three V-shaped blades with two symmetrically on the front of frame beam. Subsoiling depth is 0.4–0.5 m and required power is 73.6–88.3 kW.
1S-210B[c]/ QINGDAO LU GENG	Bent-leg tine type subsoiler		Four curved shanks carry deep loosening tines arranged to reduce energy consumption. Operating depth is below 0.25 m and working width is 2.1 m. Required power is 66.2 kW.
1S-250[d]/ HAOFENG	Straight shank chisel-type subsoiler		Vertical tine shank and chisel tine with wings in the middle of the shanks to increase soil loosening. Distance between tines is 0.4 m. Working depth is 0.25–0.3 m and required power is 66.2–88.2 kW.
1S-250A[e]/ SHANXI	Curved shank chisel-type subsoiler		Chisel tines are mounted on C-shaped shanks and spaced to increase residue and soil flow. Working depth is 0.30–35 m and required power is over 66.2 kW.

[a] www.kawabenoken.co.jp/en/products01/.
[b] http://www.nongji360.com/company/shop2/product_1814_1439.shtml.
[c] https://www.nongjitong.com/product/lugeng_1s-200b_deep_loosening_machine.html.
[d] http://www.hfjtworld.com/.
[e] http://hedongxiongfeng.cn/ProductsShow.asp?d_id=175.

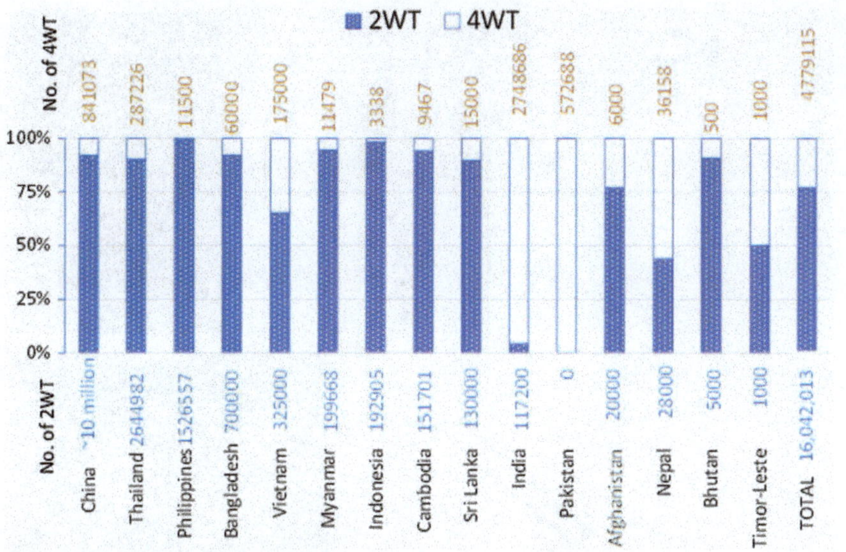

Figure 23 Tractor (four and two wheel) populations in selected Asian countries. Source: adapted from Haque and Bell (2015).

(FACASI Technical WP, 2016; Mrema et al., 2018) and elsewhere, indicating considerable interest in African countries among small farmers in mechanized farming.

Worldwide, considerable research and development has been conducted on handheld manual planters, for example Li seeder, jab planter and star-wheel planter (Fig. 24). These handheld planters are cheap (US\$3–US\$100), but have low work rates (up to 25 m^2 h^{-1}), and are best suited to widely spaced crops (e.g. maize).

Since 1990, Bangladesh has led Asian 2WT-based planter R&D as reported in Haque et al. (2013). However, most of the machines developed are not CA planters due to high levels of soil disturbance (Haque et al., 2017b). There are only a few 2WT-based planters suitable for CA systems.

A 2WT-based zero-tillage (ZT) drill was first fabricated in Bangladesh in the late 1990s (Fig. 25) (Haque et al., 2004); further development was reported by Hossain et al. (2009) and by Johansen et al. (2012) (Fig. 26). Development of the Bangladesh Agricultural Research Institute (BARI) strip-till drill (Fig. 27) was reported by Justice et al. (2004) and Roy et al. (2009).

The Versatile Multi-crop Planter (VMP) (Fig. 28) was developed during the period 2009–17 with the capacity for flexible setting-up in the field and rapid adjustment of seed rate, fertilizer rate, row spacing, seed size and planting depth in several crop establishment modes (Haque et al., 2017a; Bell et al., 2017).

Figure 24 Examples of handheld manual planter and tools for CA (a = hand planter, b = wood-based jab planter, c = plastic-based jab planter, d = punch planter, e = Li seeder, f = rolling star-wheel seed and fertilizer drill). Photos: E. Haque.

Figure 25 First version of 2WT-based zero-tillage drill. Photo: E. Haque.

Figure 26 The improved ACIAR (Australian Centre for International Agricultural Research) - Rogro 2WT-based zero-tillage drill. Photo: R. Esdaile, Johansen et al. (2012).

The Versatile Strip Seed Drill (VSSD) for 2WTs was designed as a variation of the VMP (Haque et al., 2016a) (Fig. 29). The effective field capacity of these planters ranges from 0.11 to 0.20 ha h^{-1} and the ex-factory prices range from US$600 to US$1000 in Bangladesh. Except for the VMP, none of the above CA planters is commercialized.

Figure 27 BARI (Bangladesh Agricultural Research Institute) Strip-till drill working in the field to sow wheat. Photo: I. Hossain.

Figure 28 The Versatile Multi-crop Planter (VMP) attached to a two-wheel tractor. Photo: E. Haque.

The adoption of CA is challenging in puddled rice-based cropping systems. Transplanting rice seedlings into non-puddled soils with minimum soil disturbance is an opportunity to expand the adoption of the CA in rice-based cropping systems (Haque et al., 2016b, 2018). Non-puddled transplanting can

Figure 29 A Versatile Strip Seed Drill (VSSD) attached with Dongfeng-type two-wheel tractor. Parts specification: a = seed box, b = fertilizer box, c = vertical disk seed metre, d = toolbar frame, e = seed delivery tube, f = pressing roller, g = cover of rotary shaft and tines, h = driving seat for transportation, i = depth control bar, j = furrow opener, k = seed and fertilizer off/on lever, l = seed rate adjustable lever for fluted roller type seed metres. Photo: adapted from Haque et al. (2016a).

be mechanized with the use of a rice transplanter (Hossen, 2018, Fig. 30). Dry seeding can be done using the VMP according to CA principles. Rice seed can be directly sown in strips by the VMP while maintaining crop residues and minimum soil disturbance as required for CA (Haque et al., 2018).

4 Mechanization of large-scale Conservation Agriculture systems

4.1 North America

Mechanization of CA systems in North America is in continuous flux as new technology is being developed and becoming more affordable. Major trends are highlighted in data collection, storage and analysis; development of faster and better farm machinery; harvesting equipment to handle large quantities of crop residue; and planting equipment to plant through heavy crop residue, cover crops and into standing crops.

Global Navigation Satellite Systems (GNSS) are linked with monitoring and sensing equipment that continuously gather data. Georeferenced data with up to 1–10 cm accuracy are collected from the planter, sprayer, combine, satellite, weather stations, drones, crop sensors and crop scouts, and combined with publicly available data such as soils and aerial imagery. New developments in cloud and computation technology allow the huge quantities of data to be stored remotely and accessed from multiple platforms (tablets, smart phones, computers). Data mining software allows integration of numerous types of data

Figure 30 The riding type rice transplanter with non-puddled transplanting mechanism. Photo: A. Hossen.

and contains analysis and visualization of the data allowing farmers to improve their decision-making. Uses of data include the mere ability to store and access field records easily; detailed harvest maps showing yield variability across fields enabling various applications such as varying crop varieties and input use by management zone; self-guidance systems to reduce operator fatigue increasing efficiency and work-days; automatic sprayer shutoff sections to avoid sprayer overlap, saving on product expenses and environmental impact; automatic planter unit shutoffs help reduce overplanting, saving on seed costs and improving yields on irregularly shaped fields; on-farm evaluation of inputs and technologies such as yield potential of different corn (maize) hybrids planted in side-by-side strips; or planting corn into 0.76 m spaced narrow strips that were prepared by a strip-till machine or by planting alternating winter-killed and winter-hardy cover crops the previous fall.

Farm size continues to grow in much of North America, and farmers rely heavily on larger, faster and better machinery to avoid having to hire labourers as most farms are still family operated. Tractors are available that provide more than 600 hp (448 kW) and are able to be driven at road speeds of up to 70 km h^{-1}. New corn and soybean planters deliver seeds to the seed trench with a brush-belt trench delivery system that allows planter operation at 16 km h^{-1}. These systems can be mounted on 36-metre wide planters allowing more than 400 ha to be planted in one day. Twenty-metre wide air drills to plant small-seeded crops are available that can plant almost 180 ha in one day. However, there are disadvantages to the larger equipment too, such as the transportation challenge, soil compaction concerns and difficulties to operate in small or irregularly shaped fields. The advent of driverless farm equipment could enable a new trend towards smaller equipment.

Combine heads gather grain and crop residue from a wide strip necessitating the spreading of the residue back over the entire width of the harvested swath to create a uniform seedbed for no-till planting. The first strategy to deal with residue spreading is to limit the crop residue fed through the combine. Stalk rollers pull corn stalks and leaves down to the ground, snapping the corn cob off and feeding it into the auger. Chopping knife-rollers, shredding systems and rotary flail shredders aid in speeding up stalk decomposition. Stripper heads are used to harvest small-grain cereals, such as wheat, while leaving the stalk and leaves standing. Chaff (fine material) and straw that goes through the combine is spread with spreaders and residue choppers. Light material is difficult to throw far so it is better if chaff and straw are mixed to spread it over the entire width of the combine head. But on some combines, it is necessary to install a chaff spreader to spread the fines (Fig. 31). Disk spreaders and straw choppers are available to spread straw. Some choppers create additional airflow which aids in spreading dry, fine material.

Planting equipment for CA continues to be improved in North America. Most innovations concern planters for large-seeded crops such as corn and soybeans planted at wide spacing (>0.5 m). To enable planting through tall cover crops, cover crop roller/crimper attachments have been developed that can be mounted on planters. The rollers come in sections and dynamically follow the soil surface, employing flanges that roll the cover crop away from the row, while concave disc row cleaners part the cover crop to enable accurate seed placement. This equipment facilitates 'planting green', or planting into growing green cover crops that are only killed after planting the main crop (Fig. 32). This practice helps to dramatically increase cover crop residue to increase mulch cover and longevity. The green cover crop removes excess soil water, which is often a concern at planting time. At the same time, the increased mulch conserves moisture in the summer when crops frequently suffer from drought stress. The practice has been shown to increase natural enemies of certain pests and helps

Figure 31 Combine equipped with a bat spreader and a chaff spreader. Both chaff and coarse residue should be spread over the entire harvested width to obtain a uniform seed bed. Photo: S. Duiker.

Figure 32 Mini-rollers and non-wrapping row cleaners on planters facilitate planting into green, growing cover crops that are only killed after planting. Photo: S. Duiker.

to improve soil health. Many new attachments are being developed for planters, including a wide range of row cleaners to clean residue off the row so double-disc seed slot openers can effectively open the seed slot and the seed can be placed at the proper depth. Seed firmers press the seed down to the bottom of the seed trench and can be fitted with liquid fertilizer applicators. Closing wheels come in many different models adapted to different soil and residue conditions - fingered, solid steel, rubber and posi-close are examples (Fig. 33). Interseeders have been developed also to facilitate interseeding of cover crops in standing crops. Some of these high-clearance interseeders are dual purpose and can be used as drills too, expanding the number of weeks of the year they can be used (Fig. 34).

Figure 33 A solid rubber and a curvetine closing wheel to close the seedslot while avoiding compaction above the seed and breaking sidewall compaction. Photo: S. Duiker.

Figure 34 Interseeders have been developed to accommodate planting cover crops with no-till practices into standing crops. Interseeders can also be equipped with fertilizer and herbicide applicators. Photo: C. Houser.

4.2 Latin America and the Caribbean (LAC)

Latin America and the Caribbean (LAC) were responsible for 16% of global agricultural exports between 2012 and 2014 (Duff and Padilla, 2015), comprising a range of products with emphasis on grain crops in South America.

South America has the biggest no-tillage area in the world reaching 69.9 million ha (FAO, 2016). More than 90% of cropland in Argentina and Brazil is under no-tillage system (Pac, 2017; FEBRAPDP, 2013). In Paraguay and Uruguay, a similar pattern of adoption of the system has evolved and in Bolivia, the no-tillage area has also increased over the last decade.

Throughout the last 40 years, seed drill and planter manufacturers in Argentina played a central role in system development, adoption and the scaling-up process of no-tillage agriculture. An entirely new generation of planters and drills for different crops and agro-ecosystem conditions was designed to meet the specific requirements of planting through high levels of crop residues and cover crops, into undisturbed soil.

Mutual collaboration and reciprocal feedback between leading farmers, designers and manufacturers was largely practised and the machinery needed for the evolution and adoption of the new farming system underwent a continuous improvement process.

The most successful design configuration for seeding units in Argentina consists of a flat disc opening coulter followed by separate double discs set in a 'V' form.

Development work has been done on fertilizing units which normally are attached to a separate toolbar and their position in relation to the seeding units is adjustable. There are different coulter designs for fertilizer units, mainly with double or single discs and in some cases they are linked to depth control wheels.

The use of depth control wheels for seeding units is now common practice, and they are followed by seed slot closing devices and press wheels (or other compaction or firming devices) for better soil-seed contact.

Tines and shanks are avoided because they do not operate well through the crop residue on the soil surface. So far, trial results have not shown any benefit of their use in terms of crop yield in different soil types. Blockages frequently occur which reduce the seeding quality as well as the speed and work rate of the machine.

For commercial large-scale operation in Argentina, there is a trend towards bigger no-till seeding machinery of up to and even greater than 14 m width. These machines have from 10 to 24 rows or more. Row spacing varies from 0.525 m down to 0.400 m and even 0.200 m. Tractor units are also correspondingly bigger and sometimes higher than 300 hp power output (Fig. 35).

Brazil also has a large and diversified no-till seeder manufacturing sector which has played a role in CA development (Casão Junior et al., 2012). In medium to large-scale mechanized CA farms in southern Brazil, the typical configuration for no-till seeder is between 7 and 9 rows (0.5 m spacing for maize and soybean) using tractors of up to 140 hp (Fig. 36). Lately, the trend is to increase to 11 or 13 rows and tractors of over 170 hp. In the Cerrado region (of central and northern Brazil), large-scale farms are predominant with 11–17 row seeders, the larger models requiring tractors of 180 hp or more. In this region, there is a further trend in size increment to 21 rows and beyond, with ever increasing power demand (Fig. 37).

Figure 35 A large model of no-till seeder operating under high crop residue levels in Argentina. Photo: R. Peiretti.

Figure 36 A typical no-till seeder model used in southern Brazil for maize and soybeans. Photo: A. Araújo.

Figure 37 A large model of no-till seeder used in Cerrado region in Brazil. Photo: R. Casão.

In both these Brazilian agricultural regions, no-till seeder configuration is basically the same: a flat vertical cutting disc, tine or double disc for furrow opening and fertilizer placement (Levien et al., 2011), double discs for seeds, one or two depth control wheels and compaction wheels (Fig. 38). A narrow tine with adjustable operational depth is common in medium-textured to clay soils and offset double-disc openers in lighter, more sandy soils. These configurations result in excessive soil disturbance and exposure, especially with the use of tines, which increases the accumulation of residues and hairpinning in the seeding slot. Consequently, uniformity of seeding depth and optimum soil-seed contact are impaired.

Figure 38 Soil engaging tool configuration commonly adopted in medium to clay soils in Brazil. Photo: A. Araújo.

Moreover, soil exposure increases the temperature and water evaporation rate in the seeding furrow (Brandelero et al., 2015). This can result in superficial crusting, surface run-off and weed infestation.

Innovations in soil engaging tools are required to improve no-till seeding in soils with high levels of residue cover. In addition, improvements which would widen the seeding window in medium- to fine-textured soils at high moisture levels are a desirable aim.

In the last decade, the Brazilian industry has focussed technological innovations on increasing operational efficiency of no-till seeders and on seed metering devices, also separation of seeding and fertilization operations. Variable rate seed and fertilizer distributors have been part of the development programme. Work has been done in separating seeding and fertilization in two different operations, even with different time intervals. There are advantages in adopting this practice, not least the increase in work rate for the seeding operation.

Cover crop mulching is a key component of CA and while chemical control is a dominant field method, its use can be reduced by association with mechanical means. The knife or crimper roller is a useful alternative for controlling cover crops since the biomass is left in large pieces and therefore decomposition rate is reduced. Various sizes and designs of these rollers are commercially available (Araújo, 2015).

The integration of crop and livestock under CA systems is growing fast in Brazil based on grass crops such as *Brachiaria* sp. and millet, in succession or rotation with maize and soybean crops. However, problems with no-till crop establishment of small-seed crops are still a challenge. It is not uncommon for no-till seeders to be equipped with a small seed hopper and a metering device for small-sized seeds but placement is poor and seeds are left on the soil surface, restricting germination and emergence. Better equipment to deliver small seeds at shallow soil depth and to establish an ideal soil-seed environment is still required.

4.2.1 Conservation Agriculture in sugarcane cropping systems

Brazil is the world's largest sugarcane producer with a cultivated area of 10.2 million ha in 2017/18, mainly in the southern-central region (85% of production). Over the past decade, pre-harvest burning has been gradually eliminated from the sugarcane crop and today more than 98% of the cane in São Paulo state (representing 55% of the national area) is under the mechanical green harvesting system.

A large quantity of residue (between 9% and 19.2% of the total biomass) is retained on the soil surface after harvesting, representing on average 15 t ha^{-1} of dry matter with 12% of fibre content (Landell et al., 2013; Franco et al., 2013). This is a recalcitrant residue due to the high carbon/nitrogen ratio, which decreases from 97 to 68 C/N after 12 months (Oliveira et al., 1999).

Despite the benefits of green cane residue retention, the conventional tillage system causes soil compaction due to intensive traffic of heavy machinery and the mismatched wheel and row spacing. This is the main cause of sugarcane yield decline witnessed over the last decade. The chopper harvester and the transport equipment can weigh around 16–20 and 15–30 t, respectively. The resulting ground pressure in the traffic lanes causes high levels of soil compaction, especially when combined with high soil moisture content.

A great number of tillage operations are required to incorporate the green cane residue, which increased production costs by up to 30% (Conde and Donzelli, 1997). Moreover, run-off increased during the renewal period since terrace spacing is commonly higher than the recommendation to improve manoeuvrability of the harvester.

Cane stalk yields in Brazil almost doubled between 1975 and 2000 but, after the adoption of intensive harvesting mechanization, yields plateaued at 78–80 t ha^{-1} and then declined to 68 t ha^{-1}. The same effect was experienced in Australia at the beginning of the 1990s, but in their case, the adoption of crop rotations with legumes associated with reduced tillage increased yields by 20% improved the timeliness of operations and provided savings in fuel and labour costs (Garside and Bell, 2011).

Soil losses during the conventional tillage sugarcane cropping cycle can be higher than 49 t ha^{-1} $year^{-1}$, while in the first and second ratoons, may be lower than 0.2 t ha^{-1} $year^{-1}$ (Lombardi Neto et al., 1982). An appropriate strategy to reduce run-off and soil erosion in sugarcane is the adoption of no-tillage and crop rotation with legumes. It is important to emphasize that at least 15% of sugarcane fields should be renewed every year, since this is the best practice to increase cane yield. In Brazil, the predominant legumes used for rotation in sugarcane conventional systems are soybean (550 000 ha), peanut (*Arachis hypogaea*) (122 000 ha) and sunn hemp (*Crotalaria juncea*) (300 000 ha).

Pioneering experiences on the adaptation of no-till seeders for sowing soybean under sugarcane residue were led by the agronomist Osvaldo Tanimoto, from São Paulo state. He increased the diameter of the disc coulter in no-till seeders (usually between 430 and 510 mm) to 610 mm. The original discs didn't effectively cut the sugarcane residues, resulting in frequent hairpinning and planter blockages. The double disc for fertilizer placement was also changed to a chisel tine for ease of penetration in hard soils (Fig. 39). A partnership between farmers, researchers and private companies was fundamental to develop a new crop system for sugarcane, based on CA principles, and to develop and deliver information on aspects such as the use of herbicides to manage the ratoon, biological N-fixation, the influence of soil resistance on machine configuration and the best adapted sugarcane cultivars (Fig. 40).

Figure 39 Left: a 610-mm diameter disc coulter (Tatu). Right: a chisel-tine opener for fertilizer placement (Jumil). Photo: D. Bolonhezi.

Figure 40 Sugarcane field before (left) and after soybean seeding (right). Photo: D. Bolonhezi.

In 1998, the Agronomic Institute of Campinas (IAC) started a long-term experiment to study the interaction of lime rates and soil tillage (conventional and no-till), with soybean as the main crop for rotation with sugarcane. The results showed no statistical differences in soybean yield when the no-till seeder was properly set up (Tanimoto and Bolonhezi, 2002). Seeding soybean was viable before spraying herbicides to kill the ratoon when genetically modified cultivars were adopted (Finoto and Bolonhezi, 2012). It was also observed that no-till soils accumulated almost 1.0 t C ha^{-1} $year^{-1}$ more than conventional tillage after 7 years (Segnini et al., 2013) with increasing sugarcane stalk yield (Bolonhezi et al., 2011).

Today, soybean is sown under sugarcane residue on at least 300 000 ha in the southeast of Brazil. The mean yield is around the 3.6 t ha^{-1} with a potential of 4.8 t ha^{-1} and a 32% reduction in production costs.

The use of peanut for crop rotation with sugarcane in CA systems is less important in Brazil because conventional tillage is a tradition for this crop. Furthermore, peanut producers usually are not landowners and have to pay rent for the land (90% are partners and pay US\$350 ha^{-1}) and, consequently, there is resistance to change. Another constraint is that peanut planters are not suitable for no-till. However, results have shown that peanut yields are not significantly reduced with no-till under sugarcane residue (Bolonhezi et al., 2007).

A strip tillage system (Fig. 41) has been evaluated with benefits for peanut in terms of thrips (*Enneothrips flavens*) control, pod yield, economic return, crop establishment and less soil compaction (Siri-Prieto et al., 2009; Bolonhezi et al., 2017) under sugarcane cropping system rotation.

The adoption of CA in sugarcane cropping systems must start after the last cut of sugarcane and must consider the recommended cash crops and green manures for rotation in the region. For Brazilian and Australian conditions, results have shown an average 20% yield increase when sugarcane is no-till planted after legumes (Bolonhezi et al., 2014; Garside and Bell, 2011), with benefits measured up to the third ratoon (Fig. 42).

An alternative CA planting system, known as pre-sprouted buds, has been increasingly adopted. In this system, billets germinate on appropriate substrates in greenhouses and after 60 days the seedlings are transplanted to the field with one or two rows of cane intercropped with a strip of leguminous crop. Sugarcane and legume are planted at the same time and the stalk produced

Figure 41 Soil erosion in peanut crop (left) and no-tillage and strip tillage (right). Photo: D. Bolonhezi.

Figure 42 Sugarcane conventional planter (left) and no-tillage over sunn hemp straw (right). Photo: D. Bolonhezi.

from the intercropped rows is planted after legume harvesting. The objective is to reduce costs, save time and increase the quality of seed cane but there are still problems related with seedlings' sensitivity to water deficit and herbicides (Fig. 43).

In a long-term trial (started in 2003), Bolonhezi et al. (2018) showed that direct transplanting of sugarcane seedlings through mucuna residue (Fig. 44) resulted in a cane yield increase of 16 t ha^{-1}, better development of the crop root system, reduced weed infestation, improved soil structure and an increase of 18 t ha^{-1} in soil organic carbon (SOC). For each 1 t ha^{-1} increase in SOC, there was a 1.2 t ha^{-1} increase in cane yield.

4.3 Sub-Saharan Africa

Agriculture in Africa is facing serious challenges, with cereal yields having stagnated at 1 tonne ha^{-1} over the past 40 years (Word Bank, 2008), in parallel with a projected doubling of the population by 2050. The sector needs to provide income and employment for the rural population, where 70% of the poor people live, in order to reduce rural-urban migration and combat inequalities and poverty. Other demands from agriculture are environmental

Figure 43 Sugarcane seedling transplanter (left) and transplanting under soybean straw (right). Photo: D. Bolonhezi.

Figure 44 Sugarcane seedlings transplanted in conventional tillage (left) and direct-transplanted under mucuna residue (right). Photo: D. Bolonhezi.

conservation, using less fossil fuel and fewer purchased inputs, promoting biodiversity, efficiency, resilience and climate change adaptation and mitigation.

Africa as a whole is a recent comer to the implementation of advanced technological, organizational and institutional innovations that increase the competitiveness and productivity of environmentally friendly farming systems. Led by the large-scale commercial farmers in South Africa, there are now several countries in Africa, such as Zambia, Malawi, Mozambique and Zimbabwe, which have made strategic advances in scaling up smart farming based on CA. Specialized machinery is required with CA for successful seeding in unploughed soils covered with residues, and for the management of cover crops, crop residues and weeds. The role for mechanization is fundamental on large-scale farms to cope with the large areas typical of such farms in Africa and the correspondingly larger volumes in the input and output value chains of the farms. It is presumably due to this mechanization requirement that close to 85% of the 180 million ha under CA worldwide is large-scale commercial and mechanized farming. On the other hand, 80% of the global CA farmers are smallholders in Africa, Asia and Latin America, with limited access to mechanization services, and cultivating some 10 million ha (15% of the total).

The modernization of agriculture by African countries in the 1960s through large-scale farming failed primarily due to policy distortions militating against export agriculture, low public investment in rural areas and low productivity and sustainability due to the tillage-based agriculture promoted. The advent of new, productivity- and resilience-inducing no-till/CA technologies, coupled with the introduction of investor-friendly policies, allowed accelerated growth in large-scale mechanized farming in Africa in the last two decades. Other drivers of progress include stronger produce markets, increased public investment in infrastructure, land governance and farm institutions and access to information technology that makes labour supervision easier. To date, large-scale mechanized farms in Africa constitute 10% of the farmed land and contribute an estimated 10% of the food production. Invariably, the majority of the successful large-scale farms are those switching from conventional tillage-based systems to CA or new entrants using CA-based mechanization technologies (see Case study in Box 1).

With the exception of sugarcane, the level of mechanization for other large-scale commodities is mainly on the production side, with minimal vertical integration into processing and preservation chains. Such skewed introductions of mechanization diminish the sector's ability to productively absorb labour; in the sense that mechanization reduces the labour demand on the production side and should create opportunities on the agroprocessing side. Large-scale farming can operate under two primary models of large plantations or estates, or contract farming by out-growers. The latter can present more opportunities for smallholder engagement and overall poverty reduction.

Box 1 Case study: What are the implications of switching to CA for large-scale mechanized farmers? The case of Madrugada Farm, Nakuru, Kenya

Parameter	Before CA	After 10 years with CA
Number of tractors	8	1
Fuel consumption (L ha^{-1})	25	8
Number of crops per year	1	2
Maize crop yields (tonnes ha^{-1})	5	7

4.3.1 Mechanization for enhancing productivity and resilience to climate change

Under mechanized no-till direct seeding and weed management systems, CA becomes superior to conventional tillage systems in a number of ways. These include (a) emergence of opportunities to rehabilitate and reclaim degraded lands; (b) improved rainfall capture, infiltration and *in-situ* soil moisture retention; (c) timeliness of mechanized farm operations; (d) increasing the cropping intensity with respect to time by margins of 1.5–2; and (e) cost savings from less wear and tear on machinery and reduced fuel consumption.

4.3.2 Mechanized Conservation Agriculture: technologies research and development

The key innovations behind the large-scale mechanized farming in Africa have been the development of, and access to, no-till seeders coupled with the availability of effective and affordable herbicides. The research and development work on seeders, soil cover crops and herbicides has mainly been conducted outside of Africa, by private sector and public research institutions in countries such as the United States, Brazil, Australia and China (Fig. 45).

The progress to enable farmers and society to benefit from CA systems has been uneven, and Africa has yet to take full advantage of their potential. The design capacity for the large-scale mechanized machinery is for farms of around 1000 ha, normally comprising intensive crop production of cereals, pulses and oilseeds, often associated with pastures and other fodder crops for intensive livestock production. Innovations are also demanded in the design and/or adaptation of CA machinery to ensure compliance with controlled traffic GPS-guided requirements, by aligning the wheel spacing of combine harvesters to those of tractors powering sprayers, direct seeders and grain tanks (Fig. 46).

Figure 45 An Australian-made pneumatic direct seeder (Rogro) sowing wheat on a Kenyan large-scale farm. Photo: B. Sims.

Figure 46 A grain-chaser bin follows the GPS-guided combine on a 3 m permanent controlled traffic track in a large-scale South African sorghum crop. Photo: B. Sims.

The private sector has made significant contributions in the manufacture of machinery (some appropriate for CA) better suited to local conditions while remaining competitive with foreign imports. Such manufacturers include Ndume Ltd of Kenya (e.g. no-till tine-press wheel air seeders), William Bain & Co. Holdings (rippers and sprayers) and Hästt (rippers, direct seeders) of Zimbabwe. South Africa has the highest number of large-scale CA equipment manufacturers that include BPI Manufacturing Pty Ltd (stalk chopper and deep ripper), Falcon Agricultural Equipment Pty Ltd (slashers and mulchers), Agrico (rippers and direct seeders) and GC Tillage (rippers and weed wipers) (Mkomwa, 2011).

4.3.3 Integrating mechanization with agroprocessing

On the production side, the farm operations commonly mechanized are the alleviation of plough pans using subsoilers or chisel ploughs mainly at the introduction stage of CA, agrochemical spraying, direct seeding, harvesting (by combine) and transport.

Large-scale farmers do innovate on institutional collaboration such as sharing machines and thus increase their utilization capacity while spreading the investment costs. Agventure Ltd, a cooperative of seven large-scale farmers in Kenya and based in Nakuru, for example, has invested in the purchase and running of a joint oil mill that processes canola seed – a product of their CA crop rotations – into value-added products with ready markets, namely seed cake and oil.

Empowered by their good economies of scale, large-scale farmers are generally able to negotiate for produce markets and prices with outlets including off-takers and processors. Sometimes produce is processed to add value and create demand, but an increasing number of farmers are also diversifying and integrating crops with livestock production, whereby purposely produced cereals, oil seeds and pastures are processed into livestock feeds. What is marketed by such farms are value-added livestock products such as poultry, beef, processed meat and dairy products. The diversification provides opportunities for mechanization (and employment creation) in the processing, packaging, preservation and handling of feed and livestock products.

4.4 East Asia

Many countries in Asia have significant areas under CA, such as China with 9.0 Mha and India with 1.5 Mha (Kassam et al., 2015, 2019). The main categories of large-size CA machinery introduced and developed during the past 20 years are described below.

The main type of machinery used for residue management is a large-size straw chopper with rotating blades to chop or cut residues and driven by a tractor or by a harvester. Blade design has a significant effect on chopping quality, power consumption and extension of the service life (Zhang et al., 2017). For rotary chopping, the hammer blades hit, cut and rub the straw at high speed. For cutting blades widely used in maize stubble, the cutters are driven by an eccentric to reciprocate and cut the stubble into segments. Some models are shown in Table 5.

The no-till seeders employed in the region can be divided into two types according to the method used to avoid blockage, that is disc blade and tine shank seeders. In the first type, with vertical cutting, the disc coulter is the key component, relying on the weight of the seeder to cut surface residues and soil. In Asia, the large-size no-till seeder is mainly used for maize and wheat. For instance, maize-no-till planters of six or more seed rows are used in North-East China and wheat-no-till seeders with 12 or more seed rows in the Punjab region of India. High-speed, precision and intelligent control no-till seeders are expected to be developed in Asia in the future, that is intelligent monitoring and control system, GPS, self-adjusting seeder and self-navigation technology. Some models are shown in Table 6.

The most common models of subsoilers are equipped with chisel-type or bulk-type deep looseners made up of a curved or slanted shank and a tine. These latter types of subsoilers should be employed in CA systems only after a careful analysis of the occurrence of deep compacted layers (below 0.3 m), verification of their maximum depth and planning the operation for soil recovery. When used inappropriately, for instance with too high soil moisture, operation speed, distance between tines or tine depth, soil compaction may be worsened in addition to other problems resulting from excessive soil disturbance. The curved shanks have a rake angle of less than 90°, forcing the soil to move upwards; while the slanted shanks are inclined sideways, to minimize soil surface disturbance while also requiring slightly less power (Weill, 2015). Due to the diversity of seeding patterns and soil types in Asia, the large-size subsoiler has poor adaptability, high power consumption and can suffer frequent blockage. Innovative technologies to improve subsoiling quality, reduce energy consumption and enhance working efficiency (with multiple operations) are required. Some models are shown in Table 7.

4.5 Southern and Western Australia

Australia lies among the top five countries worldwide which have adopted *en masse* CA farming systems (Kassam et al., 2015, 2019), and where no-tillage adoption is still growing and now reaching 80–90% of crop area in many regions (Llewellyn et al., 2012). This section provides an overview of the status

Table 5 Typical high-output straw choppers used in Asia

Model/ manufacturer	Type	Framework	Technical features
1JH-360/YTO Group Corp.[a]	Tractor-mounted straw chopper		Suitable for several kinds of crop straw chopping, such as maize, sorghum, wheat, rice and so forth. Working width is 3.6 m. Required power is 117.7–132.4 kW.
GEMELLA 620/ MASCHIO Gaspardo China[b]	Tractor-mounted straw chopper		Robust rotor with cutting blades in a spiral layout and notched counterblades to deal with tougher residues. Working width is 6.2 m. Required power is 110–220 kW.
RM610R/Tianjin Kuhn Agricultural Machinery Co., Ltd[c]	Tractor-mounted straw chopper		Applicable to wheat, maize and other crop straw. Two types of blades (the hammer blade and the Y-type blade) can be chosen. The hammer and Y-type blades are suitable for tough and hard residues (maize, cotton, sorghum), and made of wear-resistant cast steel with high strength. Working width can be up to 6.1 m. Required power is 140–191 kW.
GK100 (4LZ-10)/ LOVOL HEAVY INDUSTRIES CO. Ltd (Revo Valley Harvesters)[d]	Harvester rear-mounted straw chopper		Widely used in wheat, maize, soybean and other crops. Working width is up to 5.34 m. The minimum ground clearance is 0.39 m.
4YZB-3/KUBOTA[e]	Harvester underside-mounted straw chopper		Mainly used in maize stover chopping. The cutter blades have a self-sharpening function, which leads to more effective chopping performance. Located between the harvester axles, the working width is 2.028 m. Machine power is 103.3 kW.

[a] http://www.yituo.com.cn/cpzx/jjcp/jgj/201706/t20170620_156690.html.
[b] www.maschio.com/catalog/product/gemella/en_GB.
[c] http://www.kuhn.cn/internet/webcn.nsf/0/C66735C502B8AD47C1257B48002FD67D?OpenDocument&p=8.6.1.1.10.
[d] http://am.lovol.com/wheat/gn60.htm.
[e] http://www.kubota.com.cn/kams/.

Table 6 Typical large-size no-till seeders and planters used in Asia

Model/Manufacturer	Type	Design	Technical features
NMCP-ZT/ NATIONAL AGRO-INDUSTRIES[a]	Tine shank type		For seeding different crops, such as maize, peas, soybean, grams, cotton and so forth, in unprepared pre-harvested fields. Adjustable row spacing.
2BFX-12/12(264)/ SHANDONG DAHUA Machinery Co. Ltd[b]	Roto-tiller type		With strip rotary tiller in front of openers. It can complete stubble handling, furrow opening, fertilizing, seeding and compacting at one pass in heavy residue covered fields. Working width is 2.64 m. Minimum required power is 73.5 kW.
2BMZF-6/ JILIN KANGDA Agricultural Machinery Co. Ltd[c]	Double disc with residue manager type		Seven-row for seeding through maize stover. Minimum required power is 66 kW.
1590/JOHN DEERE[d]	Single disc type		Because of its higher mass and good ground clearance of the frame, the seeder can deal with high volumes of crop residue and avoid blocking. Working width is 6.1 m. Required power is at least 75 kW.
21205(2BM-12)[e]/ DEBONT Corp.	Triple disc with residue manager type		Used for seeding maize, soybean, sorghum, sugarbeet and other crops. Furrow opening, side fertilizing, seeding and compacting in one operation. High-performance air-suction seed metering device and four-bar profiling mechanism for seeding depth control. Working width is 6.6–8.4 m.

[a] https://www.indiamart.com/proddetail/paddy-grain-planter-4623611030.html.
[b] https://dhblcultivator.en.ecplaza.net/products/new-condition-wheat-and-rice-drill_3715820.
[c] http://www.jlkdnj.com/products_detail/productId=39.html.
[d] https://www.deere.com.cn/zh.
[e] http://www.debont.com.cn/cn/product/23.html.

of CA seeder mechanization found in the Southern and Western rainfed cropping regions of Australia, where winter cereals dominate under low and variable rainfall (250–600 mm annual), in often low fertility soils, exacerbated by a number of physical, chemical and biological constraints. The average farm size is large, with 23% of farms now cropping more than 1200 ha (ABARES,

Table 7 Typical large-size subsoilers used in Asia

Model/manufacturer	Type	Framework	Technical features
1S-350A/SHANXI HEDONGXIONGFENG[a]	Non-vibrating type with chiselled subsoiling shovel		Seven subsoiling tines with 0.5 m spacing, operation depth between 0.25 and 0.40 m and 3.5 m of working width. Minimum power requirement is 117.6 kW.
1SQZ-420/SHANDONG AOLONG[b]	Hydraulic folding type with curved subsoiling shovel		Eight subsoiling tines with operation depth between 0.3 and 0.5 m and 4.2 m of working width. Minimum power requirement is 154.5 kW.
1SS-370/ HEILONGJIANG DEWO[c]	Non-vibrating type with curved subsoiling shovel		Six subsoiling tines spaced between 0.55 and 0.65 m and working width variable between 3.3 and 3.9 m. Minimum power requirement is 88 kW.
SS1300A/GREAT PLAINS MFG Inc. (CHINA)[d]	Straight shank with low lift wings, following disc coulter		With 5 or 6 tines requiring a minimum of 36.7 kW for each tine. Subsoiling depth is between 0.20 and 0.45 m.

[a] http://hedongxiongfeng.cn/ProductsShow.asp?d_id=226.
[b] http://www.sdaljx.com/zhsh.asp?id=463.
[c] http://www.dewokj.com/product_er.php?class_id=13&id=59.
[d] https://www.greatplainsint.com/en-gb/cdn/farfuture/RrzyMhijsYKO1969FcmTpjyW4gxv0XRJ1op77UKIFkk/mtime%3A1433858094/sites/default/files/sub-soiler_cn.pdf.

2018), and labour productivity is high due to the prominence of labour-saving technologies and large-scale machinery. While 25% of cropping farmers are rated as high innovators, CA cropping equipment is shown to be the most common type of innovation adopted by Australian farmers (Nossal and Lim, 2011).

The mechanization of CA systems in Australia is of large scale and heavily reliant on integrated technologies to maximize crop performance. CA relevant technologies include precision seeder equipment able to accurately deliver a large variety of inputs into precise furrow locations (Fig. 47), and dedicated sensors, guidance and control systems for monitoring and responding to field

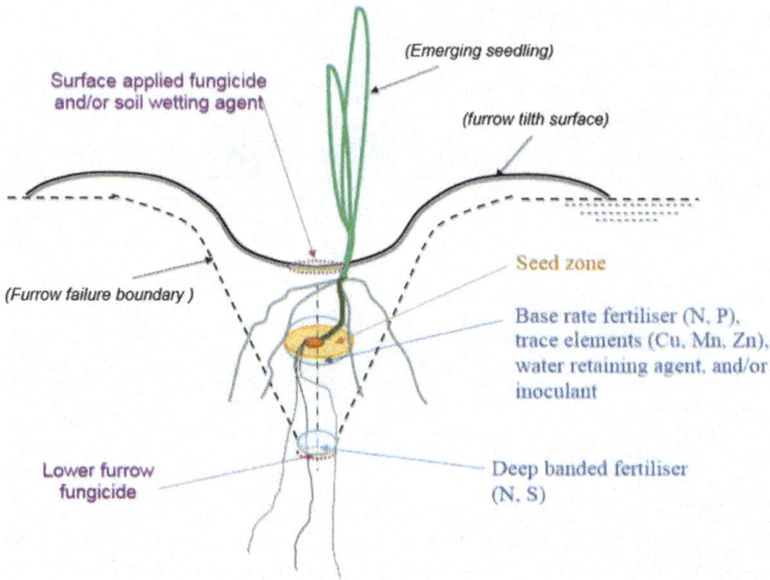

Figure 47 Furrow cross-section showing application zones for potential inputs applied under single pass crop seeding operations. These inputs reflect best practice agronomy and are enabled by advanced CA seeding systems including liquid application technology. Image: J. Desbiolles.

variability as part of precision agriculture strategies. CA mechanization has been optimized over the last 40 years and has benefited from research dedicated to optimizing seeding system performance.

The key features of large-scale CA seeding equipment include:

- Air-seeder style technology (Fig. 48) able to centrally meter up to seven products (e.g. seeds, liquid/granular fertilizers, inoculant, amendments) conveyed by air to furrow openers across the seeding implement. These air-carts are powered by the tractor hydraulics and towed either behind the seeding implement or between the tractor and the seeding implement.
- The common use of tine and press wheel seeding systems (Fig. 49a) which have the ability to till below the seed zone and separate seeds from fertilizer within the furrow (Fig. 47) to improve seedling vigour and control risks of fertilizer toxicity (GRDC, 2011). The tine-press wheel system can be adapted to maximize the water harvesting capabilities of no-till furrows, to achieve deeper moisture seeking and to increase the seedbed utilization by the use of paired row or band sowing attachments (Moodie and Desbiolles, 2016).

Figure 48 Large-scale air-seeder CA equipment are commonly 12–24 m wide, set at 220–300 mm seed row spacing and can sustain work rates in the range of 15–25 ha h^{-1}. Photo: J. Desbiolles.

- The increasing use of disc seeding systems (Fig. 49b) for reducing soil and residue disturbance, maximizing work rates and facilitating full crop residue retention (Desbiolles, 2011). These include single, double and/or triple disc blades as well as hybrid tine-disc systems and their benefits relative to tine systems have been reviewed by Ashworth et al. (2010).
- Tractor GPS guidance with Real-Time Kinematic (RTK) differential correction delivering repeatable 2 cm auto-steering accuracy across the farm, season after season, and enabling a number of 'precise row sowing' strategies, such as (1) 'Inter-row' sowing (Fig. 50a): centring sowing between existing stubble rows to solve implement blockage (tine seeders) and residue hairpinning (disc seeders) issues. This technique also provides a number of highly valuable agronomic benefits such as improved crop establishment and grain yield, lower crown rot disease (*Fusarium* pathogens) incidence, better harvestability of low-lying pulses and greater efficacy of pre-emergence herbicides (McCallum, 2007); (2) 'Edge row' sowing (Fig. 50b): close to the side of existing stubble rows to benefit from the improved soil moisture, nutrition and biology present within the old furrow, while keeping the standing stubble intact. This technique uses side-banding seeding systems and is critical for establishing uniform crops in low fertility, water repellent sands. Where CA seeder tracking behind the tractor is not

sufficiently stable, a number of options exist to more positively guide the implement on to the intended path (Desbiolles, 2017).

- The integration of pre-emergence soil-applied herbicides immediately prior to sowing and mechanically incorporated by the no-till seeding system. This Australian technique requires physical separation of the herbicide from the seed zone, obtained in the process of furrow opening and associated soil throw (Haskins, 2012). Such requirements emphasize the need to control the furrow opener soil disturbance characteristics in order to achieve reliable weed control and maintain high levels of crop safety. The latest Australian research on tine seeding systems has combined the benefits of higher speed and low disturbance sowing in a so-called bent-leg furrow opener (Barr et al., 2016, 2018).
- The mainstream implementation of various degrees of permanent tramlining, up to full controlled Traffic Farming (CTF) systems, over an estimated 22% (6.75 Mha) of grain cropping area in 2015–16 (ABS, 2016). The pillars of a CTF system are well known: permanent wheel tracks, synchronized CA seeder/sprayer/spreader and harvester widths, matching of tractor and machine wheelbases.

Their benefits comprise improved soil structure (compaction restricted to 10–15% of the field area), enhanced timeliness of operation and operational efficiency and significant energy savings (Tullberg et al., 2007).

An innovative approach to crop residue handling at sowing has seen the commercial development of the *Aqua-Till™* 'liquid coulter', which uses ultra-high-pressure waterjet cutting technology to slice very large amounts of surface residue (Taki et al., 2018) ahead of disc or tine seeding systems. Liquid herbicides have also been successfully applied with *Aqua-Till™* for the termination of cotton ratoons after harvest (Butler, 2018), enabling quicker follow-up winter grain crops under no-till systems.

Figure 49 (Left) A majority of CA seeders in Australia use tine seeding systems (right). Recent interest in disc seeding systems is especially increasing among long-term CA farmers. Photos: J. Desbiolles.

Figure 50 (Left) Inter-row sowing with RTK GPS tractor guidance is an integral part of CA systems in Australia; (right) 'Edge-row' sowing strategy is used to improve crop establishment, vigour and yield in water repellent sands, and additionally requires GPS steering of the implement. Photos: J. Desbiolles.

The precision management of seeding operations for winter grain crops is an innovation which involves extending and adapting precision planter technologies used with summer-crops (Fig. 51). The resulting uniform crop establishment at a consistent plant spacing on the row (Fig. 52) generates seed and fertilizer savings, increased grain yield and higher efficiency of inputs.

These innovative developments over the conventional CA seeder technology can be summarized as follows:

1 Singulation metering of winter grain crops coupled with direct path delivery into the furrow to achieve consistent plant-to-plant spacing on every row.
2 Low soil/residue disturbance disc seeding with controlled down pressure to achieve constant furrow depth across variable soil types and residue-free furrows.
3 Electronic ground-drive with geolocation to maintain individual row performance along curve sowing and on-off control for zero-overlap on headlands.
4 On-the-go sensing of in-furrow data: such as organic matter, moisture, temperature and residue content to optimize – in true-time – furrow uniformity, seed placement depth and seed/fertilizer rates matched to soil zone potential.
5 Synchronized seed to fertilizer placement along the seed row for maximum nutrient-use efficiency.

Figure 51 Precision planting equipment adapted for winter grain crop sowing is the new frontier of CA mechanization in Australia. Photo: J. Desbiolles.

Figure 52 Faba bean (top) and wheat (bottom) uniform crop establishment using precision planter technology adapted for winter grain crops. Photos: J. Desbiolles.

6 Bulk feed system linking a high-capacity centralized seed hopper to the singulation units on each row.

CA mechanization continues to evolve in line with the expanding knowledge of successful CA cropping systems in Australia, and the ability to improve operational and agronomic efficiencies by greater integration of automation and precision agriculture systems.

5 Future trends

Agricultural mechanization and sustainable agriculture in the view of the public often seem to be antagonistic. In fact, the replacement of animal traction systems by tractors enabling more intensive agriculture has accelerated the soil degradation processes and led to widespread soil compaction. Yet, it is not the principal cause of soil degradation. On the other hand, CA as a new farming concept only became possible with technological advance and development of new types of agricultural equipment, which started in the 1940s with the evolution of 'trash-handling' reduced tillage implements and then no-till seeders. In some cases, new technologies such as the use of big data and geospatial information systems, precision farming or autonomously working machines using artificial intelligence are seen as solutions for today's problems or alternatives to new production paradigms such as CA. In reality, most of these technologies are complementary to CA, some even make better sense in CA systems than in conventional tillage-based farming, while others facilitate CA operations. This section analyses some of these new technologies for their relevance to CA and discusses further changes, which might become relevant or necessary, when CA becomes mainstream agriculture.

5.1 Forward-looking, non-chemical, weed control

Weed management in CA systems differs from that in conventional tillage farming. While conventional farming controls weeds by mechanical destruction with tillage supplemented with chemicals as and when weeds appear, CA aims at reducing the weed presence by avoiding weed propagation and suppressing emergence.

Technologies, which have been developed, or are under development, to control the spreading of weed seeds at harvest time are just starting to influence mainstream agriculture in mechanized countries such as Australia. Such systems, which either mechanically destroy the germination capacity of weed seeds taken in by the harvester, or collect them within the chaff portion for subsequent use or destruction, exist commercially (Walsh et al., 2017). The rapidly increasing levels of adoption in Australia are the outcome of 20 years of dedicated research aiming to counter the rise of chemically resistant weeds under CA systems (Walsh et al., 2018). Their now-proven efficacy and emerging adoption elsewhere such as in the United States (Schwartz-Lazaro et al., 2017a,b) suggests harvest weed seed control techniques are likely to become increasingly relevant in CA systems globally.

The same applies to non-chemical and non-soil engaging weed control techniques, such as the use of electroshocks, microwaves or thermic energy (O'Brien, 2017; Brodie et al., 2012), which so far have had the highest adoption in organic farming systems (Smith, 2008). However, they could benefit from technologies, such as autonomous machines and swarm technologies, since most of them require very precise application to reduce the energy requirements. Techniques for interpretation of imaging can guide these vehicles to distinguish weeds from crops (Strothmann et al., 2017). Those technologies are likely to become increasingly relevant in CA systems and thus CA systems could facilitate further development of these technologies.

5.2 Conceptual design of agricultural machinery

The change of agricultural systems from tillage-based to CA might stimulate the rethinking of some agricultural machinery concepts. The harvesting of crops, which have their harvested parts below the soil surface, such as roots and tubers, are a particular challenge in CA systems. While harvesters, which pull the root crops out of the soil, are available for most crops, they have been replaced in practice mostly by digger-type harvesters, with much higher soil disturbance, but with a greater versatility for different crops. For some crops, such as cassava, the development of puller- or lifter-type harvesters seems to have stopped, while for others, such as peanut, only very few commercial companies offer puller-type harvesters. This trend might be reversed, as the soil disturbance and degradation become an issue for the harvest of root and tuber crops, resulting eventually in completely new harvesting machinery concepts.

Since the very beginning of the introduction of tractors in agricultural field operations, the ploughing operation has determined the design of the power sources, first the steam ploughs, later the agricultural tractors. The main feature of a ploughing tractor is high mass, since axle weight and ground pressure determine the pulling capacity (Edwards, 2000). Tractor draught-control systems were developed to regulate the implement pull requirements to within the capacity of the tractor, and help optimize the dynamic weight distribution of the tractor (Hesse, 1993).

Tractor engines were optimized in their performance under high and constant loads as for the ploughing task. The elimination of high-draught tillage operations, such as ploughing, from standard agricultural operations raises the question, whether such tractor design is still appropriate. For a CA cropping system, we need more versatile machines, with higher speeds, lighter weights and more flexible engine performance. Items such as draught control in the tractor hydraulic system for implements mounted on the three-point linkage become obsolete. While some of the modern tractor developments point in this direction, such as high-speed tractors with suspension for road transport work, the general mainstream tractor design is still focussed on the ploughing tractor (Edwards, 2018). Yet, as the general requirements for draught and overall tractor power are reduced under CA systems, making way for such different tractor concepts to better reflect the multiple tasks of a general farm tractor, there will still remain the requirement for high-power, high-draught tractors on large farm operations to pull wide equipment at lower speeds, such as no-till seeders. In those cases, the issue of soil compaction under no-till systems has to be addressed, which is best achieved with controlled traffic systems using permanent tram-lines and harmonized track width across all farm equipment (Kerr, 2001).

Another very relevant technology for CA in this context is the GPS-guided auto-steer function of tractors and agricultural machinery in combination with controlled traffic systems. Controlled traffic systems, as such, make more sense in CA than in conventional tillage-based farming, since mechanical repair of soil compaction has to be avoided in CA, but the machinery also requires some conceptual changes in design.

5.3 Autonomous vehicles and machines, swarm technologies

While the future of the farm tractor under a CA system is still subject to discussions and concerns, the big tractor manufacturers are already working on completely new concepts, resulting from the ideas of autonomous vehicles. The trend in these technologies is diametrically opposed to conventional agricultural machinery design, which tries to increase work rate by increasing machinery size. Autonomous vehicles allow downsizing, and applying swarm technologies, whereby high field performance can be achieved by increasing the number of these autonomous machines. Already, such technologies are applied to drones, but airborne equipment has limitations regarding the payload and energy requirements. Solutions that are more promising in agriculture seem to be small field robots, autonomous vehicles, which drive through the crops to perform different crop husbandry tasks (Daniels, 2016). These vehicles obviously will have neither the power, nor the weight, to perform tillage tasks, which makes CA the cropping system of choice for these future technologies.

5.4 Information technology in agricultural machinery and remote sensing

Precision farming concepts, linking geospatial data to crop management decisions and actions, connecting different machines of the cropping cycle, such as harvesting machines with yield monitors, soil analysers, agricultural input application equipment, including seeders, have an important role in improving productivity and reducing environmental impacts in all cropping systems, including CA. The economic potential of these technologies is probably, with some exceptions, not yet fully explored (Griepentrog, 2015). Particularly the aspect of 'big data', the connection of data in the 'cloud', so far has more relevance for marketing decisions, than for crop husbandry. But linking meteorological data with management decisions, or managing and even adjusting working parameters of machines such as harvesters through the cloud is already, today, state of the art (Klein et al., 2013; Liggesmeyer et al., 2014). Most of these technologies have a place in all cropping systems, and have no specific relevance to CA. However, GPS or sensor-controlled herbicide application is of relevance to well-managed CA systems, whereby weed incidence is reduced and a full application of herbicide to the entire field would result in product waste and unnecessary environmental loads (Adamchuk et al., 2008).

Another potential application of information technologies, particularly remote sensing with satellites in combination with big data, could lead to monitoring systems for CA adoption. Considering today's technologies and possible advances, it might be possible not only to monitor by satellite cropping areas with year-round permanent soil cover, but even improvements in soil parameters such as soil organic matter content, which then could be used for monitoring climate change mitigation potential. Another option would be logging the GPS field-data of CA-specific machinery for global mapping of CA areas. While those data would not contribute functionally to CA, they could still lead to accelerated adoption by showing the environmental impact of this concept and also facilitate monitoring for payment schemes for environmental services (Daniel et al., 2009; Siegel et al., 2005; Jin et al., 2017).

6 Conclusion

While the above displays mainly the future of the high-tech, mostly large-scale mechanization of future CA systems, there is still a very high demand for conventional, small-scale and low-cost technologies to facilitate and accelerate the adoption of CA in the small-scale family farming sectors. Also, CA opens new opportunities, such as modernization of draught animal power systems with modern no-till seeding technologies, or a wider use of single-axle tractors

(2WTs) as the entry point into motorized mechanization for small-scale farmers, converting such power source from the traditional 'power tiller' used mainly for rice puddling or transport tasks, into a versatile exclusive power source for small-scale farmers. Particularly for single-axle tractors, but also for small- and medium-size standard tractors, there is still, in many parts of the world, need for improving the local availability of affordable implements suited for CA, such as versatile multicrop no-till seeders.

7 Acknowledgement

The co-authors acknowledge with appreciation the dedication of John Wiles, who kindly devoted himself to reviewing carefully the text from Latin America and making the necessary refinements not only in the English language but also in the fluency and clarity of the text.

8 Where to look for further information

The most complete textbook on CA machinery was edited by Baker and Saxton, *No-tillage Seeding in Conservation Agriculture* (2007; see Baker et al., 2007 in the References section) which provides a broad and detailed overview of machinery for Conservation Agriculture used in different farming systems. Another important book is *Disc Seeding in Zero-till Farming Systems – A Review of Technology and Paddock Issues,* by Ashworth et al. (2010; see also References section) which despite addressing a specific theme (disc seeding) presents an in-depth description regarding components of direct seeders and its interactions with soil and residue.

FAO Conservation Agriculture websites (www.fao.org/conservation-agriculture/en/; www.fao.org/sustainable-agricultural-mechanization/en/) are good sources of information, technical literature and updates on machinery especially for small-scale farming. FAO also has organized the Global CA-CoP (Conservation Agriculture Community of Practice) which is a free service for changing information between subscribers and publishes periodic bulletins. The African Conservation Tillage Network (www.act-africa.org) has a periodic bulletin with rich information on technological advances of CA in Africa.

The Wiki page of the OSCAR project (funded by the European Union) is a relevant source of information, covering different languages, on cover crops use including data on specialized machinery and industries around the world (web5.wzw.tum.de/oscar/wiki/index.php/Machinery #Machinery_producers).

Scientific articles on machinery for CA can be found in different journals but the reader will find relevant material in ASABE´s publications (https://elibrary.asabe.org/toc.asp) and in Elsevier's journals such as *Soil and Tillage Research* (www.journals.elsevier.com/soil-and-tillage-research) and *Biosystems Engineering* (www.journals.elsevier.com/biosystems-engineering). An interesting electronic magazine with updated information on machinery and other subjects related to CA is the *No-Till Farmer* (https://www.no-tillfarmer.com/).

Important events related to the subject are the World Congress on Conservation Agriculture (WCCA) and the Africa Congress on Conservation Agriculture (ACCA).

9 References

AAB. 2006. *International Advances in Pesticide Application 2006*. Association of Applied Biologists, Warwick, UK (*Aspects of Applied Biology* 77, 524p.).

ABARES. 2018. *Australian Farm Surveys and Analysis: Report of the Farm Physical Characteristics for the Grains Industry 2015–17*. Australian Bureau of Agricultural and Resource Economics and Sciences, Department of Agriculture and Water Resources. Available at: www.agriculture.gov.au/abares/research-topics/surveys (accessed on 28 July 2018).

ABS. 2016. *Land Management and Farming in Australia 2015–16*. Catalogue No 46270. Australian Bureau of Statistics.

Adamchuk, V. I., Bernards, M. L. and Meyer, G. E. 2008. *EC08-708 Precision Agriculture: Weed Targeting Herbicide Management*. Historical Materials from University of Nebraska-Lincoln Extension 4871, pp. 1–5. Available at: https://digitalcommons.unl.edu/extensionhist/4871.

Agoumi, A. 2003. *Vulnerability of North African Countries to Climatic Changes: Adaptation and Implementation Strategies for Climate Change*. International Institute for Sustainable Development (IISD), Winnipeg, Canada.

Araújo, A. G. 1990. Definindo prioridades para a pesquisa em máquinas e mecanização agrícola adaptadas aos sistemas predominantes do centro-sul do Paraná. In: *XIX Congresso Brasileiro de Engenharia Agrícola*, Piracicaba, Brasil, pp. 631–42.

Araújo, A. G. 2015. *Is the Roller a Reliable Choice for Cover Crop Management?* Available at: http://orgprints.org/29480/1/IS%20THE%20ROLLER%20A%20RELIABLE%20CHOICE%20FOR%20COVER%20CROP%20MANAGEMENT.pdf (accessed on 10 May 2018).

Araújo, A. G. 2016. *Manual Seeders for No-Till*. Available at: http://orgprints.org/29736/1/MANUAL%20SEEDERS%20FOR%20NO-TILL.pdf (accessed on 7 May 2018).

Ashworth, M., Desbiolles, J. and Tola, E. K. 2010. *Disc Seeding in Zero-till Farming Systems – A Review of Technology and Paddock Issues*. Western Australian No-Tillage Farmers' Association (WANTFA), Northam, Australia.

Bahri, A. 1992. *Evaluation of Opener and Press-wheel Combination on a No-till Grain Drill When Seeding Wheat*. Master of Science thesis. University of Nebraska, Lincoln, NE.

Baker, C. J., Saxton, K. E., Ritchie, W. R., Chamen, W. C. T., Reicosky, D. C., Ribeiro, M. F. S., Justice, S. E. and Hobbs, P. R. (Eds). 2007. *No-Tillage Seeding in Conservation Agriculture* (2nd edn.). Food and Agriculture Organization, Rome and CAB International, Wallingford, UK.

Barr, J. B., Desbiolles, J. M. A. and Fielke, J. M. 2016. Minimising soil disturbance and reaction forces for high speed sowing using bentleg furrow openers. *Biosystems Engineering* 151, 53–64. doi:10.1016/j.biosystemseng.2016.08.025.

Barr, J. B., Ucgul, M., Desbiolles, J. M. A. and Fielke, J. M. 2018. Simulating the effect of rake angle on narrow opener performance with the discrete element method. *Biosystems Engineering* 171, 1-15. doi:10.1016/j.biosystemseng.2018.04.013.

Bell, R. W., Haque, M. E., Johansen, C., Vance, W., Kabir, M. E., Musa, M. A., Mia, M. N. N., Neogi, M. G. and Islam, M. A. 2017. Mechanized minimum soil disturbance establishment and yield of diverse crops in paddy fields using a two-wheel tractor-mounted planter suitable for smallholder cropping. *Experimental Agriculture* 54, 755-73.

Bolliger, A., Magid, J., Amado, J. C. T., Skora Neto, F., Ribeiro, MdeFdosS., Calegari, A., Ralisch, R. and de Neergaard, A. 2006. Taking stock of the Brazilian 'zero-till revolution': a review of landmark research and farmers' practice. *Advances in Agronomy* 91, 47-110. doi:10.1016/S0065-2113(06)91002-5.

Bolonhezi, D., Mutton, M. Â and Martins, A. L. M. 2007. Sistemas conservacionistas de manejo de solo para amendoim cultivado em sucessão à cana crua. *Pesquisa Agropecuária Brasileira* 42(7), 939-47. doi:10.1590/S0100-204X2007000700005.

Bolonhezi, D., Rossini, D. B., Costa, N., Marconato, M. B., Cantarella, H., Gentilin Junior, O., Garcia, J. C., de Sant'ana, S. A. and Bolonhezi, A. C. 2011. Surface application of lime for sugarcane production under no-tillage system. In: *IV International Sugar Conference*, New Delhi, India, pp. 140-4.

Bolonhezi, D., Rossini, D. B., Frizzas, A., Furlani, C. E. A. and Bolonhezi, A. C. 2014. Conservation agriculture principles applied for brazilian peanut crop system. In: *VI World Congress on Conservation Agriculture*, Winnipeg, Canada, pp.140-4.

Bolonhezi, D., Ambrosio, L. M. S., Barini, R. T., Betiol, V., Gonçalves, J. H. O., Scarpellini, J. R. and Bolonhezi, A. C. 2017. Peanut pod yield and soil compaction in Brazilian conservation agriculture system. In: *VII World Congress on Conservation Agriculture*, Rosario, Argentina, pp. 72-6.

Bolonhezi, D., Noronha, R. H. F., Furlani, C. E. A., Xavier, M. A., Garcia, J. C. and Abreu, M. F. 2018. Conservation agriculture principles applied for sugarcane crop propagated by pre-sprouted buds system in Brazil. In: *XXI World Congress on Soil Science*, Rio de Janeiro, Brazil.

Brandelero, E. M., Araújo, A. G. D. and Ralisch, R. 2015. Coverage mobilization in the sowing line and its influence on temperature and water content and on maize emergence. *Journal of the Brazilian Association of Agricultural Engineering* 35(1), 98-108. doi:10.1590/1809-4430.

Brodie, G., Ryan, C. and Lancaster, C. 2012. Microwave technologies as part of an integrated weed management strategy: a review. *International Journal of Agronomy* 2012, 1-14 (Article ID 636905). doi:10.1155/2012/636905.

Butler, G. 2018. Cotton ratoon termination with AquaTill injecticide. *The Australian Cottongrower*, December 2018 - January 2019, pp. 20-2. Available at: https://www.cottongrower.com.au/images/articles/97cf0f4a84e31ac5bd4c27f8fb6246e8.pdf (accessed on 24 December 2018).

Calegari, A., Araújo, A. G., Costa, A. C., Llanillo, R. F., Casão Junior, R. and Santos, D. R. 2014. Conservation agriculture in Brazil. In: Jat, R. A., Sahrawat, K. L. and Kassam, A. H. (Eds), *Conservation Agriculture - Global Prospects and Challenges*. CABI Publishing, Oxfordshire, UK, pp. 54-88.

Casão Junior, R., Araújo, A. G. and Llanillo, R. F. 2012. *No-till Agriculture in Southern Brazil*. FAO and IAPAR, Londrina, Brazil.

Conde, A. J. and Donzelli, J. L. 1997. Manejo conservacionista do solo para áreas de colheita mecanizada de cana queimada e sem queimar. In: *VII Seminário Copersucar de Tecnologia Agronômica*, pp. 193–205.

Conservation Agriculture for North Africa (CANA) project. 2011–15. *Adapting Conservation Agriculture for Rapid Adoption by Smallholder Farmers in North Africa*. Available at: http://www.cana-project.org.

Daniel, K. W., Tripathi, N. K. and Honda, K. 2009. An approach for estimating soil organic matter content using synthetic IRS satellite data in tropical soils of Lop Buri, Thailand. *Geospatial World*, September 1, pp. 1-9.

Daniels, J. 2016. *Future of Farming: Driverless Tractors, Ag Robots*. Available at: https://www.cnbc.com/2016/09/16/future-of-farming-driverless-tractors-ag-robots.html.

Desbiolles, J. 2011. Disc seeders in Australian agriculture: an Australian survey. In: *5th World Congress on Conservation Agriculture*, Brisbane, Australia.

Desbiolles, J. 2017. Seeder tracker and guidance for precise row sowing. *Ag Contractor and Large Scale Farmer* 99(March/April), 16–8.

Duff, A. and Padilla, A. 2015. Latin America: agricultural perspectives. *Rabobank*. Available at: https://economics.rabobank.com/publications/2015/september/latin%2Damerica %2Dagricultural%2Dperspectives/ (accessed on 30 April 2018).

Edwards, G. A. B. 2000. From conventional 'ploughing first' tractors to 'transport-first' Trantors – two decades of radical design. *Landwards*, Autumn 2000, pp. 19–24.

Edwards, G. A. B. 2018. *The Implications of Conservation Agriculture (CA) Indicate that Farm Tractors Need to be Re-designed from First Principles*. Trantor International Limited (www.trantorinternational.co.uk), Cheshire, UK.

El Gharras, O., El Mourid, M. and Boulal, H. 2017. Conservation Agriculture in North Africa: experiences, achievements and challenges. In: Kassam, A., Mkomwa, S. and Friedrich, T. (Eds), *Conservation Agriculture for Africa: Building Resilient Farming Systems in a Changing Climate*. CAB International, Wallingford, UK, pp. 127–38.

Esdaile, R. J. 2011. Conservation farming implements for two wheel tractors. In: *5th World Congress of Conservation Agriculture – 3rd Farming Systems Design Conference*, Brisbane, Australia. Available at: http://www.wcca2011.org/.

Evans, S. D., Lindstrom, M. J., Voorhees, W. B., Moncrief, J. F. and Nelson, G. A. 1996. Effect of subsoiling and subsequent tillage on soil bulk density, soil moisture, and corn yield. *Soil and Tillage Research* 38(1–2), 35–46. doi:10.1016/0167-1987(96)01020-3.

FACASI Technical WP. 2016. *Business Models of Two Wheel Tractors in Southern and Eastern Africa*. Technical Working Papers. Available at: http://facasi.act-africa.org/file/20160727 _business_models_of_br_two_wheel_tractors_in_br_southern_and _eastern_africa.pdf.

FAO. 2016. *AQUASTAT Website*. Food and Agriculture Organization of the United Nations. Available at: http://www.fao.org/nr/water/aquastat/data/query/index.html?lang=en (accessed 24 May 2018).

FAO. 2018. http://www.fao.org/sustainable-agricultural-mechanization/overview/what-is-sustainable-mechanization/en/ (accessed on 8 May 2018).

FAO and CIMMYT. 2018. *Hire Services as a Business Enterprise. A Training Manual for Small-scale Mechanization Service Providers*. Food and Agriculture Organization of the United Nations and International Maize and Wheat Improvement Centre, Rome, Italy, 142pp.

FEBRAPDP. 2013. *Expansão da área cultivada em plantio direto*. Available at: https://febrapdp.org.br/download/PD_Brasil_2013.I.pdf.

Finoto, E. L. and Bolonhezi, D. 2012. Cultivo de soja RR sobre soqueiras de cana crua. *Revista Cultivar* 155, 8–12.

Franco, H. C. J., Pimenta, M. T. B., Carvalho, J. L. N., Magalhães, P. S. G., Rossel, C. E. V., Braunbeck, O. A., Vitti, A. C., Kölln, O. T. and Rossi Neto, J. 2013. Assessment of sugarcane trash for agronomic and energy purposes in Brazil. *Scientia Agricola* 70(5), 305–12. doi:10.1590/S0103-90162013000500004.

Gan, S. B., Zhang, X. R. and Yue, L. 2014. Overview of current research situations of subsoiler. *Asian Agricultural Research* 6(8), 93–7.

Gardi, C., Angelini, M., Barceló, S., Comerma, J., Gaistardo, C. C., Rojas, A. E., Jones, A., Krasilnikov, P., Brefin, M. L. M. S., Montanarella, L., Ugarte, O. M., Schad, P., Rodríguez, M. I. V. and Vargas, R. (Eds). 2014. *Atlas de suelos de América Latina y el Caribe*. Comisión Europea – Oficina de Publicaciones de la Unión Europea (L-2995), Luxembourg, 176pp.

Garside, A. L. and Bell, M. J. 2011. Growth and yield responses to amendments to the sugarcane monoculture to crops, pasture and bare fallow breaks and soil fumigation on plant and ratoon crops. *Crop and Pasture Science* 62, 396–412.

GRDC. 2011. *Fertiliser Toxicity Factsheet: Care with Fertiliser and Seed Placement*. Grains Research and Development Corporation. Available at: https://grdc.com.au/data/assets/pdf_file/0019/203635/fertiliser-toxicity.pdf (accessed on 12 July 2018).

Griepentrog, H. W. 2015. Vom Precision zum Smart Farming – Die Potenziale und Herausforderungen für eine zukunftsorientierte Landwirtschaft. *BLT-Fachtagung 'Innovative Agrartechnik'*, 26. Februar. Universität Hohenheim, Germany.

Haque, M. E. and Bell, R. W. 2015. Development of machinery for smallholders in S & SE Asia. In: *Workshop on Modern Conservation Farming Systems* and *CIMMYT Science Week*, Lanzhou, Gansu, China.

Haque, M. E., Meisner, C. A., Hossain, M. I., Justice, S. E., Rashid, M. H. and Sayre, K. D. 2004. Two-wheel tractor operated seed drill: a viable crop establishment and resource conservation option. In: *CIGR International Conference*, Beijing, China, pp. 203–9.

Haque, M. E., Rahman, S. N. and Bell, R. W. 2013. Smallholders' minimum tillage planter adoption in Bangladesh: a successful case of private sector involvement for technology commercialization. In: *1st CIGR Inter-regional Conference on Land and Water Challenges – Water, Environment and Agriculture: Challenges for Sustainable Development*, Bari, Italy, pp. 1–14.

Haque, M. E., Bell, R. W., Kassam, A. and Mia, M. N. N. 2016a. Versatile strip seed drill: a 2-wheel tractor-based option for smallholders to implement Conservation Agriculture in Asia and Africa. *Environments* 3, 1–13. doi:10.3390/environments3010001.

Haque, M. E., Bell, R. W., Islam, M. A. and Rahman, M. A. 2016b. Minimum tillage unpuddled transplanting: an alternative crop establishment strategy for rice in Conservation Agriculture cropping systems. *Field Crops Research* 185, 31–9. doi:10.1016/j.fcr.2015.10.018.

Haque, M. E., Bell, R. W., Islam, A. K. M. S., Sayre, K. D. and Hossain, M. M. 2017a. An innovative versatile multi-crop planter for crop establishment using two-wheel tractors. *Agricultural Mechanization in Asia, Africa and Latin America* 48, 34–9.

Haque, M. E., Bell, R. W., Menon, R. K. and Hossain, M. M. 2017b. Comparative levels of soil disturbance under reduced and minimum tillage types with two-wheel tractor planting operations. In: *2nd Conference on Conservation Agriculture for Smallholders (CASH-II)*, Mymensingh, Bangladesh, pp. 62–3.

Haque, M. E., Bell, R. W., Jahiruddin, M., Hossain, M. M., Rahman, M. M., Begum, M., Hossen, M. A., Salahin, N., Zahan, T., Hossain, M. M., Hashem, A., Islam, M. A., Vance, W. H., Hossain, M. I., Esdaile, R. J. and Kabir, M. E. 2018. *Manual for Smallholders' Conservation Agriculture in Rice-Based Systems*. Murdoch University, Australia.

Haskin, B. 2012. *Using Pre-emergent Herbicides in Conservation Farming Systems*. Weed Management report. NSW Government Department of Primary Industries, 20p. Available at: https://www.dpi.nsw.gov.au/__data/assets/pdf_file/0003/431247/Using-pre-emergent-herbicides-in-conservation-farming-systems.pdf (accessed on 16 July 2018).

He, J., Zhang, Z. Q., Li, H. W. and Wang, Q. J. 2014. Development of small/medium size no-till and minimum-till seeders in Asia: a review. *International Journal of Agricultural and Biological Engineering* 7(4), 1–12.

Hesse, H. 1993. *Tractor Hitch-Control – History and Future*. Ingenieurbüro für Hydraulik, Stuttgart. Available at: https://prog.lmu.edu.ng/colleges_CMS/document/books/ABE322%20-%20Tractor%20Hitch%20Control.pdf (accessed April 2018).

Hossain, M. I., Esdaile, R. J., Bell, R. W., Haque, M. E. and Johansen, C. 2009. Development of a low cost two wheel tractor operated no-till seeder for better establishment of upland crop. *Eco-Friendly Agriculture Journal* 2, 915–9.

Hossen, M. A., Hossain, M. M., Haque, M. E. and Bell, R. W. 2018. Transplanting into non-puddled soils with a small-scale mechanical transplanter reduced fuel, labour and irrigation water requirements for rice (*Oryza sativa* L.) establishment and increased yield. *Field Crops Research* 225, 141–51. doi:10.1016/j.fcr.2018.06.009.

Jin, X., Song, K., Du, J., Liu, H. and Wen, Z. 2017. Comparison of different satellite bands and vegetation indices for estimation of soil organic matter based on simulated spectral configuration. *Agricultural and Forest Meteorology* 244–245, 57–71. doi:10.1016/j.agrformet.2017.05.018.

Johansen, C., Haque, M. E., Bell, R. W., Thierfelder, C. and Esdaile, R. J. 2012. Conservation agriculture for small holder rainfed farming: opportunities and constraints of new mechanized seeding systems. *Field Crops Research* 132, 18–32. doi:10.1016/j.fcr.2011.11.026.

Justice, S. E., Haque, M. E., Meisner, C. A., Hossain, I., Sah, G., Tripati, J., Rashid, M. H. and Amin, M. R. 2004. Giving South Asia farmers a choice: a single drill for reduced and strip till crops for 2-wheel tractors. In: *International Commission on Agricultural Engineering (CIGR) International Conference*, Beijing, China, pp. 230–7.

Kassam, A., Friedrich, T., Derpsch, R. and Kienzle, J. 2015. *Overview of the Worldwide Spread of Conservation Agriculture*. Field Actions Science Report, p. 8.

Kassam, A., Gottlieb, B., Friedrich, T., Gonzalez, E., Trivino, P. and Mkomwa, S. 2017. Mobilizing greater crop and land potentials sustainably. *Hungarian Geographical Bulletin* 66(1), 3–11. doi:10.15201/hungeobull.66.1.1.

Kassam, A., Friedrich, T. and Derpsch, R. 2019. Global spread of Conservation Agriculture. *International Journal of Environmental Studies* 76(1), 29–51. doi:10.1080/00207233.2018.1494927.

Kerr, P. 2001. *Controlled Traffic Farming at the Farm Level*. GRDC Research Update, Finley NSW, Australia.

Klein, D., Tran-Gia, P. and Hartmann, M. 2013. Big data. *Informatik Spektrum* 36(3), 219–20. doi:10.1007/s00287-013-0702-3.

Landell, M. GdA., Scarpari, M. S., Xavier, M. A., Anjos, I. Ad, Baptista, A. S., Aguiar, C. Ld, Silva, D. Nd, Bidóia, M. A. P., Brancalião, S. R., Bressiani, J. A., Campos, M. Fd, Miguel,

P. E. M., Silva, T. Nd, Silva, V. H. Pd, Anjos, L. O. S. and Ogata, B. H. 2013. Residual biomass potential of commercial and pre-commercial sugarcane cultivars. *Scientia Agricola* 70(5), 299–304. doi:10.1590/S0103-90162013000500003.

Levien, R., Furlani, C. E. A., Gamero, C. A., Conte, O. and Cavichioli, F. A. 2011. Semeadura direta de milho com dois tipos de sulcadores de adubo, em nível e no sentido do declive do terreno. *Ciência Rural* 41(6), 1003–10. doi:10.1590/S0103-84782011000600014.

Li, H. W. 2005. *The Current Status of Conservation Tillage in China, Beijing, Peoples'*. Conservation Tillage Research Centre of the Ministry of Agriculture, Republic of China.

Liggesmeyer, P., Dörr, J. and Heidrich, J. 2014. Big data in smart ecosystems. *Informatik Spektrum* 37(2), 105–11. doi:10.1007/s00287-014-0772-x.

Llewellyn, R. S., D'Emden, F. H. and Kuehne, G. 2012. Extensive use of no-tillage in grain growing regions of Australia. *Field Crops Research* 132, 204–12. doi:10.1016/j.fcr.2012.03.013.

Lombardi Neto, F., Dechen, S. C. F. and Castro, O. M. 1982. A cultura da cana-de-açúcar e as perdas por erosão. In: *IV Congresso Brasileiro de Conservação de Solo*, Campinas, 24.

McCallum, M. 2007. Multiple benefits from Inter-row sowing with 2 cm RTK GPS. In: *5th Controlled Traffic Farming Conference*, Perth, Western Australia. Available at: https://www.actfa.net/actfa-conferences/ctf07-conference/ (accessed on 20 July 2018).

Mkomwa, S. 2011. *Conservation Agriculture Equipment Supply Chain in Southern Africa: Lessons from Mozambique, South Africa, Zambia and Zimbabwe on Key Players, Services Provided Challenges and Opportunities*. African Conservation Tillage Network (ACT) and Sub Regional Emergency Office for Southern Africa, Food and Agriculture Organization of the United Nations, Nairobi, Kenya.

Montiel, K. and Ibrahim, M. 2016. *Manejo integrado de suelos para una agricultura resiliente al cambio climático*. Instituto Interamericano de Cooperación para la Agricultura (IICA), 30p. Available at: http://www.iica.int.

Moodie, M. and Desbiolles, J. 2016. *Seeding Systems for Variable Mallee Soils*. FarmTalk factsheet No. 45. Mallee Sustainable Farming. Available at: http://www.msfp.org.au/farmtalk-45-seeding-systems-variable-mallee-soils (accessed 20 July 2018).

Mrema, G. C., Kienzle, J. and Mpagalile, J. 2018. Current status and future prospects of agricultural mechanization in sub-Saharan Africa (SSA). *Agricultural Mechanization in Africa, Asia and Latin America* 49(2), 13–30. Available at: www.fao.org/fileadmin/templates/library/pdf/Ama.pdf.

Murray, J. R., Tullberg, J. N. and Basnet, B. B. 2006. *Planters and Their Components: Types, Attributes, Functional Requirements, Classification and Description*. The University of Queensland and Australian Centre for International Agricultural Research, Canberra.

Nossal, K. and Lim, K. 2011. *Innovation and Productivity in the Australian Grains Industry*. Australian Bureau of Agricultural and Resource Economics and Sciences (ABARES) research report 11.06. Australian Bureau of Agricultural and Resource Economics and Sciences, Canberra, Australia.

O'Brien, M. 2017. Microwave – a fresh look at old technology for weed control. *The Australian Cottongrower*, April–May, pp. 26–8.

Oliveira, M. Wd, Trivelin, P. C. O., Penatti, C. P. and Piccolo, MdC. 1999. Decomposição e liberação de nutrientes da palhada de cana-de-açúcar em campo. *Pesquisa Agropecuária Brasileira* 34(12), 2359–62. doi:10.1590/S0100-204X1999001200024.

Otsuka, K., Liu, Y. and Yamauchi, F. 2016. The future of small farms in Asia. *Development Policy Review* 34(3), 441–61. doi:10.1111/dpr.12159.

Pac, S. N. 2017. *Evolución y retos de la Siembra Directa en Argentina*. Asociación Argentina de Productores en Siembra Directa (Aapresid), 9p.

Ribeiro, M. F., Justice, S. E., Hobbs, P. R. and Baker, J. 2007. No-tillage drill and planter design - small-scale machines. In: Baker, C. J. and Saxton, K. E. (Eds), *No-Tillage Seeding in Conservation Agriculture* (2nd edn.). FAO, Rome and CAB International, Wallingford, pp. 204–25. Chapter 14.

Rigg, J., Salamanca, A. and Thompson, E. C. 2016. The puzzle of East and Southeast Asia's persistent smallholder. *Journal of Rural Studies* 43, 118–33. doi:10.1016/j.jrurstud.2015.11.003.

Roy, K. C., Haque, M. E., Justice, S., Hossain, I. and Meisner, C. A. 2009. Development of tillage machinery for Conservation Agriculture in Bangladesh. *Agricultural Mechanization in Asia, Africa and Latin America* 40, 58–64.

Schneider, S. 2016. *Family Farming in Latin America and the Caribbean: Looking for New Paths of Rural Development and Food Security*. Working Paper 137. International Policy Centre for Inclusive Growth (IPC-IG), Food and Agriculture Organization and UN Development Program, Brasilia.

Schwartz-Lazaro, L. M., Norsworthy, J. K., Walsh, M. J. and Bagavathiannan, M. V. 2017a. Efficacy of the integrated Harrington seed destructor on weeds of soybean and rice production systems in the Southern United States. *Crop Science* 57(5), 1–7, doi:10.2135/cropsci2017.03.0210.

Schwartz-Lazaro, L. M., Norsworthy, J. K., Barber, L. T. and Scott, R. C. 2017b. *Harvest Weed Seed Control - An Alternative Method for Measuring the Soil Seedbank, Agriculture and Natural Resource*. University of Arkansas FSA2180. Available at: https://www.uaex.edu/publications/pdf/FSA-2180.pdf

Segnini, A., Carvalho, J. L. N., Bolonhezi, D., Milori, D. M. B. P., Silva, W. T. Ld, Simões, M. L., Cantarella, H., Maria, I. Cd and Martin-Neto, L. 2013. Carbon stocks and humification index of organic matter affected by sugarcane straw and soil management. *Scientia Agricola* 70(5), 321–6. doi:10.1590/S0103-90162013000500006.

Sharma, V. K. 2014. *Design, Development and Performance Evaluation of a Multi-toolbar No-till Seed Drill for Surface Managed Loose Straw Conditions after Combining*. Govind Ballabh Pant University of Agriculture and Technology, Pantnagar, India.

Siegel, D. A., Maritorena, S., Nelson, N. B., Behrenfeld, M. J. and McClain, C. R. 2005. Colored dissolved organic matter and its influence on the satellite-based characterization of the ocean biosphere. *Geophysical Research Letters* 32(L20605), 1–4. doi:10.1029/2005GL024310.

Sims, B. and Kienzle, J. 2009. *Farm Equipment Supply Chains - Guidelines for Policy-Makers and Service Providers: Experiences from Kenya, Pakistan and Brazil*. Food and Agriculture Organization, Rome, Italy.

Sims, B. G., Thierfelder, C., Kienzle, J., Friedrich, T. and Kassam, A. 2012. Development of the conservation agriculture equipment industry in sub-Saharan Africa. *Applied Engineering in Agriculture* 28(6), 813–23. doi:10.13031/2013.42472.

Siri-Prieto, G., Reeves, D. W. and Raper, R. L. 2009. Tillage requirements for integrating winter-annual grazing in peanut production: plant water status and productivity. *Agronomy Journal* 101(6), 1400–8. doi:10.2134/agronj2009.0116.

Smith, K. 2008. *Another Look at Thermal Technology Weed Control*. Western Farm Press, Fresno, CA. Available at: http://www.westernfarmpress.com/another-look-thermal-technology-weed-control (accessed April 2018).

Strothmann, W., Ruckelshausen, A., Hertzberg, J., Scholz, C. and Langsenkamp, F. 2017. Plant classification with in-field-labeling for crop/weed discrimination using spectral features and 3D surface features from a multi-wavelength laser line profile system. *Computers and Electronics in Agriculture* 134(March), 79–93. doi:10.1016/j.compag.2017.01.003.

Taki, O., Desbiolles, J. and Butler, G. 2018. Waterjet cutting of crop residue using the aqua-till liquid coulter – a laboratory evaluation. In: *Paper #207 of the 21st International Conference of the International Soil Tillage Research Organisation*, 24–27 September 2018. Cite Universitaire, Paris, France.

Tanaka, H., Oida, A., Daikoku, M., Inooku, K., Sumikawa, O., Nagasaki, Y. and Miyazaki, M. 2007. DEM simulation of soil loosening process caused by a vibrating subsoiler. *International Commission of Agricultural Engineering (CIGR) Ejournal Volume 9*, Manuscript PM 05 010.

Tanimoto,O. S. and Bolonhezi, D. 2002. *Plantio direto de soja sobre palhada de cana-de-açúcar*. CATI, Campinas, 18pp.

Tullberg, J. N., Yule, D. F. and McGarry, D. 2007. Controlled traffic farming – from research to adoption in Australia. *Soil and Tillage Research* 97(2), 272–81. doi:10.1016/j.still.2007.09.007.

Walsh, M., Ouzman, J., Newman, P., Powles, S. and Llewellyn, R. 2017. High levels of adoption indicate that harvest weed seed control is now an established weed control practice in Australian cropping. *Weed Technology* 31(3), 341–7. doi:10.1017/wet.2017.9.

Walsh, M. J., Broster, J. C., Schwartz-Lazaro, L. M., Norsworthy, J. K., Davis, A. S., Tidemann, B. D., Beckie, H. J., Lyon, D. J., Soni, N., Neve, P. and Bagavathiannan, M. V. 2018. Opportunities and challenges for harvest weedseed control in global cropping systems. *Pest Management Science* 74(10), 2235–45. doi:10.1002/ps.4802.

Weill, A. 2015. *A Guide to Successful Subsoiling*. CETAB, Victoriaville. Available at: www.cetab.org/publications.

Wenying, L., Hongwen, L. and Shi, C. 2009. *Conservation Tillage Technology*. Heilongjiang Science and Technology Press, Harbin, China.

World Bank. 2008. *World Development Report 2008: Agriculture for Development*. Oxford University Press for the World Bank, Washington, DC.

Zhang, Z., McHugh, A. D., Li, H., Ma, S., Wang, Q., He, J. and Zheng, K. 2017. Global overview of research and development of crop residue management machinery. *Applied Engineering in Agriculture* 33(3), 329–44. doi:10.13031/aea.12167.

Zhiqiang, Z., Jin, H. and Hongwen, L. 2017. Design and experiment on straw chopper cum spreader with adjustable spreading device. *Transactions of The Chinese Society of Agricultural Machinery* 48(9), 76–87. Available at: http://www.j-csam.org/jcsam/ch/reader/view:abstract.aspx?flag=1&file_no=20170910&journal_id=jcsam.

Zhu, Z. L., Xiong, Z. Q. and Xing, G. X. 2005. Impacts of population growth and economic development on the nitrogen cycle in Asia. *Science in China, Series C: Life Sciences* 48, 729–37.

Chapter 11

Certification schemes for Conservation Agriculture systems

Juliana Albertengo, Iowa State University, USA

1 Introduction

Primary production is the first step in the whole food, fibre, energy and economy chain. Contrary to industries, farmers have always stayed away from certification schemes because there were not certification schemes developed for agricultural production, and even today not all of them Conservation Agriculture (CA). Consumers nowadays are willing and sometimes demanding to get more information on what they are buying. Thus, certification schemes appear to be the solution to this problem, creating trustworthy information on quality assurance and incorporating CA principles that include:

- No or minimal mechanical soil disturbance with no-till seeding and weeding,
- Maintaining soil mulch cover with crop residues and cover crops and
- Crop diversification with rotations, sequences and association involving annuals and perennials, legumes and cover crops.

Currently, there are various certification schemes for the agricultural sector. Many of them focus on primary products that are being produced (e.g. Round Table on Sustainable Soybean, RTRS), while others focus on the whole production process (Agricultura Certificada, AC) or on the end use (Round Table on Sustainable Biomaterials, RSB, on Biofuels). Some also include the chain of custody (specific and segregated chain of a product from the farmer to the buyer), transport and/or marketing. Certification schemes usually link

http://dx.doi.org/10.19103/AS.2019.0048.12

management practices with measurable indicators. In the following sections, various certification initiatives are described as examples. Organic certifications have been excluded since they do not comply with CA principles, because they allow tillage.

2 Certification schemes with a focus on Conservation Agriculture

Certification is a simple way to ensure that standard directives are fulfilled. Certification is the way to enhance trust with the consumer. The standard can be private or public. Certification audits are carried out by an independent third party who verifies the fulfilment of the standard requirements. When the external audit (certification audit) is completed the auditing company recommends the certification of that farmer or farm to the standard or scheme developer. Since certification systems need to be measurable, they all have a set of mandatory or self-chosen indicators. There are four different ways of selling certified products in the market that are as follows:

- Identity preserved: This model requires a full separation and full traceability from the producer to the end user.
- Segregation: This method assures that all of the physical products have originated from a mix of certified sources. However, the end user does not know the exact origin since it is a blend.
- Mass balance: This structure comprises a known amount of mixture of certified and non-certified products.
- Buck and claim: This system allows for the transfer of volume credits from the owner to its supplier and end user independently from the physical chain.

The purpose of this chapter is to explain different certification schemes linked to Conservation Agriculture. First, two certification schemes related to extensive agricultural production will be described. Then some other examples of different kinds will be given.

2.1 Round Table on Responsible Soy Association (RTRS)

The Roundtable on Responsible Soy Association (RTRS) is a civil organization that promotes responsible production of processing and trading soybean on a global level. The RTRS initiative began in Switzerland in 2006 and has expanded on a global level. Four years later, they launched the first version of the RTRS Standard and by June 2011, the first producers were certified in Argentina, Brazil and Paraguay. Those pioneers sold their product (via credits) to Europe, which was proof of the success of certified soybeans in the marketplace (RTRS, 2014a).

RTRS members include representatives of the soybean value chain and of the civil society from around the world. As stated by the Round Table philosophy, participants from different areas and with diverse interests obtain equal rights and join with a common purpose, thus assuring dialogue and decision-making by consensus (RTRS, 2014a,b). This principle of whole consensus was the reason for the creation of the RTRS Standard for Responsible Soy Production.

This standard is applicable on a worldwide level assuring that soybean production is socially equitable, economically feasible and environmentally sound; even though every country has its 'national interpretation' to adapt to the site-specific realities (RTRS, 2014c). To date, RTRS has national interpretations in Argentina, Brazil, China, India, Uruguay, Bolivia and Canada, and is currently working on an interpretation in Paraguay (RTRS, 2014c). Currently, RTRS has more than 2 937 268 tons of soybean certified around the world that is produced on 923 000 ha (RTRS, 2014d).

The RTRS Standard is based on a set of principles and criteria for certifying soybeans as a responsible crop. The pillars are (RTRS, 2014b):

- Legal compliance and good business practices,
- Responsible labour conditions,
- Responsible community relationships,
- Environmental responsibility,
- Good agricultural practices (GAPs).

Those principles and criteria must be linked to self-chosen indicators.

RTRS, in its principle N° 5, GAPs, establishes that it is necessary to demonstrate and apply known practices to maintain soil quality as: Conservation Agriculture, crop rotations and balances fertilization. These practices can also help to reduce erosion when applied.

The RTRS certificate lasts for 5 years; however, follow-up or surveillance audits are carried out during each of the intermediate years. This is a product certification since soybean product is certified. Small producers can choose to certify their soybean as individuals or as a group, thus, sharing assessment costs, both for certification and for annual surveillance.

In addition, the producer or farmer that owns or hires many sites can ask for a multi-site certification. Under this modality, all the farms under the same ownership, including leased land, and under the same management and control systems, may be granted a single certification without any need for applying for an individual certification for each one of the sites (RTRS, 2014b).

RTRS also offers other annexes or options to producers, such as the possibility to apply for a step-by-step certification (progressive entry level) that includes a continuous improvement approach (Fig. 1). In addition, there is a biofuel annex for the production of biofuels fulfilling the Renewable Energy Directive of the European Union (EU RED). Moreover, the possibility of certifying the chain of custody is also available to producers (RTRS, 2014b).

2.2 Agricultura Certificada

Agricultura Certificada (AC – Certified Agriculture) is a certification scheme developed by the Argentinean No-Till Farmers Association (Aapresid) in 2008. It is a quality management system based on good agricultural or management practices and soil quality indicators. It also includes good business and social practices. Unlike RTRS, AC certifies the whole productive process at a farm, not a determinate product. The certification lasts for 3 years with two mid-year follow-up audits. AC is based on a continuous improvement cycle. According to Aapresid, AC has more than 83 000 certified ha belonging to 47 farmers per companies (R. Belda, pers. comm., 2017).

According to the FAO Regional Office for Latin America and the Caribbean (2004), 'Good Agricultural Practice (GAP)' consists of applying the knowledge available to make a sustainable use of basic natural resources for the friendly production of agriculture-based food and non-food products that are innocuous and healthy, while focussing on economic feasibility and social stability. It is also necessary to point out that GAPs generate better working conditions and greater qualification for rural workers.

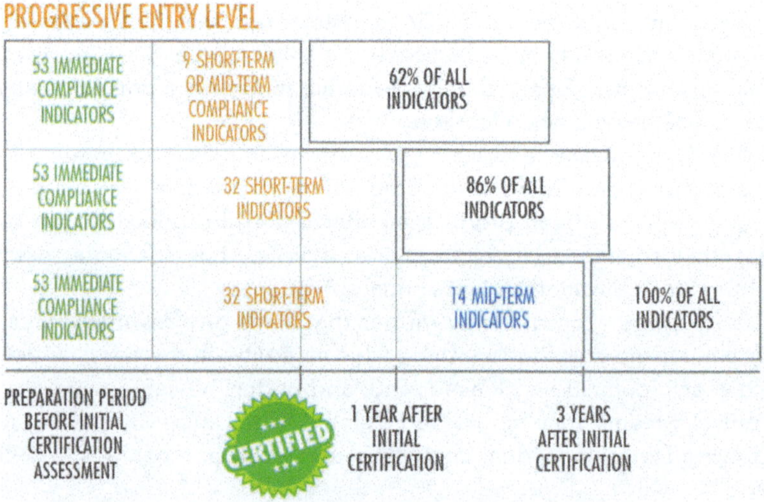

Figure 1 RTRS progressive entry level. Source: adapted from RTRS (2014b).

The main goals of the AC are:

- To provide tools for a professional agronomical management, by the ordered registry of information and the analysis of the soil quality and efficiency indicators.
- To demonstrate to the rest of society the production processes involved and their impact on the environment, allowing the capture of the value of the positive externality that the AC makes.

AC is based on two main documents (Fig. 2), namely:

- Good Management Practices Manual describes the GAPs and soil quality indicators. In addition, it refers to soil sampling, calculation methods and has some other useful tables.
- Quality Management System Protocol specifies all the documents, registry and procedures that the farmer has to fulfil in order to become certified.

AC manual mentions that, in brief, implementing GAPs requires simultaneous conditions such as:

- Investing in workers, farmers and professionals training
- Investing in adequate technologies (seeds, machinery, fertilizers etc.) and infrastructure
- Recognition by consumers/market, the implementation results should be profitable for the farmer
- Global feasibility of the agricultural enterprise in the long term – a concept where sustainability is the responsible interaction of economic, social and environmental variables

Additionally, and in productive terms, some GAP techniques assert a greater emphasis on productivity. GAPs should improve productivity within the mid-long term since their implications (registers, training, qualified personnel etc.) allow the farmer to consider with greater knowledge their productive system and, therefore, the management process is improved.

So, based on scientific studies carried out by different organizations and on Aapresid's own experience, the following GAPs have been established as the required criteria to achieve certification:

- No-tillage/Conservation Agriculture,
- Crop rotation,
- Balanced crop nutrition or fertilization,

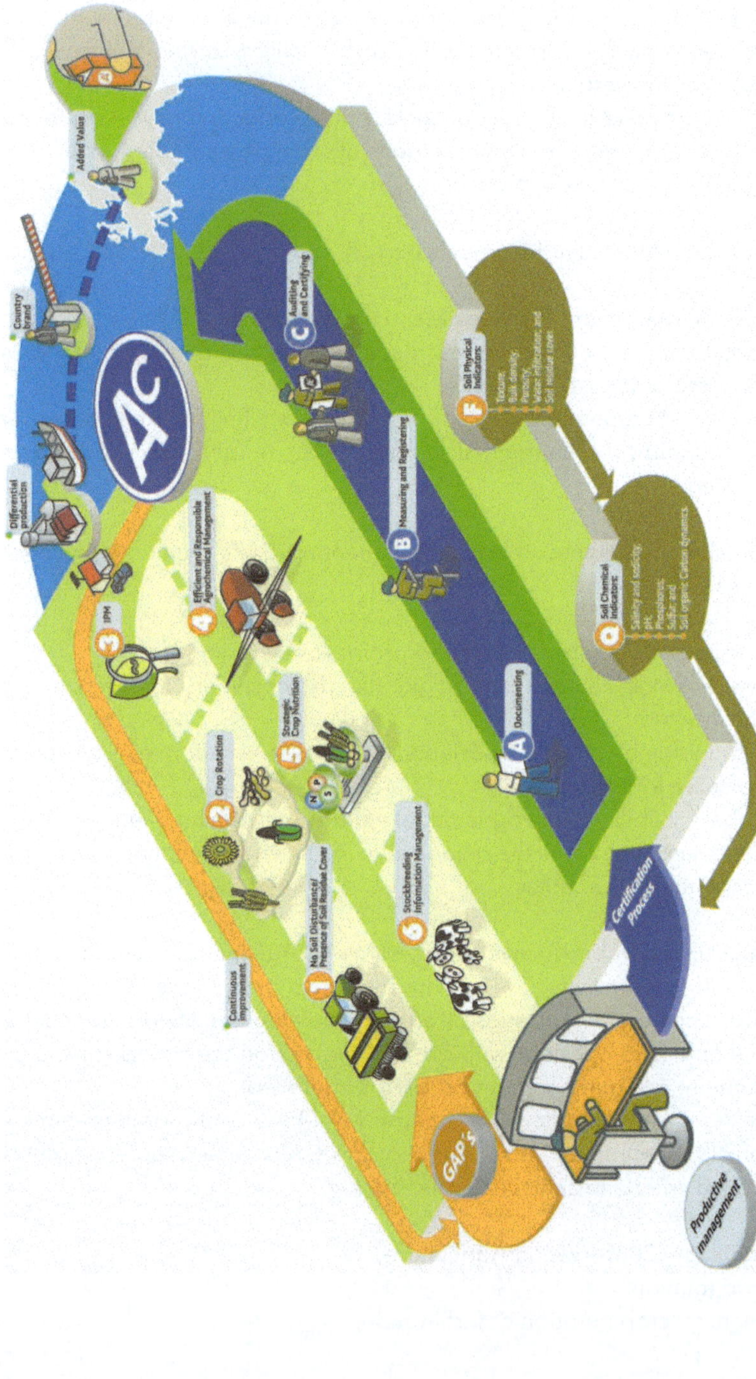

Figure 2 AC poster. Source: adapted from AC (2017a).

- Integrated pest, weed and disease management,
- Efficient and responsible use of phytosanitary products,
- Cattle management.

Contrasting RTRS, AC has preselected indicators. These indicators correlate with the GAP mentioned above. They are divided into physical, chemical, environmental and social indicators:

- Physical:
  - Soil texture
  - Density
  - Porosity
  - Crop residue cover
- Chemical:
  - Carbon content dynamics
  - pH
  - Electric conductivity
  - Phosphorus content
  - Sulphur content
- Environmental:
  - Water use efficiency
  - Fossil fuel use efficiency
  - Nutrient balance
- Social:
  - The farmer seeking certification has to choose which indicators he or she would like to measure. For example, farmers can choose indicators like gender or age percentage, home origin and number of social complaints.

However, the first question every farmer has when considering AC is: Does it have some benefits? Therefore, the following paybacks can be mentioned:

- Benefits related to a quality management system approach,
- Opportunity to generate knowledge,
- As a demand from the owner or tenant differentiation,
- Agronomic-certified history,
- Governmental benefits (tax discount),
- Input's discounts and special offers,
- Future access to certified market,
- Product differentiation,
- Differential price.

Furthermore, there are benefits that go beyond the farmer level which include:

- Leaving better soils for future generations,
- Less competition for drinkable water (because water use efficiency is improved),
- Higher water quality (lower erosion and contamination risk),
- Better atmosphere, positive impact in the climate change,
- Lower pressure on more fragile areas (by an increase in yields on better soils),
- Possibility of producing in more fragile areas without the known risks of conventional tillage.

Farmers actually have an option when it comes to certification and can ask for a joint audit for both AC and RTRS. Along with AC and RTRS, there are also further examples of certification schemes that focus on other processes or products. The following exemplify such certification schemes.

2.3 Roundtable on Sustainable Biomaterials (RSB)

The Roundtable on Sustainable Biomaterials (RSB) is an independent and global multi-stakeholder coalition which works to promote the sustainability of biomaterials. RSB's certification scheme is the strongest and most trusted of its kind. It verifies that biomaterials are ethical, sustainable and credibly-sourced. Certified entities can communicate their RSB certification to the public and shareholders to prove their sustainability to the world and strengthen their brand in sustainability issues (RSB, 2016a).

RSB bases its certification on a set of environmental, social and economic principles and criteria. They use a risk management approach, identifying the important sustainability issues for a given crop and location and focus on managing them effectively. On one hand, this ensures security and robustness and on the other, this ensures flexibility and adaptation to a wide range of products. To match the needs of operators, several 'chain of custody' options are proposed (100% segregation, mass balance etc.) (RSB, 2016b). RSB list of 12 principles and criteria is as follows:

- Legality,
- Planning and monitoring,
- Greenhouse gas emissions,
- Human and labour rights,
- Rural and social development,
- Local food security,

- Conservation,
- Soil (including CA as a must to improve soil quality),
- Water,
- Air quality,
- Technology, inputs, waste,
- Land rights.

2.4 International Sustainability and Carbon Certification (ISCC)

The International Sustainability and Carbon Certification (ISCC) is a global certification system covering the entire supply chain and various biobased feedstocks (this includes agricultural crops and their derivatives) and renewables. ISCC is a multi-stakeholder initiative governed by an association with currently more than 80 members. ISCC was developed through an open multi-stakeholder process involving approximately 250 international associations, corporations, research institutions and NGOs from all over the world in order to ensure high practicality and cost effectiveness (ISCC, 2017a). Since its beginning in 2010, more than 12500 certificates in more than 100 countries have been issued (ISCC, 2017b). Nowadays, ISCC has 2863 valid certificates (by 27 December 2016) (ISCC, 2017b).

ISCC ensures that:

- Greenhouse gas emissions are reduced,
- Biomass is not produced on land with high biodiversity and high carbon stock,
- GAPs and the protection of soil, water and air is applied (including CA),
- Human, labour and land rights are respected (ISCC, 2017a).

According to ISCC (2017a), high requirements for traceability ensure that the raw materials that are used in biofuels and other applications can be traced back to its origins. This process for traceability is called identity preserved. In addition, ISCC provides methodologies to calculate mass balances and verify greenhouse gas emissions along the supply chain. Overall, ISCC offers a 'One Stop Solution' for all crops and markets. It enables system users to access different market segments with one certificate (ISCC, 2017a). In order to meet the different requirements of each government or market, ISCC has developed annexes for each system that include: ISCC DE, ISCC DE 36th BImschV, ISCC EU, ISCC DE 36th BImSchVund, ISCC PLUS (ISCC, 2017c). Under special circumstances, National or Regional Initiatives (Technical Working Groups) can adapt the international ISCC standards to local conditions by the means of a specification of the standard (ISCC, 2017c).

The basic ISCC documents are:

- ISCC 202 (DE, EU, PLUS) Sustainability Requirements: The use of biomass can only be justified if sustainable, environmentally and socially sound production of biomass is ensured. The protection of certain areas is vital.
- ISCC 202 includes six principles and its respective criteria, not only aiming at the prevention of ecological problems but also taking care of safety and health issues of employees. These criteria are divided into 'major must' and 'minor must'. All the majors need to be complied for a successful audit, together with 60% of the minors (ISCC, 2017c).
- ISCC 203 (DE, EU, PLUS) Requirements for Traceability: Traceability does not only cover the basic requirements that products can be traced back and forth throughout the value chain from their origin to the point of final delivery but also the specification of what they are made from and how they have been processed (ISCC, 2017c).
- ISCC 204 (DE, EU, PLUS) Mass Balance Calculation Methodology: Describes the basic elements to be considered for mass balance calculation (ISCC, 2017c).
- ISCC 205 (DE; EU; PLUS) GHG Emissions Calculation Methodology: Has the goal of calculating and verifying GHG emissions along the value chain. This includes emissions from biomass production, conversion processes, warehouses and transport. GHG value is expressed in GHG emissions per ton of product (ISCC, 2017c).
- ISCC 207 (DE, EU, PLUS) Risk Management: It is based on a number of risk indicators that are continuously monitored and adjusted (ISCC, 2017c).

2.5 Better Sugarcane Initiative

The Better Sugarcane Initiative (Bonsucro) is the leading certification for sugarcane production. Bonsucro's mission is to ensure that responsible sugarcane production creates lasting value for the people, communities, businesses, economies and ecosystems in all cane-growing origins. Bonsucro is a global and multi-stakeholder NGO. It has over 450 members in 41 countries, representing more than 24% of the world's land producer in the world of sugarcane and it is a full member of the ISEAL Alliance (Bonsucro, 2016a).

The Bonsucro certification system is based on six documents:

- Certification protocol,
- Production standard (including EU),
- Chain of custody standard,
- Audit guidance for production and chain of custody standards,
- Bonsucro calculator.

Before certifying, a mill needs to be member of Bonsucro and agree to their Code of Conduct. Afterwards, Bonsucro provides training and support to achieve certification (Bonsucro, 2016b).

Membership includes access to the Bonsucro calculator, a software tool that enables self-assessment and aids mill managers to understand what corrective actions need to be implemented to ensure a successful audit against the production standard. Bonsucro also offers 'chain of custody' certification, which provides assurance for Bonsucro-certified product through its final destination (Bonsucro, 2016c).

According to Bonsucro (2016a), the intention of its certification protocol is to:

- Outline and prescribe the process and procedures of the Bonsucro certification System,
- Detail the certification requirements for organizations to demonstrate compliance to the Bonsucro certification system,
- Give criteria for the selection, qualification, monitoring and evaluation of certification bodies that wish to operate the Bonsucro standards,
- Describe the audit procedure for certification bodies to verify compliance,
- Offer useful tools and methodologies for auditors to verify compliance to individual criteria of the Bonsucro production standard and/or chain of custody standard in its appendices.

Bonsucro certification focusses on the areas shown in Table 1 (Bonsucro, 2016c):

At the moment, Bonsucro has 56 production certificates, reaching 1 052 576.08 ha and representing 4.43% of global sugarcane producing land surface. In metric tonnes it means, 64 328 493 certified tonnes of sugarcane production and 3.87% of global sugarcane production. Finally, in terms of sugar quantities, this equates to 4 571 662 tonnes of certified sugar which is 2.92% of global sugar production (Bonsucro, 2016d). These numbers are an indication of the impact of the certification scheme based on global production of sugar.

3 Other schemes

The various certification standards described in Section 2 were chosen as examples of certification schemes based on their global impact. However, in the Standards Map webpage (https://sustainabilitymap.org/login) up to 131 public and private standards for 'Agriculture Products' ranging from soybean to grapes and flower production can be found. If, for example, the list is filtered and shortened by choosing 'South America' as the producing country, 'Certification' is mandate and 'Soil Management' as the environmental key factor, 27 schemes can be found. If 'soybean' is only chosen as the product, then 19 schemes can

Table 1 Bonsucro areas of focus

Areas	Water management	Conservation	Legal compliance	Land rights	Labour rights	Salary and contracts	Health and safety	Vocational training	Efficiencies and profitability
Topics	Efficiency of water use	Biodiversity loss and high conservation value (HCV) areas	Compliance with national laws	Right to land and water usage in accordance with local law	ILO convention	Minimum wage	Drinking water		Management tool for improved performance
	Quantity of water use	Environmental impact			Minimum age of workers	Existence of contract	Safety assessments		Improved technical knowledge
	Farming yield depending on climatic zones	Environmental impact and management plan			Absence of forced or compulsory labour		First aid and emergency provision		Investment in R&D
		GHG emissions and energy usage in transport			Absence of discrimination				Use of Bonsucro calculator
		Agrochemicals: amount applied per ha per year and banned chemicals			Respect right to form and join trade unions and/or to bargain collectively				Recovery of sugar
		Impact management							Industrial efficiency
		Greenfield expansion							

Source: adapted from Bonsucro (2016c).

still be found. Although, the focus of this chapter was on four schemes, this shows how large the certification world really is.

4 Conclusion and future trends

This chapter aims to contribute to the Conservation Agriculture literature in the scientific community. Soil health or soil quality has been defined as the capacity of soil to function as a vital living ecosystem for sustaining plants, animals and humans; being the base of life (USDA – NRCS, 2017). Hence, the concept of a correct soil management as an essential tool to keep soils sustainable for future generations is not new.

Society is responsible for expressing its consciousness towards the environment and certification is a key factor for credibility. Fortunately, management practices to ensure product or process quality can be certified.

Certification schemes are a whole world by themselves. Some successful examples of certification schemes for the agricultural sector are as follows:

- Agricultura Certificada (AC) certifies the whole productive process inside a farm.
- International Sustainability and Carbon Certification (ISCC) certifies biofuel and bioliquids and its chain of custody.
- Round Table on Sustainable Biomaterials (RSB) certifies biomaterials and its chain of custody.
- Round Table on Responsible Soybean (RTRS) certifies soybean and its chain of custody.
- Better Sugarcane Initiative (Bonsucro) certifies sugar, mills and its chain of custody.

The Standards Map website allows comparison among standards. So, it is a useful guide to select the certification that best fits producer and consumer needs.

Table 2 states, as a summary, the main differences among the five selected certification schemes for this creative component.

Soils are a vital living ecosystem. Soil management is a key in future sustainability. Currently, certifications, based on Conservation Agriculture, are starting to play an important role in the farming sector. Management practices can be audited, certified and measured by those indicators. Some certifications only address the primary product, while others address the productive process and its many supply chains. All of them can be beneficial; however, which one to select will mainly depend on the target market and the producing country but without forgetting consumer's needs. In addition,

Table 2 Differences among certification schemes

Characteristics	RTRS	AC	RSB	ISCC	Bonsucro
Type of standard	General guide. Focussed on soybean.	Specific guide. Focussed on production.	General guide. Focussed on biomaterials.	General guide. Focussed on biofuels.	General guide. Focussed on sugarcane.
	Product certification.	Process certification.	Product certification	Product certification.	Product certification.
		Quality management system.			
Transparency and measuring	Flexible and self-choosing indicators.	Indicators are specified in the protocol and manual.	Flexible and self-choosing indicators.	Flexible and self-choosing indicators. ISCC GHG calculator.	Flexible and self-choosing indicators. Bonsucro calculator.
Origin	Global platform that includes different players from the whole chain.	Aapresid in consensus with scientific community.	Global platform that includes different players from the whole chain.	Global platform that includes different players from the whole chain.	Global platform that includes different players from the whole chain.
Annexes	EU RED, biodiesel, no GMO, chain of custody.	No annexes.	EU RED.	EU RED.	EU RED, chain of custody.
Scope	ILO, environmental responsibility, land use rights, responsible relationship with the local community.	ILO, environmental responsibility, land use rights, responsible relationship with the local community.	ILO, environmental responsibility, land use rights, responsible relationship with the local community.	ILO, environmental responsibility, land use rights, responsible relationship with the local community.	ILO, environmental responsibility, land use rights, responsible relationship with the local community.
Certificate period	5 years.	3 years.	1 or 2 years according to risk management.	1 year.	3 years.
	Follow-up audits in the middle years.	Follow-up audits in the middle years.	Follow-up audits in the middle years.		Follow-up audits in the middle years.
Target market	European feed and food processing companies.	Local opportunity.	European companies.	European biofuel companies.	Worldwide sugar market.
	Market recognition.	No market recognition.	Market recognition.	Market recognition.	Market recognition.

many of the requirements of certification schemes overlap, thus allowing joint certification audits to exist.

The main goal of this chapter is to help agronomists and other agricultural experts to understand the importance of Conservation Agriculture and its linkage to certification schemes which constitute a huge market opportunity for the future focussing on consumer's perception and needs.

5 References

Aapresid. (2013). *Protocolo del sistema de gestión de calidad y manual de buenas Prácticas de manejo de agricultura certificada*. Aapresid, Rosario, Santa Fe, Argentina.

Aapresid. (2017). Agricultura certificada. *Aapresid*. Available at: http://www.aapresid.org.ar/ac/ (accessed on 5 January 2017).

Albertengo, J. (2017). *Soil Health Indicators and Certification Schemes*. Iowa State University, Ames, IA.

Bonsucro. (2016a). Bonsucro production standard 4.1.1. *Bonsucro*. Available at: http://bonsucro.com/site/wp-content/uploads/2013/02/Bonsucro-Production-Standard-4.1.1.pdf (accessed on 5 January 2017).

Bonsucro. (2016b). Certification process. *Bonsucro*. Available at: http://bonsucro.com/site/certification-process/ (accessed on 5 January 2017).

Bonsucro. (2016c). Certification protocol. *Bonsucro*. Available at: http://bonsucro.com/site/certification-process/certification-protocol/ (accessed on 5 January 2017).

Bonsucro. (2016d). In numbers. *Bonsucro*. Available at: http://bonsucro.com/site/in-numbers/ (accessed on 5 January 2017).

FAO Oficina regional para América Latina y el Caribe. (2004). Las buenas prácticas agrícolas. *FAO*. Available at: http://www.fao.org/3/a-ai010s.pdf (accessed on 5 January 2017).

ISCC. (2017a). ISCC in short. *ISCC*. Available at: http://www.iscc-system.org/en/iscc-system/about-iscc/iscc-in-short/ (accessed on 5 January 2017).

ISCC. (2017b). Valid certificates. *ISCC*. Available at: http://www.iscc-system.org/en/certificate-holders/valid-certificates/ (accessed on 5 January 2017).

ISCC. (2017c). System documents. *ISCC*. Available at: http://www.iscc-system.org/en/certification-process/isccsystemdocuments/ (accessed on 5 January 2017).

ISEAL Alliance. (2016). Full members. *ISEAL*. Available at: http://www.isealalliance.org/our-members/full-members (accessed on 5 January 2017).

Papendieck, S. (2011). *Análisis comparativo de las diferentes iniciativas de estándares privados ambientales para la producción de soja sustentable y sus derivados con la normativa europea*. Sociedad Rural Argentina, Buenos Aires, Argentina.

RSB. (2016a). Certification. *RSB*. Available at: http://rsb.org/certification/about-certification/ (accessed on 15 March 2017).

RSB. (2016b). Participating operators. *RSB*. Available at: http://rsb.org/certification/participating-operators/ (accessed on 5 January 2017).

RTRS. (2014a). History. *RTRS*. Available at: http://www.responsiblesoy.org/about-rtrs/history/?lang=en (accessed on 5 January 2017).

RTRS. (2014b). Certification. *RTRS*. Available at: http://www.responsiblesoy.org/certification/production/?lang=en (accessed on 5 January 2017).

RTRS. (2014c). National interpretations. *RTRS*. Available at: http://www.responsiblesoy.org/certification/national-interpretations/?lang=en (accessed on 5 January 2017).
RTRS. (2014d). Certified volumes and producers. *RTRS*. Available at: http://www.responsiblesoy.org/mercado/volumenes-y-productores-certificados/?lang=en (accessed on 5 January 2017).
RTRS. (2016). Standard version 3.0. *RTRS*. Available at: http://www.responsiblesoy.org/wpdm-package/english-rtrs-standard-responsible-soy-production-version-3-0/?wpdmdl=11005&lang=en (accessed on 5 January 2017).
Standards Map. (2015). International Trade Centre. *Standards Map*. Available at: http://www.standardsmap.org/ (accessed on 5 January 2017).
USDA – NRCS. (2017). Healthy soil for life. *USDA – NRCS*. Available at: https://www.nrcs.usda.gov/wps/portal/nrcs/main/soils/health/ (accessed on 5 January 2017).

Chapter 12

Institutional and policy support for Conservation Agriculture uptake

Tom Goddard, Alberta Agriculture and Forestry, Canada; Gottlieb Basch, University of Evora, Portugal; Rolf Derpsh, Agricultural Consultant, Paraguay; Li Hongwen and He Jin, China Agriculture University, China; Muratbek Karabayev, International Maize and Wheat Improvement Center (CIMMYT), Mexico; Amir Kassam, University of Reading, UK; Ken Moriya, Ministry of Agriculture and Livestock, Paraguay; Roberto Peiretti, Agricultural Consultant, Argentina; and Hendrik Smith, Grain SA, South Africa

http://dx.doi.org/10.19103/AS.2019.0048.13
Published by Burleigh Dodds Science Publishing Limited, 2020.

1 Introduction

Agriculture is always under pressure to change or adapt to biophysical, social and economic conditions. Farmers are the ultimate convenors of change at the farm level in response to perceived pressures and opportunities. Those drivers may be on their farm in terms of crop, soil or livestock performance or in a top-down manner from public or private policies at regional or national and international scales.

In the first half of the twentieth century agriculture underwent a revolution based upon machinery and chemistry. The labour-saving devices of tractors, seeding and harvesting machinery allowed farms to expand their area with fewer labourers. The chemical advancements in fertilizer and herbicides allowed monoculture and yield increases along with higher yielding and disease-resistant varieties of crops. By the mid-twentieth century the assemblages of technology and science helped farmers focus on maximum production and sales or exports. Global food production more than doubled from 1960 to 2000; in China it increased five-fold in the 1980s–90s (Pretty and Bharucha, 2014). The increased production, however, was not without costs. Ramankutty et al. (2018) acknowledge the success of the Green Revolution in increasing production and the need for more areas under production to meet future food security but in doing so we will negatively impact water resources and cause soil degradation, biodiversity loss and increased greenhouse gas emissions.

Erosion and land degradation have occurred for millennia, but the intensive industrialization of farming in the twentieth century brought loss and degradation to the forefront. The conventional farming systems that established themselves also increased erosion rates and reduced the sustainability of agriculture (Montgomery, 2007; Kassam et al., 2009). Borelli et al. (2017) model current water erosion risk using the rather limited RUSLE model and found global erosion rates less than previous predictions but significant (36 Pg yr^{-1}). They single out Conservation Agriculture (CA) as the practise most adopted around the world (15% of cropland) and estimate global erosion rates have decreased 7% because of it, with remarkably larger erosion reductions in some South and North American countries (up to –33%).

International and national stimulus for CA since the mid-1990s was spearheaded by the Food and Agriculture Organization (FAO) (see Chapter 2, Volume 1) and was later joined by the World Bank and other international agencies. In addition to pioneers and champions who promoted no-till-based soil and water conservation (Kassam et al., 2014a), FAO helped identify pioneer farmers and provided support for international agreements especially in developing economies. FAO policy was to officially support CA with a website, actions in the field and through literature. FAO sponsored the first World Congress of Conservation Agriculture in Madrid in 2001, which brought farmers, specialists and policy makers together for the first time. Subsequent literature has reviewed components of CA such as No-Till implementation around the world (Goddard et al., 2008; Kassam et al., 2009, 2014b), CA prospects and challenges (Jat et al., 2014, including a foreword by the FAO Director General) and CA systems for specific regions or farm types such as smallholders in Africa (Kassam et al., 2016). FAO (2011) continued the application of the CA system in documenting the potential for sustainable intensification with the Save and Grow book series. International bodies like FAO are important to counter the entrenched policies of various countries that can be poorly developed or only inward looking and subject to many biases. The many chapters in the previously mentioned books (and others) illustrate how different societies deploy new policies and institutional support for scaling up CA.

Volume 2 of this collection is dedicated to elaborating the multiple benefits of CA systems around the world. These include the reduction in degradation of natural resources and delivering a more sustainable farming system. Consequently, it is logical and in the public interest for governments to develop policies to encourage mainstreaming of CA systems. The private benefits to the farmer have resulted in significant bottom-up drivers that has realized more than 180 M hectares of CA adoption around the world, albeit unevenly (Kassam et al., 2015, 2019). CA delivers many co-benefits to the famer such as mitigating erosion risk, drought tolerance, enhanced moisture balance, nutrient cycling, and equipment and fuel savings. Ecosystem services are enhanced for public benefit so the impetus is before governments to examine existing policies, recognize the shift in farming systems that CA represents, and move policies to enable change and address public interests.

Much of the policies in place around the world are a hang-over from the Green Revolution era of the last half of the twentieth century. Nations have always had trade and commerce as priority policy areas, and the increased production of agriculture since World War II has resulted in many policies that aligned with maximum production and international commodity trading. The farm production support for mainstream commodities and the counter-cyclical price support mechanisms have encouraged intensive monoculture production and buffered farmers from the realities of markets, thus diminishing their capacity to handle risks. Governments are awakening to the long-term unintended consequences, but it takes a lot of effort to quickly change the direction of large ships. This long process of change is underway in a few jurisdictions and the USA is undergoing a review of government support for crop insurance and the EU has started a directional change of CAP towards a greening of agriculture. After generations of protection of farms, production capacity and buffering from economic and trade realities, beliefs, biases and preferences are aligned with conventional farming systems and it will not be easy to change the thinking of the masses even though the imperative is facing society. It is helpful to look at the issues and drivers behind the imperative from different perspectives or lenses in order to see the comprehensive opportunities for policy and institutional changes.

2 Research institutions

Research institutions provide education, foster innovation, bring various players together to create knowledge and solve real world problems. They can exist as independent or government-supported universities, think tanks, experimental farms, government ministries or agencies. In nearly all cases of the emergence of CA, research institutions have been involved. In some cases, they have been more distant with a lack of interest and linkages have been created by solitary or few innovative scientists that could see the potential of CA systems.

Research stations are usually placed on high-quality, uniform and well-drained soils that represent a very small portion of farms or fields that the results are directed towards. Conventional research designs of small plots laid out in replicated blocks only account for a small amount of variance within the narrow climate-site specifics. Plot-scale No-Till equipment has been hard to source until more recently and many researchers had to modify existing or build experimental prototypes. Agronomic conditions were worse than those found at scale on CA farms yet; researchers had to publish and they did so regardless of experimental conditions. Currently, most crop breeding is carried out on conventional tillage small-plot systems which become the artefacts of ambient conditions when selecting for performance traits. That needs to be considered when evaluating varietal performance under CA conditions. This needs to be also remembered when reading literature of meta-analysis of historical journal publications. Many of the factors that impact CA systems are unaccounted for or sub-optimal; thus, meaningful results have a greater influence of chance (Derpsch et al., 2014). Unfortunately, innovative researchers that were interested in CA and willing to take career or funding risks were faced with the realities of the very limited ability of research stations and perhaps, resistance to change, or the variability of farm conditions and cycles of quick learnings and adaptations ('fast to fail' research concepts of industry). Mainstream researchers like resistant farmers were impacted by anchor and confirmation biases that prevented change from happening quickly (Awada et al., 2014; Lindwall and Sonntag, 2010). They were anchored to their conventional or historic practices and assigned more weight to results from those conditions rather than the newer, more suspicious practices.

Universities and large research establishments have the appeal of a collective of many disciplines that could work cooperatively on complex problems. That opportunity is countered by the persistence of the silo or stovepipe structure and resulting mentality (e.g. individuals 'own' labs not the collective). Working across departments and disciplines is more difficult in academic institutions that it is in the corporate world or at the field level; thus, we see a few academics involved in CA research but often not leading the agenda and as some have observed, follow along with confirming research results. Institutional funding bodies do not help with their propensity for low-funding levels and short project duration. Small, specific and short-term projects are often easier to develop, are lower cost and easier to fund (less funding at risk by the funders) as opposed to the integrative research over longer periods or multiple locations, which is helpful to CA systems. Regardless of these constraints, the small, focussed projects can solve some of the specific questions within CA. We then have to hope the assemblages of outcomes from many small projects will accurately present the answers for CA systems.

Educational or research institutions that have linkages to industry either through funding, collaboration or extension will have greater capacity to address CA issues or complex issues from other industries. An example out of the USA is the 'land grant' universities (expanded mission under the Hatch Act, 1897) that were established to provide three functions - education, research and extension, especially in the newer western states as they became settled with immigration. Today the explicit responsibility (accountability) of extension keeps those schools well connected to rural development and the agriculture industry needs. Another example from Australia is the Grains Research Development Corporation (GRDC) whose purpose is for 'Investing in research, development and extension to create enduring profitability for Australian grain growers.' It was created in 1990 and is funded by both government and a sales levy on grains and has a board of directors with significant farmer and farm industry representation. The GDRC is by design sensitive to expediently solving real-world farm problems. Both the US and Australian institutions undergo regular external reviews to ensure they are efficiently meeting operational and strategic objectives.

3 Non-government and private sector

We often focus on government as a developer and holder of policies that impact agriculture and the public. It is true that much of the public policy such as infrastructure (roads, irrigation), trade and environmental and institutional policies are developed and maintained by governments who are responsible for economic development as well as the greater public good. Increasingly though the roles of industry and non-government organizations are becoming influential and recognized as playing an important part of policy development for the agriculture sector. We have seen in the country profiles in this chapter that partnerships with industry and NGOs (with or without government involvement) have been key in overcoming barriers and moving CA forward.

Companies along the agriculture-food supply chains have undergone mergers and consolidations over recent decades. Currently there are about four large conglomerates dominating the crop input and seed industry, which have raised concerns of anti-trust and competition as concentration occurs although more recent policy analysis has found a number of other factors now playing significant roles to maintain competition in the market (MacDonald, 2017). Fuglie et al. (2011) looked at the global industries in agricultural inputs, food processing and biofuels. where they found the private expenditures on research and development were nearly US$20 billion, which accounted for about half of the total private and public research funding annual budgets. Large firms invested proportionally more in R&D. For CA to develop and adapt to local specific conditions, the support of both private and public institutions are important facilitators. Clapp and Scrinis (2017) noted that large industries have used their size to influence consumer behaviour and thinking along food supply chains and thus influence public policy in the agriculture and food sectors. Some are suspicious of anything large or powerful whether it can be countries, companies or farms and that they are only after their own short-term interests. However, large firms need to have a long-term strategy for growth and resilience, which requires knowing a wide market and investments in R&D and innovation. Clapp (2018) notes the interlinking of large agriculture industries into the farming sector and the associated faults involving environmental concerns with the size and interlayering of large industries and 'industrial farms'. Private and public institutions have both fostered public concerns with sustainable use of natural resources. If companies want to protect resources and seek new, innovative production systems, they only need to tap into the leading CA farmers or farmer organizations. Synergistic relationships result as early experiences have shown.

NGOs are also a growing influence in institutions and policy development, both public and private policies. CA farmer NGOs play in this field as well as other pro-farm, anti-farm, food, environmental, wildlife and even academic organizations as they lobby or conduct what might be termed as activist workplans in some cases. Of course, they range in size from a few people to more than ten thousand with annual budgets from zero to hundreds of millions of dollars. They also range in foundations that are strongly evidence-based to speculations and beliefs. The Internet has been a catalyst to NGOs from both good and bad perspectives. It allows people to connect themselves across great distances for common interests. It has allowed a greater reach for both lobbying activities and funding campaigns. It has also allowed extremists to assemble to critical mass levels and given loud voices to those that would not have had that in the past. Many NGOs rely on philanthropic funding and do not have much or any research capacity other than have staff or volunteers scan Internet sources and publish or re-publish short articles for the general public. Society is at a very interesting transition or evolutionary stage of how we connect and distribute and consume information.

The World Wildlife Fund is one NGO that has been involved in influencing agricultural sustainability for decades. They have been initiators or supporters of all the global round tables for sustainable commodity production since the establishment of the Roundtable for Sustainable Palm Oil (RSPO) with an exploratory meeting in 2001 between various NGOs and corporations. Utilizing change theory, they identified key players to work with in order to change the masses such as large companies in the production–food supply chains as well as the continued development of various round tables. The round tables have all developed principles that are 'pull' policies of voluntary standards. Those round table principles align well with CA. Perhaps one could say that early CA developments influenced boards and in some cases CA practitioners or that private sector CA partners are board or table members. Another example of pull policy from the voluntary standards category are the various sustainable certification systems for agriculture and forestry producers. These are applied around the world and are structured for continuous improvements. Although the certification systems were born in boardrooms, they undergo a reality check through farmers that adjust them to meet their needs, abilities and pragmatism at the field level.

Industry and NGOs operate within and influence both public and private policies. Governments' favourite policy tool is from the regulatory toolbox. Regulations are good for creating a backstop or defined point from which we want to stay on the good side usually for the sake of public interest or safety. Examples are policies for food safety, speed limits and interest rates. Industry will utilize regulatory policy (e.g. human safety) but also utilize voluntary standards (e.g. standard operating procedures, certifications). The population of farmers, like industry, operates with several types: ones that governments dictate, ones that buyers or the market dictates and ones the farmer or farm organization may prefer. The regulatory backstop usually sets an acceptable minimum level (most farmers exceed this level), while voluntary standards are often pull policies that acknowledge best practices and encourage the bulk of farmers to move in that direction (Fig. 1). What is the role of governments? For governments to have effective and efficient policy, they need to recognize the suite of policies and establish a framework and strategy to drive change at several points with all stakeholders and partners.

Agriculture policies tend to be a mixture or blend of private and public policy and the amount of one or the other will be dependent upon the country, level or type of policy and the amount of integration or intersection that exists between the private and public entities. When working with

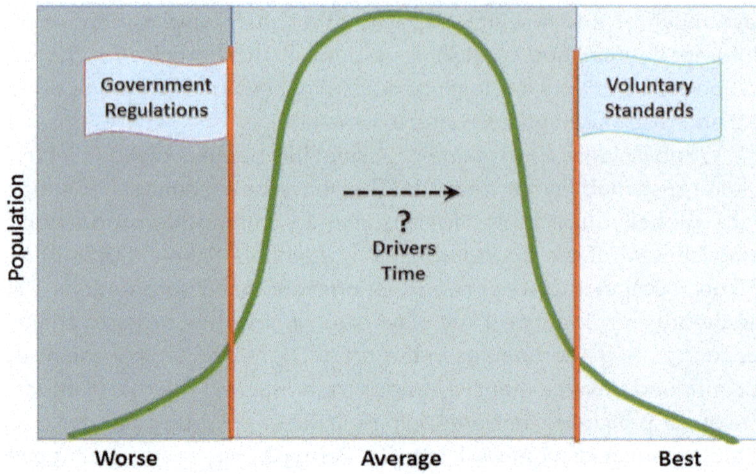

Figure 1 Public and private policy can both be used for a shift change. Source: adapted from WWF.

governments, industry or NGOs, we must be aware of the contexts in which they develop policy and the sometimes-opposite characteristics they employ (Fig. 2). Governments often go through a long-structured development process with one public consultation process up front. Industry can develop policy in a fraction of the time and may have ongoing cycles of consultations to adapt policy or anticipate additional needs. Industry can be nimbler and

Public Policy	Private Policy
• Long development process	• Short development process
• Structured public consultation	• Informal consultation (unseen, private) as well as public
• All interests encompassed	• Can be targeted/specific interests
• Cautious, conservative	• Radical, risky, pushing the envelope
• Forcing collaboration	• Collaborative willingness (pre-competitive)
• Public client	• Public client
• Long approval process	• Short approval process
• Large governance	• Small (CEO or group) governance
• Handles slowly	• Handles flash points quickly
• Plentiful resources	• More limited resources

Figure 2 Some characteristics at the ends of the public – private policy continuum.

more dynamic in policy development as they are aware, like farmers, that time costs money. They will find the optimal mix of evidence and structure, then move forward, ready to re-evaluate and adapt. Agriculture does need the assurance of government regulations as a backstop and governments need to be involved for the interests of the public and natural resources (public property). CA organizations need to work with all parties and recognition of the characteristics and needs of the various players will provide robust policies.

Co-authors from seven countries around the world were asked to provide an overview that included early movements of CA; barriers and how they were overcome; and synergisms and integration factors. A synthesis will follow their individual stories.

4 Conservation Agriculture (CA) development in Canada

The northern extent of the Great Plains of North America constitutes the prairie region of Canada and comprises nearly 90% of the farmland in the country. Similar to the USA, this region was intensively cultivated in the 1900s and experienced soil degradation since the 1930s. The national government created a special agency (Prairie Farm Rehabilitation Administration) devoted to restoration and protection of the 'breadbasket' of Canada. However, intensive cultivation, summer fallow and monocropping allowed degradation issues to persist. Thus, the concerns of land degradation and the hunt for better farming practises continued and started a shift to No-Till and CA in the late 1970s. Now 60% of cropped land in Canada is done with No-Till, with higher adoption rates on the prairies (Statistics Canada, 2016).

Innovative farmers in the 1970s learned of No-Till planters and imported some from the USA and UK for on-farm trialling. Luckily there were some champions in a few research institutions that stepped away from conventional thinking and supported, collaborated and experimented with farmers from the early days onwards. Farmers self-organized into groups at the provincial level and in one case a Canada–USA farmer organization, the Man-Dak No-Till Association between Manitoba, Canada and North Dakota, USA. The organizations produced newsletters, conferences, field days, and tours to research stations and other provinces (Lafond et al., 2014).

Farmers were not satisfied with the lack of suitable equipment and the large machinery companies ranged from uninterested to hostile to the paradigm shift and anti-tillage sentiments. The innovative pioneers ultimately invested in start-up companies that produced No-Till equipment. Those became very successful and were ultimately bought up by the larger machinery companies as they realized the market opportunities and needed to catch up on knowledge and expertise (Lindwall and Sonntag, 2010).

Researchers early in their careers were the ones that linked up with farmers to develop synergisms to extend and speed up their research with limited support. They realized that the conventional agronomic research results were not applicable to CA systems and gradually moved the research institutions to adopt No-Till practises at plot level so that everything from breeding to fertility and pathogen research was applicable to CA.

4.1 Barriers and hurdles

- The momentum of the Green Revolution that was focussed on maximum commodity production under intensive tillage monoculture.
- Mindset of farmers – both cognitive and anchor biases dominated as well as fear of ridicule by neighbours (Awada et al., 2014).
- High costs of equipment and costs of experimentation (crop failures or weed issues).
- The invalidity of historic research results from intensive tillage agronomic systems.
- The reduction of extension services from government agencies across the prairie region.

4.2 Catalysts

- Farmer leadership in self organization and influencing others.
- Government support for farmer organizations, field days, equipment grants or sharing agreements.
- An early Senate report underlining the soil conservation crisis (Sparrow, 1984).
- Researcher collaborations, extension materials and reviews of CA systems benefits (Baig and Gamache, 2011).
- Internet-based directory of CA farmers with their farm and equipment characteristics so new farmers could easily find coaching for their circumstances. A CA farmer match-making service!

4.3 Integration and synthesis

- Enabled a soil carbon sequestration protocol for the Alberta greenhouse gas offset system to show greenhouse gas mitigation from the agriculture sector (see Chapter 13, Volume 2 and Reicosky et al., 2012).
- Federal research stations have moved to No-Till as a basis for all plot research. Some universities and colleges as well.
- Federal-provincial five-year development agreements are funding CA and associated activities.

- Life Cycle Analysis and Environmental Footprints of agriculture commodities are based upon CA principles as a baseline and align well with international markets and conventions (e.g. Sustainable Development Goals).

5 Conservation Agriculture (CA) development in Kazakhstan

Kazakhstan is characterized by a rich diversity of climatic conditions and soil types. The existing production systems for grain crops are typified by:

- rainfed agriculture dominated by spring grain crops (northern and central parts of the country);
- rainfed agriculture dominated by winter grain crops (southern and south-eastern parts of the region);
- irrigated agriculture (mostly in the southern parts of the region).

The climate of Kazakhstan is distinctly continental, and a large part of the territory is affected by wind and water erosion (improper irrigation). Concerns of drought and salinity are growing. Soil fertility has decreased dramatically throughout the region associated with a loss of soil carbon and, weed infestations of cereal fields have increased. Water deficiency and insufficient soil moisture levels remain among the major problems at a national level. It has become increasingly clear that under conventional systems, improvements in plant production should be achieved through a production system based upon the conservation and sustainable use of water, soil, energy and other natural resources in the context of the whole environment. Today, such a system is the key to survival of Kazakhstan's farmers, especially those producing wheat, which is the major export food commodity.

The world experience has shown that traditional farming systems and their focus on high productivity has led to soil degradation and reduced input use efficiency. Negative components of the traditional farming systems are intensive tillage, returning little organic matter to the land and monoculture. Essentially, CA is the further development and improvement of conventional systems by including all of the other principles of sound soil and crop management, but avoiding the negative components. However, we must recognize that even with the apparent simplicity of a focus on the three principles, CA remains a complex technology and implies changes in a number of technological components of the existing traditional systems. It is necessary to change two basic paradigms: the paradigm of soil tillage and the paradigm of linear knowledge flow.

Many agricultural research and extension systems are based upon a linear model of knowledge flow, with new knowledge being developed in research organizations, passed on to agricultural extension agents who in turn pass on the new knowledge and information to farmers. While this model may be applied to simple technologies, it does not always effectively work with complex technologies, especially when research institutes do not have the capacity to develop a systems approach to functional packages of multiple technological components for all farmer situations. Development and adoption of innovation systems (platforms) are based upon collaborative networks of multiple agents, including farmers-innovators and decision-makers, all utilizing their own knowledge, external information and policy support to help overcome problems and develop functional systems for local farming conditions and farmer circumstances. Successful adaptation and promotion of CA in Kazakhstan can be considered as an example of this approach.

Adoption of CA technologies in Kazakhstan was initiated in 2000 and comprised zero/minimum soil tillage, leaving crop residues in the fields, direct seeding with narrow chisel and disk openers, permanent bed-planting and furrow irrigation, etc. Thanks to the combined efforts of scientists and farmers, international organizations (CIMMYT, FAO, ICARDA, World Bank, etc.) and support by the state and government bodies, the areas under No-Till have been increasing from virtually zero to an estimated area of 2.9 M ha in 2018 (Fig. 3). The utilization of CA-based technologies has become an official state policy in agriculture in Kazakhstan. Since 2008, the government of Kazakhstan has been subsidizing farmers who are adopting CA-based technologies. The result is that Kazakhstan is now included among the top ten countries with the largest areas under No-Tillage in the world.

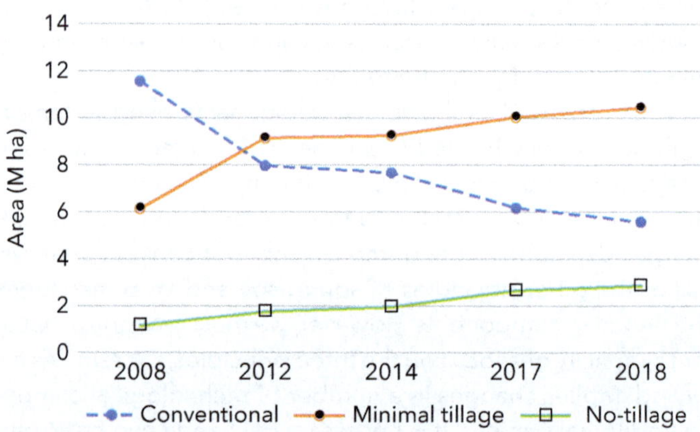

Figure 3 Growth in areas under different management in Kazakhstan. *Conventional*: Multiple tillage trips with blades and sweeps (mostly monocropping). *Minimum*: Direct seeding with V-shaped openers or with narrow chisels regularly combined with shallow harrowing/cultivation (little crop diversification). *No-till*: Direct seeding with narrow chisels or double discs openers (crop diversification).

CIMMYT, FAO and the World Bank experts analysed the current state of CA adoption and wheat production in Kazakhstan and made the following conclusions:

- Yield of spring wheat under No-Till, in average, 48% to 58% higher in comparison with the conventional technologies. The advantages of CA/ No-Till technologies are especially evident in years of drought.
- In 2012, Kazakhstan was ranked first in Europe and the Central Asia region and 9th in the world for No-Till adoption.
- Example of CA advantage: In 2012, the driest year for the last decade, wheat production was estimated at 11.0 million tonnes. Wheat No-Till (80% of the No-Till area) produced an estimated 1.8 million tonnes of wheat. The incremental wheat yield advantage from this No-Till area is estimated at 0.72 million tonnes, equivalent to around 200 million dollars.
- Increased income and food security during the last 3 years:
  - An estimated 580 million dollars incremental income;
  - Satisfied cereal requirements of about 5 million people annually.
- Climate change mitigation: Kazakhstan contributes to the annual sequestration of about 1.3 million tonnes of CO_2e.
- The CA technologies adopted fully align with the concepts and requirements for innovative technologies and systems.

Despite the progress of Conservation Agriculture in the country, it is important to further concentrate joint efforts to address the following needs in the field of agricultural science and production:

- Weed control (infestation management) remains a major issue, which needs to be addressed. It is important to develop efficient methods to combat weeds, especially for the early stages of introduction of zero tillage technologies. Selection of herbicide types and identification of efficient combinations, concentrations and application schedules for the particular weed species present are the focus of this research. The research should provide recommendations on the most efficient and cost-effective measures to combat weeds under zero tillage. Experience shows that ultimately zero tillage helps reducing the weed infestation level significantly; however, in the transition period weed control is a serious issue.
- Development and introduction of economically viable crop rotations and diversification of production are important pre-requisites for successful implementation of CA in Kazakhstan. Besides obvious economic advantages, the introduction of diversified crop rotations under zero tillage technologies will significantly reduce weed infestation levels and plant

vulnerability to diseases as well as improve soil fertility. Newly introduced crops should also receive attention in processing and marketing.

- There is a need for more thorough research on the application of mineral fertilizers and the balance and dynamics of soil nutrients under zero tillage and crop rotation technologies. It is important to remember that application of fertilizers is important during the transition period not only to ensure high grain yields, but also to produce sufficient biomass for soil cover, which is an important principle of CA.
- An important component of the strategy for promotion of zero tillage technologies for soil conservation is the building of scientific and technical capacity: teaching new technologies and agricultural methodologies to specialists and farmers, conducting various training courses and programmes for scientists, drawing from international experience, providing consulting services and field days led by highly qualified specialists and building public awareness of the appropriate agricultural technologies.

At present, Kazakhstan is considered to be one of the world's most important regions for global food security with its rich land resources, good research capacity and well-developed economic infrastructure. As per official data, by 2025, the global level of cereal production should reach 3 billion tonnes in order to feed the 8 billion people. In order to achieve that goal, wheat, the most important food crop should achieve an annual increase in production of 2% (versus the current annual increase of 1.3%). The increase is followed with the threat of the growing influence of negative factors such as constrained water supply, drought, temperature increase, land degradation, emergence of new dangerous diseases, increasing use of plant products for biofuels and livestock production. Successfully overcoming these factors and achieving sustainable growth of agriculture in Kazakhstan, and the world, will depend on new technologies and development of research and science.

6 Conservation Agriculture (CA) adoption in Europe

In Europe, the first steps towards CA in the form of No-Till were driven by the attempt to reduce production costs associated with machinery, fuel, time and labour. Despite intensive research on the different aspects of conservation tillage after the invention of Paraquat in 1955 and its commercial release in 1961, No-Tillage and even reduced tillage did only receive minor attention until the end of the last century. Throughout Europe, the history of CA varies considerably from country to country. In the UK, the first very positive results made the area under No-Till grow to almost 300000 ha in the early 1980s. However, the straw burn ban caused farmers to abandon this technique due to increasing problems of weed control and volunteer cereals (Christian, 1994).

On the contrary, the first experiences carried out by Bakermans and de Wit in the Netherlands were far from being successful, which made Van Ouwerkerk and Perdok (1994) conclude that no-tillage was not feasible in Dutch arable farming. Already in the early 1980s, very positive results were reported from Spain and Portugal where both water scarcity during spring and summer and the risks of severe water erosion during winter encouraged the use of soil conservation measures.

Today, the area under CA in Europe amounts to 3.1 M ha, which according to Kassam et al. (2019) correspond to around 4.3% of the annual ('arable') cropland area. The countries with the biggest areas under CA in Europe are Spain, Romania, UK, France, Italy, Finland and Germany with 900, 584, 362, 300, 284, 200 and 146 kha, respectively. Surprisingly, there is no uniform evolution of CA within Europe. However, some countries are showing a rapid increase in the CA area over the recent years (e.g. UK increased from 25 kha in 2008/09 to 362 kha in 2015/16), most countries are stagnating, while others lost area under CA (e.g. Germany decreased from 354 kha in 2008/09 to 146 kha in 2015/16). The reasons for a smaller or greater uptake of CA in the different regions of Europe are manifold ranging from financial incentives, timeliness of planting, natural conditions and the smaller or greater need to cut production costs, to the availability of institutional support, especially research and dissemination, etc. Another important driver that can be considered is the establishment of nuclei of successful CA farmer or farmers' groups in several regions throughout Europe. Several of the national associations of the European Conservation Agriculture Federation (ECAF) succeeded in establishing and supporting such farmers or groups. A recently established yearly event by a farmer's family in the UK (Groundswell No-Till Show) may explain the renaissance and tremendous spread of CA in the UK over the past few years.

Less well documented is the area of perennial crops where the principles of CA, based on the permanent cover of the inter-row space, are applied. In Europe, this is of special relevance as around two-thirds of the perennial cropland are under olives and grapes, whose cultivation is often on hilly and rolling landscapes where soil conservation measures would be crucial. Only in Spain this area covered with olive orchards mainly amounted to 1.3 M ha in 2018.

While in many countries the initial adoption of CA on a small area was driven by research institutions, in Denmark in the 1960s and 1970s and in Finland at the very end of the last century, the initial adoption was clearly farmer-driven. In Denmark, however, the No-Till practice was replaced by reduced or minimum tillage before seeding, whereas the spread of No-Till among farmers in Finland was rather fast, reaching 13% of the total area of cereals and oilseed crops by 2008 (Soane et al., 2012).

Another more important player in the orientation of agriculture and farming practices has become the European Common Agricultural Policy (CAP). Until the end of the last century, the income support to the agricultural sector was clearly production oriented favouring high productivity levels, obtained with massive external inputs, instead of promoting competitiveness and sustainability. Despite a shift from a production-oriented income support to an 'area-based' support through so-called direct payments since the dawn of this century, there continues to be an enormous welfare transfer from consumers and taxpayers to the farming sector, justified by the high standards (environmental and animal welfare) European farmers have to comply with. As a matter of fact, these standards apparently have been increased over the programming periods, from Statutory Management Requirements (SMRs), Good Agricultural and Environmental Condition (GAECs), Cross-Compliance to the Greening Measures that will transform into the new 'Enhanced Conditionality' in the next CAP. Although mandatory for farmers, the EU member states (MS) have a high discretionary power to interpret and implement these standards and measures in order to satisfy the EU requirements for climate action and the preservation of water, soil, biodiversity and landscapes. However, the CAP created possible synergies between EU and national policy goals by incentivising MS to propose and implement specific measures for soil protection within the Rural Development Programmes, also known as the 2nd Pilar of CAP, which could be defined but had also to be co-financed by each MS. Therefore, concrete policies to promote CA were implemented in those MS that included CA in their specific agri-environmental measure schemes to achieve the goals of soil conservation (e.g. Spain, Switzerland, Italy, Germany, and Portugal). Even in these MS the respective policies and financial support schemes vary greatly between regions within the MS.

While financial incentives in whatever form were capable to enhance the initial adoption of CA practices, the continuity of CA after reducing or even ceasing the incentives was only observed where the benefits were perceived as a real added value when compared to the situation before. In addition, the level of support will influence farmer's level of willingness to break with the conventional, well-known system and run the risk of trying a new, more knowledge demanding system. In Portugal, for example, a considerable uptake of CA could be observed in the middle of the last decade, reaching around 8% of annual cropland. After the change of the government and the end of the support for CA, the area under CA dropped drastically and has stabilized today at around 4% of the cropland. This clearly shows that financial support alone cannot sustain initial uptake. Without further support to address the manifold challenges of a system change, the adoption of CA in Europe will continue facing strong restrictions. The additional support necessary to harness the full potential of CA would need to come from institutions (research, extension, education, training), from the private sector (equipment and other technical solutions, seeds, etc.), from policy (recognition in whatever form of the benefits of CA) and through the public awareness of the unsustainability of conventional farming.

Adoption of CA throughout many regions in the world has been driven mainly by economic reasons or severe soil degradation processes. Despite existing and identified major soil threats (EU – Soil Thematic Strategy) (e.g. soil loss through water erosion in Europe has been reported to amount yearly up to 970 Mt (Panagos et al., 2015)), soil degradation and the need to improve soil health and quality in Europe appear to be a more important subject for research rather than a major concern for the public in general and often not even for the farming community. However, stagnating cereal yields in Europe (Brisson et al., 2010) although attributed by some to the effects of climate change may indicate that soils are no longer sustaining yield increases that could be expected due to improved inputs and equipment. In addition, the constant support to farms in the form of direct payments reduces the economic pressure for change, thus maintaining the arable farming sector within acceptable economic thresholds (comfort zone).

Another important barrier for the adoption of CA in Europe, much more in some countries than in others, is the 'glyphosate' issue. Public opinion, driven by emotion and misinformation, and analytical capacities able to detect 'pico' levels of residues of this herbicide almost everywhere, have put enormous pressure on farmers, above all, CA farmers who supposedly not only depend on this chemical but use it in much higher quantities, which does not stand up to reality. The potential ban of glyphosate or restrictions for its use will certainly have an impact on the further adoption of CA or even the continuity of areas already under CA.

Since the launch of the EU Soil Thematic Strategy in 2002, awareness about the importance of healthy soils for the delivery of almost all ecosystem services rose steadily. European research funds for soil-related topics and sustainable soil management grew exponentially in the EU Framework programmes 6 & 7, and lately under Horizon 2020. The more the EU Commission, rather than the EU politicians, understands the potential contribution of CA to soil conservation, to mitigate climate change and to help adapt to its effects, the greater recognition might be accorded to CA as a suitable agri-environment measure deserving the CAP support. Yet, this recognition is somehow overruled by agro-ecological concerns that appear to be much bigger regarding CA when compared to conventional farming.

6.1 Factors influencing CA adoption in Europe

1 CA has been driven by both top-down and bottom-up policies or movements. There are both top-down initiatives through universities

or mid-level polytechnical schools as well as farmer driven bottom–up pushes such as the Groundswell initiative or the various national farmer CA organizations.

2 Research and extension help. Where research in CA has taken place, adoption levels are higher. Extension services are nearly non-existent in Europe and farmer organizations or private sector must strive to acquire and disseminate knowledge and skills (for any new farm practises).
3 Service and machinery providers specializing in CA are rare. This is a barrier for farmers seeking equipment options or customizing CA to local, specific conditions. Service providers earn more with conventional, high input agriculture. Private sector and non-government organizations remain distant due to fears of pesticide use or other perceived dangers of the unknown. This is expressed to a different degree depending upon the specific country.
4 Farmer organizations needed an incubation or gestation period to self-organize and move forward. Today because of the greater ease at connectivity and transfer of information, more organizations are appearing. As more appear, they support one another.

7 Conservation Agriculture (CA) development in China

In China, CA has been extended and adopted due to its positive effects on soil, the agricultural environment. The development of CA is mainly divided into four stages:

7.1 The first stage: initiation (1960s–1990)

In the 1960s, the introduction, experimentation and demonstration of single technology of CA began to develop in China, for instance, No-Tillage, subsoiling, straw cover, etc. However, due to limited technology, machinery, social and economic development, these single technologies were only used for small-scale demonstration and experiments in some areas. From the 1960s to the early 1970s, wheat was seeded without tillage in Heilongjiang and Jiangsu province. In 1980s, research on dryland agricultural farming system had a development tendency to crop residue cover and minimal tillage. However, due to the lack of supporting equipment for conservation tillage, especially No-Tillage seeders, made it difficult for large-scale CA extension in China.

7.2 The second stage: systematic research (1991-2001)

In 1991, the current China Agriculture University started systematic experimental CA research with combinations of machinery and agronomy. Wheat and maize

Published by Burleigh Dodds Science Publishing Limited, 2020.

were research targets, at Linfen and Shouyang, respectively, which are one-year cropping areas, and a long-term experimental observation site was established on the Shanxi Loess Plateau. The agricultural machinery bureau in Shanxi province took the lead in the demonstration and production of CA in China. The China Agriculture University cooperated with the Agricultural Machinery Bureau in Shanxi province, University of Queensland, Chinese Academy of Agriculture Sciences, Agricultural Machinery Bureau in Hebei province and Agricultural Mechanization Promotion Station in Liaoning province. Eleven experimental sites for CA were successively established in Shanxi, Hebei, Gansu and other provinces or province-level municipalities. In May 1999, the Conservation Tillage Research Center of the Ministry of Agriculture was formally established. Research results at this stage proved the adaptability of CA in China, and a series of small- and medium-sized conservation tillage machines were developed.

7.3 The third stage: demonstration, extension and key research (2002–2008)

In 2002, The Ministry of Agriculture held the First National Conservation Tillage Site Meeting in Shanxi that reviewed research completed to date. The development of CA turned to key research areas and the scope of demonstration and extension was expanded. In 2005, the government's 2005 No 1 Document first underlined 'Reform traditional tillage and develop conservation tillage'. Furthermore, the CA Expert Group of the Ministry of Agriculture was established. In 2007, the Ministry of Agriculture issued opinions on striving to develop CA. From 2002 to the end of 2008, the national and local governments had continuously issued relevant documents to require the application and extension of CA. The area of CA grew from more than 6.667×10^4 ha to more than 2.667×10^6 ha.

7.4 The fourth stage: rapid development (2009–present)

In 2009, the State Council issued a Conservation Tillage Project Construction Plan (2009–2015) to strengthen research and extension of CA. Since 2014, subsoiling has been subsidized by the central government. In China, the technical support system for the popularization and extension of CA has been gradually strengthened, and the CA technical route suitable for different regions has been formed, including areas harvested twice a year (North China), areas harvested once a year (Loess Plateau), northwest frigid wind-blown sand areas, west oasis farm area, northwest high latitude cold ridge cultivation area and south paddy area. Moreover, No-Tillage seeders now perform better. By the end of 2017, the CA area in China was 7.584×10^6 ha.

7.5 Barriers

7.5.1 Traditional paradigm

Farmer's traditional views and beliefs were ingrained by the long-term traditional tillage in China. Because of the large differences between CA and conventional tillage, it was difficult to change the farmer's traditional beliefs. During the first and second stages in China, CA experiments and demonstrations were held in some areas; however, farmers in many other regions could not obtain substantive benefits brought by CA or access to research information. Thus, publicity and education needed to be strengthened.

7.5.2 Lack of continuous systematic policy and financial support

During the first and second stages, there were no national support policies in China. By the third stage the Chinese government began to issue the relevant policies in succession to support the development of CA, including certain subsidies for farmers and project funds for research. However, there is still a lack of long-term systematic policies and funds to ensure continued development of CA.

7.5.3 Shortage of suitable and high-performance machines

Poor-quality No-Till seeders seriously limited the development of CA in China. Furthermore, the differences in soil properties, crop species, planting patterns and technical demands in different regions necessitated the development of planters suitable for specific regional circumstances.

7.5.4 Incomplete technology model

Some areas had adopted parts of CA and ended up with an incomplete CA system that resulted in sub-standard performance. Interpretation of incomplete CA systems led to incorrect conclusions or beliefs.

7.5.5 Inadequate training for farmers and technicians

Currently, the farmers and technicians aren't trained timely and adequately, and don't understand CA sufficiently, which leads to the incorrect selection and use of suitable machines and agronomic models.

7.5.6 Lack of a systematic extension approach

The extension of CA needs a systematic extension approach. There is a lack of relevant farmers' demonstration households and subsidies for the purchase of CA machines and application of CA.

7.6 Catalysts

Governments, producers (scientific research institutes, enterprises) and users (farmers, production service organizations) play an important role in the development of CA. Barriers can be overcome with the integration and coordination among the above three players. The catalysts (technical measures) that have been or should be adopted are as follows.

7.6.1 Propagation of CA in multiple ways

To reach more farmers with CA information and to motivate adoption it is necessary to promote it in various ways. Experiments and demonstrations were developed for each unique area and television shows and broadcasts related to CA were produced to portray the importance, advantages and techniques. Digital media such as mobile phones or computers are an effective means for publicity using text, pictures, audio and video. Some live interactive events were developed to answer farmer questions and learn farmer's needs, thus facilitating further specific or targeted extension of CA.

7.6.2 Government policy and financial support

In 2005, the No. 1 Central Document issued by the national government first proposed 'Reform traditional tillage and develop conservation tillage'. Since then, this technology was mentioned as a specific development area for eight consecutive years. In addition, the various central and local governments have continuously issued documents that require the application and extension of CA. Since 2014, the central government has subsidized subsoiling. The implementation of the subsidy policy has greatly mobilized the enthusiasm of farmers to adopt CA. Moreover, CA research is supported by various funds. We suggest a long-term systematic policy should be created for sustained development of CA with continuous financial support for research and farmer purchase (as a subsidy) of CA equipment.

7.6.3 Applicable innovative research on machinery

The government support for conservation tillage led to more and more factories to produce conservation tillage machines, such as Kangda in Jilin,

Nonghaha in Hebei, and Haofeng and Xinle in Henan. Research institutes and manufacturing enterprises cooperated to develop high-performance machines including No-Tillage seeders, straw returned machines, subsoilers, etc. In order to improve seeding quality, the China Agriculture University developed some No-Tillage seeders such as straw-flowing and anti-blocking No-Tillage seeder, powered anti-blocking No-Tillage seeders, etc. At present, machines suitable for regionally specific circumstances is a research focus for CA development.

7.6.4 Improvement of the technology model

Through long-term research, the technology models of conservation tillage for specific areas are gradually becoming formed (e.g. single or double cropped areas, northwest frigid wind-blown sand areas, west oasis farm area, northwest high latitude cold ridge cultivation area and south paddy area). However, technical standards and specifications suitable for different regional characteristics have not yet been fully established and impacts further expansion of CA. The perfection of conservation tillage technology models should receive more attention.

7.6.5 Better training for farmers and technicians

There are various modes for training in China such as conducting training courses, holding live meeting, tour guidance and field teaching, establishing a technology information market, etc. Training for farmers is aimed at understanding correct methods to choose suitable machines and the appropriate planting technique for local conditions. The main instructional methods are either intensive presentations through training courses or instruction in the field. Training for technicians is done with agricultural machinery experts who are invited to teach theoretical knowledge and the latest developments along with practical operations demonstrated in the field. Long-term contact between farmers and technicians should be gradually established to facilitate training.

7.6.6 Improvement of extension

Promotion and adoption of CA require strong support from multiple departments. Initially, a network of extension services for experts, technicians, technology demonstration households and farmers should be built along with establishing an administrative and technical interaction system to link the departmental, provincial and regional field levels. Moreover, it is necessary to establish a new technology extension mechanism for farmer, finances and materials to enter a village directly and creating a stable connection between CA technicians and farmers. Furthermore, more precise technical guidance can be guaranteed through the establishment of core demonstration areas and demonstration farms, which can overcome problems like poor technical extension channels and weak connections between the various links. To increase the extension of CA, it is necessary to make full use of existing agricultural subsidy policies and various project funds such as agricultural machinery subsidies and scientific and technological training.

7.7 Synergies and integration

Synergies and integration of policies occurred over time as CA became a focus of research and development at institutions with government policy and financial support. As more became known and understood about CA, ministries and institutes other than those in the ministry of agriculture developed complementary and aligned policies. CA involves a comprehensive agricultural technology system with various components such as planting patterns, planting techniques, supporting equipment, implementation mechanisms and policy guarantees. With the advancement of CA, synergies and integration will continue to expand and develop. Due to the vast territory of China and the wide variety of natural conditions, the different regions have different CA modes. Therefore, it is necessary to adopt CA forms tailored to local conditions to ensure success.

With the continuous implementation of CA, various departments and governments at all levels have introduced corresponding promotion policies.

The government's No 1 document for eight consecutive years mentioned the need to strengthen and vigorously carry out conservation tillage and related agricultural projects. The *No. 1 Central Documents of State Council* have carried these policy statements of support:

2005, 'Reform traditional tillage methods and develop conservation tillage'.
2006, 'Long term of implementation for conservation tillage demonstration'.
2007, 'Establish subsidies pilots for No-Till planting technology'.
2008, 'Implement project of conservation tillage projects continuously'.
2009, 'Extension for conservation tillage vigorously'.
2010, 'Extension of conservation tillage techniques'.
2011, 'Using techniques such as plastic mulching, deep tillage and conservation tillage, etc'.
2012, 'Extension of techniques such as precision planting, depth fertilizer application and conservation tillage techniques'.

Documents approved by the State Council have carried these statements:

2005, 'Decision on further strengthening of preventing desertification' was proposed, which emphasized implementing conservation tillage measures such as No-Till planting.

2006, 'Notice on Strengthening Relevant Issues about Land Regulation' was proposed, which suggested implementing CA to improve soil fertility.

2007, 'Planning of project construction for conservation tillage' was proposed, which suggested implementing CA in 600 sites in 15 provinces in China, and which also established the National Conservation Tillage Technology Research Center.

2012, 'Program for National Agricultural Water Saving (2012–2020)' was proposed, which suggested accelerating the extension of conservation tillage techniques in areas with higher soil erosion and water erosion.

April, 2013, 'Plan for National Defense and Control in Sand (2011–2020)' was suggested, which mentioned using conservation tillage to prevent sandstorms in three different areas.

July, 2013, 'Plan Outline of National Medium and Long-Term Science and Technology Development' was proposed, which proposed the research of appropriate machines for CA for different local conditions.

May, 2015, 'Plan for National Agricultural Sustainable Development (2015–2030)' was proposed, which proposed the implementation of conservation tillage in the northeast of China and other regions and provided subsidies for related machinery – this policy is going to solve the over-exploitation problem of groundwater.

July, 2015, 'Opinions on accelerating the transformation of agricultural development methods' was proposed, which mentioned expanding the subsidy for deep-ploughing in preparation of planting.

2016, 'Plan for National Agricultural Modernization' was proposed, which suggested optimizing a subsidy policy for purchase of agricultural machinery and increase extension of conservation tillage.

2017, 'Planning Outline of National Land' was proposed, which suggested strengthening of conservation tillage in dryland in northern China and increasing the protection of cultivated land resources.

In June, 2018, 'Win defend battle of blue sky: three's plan' was suggested to promote CA to suppress farmland dust.

At the end of 2018, the State Council proposed to strength innovation and development of sustainable farming equipment.

Documents approved by The National People's Congress

2009, a resolution about 'respond actively to climate changes' was proposed, to increase the soil fertility by implementation of CA and the construction of grassland.

2015, a report was formed in the process of examination of 'Agricultural Law of the People's Republic of China' by an inspection team, which mentioned supporting the innovation method for conservation tillage.

Documents approved by National Development and Reform Commission

2011 to 2018, conservation tillage technology was included in the proposed 'Policies and Actions response to climate change in China' by National Development and Reform Commission.

2014, conservation tillage was included in the proposed named 'National Climate Change Planning (2014–2020)'.

Documents approved by a number of ministries and commissions

2005, 'China Water Saving Policy and Technical Outline' was proposed by National Development and Reform Commission, Ministry of Water Resources and Ministry of Environmental Protection, which mentioned promotion of advanced technology of CA such as No-Till planting.

2012, 'Opinions on further job in basic farmland protection' was proposed by the Ministry of Land and Resources, National Development and Reform Commission, Ministry of Water Resources, and Ministry of Environmental Protection and Ministry of Construction, which mentioned promotion of conservation tillage, formulated fertilization and returning farmland to forest.

2016, 'Action Plan for Agricultural Machinery Development (2016–2025)' was proposed by the Ministry of Industry and Information Technology, Ministry of Agriculture and National Development and Reform Commission, which mentioned the promotion for conservation tillage, straw return to the field and water irrigation.

2017, 'Outline of Northeast Farmland Protection Plan (2017–2030)' was proposed by the Ministry of Industry and Information Technology, Ministry of Agriculture, Ministry of Land and Resources, and Ministry of Finance and Ministry of Water Resources, which suggested the technology of No-Till planting, straw return to the field and innovation pattern for planting.

Documents approved by Ministry of Environmental Protection

2011, Science and Technology Development Plan for National Environmental Protection in 'Twelfth Five-Year' program was proposed, which suggested the research of pollution prevention and control technology under conservation tillage.

2013, comprehensive technology policy for prevention and control of fine particulate pollution in air was proposed, which suggested the adoption of No-Till planting and straw returning to field.

2015, a report named 'Taking reform and innovation as the driving force to accelerate the agricultural modernization' was signed by Prime Minister Li Keqiang, which suggested the adoption of CA such as straw returning to field and using organic fertilizer to increase the soil fertility.

8 Conservation Agriculture (CA) adoption in South Africa

No-Till trials started in 1977 in the KwaZulu-Natal (KZN) Province of South Africa and in the early 1980s, a few farmers in KZN and the Western Cape Province started planting limited areas using direct drilling. At the moment the highest adoption among commercial farmers (around 50%) is in the Western Cape (winter rainfall) and KZN (summer rainfall), followed by the Highveld summer rainfall areas (mostly maize-based systems) with an adoption of around 15%. A very small percentage of smallholders have adopted CA (below 5%).

The first No-Till plantings at the Cedara Agricultural Development Institute (current name) in KwaZulu-Natal were made by Dr John Mallett under the auspices of the (then) Grain Crops Research Institute (now part of the Agricultural Research Council (ARC)), and the first tillage comparison planted in 1977. Dr Mallet made contact with No-Till developments in the USA through the South African agricultural attaché in the USA, Dr Aubrey Venter, which catalysed local research and development. In the early 1980s, Natal farmers, led by Messrs Anthony Muirhead of Winterton, and Charles Shepherd and John Jackson of Bergville, started planting limited areas using direct drilling, and soon after so did Western Cape farmers such as Messrs Jack Human and Piet Gildenhuys near Heidelberg. In all cases, these farmers started in partnership with various technicians at Cedara and the Western Cape Department of Agriculture, respectively. Other forms of 'conservation tillage' also started in predominantly maize-producing summer rainfall areas on the Highveld. In 1973, the rip-on-row system started on the very sandy soils in the western production areas of the country to alleviate soil compaction and prevent wind erosion, which was done collaboratively between farmers and key academics and technicians such as Mr Carl Koch and Professor Alan Bennie from the University of Free State. From the 1990s, a small number of farmers in maize-producing areas on the Highveld started CA, but mostly on their own, or in association with farmer groups such as the Ottosdal No-Till Club in the North West Province; many of them were inspired after cross-visits to more experienced CA farmers in the Western Cape or KZN, or countries like Argentina, Brazil, USA or Australia.

The key motivation for these early actions was wide-scale soil depletion due to tillage, giving rise to widespread erosion and desiccation.

At the beginning, the major barrier was the unavailability and affordability of suitable planting equipment and proper weed control practices. Commercial planters were available from established importers and manufacturers, but costs were high and there was often a reluctance to bring in implements which might reduce tractor sales. Collaborative initiatives between farmers and researchers or governments and industry were necessary to overcome some of the barriers. For example, the Western Cape Department of Agriculture, Martin Adams of Rovic Leers and John Jackson, a Bergville farmer, designed conventional planter conversions which enabled many commercial farmers to test the system at minimal expense. A similar initiative led by the ARC-Grain Crops Institute in association with the School of Bio-Resource Engineering of the University of Natal and the KwaZulu-Natal Department of Agriculture resulted in a No-Till adaptation being made commercially available for the SAFIM-animal-drawn planter for smallholders. Presently a wide range of No-Till planters are available for commercial farmers, while options are also available for hand-operated and animal-drawn No-Till planters.

Other barriers have been a paucity of information or understanding (knowledge) of the biological requirements, especially of soil, the processes and threat of soil degradation, as well as the implications and potential benefits of CA. In some areas, advisors and researchers have been reluctant to recommend what is viewed as a high-risk production system, and many of them recommend that farmers do NOT adopt 'No-Till' and, for example, claim 'No-Till' is responsible for disease outbreaks, soil acidification, compaction and loss of yield and income.

A few key initiatives have been launched in the last 20 years working with farmers to raise their awareness, knowledge and skills to improve the adaptation of CA in their own specific situations. Regional initiatives, such as the KZN No-Till Club and the CA movement in the Western Cape, have consistently made positive impacts through awareness events, research and publications. The most recent initiative has been the CA Farmer Innovation Programme (FIP) funded by The Maize Trust and implemented through Grain SA that started in 2014 with the aim to research, develop and adapt CA systems under commercial and smallholder farmers in different regions.

These initiatives have been successful because they essentially follow a farmer-centred Innovation Systems approach, including the creation and facilitation of multi-stakeholder platforms (on all levels) for improved integration, interaction, communication, learning and adaptation. A similar approach has been followed to some extent by the National LandCare programme in collaboration with the ARC from the late 1990s to around 2014. The new CA policy has included these successful ideas, which will eventually guide the implementation thereof.

The recent national CA policy (DAFF, 2017) could be seen as the key government initiative that was prompted by the urgent need to adapt conventional farming systems which are compromising agricultural production and resulting in alarming natural resource degradation. This policy was jointly requested and drafted by key stakeholders such as government departments, the ARC and the agricultural industry. This policy will hopefully help to create an enabling environment for South African producers to curb increasing production costs and mitigate threats to soil health, water quality and biodiversity whil increasing productivity and efficiency.

CA is also aligned with various other agricultural policies and programmes of government, especially the recent policy initiatives developed and adopted during this last decade, such as the National Development Plan (NDP), New Growth Path (NGP), Industrial Policy Action Plan (IPAP) and NEMA Act CA aligns perfectly with the LandCare programme, which focusses on community-based land resource management as well as the national white paper on climate change and climate-smart agriculture.

8.1 Factors important in influencing adoption of Conservation Agriculture (CA) in South Africa

8.1.1 Education support

All universities (with agriculture departments) have started to focus their training on CA, especially at the Masters and PhD levels. The CA FIP, LandCare and Western Cape follow a farmer-centred or bottom-up training and learning approach aiming on farmer empowerment, i.e. the development of knowledge and skills on farm level.

8.1.2 Research and extension

The national Agricultural Research Council (ARC) has supported CA since the start in the 1970s and 80s. Between 1997 and 2015, they implemented a range of government-funded LandCare-related projects among smallholders, but are at the moment restricted by capacity and funding shortages. At the moment a substantial portion of the research is funded by the industry (i.e. The Maize and Winter Cereal Trusts) and coordinated by the CA FIP (summer rainfall areas), and some provincial departments of agriculture, such as KZN and Western Cape. Agricultural businesses are also very active in providing technical research and extension services to their farmer clients, but not all of them are convinced or capacitated to support CA in the process. Many governmental extension officers are involved with CA projects, mostly with smallholders, but they generally lack the basic knowledge of CA systems and the development thereof.

8.1.3 Service providers

A number of machinery companies support CA and provide quality implements for different scales; however, there seems to be resistance from some machinery companies (which mostly sell tillage implements) and agro-chemical companies that sell fertilisers and pesticides.

The demand for cover crop seeds has increased drastically and a number of seed companies have responded quite well in providing this service. The

Published by Burleigh Dodds Science Publishing Limited, 2020.

integration of livestock within CA cropping systems using multi-species cover crop practices has been driving this trend.

8.1.4 Farmer groups

Most CA activities among commercial farmers are driven by farmers themselves. Following similar international trends, innovative and successful local CA farmers tend to act as role models, inspiring those who witness and experience their success. In most cases, these farmers organize themselves into groups (Innovation Platforms or IPs), leading towards enthusiastic and successful CA promotion and adoption initiatives, characterized by activities such as on-farm trials, conferences, field days, workshops, publications and other awareness-creating activities. Several of these initiatives have sprung up among farmers in the last decade such as the Ottosdal No-Till Club in the North West Province, Riemland and Ascent study groups both in the north-eastern Free State Province and the Sandy Soils Development Committee in the north-western Free State Province, which are all commercial CA farmer groups. The CA FIP, Western Cape CA movement and the KZN No-Till Club have been supporting these organizations quite well, although there is a big need to scale out and establish these IPs in more regions. The work of the Mahlathini Development Foundation (MDF), implementing smallholder CA projects under the CA FIP, using micro-financing groups as platforms to scale out CA, has been extremely successful and received national and international recognition.

8.1.5 Enabling public policy or policy barriers

Policy has followed rather than led CA adoption up to the current period. Limited funding currently flows through the governmental LandCare programme (about 10% of its allocation), mostly directly benefiting smallholder and new commercial farmers. The ARC does not get dedicated government funding for CA specifically and is dependent on external funding to do CA research.

The new CA policy should be implemented soon and will hopefully facilitate access to funding and capacity and hopefully obtain a much better response from provincial governments to support CA.

8.1.6 Private policy influences

For the last decade or so, The Maize and Winter Cereal Trusts of South Africa invested financially, e.g. through the Agricultural Research Council (ARC) and other stakeholders, in CA research projects. From 2013, The Maize Trust, in cooperation with Grain SA, supported the development of the CA FIP, primarily using farmer-led Innovation Platforms to launch activities such as on-farm trials and awareness events. The main aim of the CA FIP is to develop and research CA in collaboration with grain farmers and hence follows a farmer-centred integrated system philosophy. Several major CA Innovation Platforms or projects with commercial farmers and with smallholder farmers in different regions have been established and coordinated by the CA FIP. These successful concepts, approaches and technologies have all been included in the national policies on CA and Climate-smart Agriculture.

8.2 Conclusion

Feedback and observations during recent major outreach and awareness activities organized in South Africa (i.e. CA conferences and farmers' days attracting thousands of farmers and other stakeholders) showed that an enormous interest has been developed among grain producers and other key stakeholders during the last 5 years. It is becoming clear that South Africa is in the early stages of a genuine agricultural revolution towards a sustainable crop–livestock industry, where it is the innovators and early adopters (farmers, innovative researchers and scientists, journalists, etc.) who are leading the way. There were several key initiatives that influenced this transformation, such as the KZN No-Till Club, the CA movement in the Western Cape and the CA FIP, and for the sake of the whole farming sector in South Africa, these structures, or other similar structures, will have to be very prominently positioned in the next 5 years, as this period will be crucial to sustain this momentum. Already, a significant proportion (around 20%) of grain producers in South Africa are using CA and all indications and expectations are that the numbers will grow exponentially into the future.

9 Conservation Agriculture (CA) development in Paraguay

No-Till adoption in Paraguay started at about 1982/83 by agricultural cooperatives located near the Brazilian border (mainly of German and Japanese descent), which had heard about this new method of seeding from their peers in neighbouring Brazil. At this early stage there was little or no influence from outside. Ten years later, about 20 000 ha of No-Till was being trialled by farmers in Paraguay, which had a lot of problems because there was no support for the technology. In 1990, the Paraguayan Ministry of Agriculture formally requested the German Government for a project to control erosion and diffuse No-Tillage system practices to farmers. In 1993, a joint venture was signed between the Paraguayan Ministry of Agriculture and the German Development Agency GTZ. One of the biggest problems at the beginning was the national mindset of soil preparation for any crop was 'you do one or two ploughings followed by two diskings with a disk harrow'. This attitude was institutionalized across the extension service, universities, agricultural schools, libraries, journalist, etc. and provoked resistance among farmers and technicians for the new method of seeding.

To break this well-rooted paradigm, a series of strategies were developed. One of them was to create a public–private partnership (PPP) alliance between the Paraguayan Grains and Oilseed Trader Association (CAPECO) with the Ministry of Agriculture and GTZ. The interesting thing is that no official documentation was signed for this alliance and the same type of agreement happened with other stakeholders later. With CAPECO´s influence, it was possible to invite President Wasmosy and the Minister of Agriculture Torres and staff to a conference on No-Tillage agriculture in Asunción in 1994. It was crucial for the technology to be accepted by these high-ranking officials in order to break the general resistance against No-Tillage. Remember that Paraguay was ruled by a dictator until 1989, which meant that authority's words and official statements were respected. At that time farmer associations were formed, which had an important role in spreading the No-Tillage system in Paraguay. They united together into a federation – Federación Paraguaya de Siembra Directa para una Agricultura Sustentable (FEPASIDIAS) – which continues to be active and in 2018. It organized the second Paraguayan Congress on CA with about 20 high-ranking international speakers.

The Soil Conservation Project believed that 'farmer to farmer extension' would be an effective way of getting producers interested in the new technology. For this, the Project invited the three top pioneers in CA from Brazil (Nono Pereira (deceased), Franke Dijkstra and Herbert Bartz), each with almost 20 years of experience in CA on their farms, to discuss their experience in No-Tillage with Paraguayan farmers. Initially with the help of the CAPECO and local Cooperatives, farmers from the southern region were invited for presentations about CA, giving them the opportunity for discussion with one of the pioneers. Moreover, they had the whole afternoon open for exchanging ideas and experiences with them informally. Then a local farmer-pioneer that had already adopted the technology was selected to discuss all practical problems and experiences. This was repeated in two other nearby regions during the same week and then with two other farmer-pioneers in other areas in Paraguay over the subsequent 2 weeks. After the 3 weeks, 1 500 farmers had learned from experienced and knowledgeable pioneers and participated in discussions with them about the new, novel way of farming. The effect of learning from peers about pragmatic experiences and knowledge was evident resulting in many farmers convinced they needed to adopt CA to control erosion and to be able to practice farming in a sustainable way. A lesson learned by the project managers was the value of informal but effective PPPs to support project activities among farmers quickly and efficiently (without bureaucratic burdens). The project ended in 2001 and the No-Till system had expanded to more than 1.1 M ha in Paraguay. This was only achieved thanks to highly motived participants and well-trained extension agents.

As already mentioned, mindset was among the main barriers for adoption as well as ignorance about the system, especially in regards to weed control and cover crop management. To remove these barriers, it was necessary to organize workshops, symposia, congresses and other events to educate farmers, extensionists and researchers in all the aspects of the No-Till system. Another way of diffusing the technology was to organize farm tours to other regions and neighbouring Brazil. To get more farmers to adopt the technology messaging with a multimedia extension approach was adopted involving newspapers, radio, television, posters and documents. The integration and leveraging through the PPP with the multi-stakeholder approach activated synergies that convinced more and more farmers to adopt. The goal from the beginning was always to get the adoption as a system: No-Tillage plus permanent cover and crop rotation. This goal has been achieved only by top farmers while the majority uses monoculture soybeans. Here it has to be clarified that farmers in Paraguay use something like diversified monoculture because they plant 2.5–3 crops or cover crops in a year trying to maintain soil cover year-round. In order to keep up with the newest technology, the project did research on crop rotations and cover crops and disseminated the findings in seminars, congresses, etc.

A key factor for rapid growth in Paraguay was the joint venture project between the Paraguayan Ministry of Agriculture (Directorate of Extension) and the German Development Agency GTZ, started in 1993, to control erosion by cover crops and No-Tillage techniques. Here the Paraguayan counterpart to the project 'Ingeniero' Ken Moriya has to be acknowledged, who did a great job in developing and coordinating all extension activities, and towards the end of the project created the National Soils Department, which had the responsibility to look for increasing the soil quality and for sustainable crop production in Paraguay. Today (2019), there are more than 3 M ha under No-Till in Paraguay, which is more than 90% of the national crop land. Thus, Paraguay ranks in the sixth place in terms of adopted hectares worldwide. This continued growth of the area under No-Till was possible because the main barriers to adoption had been overcome and enough technicians had been trained and continued to provide professional advice to farmers. The tillage paradigm had changed at all levels in the ministries as well as in the private sector, which means that the project strategy could persist in a self-supporting sustainable way without the need for international financial support.

Even though the PPP worked to eliminate barriers for farmers, industry and national government, drawbacks existed elsewhere. For instance, when the municipalities support small farmers, they think the best way of improving their life is to donate a tractor and supply them with access to a plough and disc harrow. The intensive tillage paradigm was still the mindset of some influential people. The good news is that recently (2018/19) several municipalities and other institutions are now supporting CA. In addition, funds from Itaipu (the largest hydroelectric power plant in joint venture with Brazil) as well as Yacyretá (a large power plant joint venture with Argentina) are now targeting their assets to support CA realizing the co-benefits of watershed protection. Another problem is the adoption by small farmers. While more than 90% of mechanized farmers adopted CA, only a small proportion of small farmers practice No-Tillage currently despite great efforts with this target group. This is because of difficulty in application of CA, since most small farmers need much more time of technical support (up to 10 years) until they can operate independently. As USA farmer R. Bieber once said, 'No-Tillage is not a farming practice, it is a concept of the mind'.

In summary, the following main factors induced a rapid change in Paraguay:

1. Efficient and economic erosion control.
2. Appropriate knowledge was available within the region.
3. Widespread use of cover crops.
4. All sectors involved (private and public) provided consistent, positive messaging, with little contradictions.
5. Other forms of 'conservation tillage' like minimum tillage, reduced tillage, etc. have never been recommended to farmers.
6. There has been effective farmer-to-farmer extension through farmers associations.
7. Publications with adequate, practical, useful and timely information were made available to farmers.
8. Local economic evaluations with a systems approach by highly qualified international experts showed high economic returns of the No-Tillage system, compared to conventional agriculture.
9. Paraguayan farmers have to be competitive in the global market without farm/market subsidies, which means that they have to manage cost-effective production systems to survive.
10. There have been no major forces against the system.

10 Conservation Agriculture (CA) development in Argentina

Starting during the 1960s and 1970s, a general intensification of tillage took place in Argentina and in other countries of America. The process was encouraged by the development of more powerful tractors which in turn accommodated and promoted more tillage tools. Farmers accepted that this was the modern and only way to grow crops. More crops could be grown with more horsepower to meet increasing global commodity markets.

Soils were one of the first ecosystem components to become degraded under intensive cultivation. Water and wind soil erosion became an 'easy to see physical soil loss' and an 'uncomfortable and unacceptable soil damaging reality'. As well, a decrease of the soil organic matter occurred from oxidation from repeated tillage operations. A decrease in the nutrient supplying power occurred from reduced organic matter as well as the lower than optimal level of fertilization to replace the soil nutrients exported by the harvested grains and forages.

Argentinean farmers as well as other farmers of America were strongly motivated to find and evolve to a new way to manage their soils in particular and their agroecosystem in general, in such a manner that they could stop the land degradation and keep growing profitable crops. In the 1970s, a group of Argentina farmers – most of them agronomists directly engaged with farming operations faced with these realities – developed a 'new paradigm strategy to manage their soils' for crops of any type, annual or perennial.

At the beginning of this evolutionary process, the proponents were inspired by early experiences and literature coming from other countries. From the USA came the 'Plowman's Folly' by Edward H. Faulkner (1943) and experiences by Dr Shirley Philips and H. M. Young Jr, who described their experiences both at the University of Kentucky and early No-Till practices on the farm of Mr. Young during the 1960s. Their book, entitled *No Tillage Farming* by Dr Shirley H. Phillips and Mr. H. M. Young Jr, was very impactful and motivational for farmers in Argentina providing a wide source of information for those that were starting to move to a No-Till system as a valid way to improve their farming.

Also, during the seventies, a group of leading Argentinean farmers that later founded the Argentina No-Till Farmers Association (AAPRESID) contacted Mr. Carlos Crovetto from Chile who had implemented the No-Till system on his farm. Mr. Carlos Crovetto's personal knowledge and experiences provided inspiration and helpful guidance for Argentinean (and American) farmers. It showed the 'long way' they would have to walk to design, build, adjust, test and finally scale up an entirely different farming system, changing the agricultural paradigm in the process. This new approach was primarily based upon eliminating tillage, keeping soils covered with mulch as much and as long as possible and implementing 'crop rotation and diversity'. Farmers and agronomists also grasped the associated concepts of No-Till, for example, 'soil nutrition rather than crop fertilization' and 'cover or services crop technologies'.

Similar to Argentina, other South American countries like Brazil, Paraguay and Uruguay also had pioneers that followed similar paths at evolving their farming systems on this matter. They too realized the importance of a bottom-up approach of self-organizing farmers and field agronomists. The AAPRESID was created in Argentina (Argentinean No-Till Farmers Association), FEBRAPDP in Brazil (Brazilian Federation of No-Till Farming), FEPASIDIAS in Paraguay

(Paraguayan Federation of No-Till Farmers) and AUSID in Uruguay (Uruguayan Association of No-Till Farming). These organizations were key players in the change process and provided the foundation that initiated and collaborated in the evolution of the new farming system within the southern cone countries of America. Later they founded the CAAPAS, the American Confederation of Farmers working for a Sustainable Agriculture. This last institution signified the need to further collaborate and share knowledge across borders among eleven countries of America, from Argentina to Canada, with a common goal to help each other by openly sharing knowledge and experiences and working for synergies among them and private companies. In the beginning academia as well as some governments and official institutions were not too enthusiastic about the ongoing changes and consequently not helpful. In fact, in some cases, they served as barriers with scepticism and were not proactive to address the variety of issues that needed to be resolved in order to be able to speed up the transformation process. Later, when farmers achieved progress and the farming evolution process acquired a considerable degree of momentum, the sceptics got on board and started to contribute to the process.

The process in South America was definitely led by farmers and their organizations. They started to observe and demonstrate the benefits and desirable characteristics of the new paradigm and acquired solid empirical evidence of positive results. After 40 years of farming under the new CA model, they can show not only the productivity and profit benefits but also clear signs of being positioned even beyond an immediate sustainable cropping system. They were able to improve most of the indicators that represent the three conceptual components of sustainability, the economic, environmental and social dimensions.

10.1 Barriers and hurdles

10.1.1 Absence of practical experience

At the beginning of the development and adoption of the No-Till System, there was not enough practical experience (critical mass), supporting scientific evidence or developed technologies to operate in the new farming scenario. In addition, the absence of practical experience was also reflected in the lack of equipment and properly designed machinery (basically direct seeders) to meet the needs of the new farming approach.

Since the process was predominantly a bottom-up one, farmers' communications and actions were critical to overcome these initial barriers. The early-movers connected among themselves with a generous attitude and disposition to opening and sharing knowledge and experiences and also to help each other. This was a critical and central value to collectively gain practical experience during the early phase of the development and adoption of the new system.

The synergies that developed with commercial companies who saw new opportunities in the farming approach and were interested in the potential market were helpful in many cases. Frequently, companies were keen to create synergies of practical experience that at the end of the day benefited both parties. Companies contributed in different ways but basically by supporting farmers' actions with experimental trials, field days, conferences, etc. and by developing specific products and inputs that farmers needed to move ahead with the development and scaling up of the new system.

10.1.2 Absence of adequate machinery

The absence or lack of adequate machinery (more specifically No-Till seeders) well adapted to operate properly in No-Till conditions across stubble and mulch covered soils (soils covered with five or even more tonnes per hectare of biomass covering the soil surface) was a significant and relevant barrier especially in the early days.

From the beginning some planter factories paid attention to developments with enthusiasm and a willingness to help farmers by developing new planters and drills based upon the entirely new paradigm. Under the new system a contrary concept of large-scale planting into undisturbed soil with a significant layer of mulch had to be managed. Factories that benefited the most were those that maintained communications with the leading farmers and properly understood the paradigm shift that was taking place and the potential market that was going to be developed. Again, a win–win synergism resulted from farmers and manufacturers keeping open minds and learning together.

10.1.3 Absence of or insufficient government support and resistance to change

10.1.3.1 Government financial support mostly absent

With only few exceptions within the Latin and Central American countries, generally governments did not respond properly and in a timely manner to effectively support farmer efforts to transform and evolve a farming system towards an entirely new one. In some cases, such as South Africa, government research and extension technicians effectively supported commercial farmers in the early days of CA development, while they have had a greater focus on smallholder support more recently. The task they were faced with represented a paradigm change that involved challenging the basic pillars of conventional agriculture and for the public to understand this change.

In most cases, the lack of or unsatisfactory level of governmental understanding and support was overcome by the farmers pushing for changes proactively and with conviction. Governments were also influenced by the evident support of the private sector and companies interested in the change process and in the potential benefits of successful implementation of CA.

10.1.3.2 Human resistance to change

Since its beginning, agriculture was based upon and strongly associated with soil tillage. Now the new model was proposing the opposite – to completely quit all types of tillage operations. This recent transformation is a blink in time of the thousands of years of the history of tillage-based agriculture. Understandably this movement encountered mental and cultural barriers resistant to change. Conservative agricultural communities resist taking risks and a fundamental change could only proceed when farmers (and ultimately the greater agricultural community) had a clear understanding of the need to change, the imperative to do so and the multiple benefits of the outcome. Furthermore, smallholder farmers are risk averse and need much greater support and the design, implementation and facilitation of quite complex innovation processes (social, economic and technical) to develop sustainable CA systems.

Human resistance was overcome by the leadership of the pioneers as well as the collective action of farmers and in some cases with external support and facilitation, who became convinced of the need to change, accepting the risks and understanding the potential benefits that would accrue. The path of the status quo of conventional agriculture was obviously not sustainable.

11 Subsidization of agricultural activities as a barrier to change and evolution of processes

When we observe the degree to which CA or No-Till systems have been adopted in countries around the world, there seems to be a negative correlation between the level of adoption and the level of farmer subsidies.

In general, subsidies sterilize the necessary will and determination to face the risks that the subsidies are associated with. Even if a given change indicates a relatively easy-to-achieve benefit, a person will tend to choose the option of remaining in a protected 'comfort zone' and business as usual is rewarded. On the contrary, we have seen throughout human history that necessity has been the mother of many changes.

Assistance in overcoming this barrier came from two sources. It was evident that farmers in subsidized countries were happy with status quo in their comfort zone and lacked a desire to change. In contrast, the behaviour of farmers from low-subsidy countries sought out more production efficiencies and environmentally sustainable practices. The persistence of subsidies underlined the growing differences between countries and motivated the one group of famers more than the other. Secondly, the countries with low subsidization in many cases had higher taxation pressures and farmers were exposed to the harsh realities of markets and were obliged to find better alternatives. This was the case for several of the South American countries such as Argentina and also to some extent Brazil, Paraguay and Uruguay during the early seventies and eighties resulting in farmers realizing the 'need to change' and hence change began. After an initial gestation period, change happened quickly, particularly since 1990, and these countries now have No-Till CA systems on 80–90% of their agricultural croplands.

12 Initial reluctance and scepticism of academia

With some minor differences between the four leading countries of South America in adopting the new CA farming system, the process of perceiving the 'need to change' as well as the urgency and practical implementation of the change were definitely a bottom–up process more than a top–down one.

The absence or reluctance of academia to become involved in the early stages of CA adoption was based upon a number of interacting reasons. There was a significant shock to those that supported the long-existing paradigm with their assertion that intensive tillage and bare soil were necessary for crop production. They resisted the need for change and were comfortable to maintain the status-quo. Academia was risk-averse and saw no need for alternate forms of production and preferred to 'stay the course' within the momentum and capacity of research institutions. Leaders are better prepared to face the fear of the unknown because of the potential for positive outcomes. Within institutionalized academia, even a small amount of leadership or lone individuals can be held down by the bureaucracy of the establishment.

Farmers were closest to the detrimental issues of conventional production systems and saw first-hand the degradation and threats to the sustainability of their farms. They were obliged to seek out and test new practises as they could not afford the consequences of maintaining the trajectory they were on. Academia did not experience the same urgency and compelling need for change that the farmers did; so farmers moved ahead on their own.

Another reason for the early separation of farmers and academia might have been that academia felt left behind when farmers and industry leapt forward. The expected level of scientific validation was not there for them to voice approval for the departure from normal practise. Indeed, early contributions from academia were often warnings of possible negative consequences of abandoning tillage and bare field conditions. Thankfully those predictions of doom have not materialized.

Published by Burleigh Dodds Science Publishing Limited, 2020.

Farmers overcame the lack of interest or involvement of academia by leading farmers doing farm trialling and sharing information with other leaders across regions and borders. The developing synergies between farm organizations, private sector companies and in some cases government or international development supported technicians also injected some expertise, science and resources to enhance field experimentation as well as a range of extension activities. This accelerated knowledge generation and went around or ahead of the bulk of academia. This groundswell of interest, activity and knowledge generation eventually attracted the attention of academia, which has since come on board and is supporting with research work on experimental stations and in laboratories to add depths to the scientific understanding of CA systems.

13 Commonalities and differences in the global growth of Conservation Agriculture (CA)

When examining the experience profiles of the range of countries and other experiences of the authors, one recognizes many learnings and growth of innovation in farming systems. Even though there are unique differences or special cases, it is useful to look at the collective through different lenses or from different aspects to identify commonalities or differences that become key aspects that are important when analysing existing or designing new policies.

The following four aspects proved to be important.

13.1 Early movements

All early actions were taken by farmers, as early as the 1960s and 1970s. In the cases of China, Canada and South Africa, there was also interest and support by some researchers, which fostered some early research institute collaborations.

The leading innovators identified others and self-organized with a common goal of knowledge development and exchange. These were often farmers but, in some cases, included an interested industry or researcher. Self-organization did not respect political boundaries and reached across borders.

Collaboration was key. Early movers identify others, support one another, identify issues or barriers and design pathways to overcome them.

13.2 Barriers

Technology, especially planters, was an early barrier that persisted for a few decades. The barrier was overcome by involving industry manufacturers in the iterations of designs that were required as agronomic knowledge increased and the impacts of site-specific conditions were recognized.

Published by Burleigh Dodds Science Publishing Limited, 2020.

Experience and knowledge were another early barrier that was overcome by connecting early adopters to one another, to interested scientists or change agents (facilitators) and to industry. Eventually they also connected to academia and research institutions.

Mindset barriers on behalf of farmers, industry, government and researchers were issues early on and in some cases persist until today, especially where nourished by a lacking of the need to change. Biases, paradigms, conservatism vested interests or resistance of input providers and inertia of large institutions all come to play against adoption and mainstreaming. Farmers, industry, government officers and researchers were all subject to scepticism, ridicule, and non-cooperation by their peers and superiors. Pathways to overcome these barriers were customized depending upon the target circumstances.

Existing government policy was identified as a barrier in several cases from the massive Common Agriculture Program of the EU to smaller national and subnational policies that preferentially maintained a focus (and subsidization) on maximum yields and conventional production systems.

13.3 Catalysts

Self-organizing for collaborative efforts had a strong impact in all cases early on and continuously. The social networks that develop catalyse the immediate needs and are strategic for future, unanticipated needs. Diversity within the network allowed them to tackle complex problems and be resilient.

Policy delivered both intended and unintended consequences. Government policy helped with funding institutions, supporting NGOs and extension, creating soil protection or health policies and creating and empowering advisory groups. In other cases, the persistence of government policies for conventional agriculture with large subsidies catalysed a counter movement of CA.

Multiple pathways or points of contact and influence were important. This included multi-mode and multi-media extension efforts, training for all players, site-specificity recognition and identifying all players directly and indirectly involved. In some cases, the involvement of a president or government leader catalysed action, while in other cases the public–private partnership between farmers, industry and government institutions provided the needed fuel.

Strong leadership or facilitation of individual innovators has been driving many successful CA initiatives. These innovators or catalysts are mostly farmers, but they are also facilitators, researchers or extension officers from private and government institutions. The role of highly experienced and skilful facilitators to transform complex smallholder situations into sustainable CA systems is important.

13.4 Synergies and integration

CA systems are inherently integrated and find synergies in biologic systems. The multiple benefits of CA are becoming recognized by governments and international accords. China has not only an agriculture ministry but also a number of other ministries and national institutions all recognizing CA in their policy positions. Other countries are starting to move on those pathways as are international agreements on nature, climate change, social well-being and sustainability.

Ecosystem services are recognized as a key co-benefit of CA and some countries such as Paraguay, Brazil, Canada, South Africa and the EU have or are considering payment for ecosystem services schemes from both government, private industry and NGO policy initiatives.

The holistic and integrated nature of CA aligns well with universities, research institutions and extension efforts that are restructuring organizations and educational curricula to reflect more modern concepts of sustainability and social preferences.

The collaboration and partnerships between farmers, industry and NGOs have strengthened ties and developed capacity in the parties involved. The increased capacity spans all channels: communications, leadership, research, data and knowledge. That has attracted better connections with others such as academia or other NGOs not involved early on. In some cases, the increased capacity can also allow the PPPs to exist without interaction or support from others.

14 Conservation Agriculture (CA) as a disruptive innovation

Society, business and governments are in the midst of many disruptions both destructive and constructive. The changing baselines or paradigms need to be acknowledged, examined and utilized by policy practitioners.

Communications is now nearly instant across the country side and around the world thanks to the nearly ubiquitous smart phones and Internet. Communications can be in any form of media – text, audio, video – farmers can take a photo of an insect and identify it themselves with Internet-based searches or send it to an entomologist at a laboratory thus getting expert access at the field or farm levels.

Citizen science is another important application of technology that has many benefits including cost effectiveness data collection over wide areas, access to analytical expertise, creation of valuable datasets and bringing people together over common causes. The United Nations Environment Program listed citizen science as one of the ten emerging issues in their 2014 year book (UNEP, 2014). The emergence of it had been documented 5 years earlier, but the examples since have skyrocketed. Ryan et al. (2018) called it exponential growth when they examined applications and potential in food and agriculture research. They saw more opportunities with new sensor technologies, artificial intelligence and machine learning, leveraging citizens to support local decision making and, supporting and strengthening extension services. Van Etten et al. (2019) over 4 years collected results from 12 409 farmer-managed plots (three varieties) from three countries around the world. They could show their analysis could reduce climate bias, incorporate seasonal forecasts, risk analysis and geographic extrapolation. Recommendations resulting from the citizen science methodology led to important differences from previous recommendations.

More sensors and sophisticated monitoring technology are appearing every month. Robotics, once thought to be futuristic, are now being deployed on farms. Along with the appearance of robots has come a countering of bias or a paradigm shift that farm equipment had to be larger to farm larger areas quickly with the same number of operators. Robotic farm equipment has become smaller (for now anyway) as they can work continuously and as multiple units (swarms). This may also present opportunities for smaller land holders as capital costs go down with developments.

Does all this challenge the classical thinking of experimentation and evidence-based policy? We are not constrained to wait for scientists to do experiments on a hundred small plots at research stations with good land. We can collect thousands of field measurements on all types of fields under a wider range of conditions in a shorter period of time and use advanced analytical techniques on a central server to provide reports in shorter turnaround times. Are we entering the era of the 'data by-pass' where data does not go to institutions or governments? Will independent experts and consultants working with farms generate more knowledge and evidence and, hold the data? Will industry move earlier and quicker to trust citizen science evidence than governments and be nimbler in policy development? There are big opportunities for farmers, farm NGOs and consultants to deliver more evidence that is more precise and wider reaching in short periods of time. This is also an opportunity for extension services to grow and provide value to their clientele.

15 The future of Conservation Agriculture (CA) policy

Main policy influencers of CA are research institutions and industry or NGOs which together intertwine their policies in response to government created policy or perhaps their policies illicit a policy response from governments and their agencies. Governments have always struggled with forming policy to encourage innovation, economic diversification and stability. The task has becoming more challenging (or with more opportunities) with globalization, international trade agreements, multinational agribusinesses, increasing public expectations and the disruptions of technology mentioned earlier. Policies and their effects can be complex and interconnected; however, a brief topology of policy types may help frame paths forward for CA practitioners and policy influencers.

Elgie and Brownlee (2017) provided a framework for policy with four high-level categories when looking at the problem of stimulating clean innovation (products and practices that improve environmental performance). CA is a form of clean innovation and can be framed in the same categories.

Strengthen policies make the system more effective and resilient. This could be achieved by facilitating information exchange and social capital building through conferences or collaboration among institutes. Strategies are built for CA, data are gathered and pooled and talent is built.

Grow policies help ideas become realities into new markets. This could be grants or loans to farmers to access equipment or other capital. It could be loans or business development assistance to equipment designers and manufacturers.

Push policies drive new ideas and initiatives. This would include support for academic research (e.g. targeted funds for CA research or a Chair of CA at a university) or industry research (matching grants). It might also include legislation to prevent erosion or soil degradation.

Pull policies are more market orientated to pull change in the right direction at a quicker pace or broader spread. CA certification systems or CA being a component of other markets (e.g. ecosystem services, carbon markets) will allow the private sector to develop markets and find efficient ways to supply them.

All these policy areas work together as we have seen in the national examples presented. CA, as with other clean innovations, involve a paradigm shift, as mentioned in the country profiles. This then requires a comprehensive look at all policy areas – it is not a simple task of tweaking one or two policies and maintaining the status quo. In order for CA to move effectively, all categories should be examined and analysed for appropriate policies that will foster CA as well as identify policies (existing or planned) that would be counterproductive. Some countries are examining crop insurance products, recognizing there is an increasing public cost to provide subsidized assurance to sub-optimal production systems that might be more exposed to climate change and have higher greenhouse gas emissions.

A second policy formation area is that of policies to stay relevant and to be adjustable or adaptable to changing conditions or uncertainty. CA is a practise that adjusts to locally specific conditions at the farm as well as institutional conditions such as local governments, research/educational institutions and industries. Policies should reflect that ability to adjust. Likewise, as mentioned earlier, disruptions of technology will become the norm as will diversity and volatility of markets. In order to plan for and design adaptive policy Swanson et al. (2009) researched policies in different countries and derived a series of seven principles or tools to either audit or develop policies (see Box 1).

Box 1 Adaptive policy design

Seven tools to facilitate policy development for capacity to adapt to anticipated and unanticipated conditions (used with permission from Swanson et al., 2009).

1 *Integrated and forward-looking analysis* - By identifying key factors that affect policy performance and identifying scenarios for how these factors might evolve in the future, policies can be made robust to a range of anticipated conditions and indicators developed to help trigger important policy adjustments when needed.
2 *Multi-stakeholder deliberation* - A collective and collaborative public effort to examine an issue from different points of view prior to taking a decision, deliberative processes strengthen policy design by building recognition of common values, shared commitment and emerging issues, and by providing a comprehensive understanding of causal relationships.
3 *Automatic policy adjustment* - Some of the inherent variability in socio-economic and ecological conditions can be anticipated, and monitoring of key indicators can help trigger important policy adjustments to keep the policy functioning well.
4 *Enabling self-organization and social networking* - Ensuring that policies do not undermine existing social capital; creating forums that enable social networking; facilitating the sharing of good practices and removing barriers to self-organization - all strengthen the ability of stakeholders to respond to unanticipated events in a variety of innovative ways.
5 *Decentralization of decision-making* - Decentralizing the authority and responsibility for decision-making to the least effective and accountable unit of governance, whether existing or newly created, can increase the capacity of a policy to perform successfully when confronted with unforeseen events.
6 *Promoting variation* - Given the complexity of most policy settings, implementing a variety of policies to address the same issue increases the likelihood of achieving desired outcomes. Diversity of responses also forms a common risk-management approach, facilitating the ability to perform efficiently in the face of unanticipated conditions.
7 *Formal policy review and continuous learning* - Regular review, even when the policy is performing well, and the use of well-designed pilots throughout the life of the policy to test assumptions related to performance can help address emerging issues and trigger important policy adjustments.

Steps 1 and 7 are connected and the government (or other institutions) is often weak on both of these. Time needs to be taken for thoughtful analysis and perhaps use scenario tools to anticipate what would happen if CA became mainstream. A few countries are nearly there and could act as case studies.

Published by Burleigh Dodds Science Publishing Limited, 2020.

Step 4 seems to be where CA got its start in all countries. The need is for that self-organizing to maintain itself as a *strengthening* policy category mentioned above. The collective knowledge of these networks is shared among members and perhaps with larger networks (global CA) but how many of them have an authorized channel upwards to ministers of agriculture or research station directors? An annual summary of issues, needs and achievements is valuable information for policy makers.

Step 2 can be easier if Step 4 is already in place (as happened in the early stages of CA developments) because collaboration exists, at least from the production agriculture level. More stakeholders need to be consulted to ensure CA policies align with all affected or impacted areas such as environmental impacts or supply chains and consumers. This is needed for both *push* and *pull* policies.

Step 3 is often weak in policy development. Measures are often not thought out ahead of time and tend to be a management response after the policy is in place. Are CA policies automatically adjustable to changes in governments, industry or climate?

Step 5 is another point that is often weak in policies, especially government policy where a centralized authority is the preferred default position. Empowering others in a distributed network or hierarchy can result in more adaptive policy design and implementation.

Step 6 is important for longevity and adjustment to changing conditions. It is a characteristic mirrored from ecology and the study of natural ecosystems – variation or diversity is the secret for resilience.

A final word on policy formation is that the term 'evidence-based policy' is one that is commonly waved in front of policy makers to imply that solid evidence, usually, but not always, inferring peer-reviewed science, is taken into account in policy formation. At the time that No-Till was first emerging, Caplan (1979) was studying how policy was formed in the USA and developed a Two Communities Theory that recognized there was a gap or separation between academic evidence and other evidence used by policy makers. Academic literature is only a portion of the evidence used for policy making and a significant amount of other 'evidence' is gathered from a variety of sources (e.g. media, consultations, social sources) that influences the final outcome. More recently Newman et al. (2016) re-examined the Two Communities Theory in Australia with surveys of policy makers and found results similar to previous studies in developed countries. Rather than a polarity of two separate and homogenous communities, they recognized more of a continuum between the two depending on the topic areas, individuals and the amount of political versus administrative policy components that were involved. Scarrow (2018) found that at the executive board level of a government science commission in the USA, very little time was spent on discussion of peer-reviewed science. At the technical working group level more time was spent on academic literature, but significant amounts of time was spent with grey literature and other non-published evidence, along with discussions of gaps and connecting the pieces for the mandated tasks before them.

This recognizes that modern policy makers, administrative bodies, institutions and the like do consider 'evidence' from all sources although sometimes, not much of it from the academic literature, contrary to popular belief. This emphasizes on the opportunity and compelling argument for farmer organizations and collaborative bodies (assemblages of farmers, specialists, researchers, industry and NGOs) to play a role of ensuring all forms of evidence are before institutions and policy makers. The exponential growth of all types of information continues and CA organizations have to step up to ensure correct or valid information is utilized and, they can further serve a role of synthesizing information. The categories of policies mentioned earlier indicate that size of the job and the adaptive policy tools indicate the recognition of the value and need for CA organization input and support.

16 Relevance of Conservation Agriculture (CA) at the global policy and governance level

At the UN Earth Summit in Rio in 1992, three major global environmental threats to sustainable development were recognized. These were climate change, loss of biodiversity and land degradation and desertification. To address these global challenges, three United Nations conventions were established to help understand the nature of the threats and formulate possible solutions, legally binding agreements, the Rio Conventions, which were opened for signature. They are The UN Framework Convention on Climate Change (UNFCCC); the UN Convention on Biological Diversity (UNCBD); and the UN Convention to Combat Desertification (UNCCD). In addition, the Earth Summit resulted in The Rio Declaration on Environment and Development; The Agenda 21 and The Forest Principles.

To ensure compliance with the Rio agreements, particularly the Rio Declaration on the Environment and Development and Agenda 21, delegates to the Earth Summit established the UN Commission for Sustainable Development (UNCSD). In 2013, the UNCSD was replaced by the High-level Political Forum (HLPF) on Sustainable Development that meets every year as part of the Economic and Social Council (ECOSOC) meetings, and every fourth year as part of the UN General Assembly meetings. To coordinate activities related to sustainable development, a UN Sustainable Development Group with a Development Coordination office has been established. The main contribution of the HLPF and the Sustainable Development Group has been the formulation of the 2030 Agenda for Sustainable Development with its 17 Sustainable Development Goals (SDGs), which were adopted in September 2015.

There are several SDGs that can benefit directly from agricultural land use that is based on CA systems. These are SDG 1 (no poverty), 2 (zero hunger), 6 (water), 7 (energy), 8 (economic growth and employment), 12 (sustainable consumption and production), 13 (climate action), 14 (marine resources) and 15 (terrestrial ecosystems), although it is realized that all SDGs are interconnected. Chapter 1 in volume 2 elaborates on the key benefits of CA systems and shows how they can contribute to the achievement of the above SDGs.

CA systems contribute to SDGs through increased productivity and profit, improved soil health, reduced or minimized use of pesticides and herbicides, reduced machinery, labour and energy use and costs, decreased soil erosion and land degradation, increased biomass for livestock, greater adaptability to climate change, climate change mitigation, rehabilitation of degraded lands and ecosystem services and enhanced ecosystem services to society. As every nation is committed to achieving the SDGs, agriculture based on CA can become part of the solution provided national policies and global policies can integrate CA as a key component of sustainable land use.

The three UN Conventions to address the challenges of climate change, loss of biodiversity and land desertification and degradation are being implemented through the COP processes. These COP processes have yet to embrace the CA-based land use as part of the solution. For climate change, at COP 21, the 4 per 1000 proposal from the French Government was accepted and this initiative is expected to facilitate the process of increasing soil carbon content by 0.4% per year, which is equivalent to the total amount of carbon being emitted annually. However, the 4 per 1000 initiative has yet to establish how this can be done, along with the supporting scientific evidence. Agricultural land use based on good-quality CA systems has a major contribution to make to climate change mitigation by sequestering carbon in the soil. It has a major contribution to make to climate change adaptability. Given that CA systems are acknowledged to be climate-smart and given that there are already 180 M ha of CA cropland globally, it is important that the role of CA in climate change mitigation and adaptability is recognized in national and international policies and institutional land use strategies respond to international climate change agreements such as COP 21 in Paris and COP 22 in Marrakesh. At COP22, the Moroccan Government's Triple A initiative of Adaptability of Agriculture in Africa to climate change was accepted. Again, CA land uses have a major role to play in implementing the Triple A initiative. The Moroccan Government itself has also proposed to support the transformation of 2 M ha of cropland currently under tillage agriculture to CA systems over a period of 5 years.

Similarly, the role of CA in reversing land degradation, enhancing soil health and productivity and rehabilitating soil-mediated ecosystem services is now known as elaborated in Kassam et al. (2013) with successful examples from Brazil, Canada, Spain, China and Australia. Recently, under the UNCCD, a framework for Land Degradation Neutrality has been proposed. Given that the land use systems based on CA systems are capable of reversing land degradation as well as rehabilitating abandoned agricultural lands, it is important that UNCCD's Land Neutrality Framework incorporates the CA dimensions so that solutions to combating desertification and land degradation are based on CA regenerative processes.

Equally, the role of CA in enhancing agrobiodiversity as well as biodiversity in general is being increasingly documented. CA systems are known to enhance soil health and biology. This is based on increasing soil biodiversity in terms of promoting soil microorganisms as well as meso fauna including earthworms, termites, spiders, beetles, frogs and other soil inhabiting organisms. Over time, CA systems create a soil habitat for all kinds of organisms that can play a useful role in nutrient, water and carbon cycling, as well as integrated pest management. Similarly, CA systems are known to promote biodiversity above the ground through their diversified cropping systems and mulch cover, both of which not only contribute to enhancing soil biodiversity, but also above ground biodiversity including natural enemies of pests. Through this increase in soil and crop biodiversity, CA systems can offer greater resilience to biotic and abiotic stresses. CBD and FAO have recognized the role of CA in enhancing biodiversity as seen from their recent documents (e.g. FAO, 2019).

In light of the above, it is only logical to view CA as offering a common land use management framework to achieve the objective of the three UN Conventions, but also contribute to achieving the SDGs.

17 Future trends and conclusion

CA has come a long way in the past several decades and it has a bright future ahead. The adoption journey has been uphill and the path has had numerous potholes and rocks to navigate around. In many cases the development of practice, policy and institutions has been from the bottom-up; more recently however, we have seen top-down developments as larger institutions and industry break out of old paradigms and the inertia of conventional agriculture. The CA future needs to link across stakeholders and influencers in a collaborative fashion to enable adaptive and robust policies to be implemented at all levels. Stakeholders are farmers, government, academia and industry and NGOs and all should be involved to develop CA policy. Some simple frameworks have been presented from which to frame analysis or development of policies at any institutional level.

CA is at an opportune moment. The public are environmentally conscious and in favour of production practises with a small environmental footprint and have co-benefits for the public good. International conventions are looking to agriculture for support on a number of policies from economic diversity to climate change to the environment. CA is no longer the fear of the unknown. It has found adoption around the world in a wide range of farm circumstances. Science evidence has grown to validate CA systems and support policy or institutional developments which can align and strengthen other policies. The disruptions of technology, corporate emergence as a policy influencer, globalization and climate change all invite opportunities for CA to be positioned well to align with multiple interests. We need to ensure governments and institutions are aware that elegant solutions are available.

18 Where to look for further information

A general reference for Conservation Agriculture is the UN Food and Agriculture Organization webpage which carries a library of resources on the topic (http://www.fao.org/conservation-agriculture/en/). Cornell University, USA, maintains a library of CA resources which includes academic literature as well as grey literature, CA association newsletters and the FAO hosted CA Community of Practice (CA-CoP) list of literature mailings. The Cornell site also contains a searchable Zotero database (in several languages) where one can search for policy or institutional references. (http://conservationagriculture.mannlib.cornell.edu/). Regional organizations such as the European CA Federation (ECAF) website (www.ecaf.org) and African Conservation Tillage (ACT) Network website (www.act-africa.org) offer policy-related technical publications, as do to many of the websites of the large number of national and sub national No-Till and CA organizations.

Policy for CA is not the main topic of papers or books that we are aware of. It is sometimes presented as a subtopic of global context publications as in Jat et al. (2014) and Kassam et al. (2013). Kassam et al. (2014b) review policies and policy support for CA looking specifically at necessary conditions for adoption of CA and how to design and implement policies. The UN-FAO Save and Grow publication includes a final section on policy and institutional development with a focus on smallholder farms (FAO, 2011). An examination of CA in the context of climate change in Africa and policy needs can be found in Kassam et al. (2016) and Dzvimbo et al. (2017).

19 References

Awada, L., Lindwall, C. W. and Sonntag, B. 2014. The development and adoption of conservation tillage systems on the Canadian Prairies. *International Soil and Water Conservation Research* 2(1), 47–65. doi:10.1016/S2095-6339(15)30013-7.

Baig, M. N. and Gamache, P. 2011. *The Economic, Agronomic and Environmental Impact of No-Till on the Canadian Prairies* (version 2). Alberta Reduced Tillage Linkages,

Alberta, Canada. Available at: http://www.reducedtillage.ca/ImpactnotillrtlaugFinalAug30optomized.pdf.

Borrelli, P., Robinson, D. A., Fleischer, L. R., Lugato, E., Ballabio, C., Alewell, C., Meusburger, K., Modugno, S., Schütt, B., Ferro, V., Bagarello, V., Oost, K. V., Montanarella, L. and Panagos, P. 2017. An assessment of the global impact of 21st century land use change on soil erosion. *Nature Communications* 8(1), 2013. doi:10.1038/s41467-017-02142-7.

Brisson, N., Gate, P., Gouache, D., Charmet, G., Oury, F.-X. and Huard, F. 2010. Why are wheat yields stagnating in Europe? A comprehensive data analysis for France. *Field Crops Research* 119(1), 201–12. doi:10.1016/j.fcr.2010.07.012.

Caplan, N. 1979. The two-communities theory and knowledge utilization. *American Behavioral Scientist* 22(3), 459–70. doi:10.1177/000276427902200308.

Christian, D. G. 1994. Experience with direct drilling cereals and reduced cultivation in England. In: Tebrügge, F. and Böhrnsen, A. (Eds), *Experience with the Applicability of No-Tillage Crop Production in the West-European Countries.* Proceedings of the EC-Workshop I. Wissenschaftlicher Fachverlag, Giessen, Germany, pp. 25–31.

Clapp, J. 2018. Mega-mergers on the menu: corporate concentration and the politics of sustainability in the global food system. *Global Environmental Politics* 18(2), 12–33. doi:10.1162/glep_a_00454.

Clapp, J. and Scrinis, G. 2017. Big food, nutritionism, and corporate power. *Globalizations* 14(4), 578–95. doi:10.1080/14747731.2016.1239806.

DAFF (Department of Agriculture, Forestry and Fisheries), 2017. Draft Conservation Agriculture Policy. Pretoria, South Africa. Accessed on 16 Aug 2019 (https://www.nda.agric.za/docs/media/Draft%20Conservation%20Agriculture%20Policy.pdf)

Derpsch, R., Franzluebbers, A. J., Duiker, S. W., Reicosky, D. C., Koeller, K., Friedrich, T., Sturny, W. G., Sá, J. C. M. and Weiss, K. 2014. Why do we need to standardize no-tillage research? *Soil and Tillage Research* 137, 16–22. doi:10.1016/j.still.2013.10.002.

Dzvimbo, M. A., Mashizha, T. M., Monga, M. and C. Ncube. 2017 Conservation Agriculture and Climate Change: Implications for Sustainable Rural Development in Sanyati, Zimbabwe. *J. Soc. Dev. Sci.* 8(2):38–46.

Elgie, S. and Brownlee, M. 2017. *Accelerating Clean Innovation in Canada*. Smart Prosperity Institute, University of Ottawa, Canada, 24pp.

FAO. 2011. *Save and Grow: A Policymaker's Guide to the Sustainable Intensification of Smallholder Crop Production*. UN-FAO, Rome, Italy, 112p. ISBN 978-92-5-106871-7. Available at: http://www.fao.org/ag/save-and-grow/en/index.html.

FAO. 2019. *The State of World's Biodiversity for Food and Agriculture.* Food and Agriculture Organization, Rome.

Fuglie, K. O., Heisey, P. W., King, J. L., Pray, C. E., Day-Rubenstein, K., Schimmelpfennig, D., Wang, S. L. and Karmarkar-Deshmukh, R. 2011. Research investments and market structure in the food processing, agricultural input, and biofuel industries worldwide. ERR-130. U.S. Dept. of Agriculture, Econ. Res. Serv., December 2011.

Goddard, T., Zoebisch, M. A., Gan, Y. T., Ellis, W., Watson, A. and Sombatpanit, S. (Eds). 2008. *No-till Farming Systems*. Special Publication No. 3. World Association of Soil and Water Conservation, Bangkok, 544pp. ISBN: 978-974-8391-60-1

Jat, R. A., Sahrawat, K. L. and Kassam, A. H. (Eds). 2014. *Conservation Agriculture: Global Prospects and Challenges.* CABI Publishing, Wallingford, UK, 424pp.

Kassam, A., Friedrich, T., Shaxson, F. and Pretty, J. 2009. The spread of Conservation Agriculture: justification, sustainability and uptake. *International Journal of Agricultural Sustainability* 7(4), 292–320. doi:10.3763/ijas.2009.0477.

Kassam, A., Basch, G., Friedrich, T., Shaxson, F., Goddard, T., Amado, T., Crabtree, B., Hongwen, L., Mello, I., Pisante, M. and Mkomwa, S. 2013. Sustainable soil management is more than what and how crops are grown. In: Lal, R. and Stewart, R. A. (Eds), *Principles of Soil Management in Agro-ecosystems*. Advances in Soil Science. CRC Press.

Kassam, A., Derpsch, R. and Friedrich, T. 2014a. Global achievements in soil and water conservation: the case of Conservation Agriculture. *International Soil and Water Conservation Research* 2(1), 5–13. doi:10.1016/S2095-6339(15)30009-5.

Kassam, A., Friedrich, T., Shaxson, F., Bartz, H., Mello, I., Kienzle, J. and Pretty, J. 2014b. The spread of Conservation Agriculture: policy and institutional support for adoption and uptake. *Field Actions Science Reports* (Online) 7. Available at: http://factsreports.revues.org/3720.

Kassam, A., Friedrich, T., Derpsch, R. and Kienzle, J. 2015. Overview of the worldwide spread of Conservation Agriculture. *Field Actions Science Reports* (Online) 8. Available at: http://factsreports.revues.org/3966.

Kassam, A., Mkomwa, S. and Friedrich, T. (Eds). 2016. *Conservation Agriculture for Africa: Building Resilient Farming Systems in a Changing Climate*. CABI Publishing, Wallingford, UK.

Kassam, A., Friedrich, T. and Derpsch, R. 2019. Global spread of Conservation Agriculture. *International Journal of Environmental Studies* 76(1), 29–51. doi:10.1080/00207233.2018.1494927.

Lafond, G. P., Clayton, G. W. and Fowler, D. B. 2014. Conservation agriculture on the Canadian prairies. In: Jat, R. A., Sahrawat, K. L. and Kassam, A. (Eds), *Conservation Agriculture: Global Prospects and Challenges*. CABI Publishing, Wallingford, UK.

Lindwall, C. W. and Sonntag, B. (Eds). 2010. *Landscape Transformed: The History of Conservation Tillage and Direct Seeding*. Knowledge Impact in Society, University of Saskatchewan, Saskatoon.

MacDonald, J. M. 2017. *Mergers and Competition in Seed and Agricultural Chemical Markets*. ERS-USDA, Amber Waves, 3 April 2017.

Montgomery, D. R. 2007. Soil erosion and agricultural sustainability. *Proceedings of the National Academy of Sciences of the United States of America* 104(33), 13268–72. doi:10.1073/pnas.0611508104.

Newman, J., Cherney, A. and Head, B. W. 2016. Do policy makers use academic research? Reexamining the "two communities" theory of research utilization. *Public Administration Review* 76(1), 24–32. doi:10.1111/puar.12464.

Panagos, P., Borrelli, P., Poesen, J., Ballabio, C., Lugato, E., Meusburger, K., Montanarella, L. and Alewell, C. 2015. The new assessment of soil loss by water erosion in Europe. *Environmental Science and Policy* 54, 438–47. doi:10.1016/j.envsci.2015.08.012.

Pretty, J. and Bharucha, Z. P. 2014. Sustainable intensification in agricultural systems. *Annals of Botany* 114(8), 1571–96. doi:10.1093/aob/mcu205.

Ramankutty, N., Mehrabi, Z., Waha, K., Jarvis, L., Kremen, C., Herrero, M. and Rieseberg, L. H. 2018. Trends in global agricultural land use: implications for environmental health and food security. *Annual Review of Plant Biology* 69, 789–815. doi:10.1146/annurev-arplant-042817-040256.

Reicosky, D. C., Goddard, T. W., Emerson, D., Chan, A. S. K. and Liebig, M. A. 2012. Agriculture greenhouse gas trading markets in North America. In: Liebig, M. A., Franzluebbers, A. J. and Follett, R. F. (Eds), *Managing Agricultural Greenhouse Gases: Coordinated Agricultural Research through GRACEnet to Address Our Changing Climate*. Academic Press, San Diego, CA, pp. 423–38.

Ryan, S. F., Adamson, N. L., Aktipis, A., Andersen, L. K., Austin, R., Barnes, L., Beasley, M. R., Bedell, K. D., Briggs, S., Chapman, B., Cooper, C. B., Corn, J. O., Creamer, N. G., Delborne, J. A., Domenico, P., Driscoll, E., Goodwin, J., Hjarding, A., Hulbert, J. M., Isard, S., Just, M. G., Kar Gupta, K., López-Uribe, M. M., O'Sullivan, J., Landis, E. A., Madden, A. A., McKenney, E. A., Nichols, L. M., Reading, B. J., Russell, S., Sengupta, N., Shapiro, L. R., Shell, L. K., Sheard, J. K., Shoemaker, D. D., Sorger, D. M., Starling, C., Thakur, S., Vatsavai, R. R., Weinstein, M., Winfrey, P. and Dunn, R. R. 2018. The role of citizen science in addressing grand challenges in food and agriculture research. *Proceedings of the Royal Society. Series B: Biological Sciences* 285, 20181977. doi:10.1098/rspb.2018.1977.

Scarrow, R. 2018. Policy without science. *Nature Sustainability* 1(8), 385-. doi:10.1038/s41893-018-0128-7.

Soane, B. D., Ball, B. C., Arvidsson, J., Basch, G., Moreno, F. and Roger-Estrade, J. 2012. No-till in northern, western and south-western Europe: a review of problems and opportunities for crop production and the environment. *Soil and Tillage Research* 118, 66–87. doi:10.1016/j.still.2011.10.015.

Sparrow, H. O. 1984. *Soil at Risk – Canada's Eroding Future*. Standing Senate Committee on Agriculture, Fisheries and Forestry, Senate of Canada, Ottawa, Canada, 129pp.

Statistics Canada. 2016. Tillage practices used to prepare land for seeding. Table 32-10-0408-01. Statistics Canada, Ottawa, Canada. (online open data). Available at: https://www150.statcan.gc.ca/t1/tbl1/en/tv.action?pid=3210040801.

Swanson, D., Barg, S., Tyler, S., Venema, H. D., Tomar, S., Bhadwal, S., Nair, S., Roy, D. and Drexhage, J. 2009. Seven guidelines for policy-making in an uncertain world. In: Swanson, D. and Bhadwal, S. (Eds), *Creating Adaptive Policies: A Guide for Policy-Making in an Uncertain World*. Sage Publications, Thousand Oaks, CA and IISD International Development Research Centre, Ottawa, Canada, pp. 12–24. Chapter 2.

UNEP. 2014. UNEP Year Book 2014: Emerging issues in our global environment. United Nations Environment Programme, Nairobi, Kenya. Available at: https://wedocs.unep.org/handle/20.500.11822/17716.

van Etten, J., de Sousa, K., Aguilar, A., Barrios, M., Coto, A., Dell'Acqua, M., Fadda, C., Gebrehawaryat, Y., van de Gevel, J., Gupta, A., Kiros, A. Y., Madriz, B., Mathur, P., Mengistu, D. K., Mercado, L., Nurhisen Mohammed, J., Paliwal, A., Pè, M. E., Quirós, C. F., Rosas, J. C., Sharma, N., Singh, S. S., Solanki, I. S. and Steinke, J. 2019. Crop variety management for climate adaptation supported by citizen science. *Proceedings of the National Academy of Sciences of the United States of America* 116, 4194. doi:10.1073/pnas.1813720116.

Van Ouwerkerk, C. and Perdok, U. D. 1994. Experiences with minimum and no-tillage practices in the Netherlands. I: 1962-1971. In: Tebrügge, F. and Böhrnsen, A. (Eds), *Experience with the Applicability of No-Tillage Crop Production in the West-European Countries*. Proceedings of the EC-Workshop I. Wissenschaftlicher Fachverlag, Giessen, Germany, pp. 93–104.

Index